ELECTRICITY AND ELECTRONICS

ELECTRICITY AND ELECTRONICS

SECOND EDITION

REX MILLER
State University College
Buffalo, New York

FRED W. CULPEPPER, JR.
Old Dominion University
Norfolk, Virginia

DELMAR PUBLISHERS INC.®

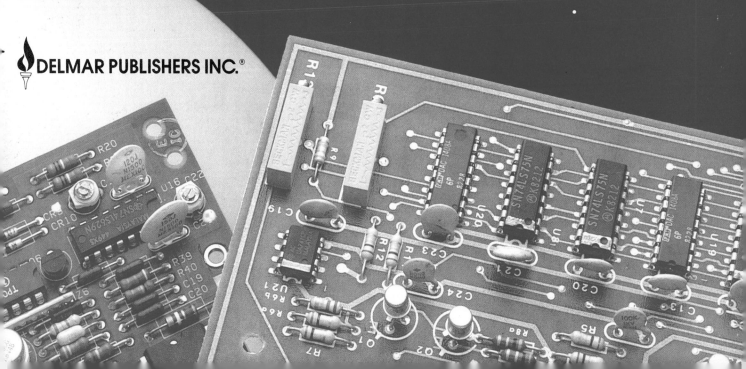

NOTICE TO THE READER

Cover photo credit: Bruce Parker photographer
Cover design by Juanita Brown

Delmar Staff
Associate Editor: Christine E. Worden
Project Editor: Laura Gulotty
Production Supervisor: Karen Seebald
Art Manager: John Lent
Art Coordinator: Michael Nelson
Design Supervisor: Susan C. Mathews

For more information address Delmar Publishers Inc.,
2 Computer Drive West, Box 15-015
Albany, New York 12212

Printed in the United States of America
Published simultaneously in Canada
by Nelson Canada
A division of The Thomson Corporation

10 9 8 7 6 5 4 3 2 1

Library of Congress Cataloging-in-Publication Data

Miller, Rex.
    Electricity and electronics / Rex Miller, Fred W. Culpepper, Jr. —
— 2nd ed.
        p.   cm.
    Rev. ed. of: Energy, electricity/electronics, © 1982.
    Includes index.
    ISBN 0-8273-4419-8
    1. Electric engineering.   2. Electronics.
I. Culpepper, Fred W.
II. Miller, Rex, 1929–
Energy, electricity/electronics.
III. Title.
TK146.M537   1991
    621.3—dc20   90-20530
                                                    CIP

CONTENTS

1.1 Development of Electricity and Electronics • 1.2 Basic Skills in Electricity and Electronics • 1.3 Tools for Electrical and Electronics Work • 1.4 Solder and Soldering • 1.5 Desoldering • 1.6 Nonsoldered Connections • 1.7 Printed-Circuit Boards

2.1 Importance of Safety • 2.2 Safety Rules • 2.3 Tool Safety • 2.4 Electrical Safety

3.1 Some Basic Electrical Terms • 3.2 Structures of Matter • 3.3 Conductors, Insulators, and Semiconductors • 3.4 Voltage (Electromotive Force) • 3.5 Current • 3.6 Resistance • 3.7 Ohm's Law • 3.8 Power

4.1 Resistance and Resistors • 4.2 Types of Resistors • 4.3 Resistor Color Coding • 4.4 Variable Resistors • 4.5 The Electrical Circuit • 4.6 Series Resistance Circuits • 4.7 Parallel Resistance Circuits • 4.8 Series–Parallel Resistance Circuits • 4.9 Kirchhoff's Laws

5.1 Nature of Magnetism • 5.2 Magnetic Theory • 5.3 Electromagnetism • 5.4 Uses of Electromagnetism

6.1 How Meters Work • 6.2 Meter Movement Characteristics • 6.3 Reading an Analog Meter • 6.4 The Ammeter • 6.5 The Voltmeter • 6.6 The AC Ammeter • 6.7 The AC Voltmeter • 6.8 The Series Ohmmeter • 6.9 The Shunt Ohmmeter • 6.10 The Multimeter • 6.11 The Wattmeter • 6.12 The Megger • 6.13 The Digital Meter • 6.14 Clamp-on Meters

7.1 The Nature of Alternating Current • 7.2 The AC Generator, or Alternator • 7.3 AC Waveforms • 7.4 Maximum and Peak Values of AC • 7.5 Transformers

ACTIVITIES AND PROJECTS

PREFACE

Electricity and Electronics is an introductory text. It is intended for use in courses that introduce students to the principles—and to the practical applications—of electricity and electronics.

On completing readings and work assignments of this integrated learning program, students should build a level of knowledge and understanding that will enable them to make meaningful decisions about further studies in electricity and electronics. Knowledge gained from the experiences within this program should also help students determine whether they are interested in careers related to the electrical or electronics fields.

For all students, whether or not they are interested in further study or in work opportunities, the background built through this program will help equip them as more discerning consumers. Another assured learning result for all students is a consciousness of household and occupational safety.

ORGANIZATION

As can be readily seen through a review of the Table of Contents, this book uses a building-block approach in the presentation of information. The organization of this book assumes no previous knowledge or experience on the part of the student.

Information presentations start with the very basics: A history of electricity and electronics, followed by a description of the basic skills in electricity, is provided in the initial chapter. To this foundation, the book adds—one increment at a time—explanations about the different properties and functions of electricity and about different types of circuits and devices.

By the time students have completed the first half of the book, they should have formed a theoretical foundation of knowledge. This understanding, in turn, is applied to a series of practical applications of electricity and electronics presented in the remainder of the text.

New to the second edition is expanded coverage of generators, electric motors, power supplies, amplifiers, oscillators, and integrated circuits. The chapter on electronic communications has been revised to include descriptions of up-to-date technologies. A chapter on robotics has also been included in this edition.

The body of the text ends with a chapter on career opportunities.

The content concludes with an extensive glossary of terms that can be used as a reference tool throughout the study program.

FEATURES

All technical terms are defined in context at the time of their first use. After that, all new terms are used in context at least twice within the first page or two of text after their initial introduction. Meanings of previously used terms are repeated for reinforcement throughout the text to assure ease of reading and comprehension.

Similar techniques are used to help assure student understanding of the mathematical content of this book. Equations and mathematical formulae are used only as necessary. In addition, the math that is used has been selected because it builds a basis for students to move ahead with additional study in this field. To make the math clear, actual examples, using figures from illustrations and the text, are provided to show how all mathematical problems are solved.

Further, all content—verbal and mathematical—is reviewed both in context and at the end of each chapter. The text itself is presented in a series of topical units. At the end of each topical unit, a series of review/discussion questions is provided. This provides an opportunity to monitor student progress on a learn-as-you-go basis.

Throughout the book, textual presentations are illustrated extensively. These illustrations are positioned and cited in the text in a manner designed specifically to promote comprehension.

Within each chapter, there is a series of reinforcement aids. Some of these features are as follows:

- Each chapter now begins with a listing of specific learning objectives that should be met, and ends with a summary. This provides both opportunity and stimulus for review if an individual student is unsure of any element of subject content.
- At the end of each chapter, there is a list of key terms taken from the text. Students should know the meanings of these terms and be able to use them. There is an incentive for review if an individual is unsure of the meaning of an important term.
- Scattered throughout the second edition are numerous hands-on activities and projects that are designed to build understanding by giving students the opportunity to apply their newly acquired knowledge. Each chapter contains a list of suggested activities to further this understanding.
- Selected chapters contain problems that enable the students to apply mathematical formulae presented in the text.
- Strong emphasis is placed—throughout this book—on safety in the use of electrical and electronic devices. Chapter 2 is devoted entirely to safety. Then, within most chapters, there is a list of safety tips that deal specifically with the content areas covered.

ACKNOWLEDGMENTS

Governmental and business organizations in electrical and electronics fields have provided many illustrations used in this book. Contributions of illustrations and information are acknowledged in context throughout the book.

We would also like to acknowledge the contributions of David Gunzel of Radio Shack and Toni Lenuzza of PCB Piezotronics, Inc.

Special acknowledgment is also in order for reviews provided in their respective areas of expertise by Dr. Robert Eversol and Ernest Ezel, both of Western Kentucky University, Richard S. Miller of Northeast Independent School District, San Antonio, Texas, and Mark R. Miller of Texas A & M University, College Station, Texas.

Dr. J. Kenneth Cerny, Oakland School District, Pontiac, Michigan, conducted the content readability analysis that helped establish comprehensibility of this book.

— Rex Miller

Reviewers

Gregory Bazinet
University of Southern Maine
Gorham, Maine

Ward Belliston
Utah State University
Logan, Utah

John A. Cappella
Kalamazoo, Michigan

John Conboy
Alton, New Hampshire

John Cooper
Liverpool Central Schools
Liverpool, New York

Michael Grimes
Chillicothe High School
Chillicothe, Ohio

Raymond Jung
San Gabriel High School
San Gabriel, California

George Legg
Ossining High School
Ossining, New York

Richard McCammack
O'Fallon High School
O'Fallon, Illinois

Dirk Mroczek
Great Bridge High School
Chesapeake, Virginia

Phillip N. Pfeiffer
J. M. Tate Senior High School
Gonzalez, Florida

Dave Pullias
Richardson Independent School District
Richardson, Texas

Dick Robinson
Rutland Area Vo-Tech Center
Rutland, Vermont

C. Guenn Williams
New Braunfels, Texas

Chapter 1

INTRODUCTION TO ELECTRICITY AND ELECTRONICS

OBJECTIVES

After studying this chapter, you will know

- *the impact of electricity and electronics on modern life.*
- *the many uses of electricity/electronics in communications systems, production systems, construction systems, and manufacturing systems.*
- *how the technologies are interdependent.*
- *the various handtools needed for electrical/electronics work.*
- *how to use handtools safely.*
- *how to solder correctly.*
- *how to read a schematic.*
- *how to make wire wrap connections.*
- *how to plan the layout for a printed circuit.*
- *how to make a printed circuit.*

INTRODUCTION

It is hard to imagine life without electricity. Electricity is needed to start a car, to keep the engine running, and to light the way when the car is on the road at night. Without electricity, there would be no television, computers, calculators, radio, radar, microwave ovens, or any number of other things we call "necessities" today.

In the days before electricity, muscle power had to be used to build things. It was also a slow and laborious task to get from one place to another or to ship things. Wind power aided in getting some things done, such as pumping water, moving ships, and grinding grain.

1.1 DEVELOPMENT OF ELECTRICITY AND ELECTRONICS

Hundreds of years passed between the Chinese discovery of magnetism in about 80 A.D. and William Gilbert's scientific analysis of the phenomenon in 1600 A.D. From 1600 on, the developments increased in frequency. In 1879, Thomas Edison was successful in creating the carbon-filament light bulb. Soon it was distributed to many people for their comfort and enjoyment.

By the end of the 1800s, a number of fundamental equations, laws, and relationships had been established in electricity. These developments made it possible for the field of electronics to flourish, Figure 1-1. Figure 1-2 lists some of the primary contributors to the field of electricity and electronics.

Impact of Electricity and Electronics on Modern Life

The impact of the electrical age was felt throughout the world. Old methods were replaced by new ones. People were relieved of back-breaking labor by electrical machines that could do the job better and at less cost. Fears that new inventions or methods would take the place of manpower and create mass unemployment were proved unfounded. Instead of unemployment, electricity brought new industries, each needing more employees than those replaced by electrical machines.

Two important developments that had a great impact were the discovery of the properties of electromagnetic waves and the invention of the vacuum tube. These led to the development of the radio.

Radio developed rapidly, soon becoming the major entertainment medium. By the 1930s, almost every U.S. household had a radio. The use of radio for communications became popular during the days before World War II. Experiments in other fields were also being conducted at that time and some progress was made in the development of television.

Before World War II, there were only two major classifications in the field of electricity. One related to power generation, distribution, and utilization, and the other related to radio transmission and reception. Television was available only on an experimental basis at that time.

World War II was the turning point in the development of electronics. Fantastic electronic devices, such as radar, sonar, and infrared detectors, were then developed. Equipment using these principles was developed for automatic aiming of artillery and for navigation of ships and aircraft.

After World War II, development continued in electronics equipment and the introduction of television dominated the home-entertainment field for years. High-quality sound reproduction gained popularity, as did high-definition color television. FM stations increased the quality of sound broadcasting and stereo records and tapes offered music lovers a greater realism than early phonographs did.

Year A.D.	Event	Year A.D.	Event
80	Chinese discover magnetism.	1915	AM radio is developed.
1600	William Gilbert states the basic principles of magnetism, experiments with static electricity, and establishes the use of the word *electron*.	1921	First commercial radio broadcast, from KDKA, Pittsburgh, takes place.
1752	Benjamin Franklin invents the lightning conductor.	1924	Zworykin obtains a patent on a TV camera tube.
1770	John Cuthbertson develops the electric battery.	1927	First TV station permit is issued.
1789	Galvani performs the frog-leg experiments.	1928	First color TV is produced.
1800	Alessandro Volta develops first electric cell.	1933	FM radio is developed.
1819	Hans Christian Oersted discovers the relationship between electricity and magnetism.	1937	First radio telescope is developed.
1826	Andre Marie Ampere develops a mathematical basis for electrodynamics.	1938	Klystron tube is invented; fluorescent lighting is developed.
1827	Georg Simon Ohm establishes the relationship between voltage, current, and resistance.	1939	TV is demonstrated at World's Fair in New York.
1829	Joseph Henry discovers self-inductance.	1940	Radar is invented.
1831	Michael Faraday explains electromagnetic induction.	1941	First commercial FM broadcasts occur.
1837	Telegraph demonstrated by Cooke and Watson in England; electric motor developed by Thomas Davenport.	1946	Printed circuit board is developed; Electronic vacuum-tube computer (EVIAC) is developed.
1838	Morse devises his code for the telegraph; first submarine cable is laid.	1948	Transistor is invented.
1842	Robert Wilhelm Bunsen demonstrates the carbon-electrode battery.	1954	Color TV broadcasts begin; transistor radio is developed.
1862	James Clerk Maxwell develops mathematical equations setting forth his electromagnetic theory of light.	1957	Citizens Band is created; stereo records are introduced; *Sputnik* is launched.
1876	Bell invents the telephone; Edison invents the phonograph.	1958	First laser is developed.
1878	Hertz experiments with radio waves; William Crookes experiments with the cathode-ray tube.	1959	Integrated circuit is developed.
1879	Thomas A. Edison and Joseph W. Swan demonstrate the carbon-filament light bulb.	1960	First weather satellite is used.
1882	The Edison effect is noted.	1962	MOA integrated circuits are developed (Hofstein).
1888	Hertz makes the first transmitter and receiver of radio waves.	1963	Products with integrated circuits are introduced; first Pulse Code Modulating Circuit is introduced.
1898	Ferdinand Braun constructs the first cathode-ray scanning device; Marconi sends and receives wireless messages.	1965	First word processor is developed.
1904	Fleming invents the rectifier tube.	1970	Computer floppy disc is developed.
1905	DeForest invents the triode amplifier tube.	1971	Liquid crystal display (LCD) is introduced.
1911	Theory of atomic structure is developed by Ernest Rutherford.	1973	First microcomputer is developed.
		1974	First video home recorder is introduced; first fiber-optic circuits are developed.
		1977	Video disc becomes available.
		1980	Satellite TV becomes popular.
		1981	Silicon 32-bit chip is introduced.
		1982	First Megabit IC is developed; flat screen for personal TV is introduced.
		1985	CD-ROM (compact-disc, read-only memory) is introduced.
		1987	Superconductivity is confirmed.

Figure 1-1. Developments in electricity and electronics time line.

Development of the transistor made equipment smaller, and, eventually, with the aid of space exploration, the integrated circuit and the microprocessor were developed. The computer dominated public consciousness and became a popular home installation. Many microwave ovens and washing machines utilizing the microprocessor in the timing of various operations brought into the home the latest in electronics systems technology.

Environmental Considerations

Electronics has been used in reducing the air pollution caused by the internal-combustion engine. Computer control of combustion has also led to the production of the catalytic converter and of engines that can perform as required and still not produce large quantities of harmful gases that pollute the air.

Electronic devices are utilized in monitoring the qual-

Scientist	Dates of Birth and Death	Contribution
Franklin, Benjamin	(1706-1790)	Kite experiments
Galvani, Luigi	(1737-1798)	Frog-leg experiment (electrophysiology)
Volta, Alessandro	(1745-1827)	Electric battery
Ampere, Andre Marie	(1775-1860)	Laws of magnetism
Oersted, Hans Christian	(1777-1851)	Magnetism–electric-current relationship
Gauss, Karl Friedrich	(1777-1855)	Unit of strength of magnetic field
Sturgeon, William	(1783-1850)	Electromagnet (solenoid)
Ohm, Georg Simon	(1787-1854)	Ohm's law
Faraday, Michael	(1791-1867)	Electromagnetic induction
Morse, Samuel F. B.	(1791-1872)	Telegraph in U.S.
Henry, Joseph	(1797-1878)	Fundamentals of electromagnetism
Wheatstone, Charles	(1802-1875)	Telegraph in England, Wheatstone bridge
Cooke, William F.	(1806-1879)	Telegraph engineering
Maxwell, James Clerk	(1831-1879)	Electromagnetic fields
Hughes, David Edward	(1831-1900)	Carbon microphone
Crookes, Sir William	(1832-1919)	Cathode-ray tube and vacuum tube
Fleming, Sir John Ambrose	(1840-1945)	Rectifier tube/Fleming valve
Bell, Alexander Graham	(1847-1922)	Telephone, wax records
Edison, Thomas	(1847-1931)	Electric light, phonograph, and other inventions
Braun, Karl Ferdinand	(1850-1918)	Cathode-ray tube as measuring device (Nobel Prize with Marconi in 1909)
Hertz, Henrich Rudolph	(1857-1894)	Electromagnetic waves
DeForest, Lee	(1873-1961)	Triode amplifier tube
Marconi, Guglielmo	(1874-1937)	Wireless telegraphy (radio) (Nobel Prize in 1909)
Armstrong, Edwin Howard	(1890-1954)	FM radio
Watson-Watt, Sir Robert Alexander	(1892-1973)	Practical radar
Zworykin, Vladimir Kosma	(1899-1982)	TV camera tube

Figure 1-2. People who contributed to the development of electricity and electronics.

ity of the air and water. Such devices have been developed to aid in many ways the fight against pollution of our environment.

Cost of Energy versus Environmental Damage

Of course, the generation of electricity has also been of concern to those involved in monitoring the quality of air and water. Low-sulfur coal has been utilized in fueling steam-powered generating stations, and electronic scrubbers have been employed in cleaning up emissions in power plants.

Atomic-power plants are also sources of environmental concern. The return of the water used to cool the reactors is of particular concern. The temperature and contents of the returned water have to be carefully monitored. Here again, electronics plays a part in the monitoring and alarm systems.

The cost of electrical energy and of the electronics products being manufactured has to be considered in any debate concerning environmental safety: What price will we have to pay for the power generated and the products manufactured in terms of environmental damage? Agencies have been established to study such questions and to protect the environment from damage.

Future of Electricity and Electronics

Any consideration of or projection into the future of electricity and electronics has to be tempered with concern for the long-term impact on the environment and on society. Satellite communications that can cover the world will make their contribution to education, entertainment, and world development as time progresses. The only limitations will be people's imaginations. It is amusing to note that in the early 1800s, the head of the U.S. Patent Office said it was time to close the office because everything had been invented! Since he made his statements, it has become clear that the frontier of unlimited electronics development is limitless. We have before us possibilities as yet unknown in the medical, space, and communications fields.

We have to consider ways to produce electricity without polluting the atmosphere and the land on which we live. We also have to think in terms of smaller products, thanks to the computer and calculator, which have changed our attitudes about the size of electronics products.

Technologies Interdependence

Technologies are interdependent, partially because of the ways in which electrical and electronic systems are developed and manufactured. As the world becomes more complex, its systems for communicating and doing business become more complex. Electricity and electronics play an important role in the expansion of any technology as it may be applied to scientific endeavors.

There is no way to separate the study of mathematics from the development of communications skills in the area of electricity and electronics. Since both of the latter involve the study of many aspects of the physical properties of nature, they can be largely reduced to mathematical concepts. Their interactions can then be stated as laws, formulas, or theories. Once a theoretical base is established and mathematical formulas are applied to the laws of physics, it becomes possible to make practical use of this base in the development of electricity control. Electronics can actually be considered the science of control of electricity by use of the vacuum tube, transistor, and chip, or integrated circuit.

Communications technology involves all aspects of the electrical and electronics means of communicating. This term can cover everything from electronic print to computer-aided drafting, television pictures, video recording and playback, radio, and space communications.

Production technology can be used in reference to electrical power inasmuch as it takes much electrical energy to produce the products we use every day. The electricity is used at the assembly line and for manufacturing processes, as well as to power robots.

Control of manufacturing processes is carried out completely by electrical or electromechanical means.

Production planning and data storage rely completely upon the ability of electricity to perform inexpensive operations and store information as electrical charges.

Transportation technology relies heavily upon electronics for control and routing. Airplanes as we know them would be useless without radar, computers, and communications equipment. Trains also use electronics for communications and control of trains and cargo. Automobiles and trucks rely upon electronics to control their engine combustion and resulting pollutants. Monitoring of engine operation is done through use of electricity and electronics. On the road and in the car, telephone and radio communications speed up the movement and handling of people and goods. The efficiency of modern transportation systems is a result of the proper utilization of electronics.

Construction technology is beginning to use computers for scheduling and for estimating costs. Electronics packages designed especially for the construction industry provide heat, light, and conditioned air for buildings, as well as power for elevators and escalators and for running water.

Today's home relies upon an electrically controlled kitchen, fans for ventilation, and furnaces and air conditioning for temperature comfort. Construction systems have to accommodate these necessities when a house is built. Lighting is provided by electricity. Television, stereo music, and telephones are all objects of the construction system's concern and must be properly planned for correct installation.

In short, energy, electricity and electronics are woven into the fabric of modern-day society. Without controlled electrical energy, we would not be able to function as we do. Our standard of living would be much lower than it is today. The future thus depends on electricity and its applications. That is why it is so important for everyone to have a working knowledge of the concepts and applications of electricity and electronics.

REVIEW QUESTIONS FOR SECTION 1.1

1. What did William Gilbert do to advance our knowledge of electricity?
2. What did Thomas Edison do to make electricity useful?
3. What did people fear about electricity and the advent of electrical machines?
4. What two important discoveries introduced the electronics age?
5. What was the turning point in the development of electronics?
6. How has electronics changed the automobile's environmental impact?
7. How is electronics used to control atomic-powered generator plants?
8. What changed our way of thinking about the size of electronics products?
9. What is electronics?

1.2 BASIC SKILLS IN ELECTRICITY AND ELECTRONICS

This chapter has introduced you not only to the essential elements of any study of electronics and electricity, but also to the basic skills you need to perform on the level expected of you in school.

You will need to learn a number of things before you can become an electricity or electronics employee. The field is always in need of high-quality persons who like to learn and keep pace with everyday developments in the field.

A good place to start is with your own project. Much can be learned by doing such a project. Learning to plan and to read schematics is also very important, as are safe work practices. Read Chapter 2, on safety, before you get underway with your first project.

To handle projects on your own, you will have to be able to read schematic drawings. Schematic drawings are plans. They show the current paths and components for circuits. The symbols presented throughout this book are used in schematic drawings.

In this chapter, you will learn how to use some basic tools safely.

You are ready to begin work on projects. The experience you gain will help you decide about continuing your studies in this field. The exposure you get will also help you decide whether you are interested in electrical or electronics careers.

Planning

It is valuable to look ahead in anything you do. Planning is a method for looking ahead. There is value in planning. It provides a chance to prepare for the work to be done. You determine what parts, equipment, and tools are needed. Potential problems that might come up are identified. Identifying problems can be valuable. Once you know what to expect, you can avoid the problems. You can prepare to deal with them. Or you can take alternative courses to avoid them.

It takes only minutes to plan a project. Planning will save time in the long run. So, planning is a good investment. To plan well, you should use a planning tool. This can be a form like the one illustrated in Figure 1-3. This form gives you an opportunity to list the materials you need. You also list the steps you will take to complete a job. The form provides space to develop a schematic of the circuits involved. As you complete these steps, you will have a chance to identify potential problems. The form then provides space to list both problems and solutions.

Reading a Schematic

In electrical or electronics work, you are often required to read schematic drawings. A **schematic** is a diagram of the circuit to be built. Schematics are drawn with stan-

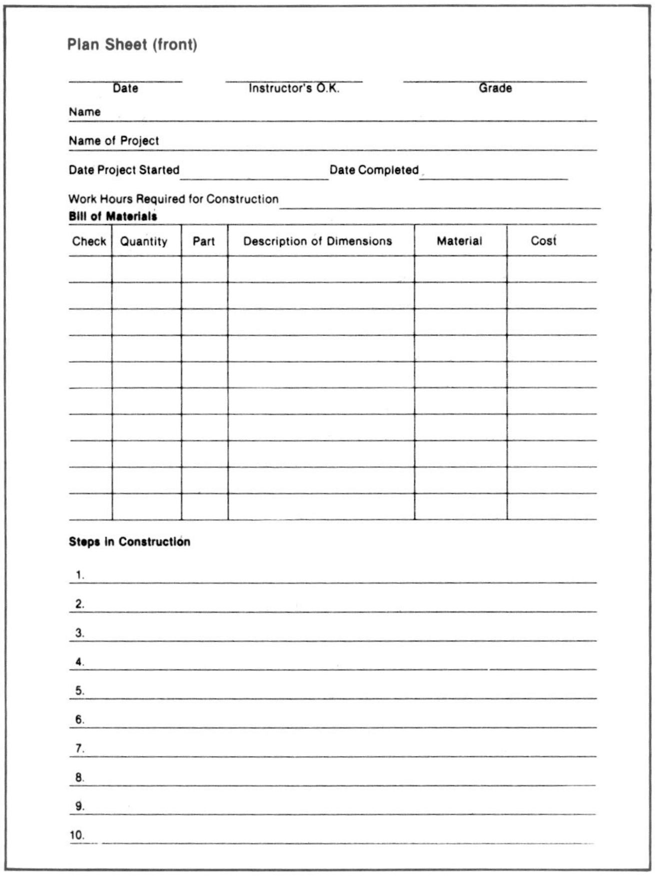
Plan Sheet (front)

Date Instructor's O.K. Grade

Name

Name of Project

Date Project Started Date Completed

Work Hours Required for Construction

Bill of Materials

Check	Quantity	Part	Description of Dimensions	Material	Cost

Steps in Construction

1.
2.
3.
4.
5.
6.
7.
8.
9.
10.

Figure 1-3A. Project plan sheet, front.

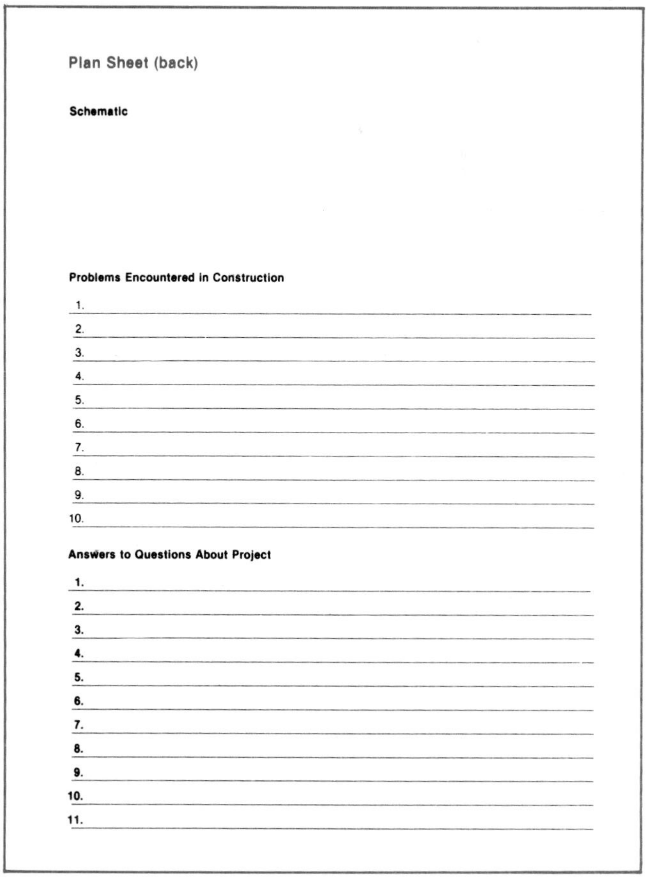
Plan Sheet (back)

Schematic

Problems Encountered in Construction

1.
2.
3.
4.
5.
6.
7.
8.
9.
10.

Answers to Questions About Project

1.
2.
3.
4.
5.
6.
7.
8.
9.
10.
11.

Figure 1-3B. Project plan sheet, back.

dard symbols. You will learn these symbols as you work your way through this book. Many of these symbols are shown in Figure 1-4.

These symbols tell you what components are needed for a project. They show how all components within a circuit or device are connected to each other. So, it is important for you to learn these basic symbols. You should

also learn to read schematic diagrams in which the symbols are used.

Look at the schematic in Figure 1-5. This is a half-wave power supply. Its function, in general, is to supply a 12-V DC power source. To do this, the power supply transforms 120-V AC current to 12-V DC.

A power supply of this type might be used in a shop

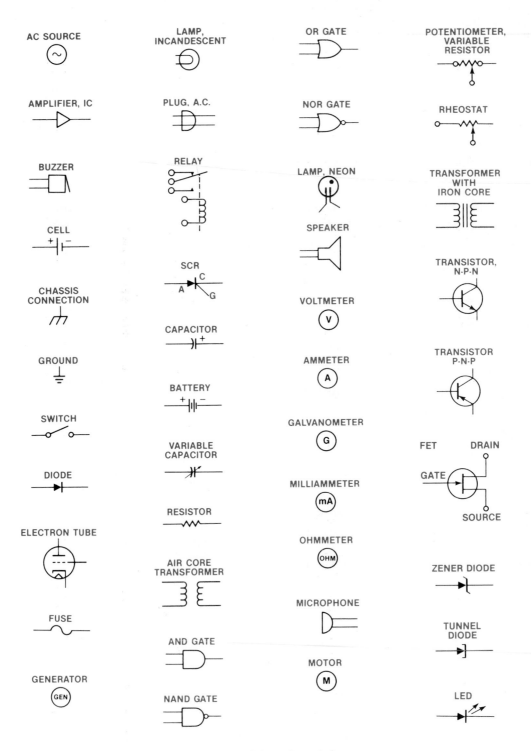

Figure 1-4. Schematic symbols.

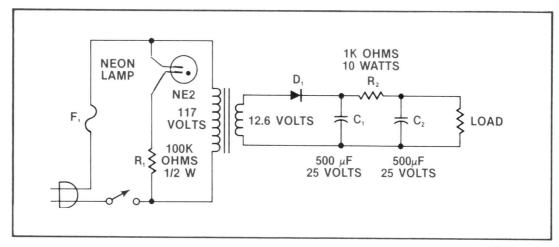

Figure 1-5. Schematic of half-wave power supply.

that repairs automotive radios. With experience and care, you could build a power supply of this type for your own use.

Check some of the important components of the circuit in Figure 1-5. First, notice the fuse and the switch. These are safety requirements. Never overlook them in your own work.

Now, look at the circuit branch with the neon lamp. The neon lamp is used as an indicator. It tells the user whether the power supply is on or off. This is also an important safety feature. If you are working with the power supply, this indicator light is useful. It tells you when the unit is on. And it helps remind you to turn the power off when the unit is not in use.

The neon lamp is connected in series with a resistor. The resistor is rated at 100,000 Ω.

Now notice the transformer. This is also a standard unit. It is called a filament transformer. It reduces voltage from 117 V to 12.6 V. For use in operating auto radios, DC is needed. The diode makes this conversion.

For use in radios, it is necessary to smooth out the flow of DC current. This is done by the two capacitors and a resistor. The capacitors and resistor are connected between the diode and the load.

The more schematics you read, the more you learn. Think of the electrical and electronic symbols as you do

letters of the alphabet. With a little practice, reading schematics will come naturally.

REVIEW QUESTIONS FOR SECTION 1.2

1. Where on the form in Figure 1-3 would you draw a schematic of your project?
2. What part of the planning form would you use to list the steps to follow in completing a project?
3. What is the function of a transformer?
4. How does diode D_1 work in Figure 1-5?
5. If fuse F_1 in Figure 1-5 blows, what may be the problem?
6. If the load is removed, will there be current flow in Figure 1-5?
7. Name the parts of the circuit in Figure 1-6.

1.3 TOOLS FOR ELECTRICAL AND ELECTRONICS WORK

To work with electricity or electronics, you need certain special tools. Each of these tools does one or more specific jobs. Some of the more important tools include:

- Diagonal-cutting pliers
- Side-cutting pliers
- Long-nose pliers
- Combination pliers
- Wire strippers and cutters
- Wire/screw cutters
- Screwdrivers
- Nut drivers
- Allen wrenches
- Scratch awls
- Chassis punches

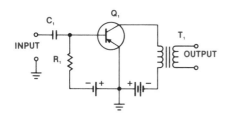

Figure 1-6. Practice schematic. Identify all components.

Diagonal-Cutting Pliers. This tool is used for cutting wire, Figure 1-7. The jaws have hardened edges. The cutting edges are at an angle to permit them to cut flush with the work surface. They should never be used as a gripping tool or to cut hardened wires. Like other pliers, they are made in many different sizes and are grouped by their overall length. The most common sizes are 4, 5, 6, and 7 inches.

Side-Cutting Pliers. These pliers are designed for both cutting and gripping. They are useful for bending, twisting, and cutting heavier wire. Some side-cutting pliers have a wire-stripping groove. They are available in 6-, 7-, and 8-inch lengths, Figure 1-8.

Long-Nose Pliers. These are used to grip and shape wires. For example, you would use long-nose pliers to form loops at the ends of wires. These are used to connect wires to screw terminals. These pliers come in both straight and curved-nose types. This type of pliers is handy when work space is limited and for holding small work. The most common sizes for these pliers are 5 and 6 inches, Figure 1-9.

Combination Pliers. These are multipurpose tools, Figure 1-10. These can be used for holding or bending

wire. They are made in 5-, 6-, 8- and 10-inch sizes. The size indicates the length of the pliers.

Wire Stripper and Cutter. This is a specialized tool. Wire strippers come in a number of shapes. They are used to strip the insulation from the ends of wire. The same tool also cuts wires to needed lengths, Figure 1-11.

Wire/Screw Cutter. This is a useful combination tool. It can be used to cut or strip flexible wire. It can also be used to cut screws to lengths needed in assembly work. It cuts the screws without ruining the threads. In addition, the same tool can be used to crimp terminals to the ends of wires, Figure 1-12.

Screwdrivers. These tools come in a wide variety of sizes and shapes. They are designed for specific purposes or jobs. Screwdrivers have two types of heads. One is the standard head. This is used for screws with slotted heads. Standard screwdrivers come with shaft lengths of 1 to 12 inches.

The other is the **Phillips screwdriver.** This is used for special screws with crossed slots in their heads. These screwdrivers come in sets. The blades are sized on a numbering system from 0 to 4. Phillips screws, generally, are used in manufacturing. The shape of the head is designed for use with power tools. Care should be taken to use a screwdriver with a blade that fits the slot in the screw snugly, Figure 1-13A. Other types of recessed screws and screwdrivers are also shown in Figure 1-13B (page 10).

Nut Drivers. These are used in assembly work. They hold the nuts in place while screws or bolts are tightened, Figure 1-14.

Figure 1-7. Diagonal-cutting pliers.

Figure 1-8. Side-cutting pliers.

Figure 1-9. Long-nose pliers.

Figure 1-10. Combination pliers.

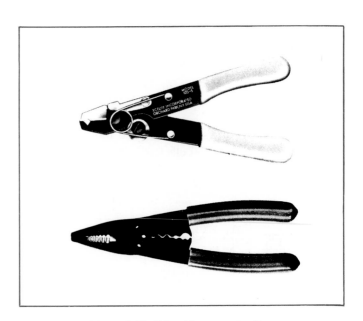

Figure 1-11. Wire strippers and cutters.

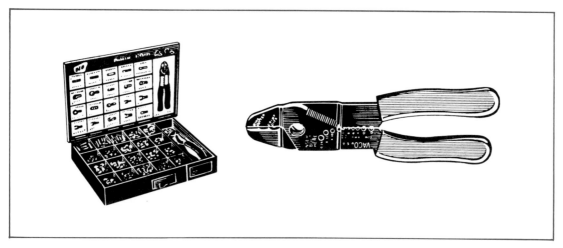

Figure 1-12. Wire/screw cutter. Also used to crimp terminals. Terminals are at left. These materials are sold as a kit.

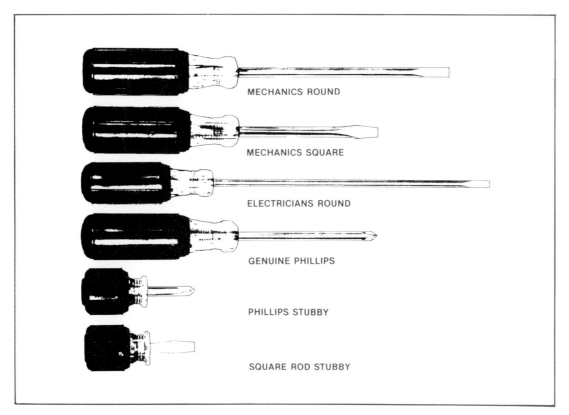

MECHANICS ROUND

MECHANICS SQUARE

ELECTRICIANS ROUND

GENUINE PHILLIPS

PHILLIPS STUBBY

SQUARE ROD STUBBY

Figure 1-13A. Screwdrivers.

Allen Wrenches. These are used to tighten or loosen special bolts. The wrenches fit into six-sided holes in the heads of the bolts. Allen wrenches are available in sets. They are made in English-system sizes, 3/32 and 1/8, or in metric sizes, 4 mm, 5 mm, and 6 mm. A set of Allen wrenches is shown in Figure 1-15.

Scratch Awls. This tool is used for marking metal parts such as sheet metal or conduit. The end of a scratch awl has a sharp point. The scratch marks usually indicate places where metal is to be cut, bent, or drilled. Because

of the sharp point, this tool should be used with care, Figure 1-16.

Chassis Punches. These are used to create holes in sheet metal. The sheet metal is usually used as a mounting for electrical parts. A set of chassis punches is shown in Figure 1-17.

These are just some of the tools necessary for safe electrical and electronics work. All hand tools must have insulated handles. Know your tools. Select and use them with care.

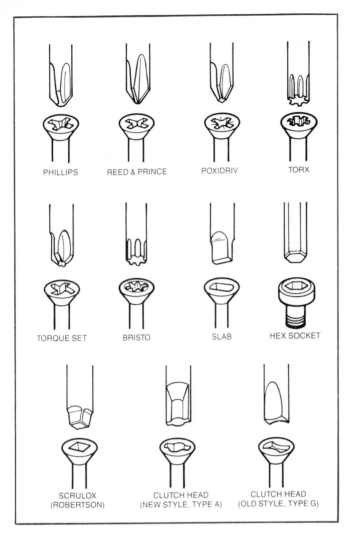

PHILLIPS   REED & PRINCE   POXIDRIV   TORX

TORQUE SET   BRISTO   SLAB   HEX SOCKET

SCRULOX (ROBERTSON)   CLUTCH HEAD (NEW STYLE, TYPE A)   CLUTCH HEAD (OLD STYLE, TYPE G)

Figure 1-13B. Recessed screws and screwdrivers to fit.

Figure 1-14. Nut driver.

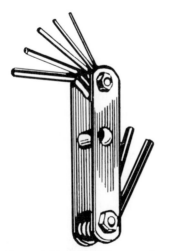

Figure 1-15. Set of Allen wrenches.

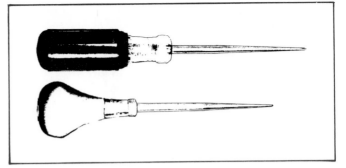

Figure 1-16. Scratch awls.

Figure 1-17. Chassis punches.

REVIEW QUESTIONS FOR SECTION 1.3

1. Why are the cutting edges of diagonal-cutting pliers at an angle?
2. What work is done with side-cutting pliers?
3. What feature of long-nose pliers makes them ideal for tight working areas?
4. What is the purpose behind the shape of the head of a Phillips screwdriver?
5. What is a scratch awl and for what jobs is it used?
6. What are nut drivers and for what jobs are they used?

1.4 SOLDER AND SOLDERING

The most common technique for joining wires, lugs, and terminals in electrical and electronics circuits is soldering.

Solder is an alloy metal. It is made from tin and lead. There are three standard alloy mixtures for solder. These mixtures are given as percentages. The tin percentage is always stated first. Thus, a 40/60 solder contains 40 percent tin, 60 percent lead. There are also 50/50 and 60/40 solders.

In general, the more tin contained in solder, the higher is the quality. In industry, 60/40 solder is used by quality manufacturers.

Safety Tips

1. Be careful when you are around hot soldering irons. Severe burns can be caused by careless use of a soldering iron.
2. Be careful with wire cutters. They can also cut flesh.
3. Wire cutters and slip joint pliers can cause painful injuries at the point where they are adjusted. Make sure your hands are free of these areas.
4. Hot solder can burn. Be careful when using the soldering iron to desolder. Make sure the hot solder is not thrown onto your neighbor or on clothes. Today's miracle fabrics contain plastics. Hot solder will melt them. Holes caused by solder burns are almost impossible to mend.

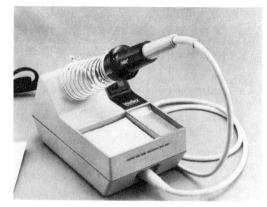

Figure 1-18. Complete soldering station. (Weller)

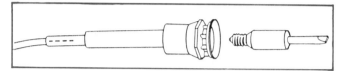

Figure 1-19. Drawing of soldering iron.

Soldering is the technique of joining electrical or electronic connections. In soldering, solder is melted to form a coating over the connection point, forming a joint. Standard solders melt at between 450 and 600 degrees F [232.2 and 315.5 degrees C].

Most solder contains a chemical. This is called **flux**. The purpose of the flux is to clean the area of the connection. This allows the melted solder to flow easily. Flux also prevents oxidation. Two types of flux are available. One is an **acid-core flux**. Acid flux is used primarily in sheet-metal soldering. Never use acid-core flux when soldering copper.

For electrical and electronics work, only rosin flux is used. Rosin flux may be purchased as a paste. However, the most common form is as part of the solder. Flux is built into soldering wire. It is then known as **rosin-core solder**.

There are three methods of applying solder:

- Contact
- Dip
- Wave.

Contact Soldering. This method uses devices called **soldering irons** or **soldering pencils**. Figure 1-18 shows a complete soldering station. It includes a soldering iron. In addition, the same unit has a power supply that controls the current. The current heats the iron. There is also an area containing a wet sponge. This sponge is used to clean the tip of the iron.

Figure 1-19 is a drawing of a soldering iron. This is a self-contained iron. It is simply plugged into an outlet for use. The heating elements are within the end of the iron, which screws into a base.

Soldering irons come in a variety of power ratings. For work on electronics, choose a low-wattage iron. The rating should be between 20 W and 30 W. For heavy-duty electrical devices, the wattage may be higher. Soldering irons used on heavy-duty electrical jobs are rated up to 500 W.

A self-contained soldering iron is heavier than one built into a soldering station. In addition, the soldering station has sensing circuits. These control the temperature of the soldering iron.

Figure 1-20 is a drawing showing a soldering pencil. This is similar in construction to a self-contained soldering iron. However, the pencil is smaller and is used for finer work.

Figure 1-21 is a drawing of a portable soldering pencil. Power for this unit comes from a rechargeable cell within the handle. There is no electric cord. Thus, the portable units are convenient and easy to use. However, they must be recharged periodically.

Figure 1-22 illustrates three common types of tips used for soldering irons and pencils. Any of these tips can be used for any soldering job. The choice depends on the preferences of the user.

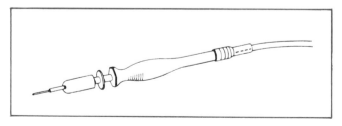

Figure 1-20. Drawing of soldering pencil.

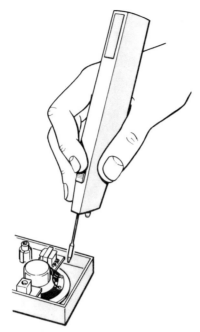

Figure 1-21. Portable soldering pencil. (Weller)

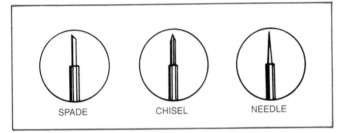

SPADE   CHISEL   NEEDLE

Figure 1-22. Three commonly used soldering tips.

Figure 1-23 pictures a soldering gun. Many hobbyists use units of this type. The trigger of the gun activates a transformer within the unit. Heat is generated through induction. The heat is actually created at the tip of the gun.

To form a solder connection, follow a series of steps:

1. Clean the soldering iron or pencil. If you have a work station, use the wet sponge. If not, portable sponges are available.
   NOTE: To be cleaned properly, the iron must be hot. An iron or pencil should always be at full heat before you begin to use it.
2. Make sure the connection to be soldered is clean.
3. *Tin* the iron. To do this, touch the tip of the iron with solder wire. This assumes you are using rosin-core solder. If not, dip the tip of the wire into flux first. A small spot of solder will form on the tip of the iron or pencil. Wipe the tip with a rag to cause the solder

Figure 1-23. Soldering gun. (Weller)

to coat or "tin" the tip. With the iron tinned, you are ready to solder.

4. Touch the connection with the tip of the iron or pencil. Keep in mind what you are doing. You are heating the connection so that the solder will flow naturally. Simply allow the solder to flow into the connection. The flowing solder makes the connection.
5. Check the quality of your work. A good solder connection should be clean and shiny. There should be no cracks. If the solder is cracked or dull, this indicates a cold solder connection. The iron was not hot enough. Or the connection was not preheated properly. A cold solder connection is unsatisfactory. The connection should be reheated. Also, be careful that the solder does not bridge. *Bridging* occurs when solder runs across copper strips along a printed-circuit board. This causes a short circuit. Figure 1-24 illustrates bridging between connections.

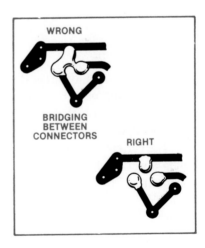

WRONG

BRIDGING BETWEEN CONNECTORS

RIGHT

Figure 1-24. Correct and bridged solder connection.

Soldering Tips

1. Whenever you solder, make sure you use an iron with the proper wattage rating.
2. Make sure you use the proper solder.
3. When soldering a printed circuit, it is very important to choose the right tip for the iron.
4. Make sure you don't hold the hot tip to the printed circuit board too long. The strips of copper will lift from the board if overheated.
5. Apply enough heat to the metal surfaces you are joining.
6. Keep the iron tip brightly coated with solder.
7. A soldering iron in the 20- to 30-W range is recommended for most printed-circuit-board work.

Dip Soldering. For mass production, it is possible to solder a number of connections in one operation. All of the components are put into position first. Then they are lowered into liquid solder contained in a pot, Figure 1-25.

The leads of the connecting wires pick up solder. When the connection is removed from the pot, the solder cools and components are held in place.

Wave Soldering. Wave soldering is a faster, more advanced, mass-production technique. Boards are prepared in the same way as described above for dip soldering. However, there is a pressure mechanism in the pot. This causes the hot solder to flow upward in waves. Circuit boards with components in position are placed just above the level of the hot solder. The waves cause the solder to touch the connection. Soldering is clean and fast. The wave-soldering process is illustrated in Figure 1-26.

___ **REVIEW QUESTIONS FOR SECTION 1.4** ___

1. What metals are used in making solder?
2. What is flux?
3. On what kinds of jobs should acid flux be used?
4. On what kinds of jobs should rosin flux be used?
5. Why should a low-wattage iron be used for electronic soldering?
6. What heat should an iron reach before you begin soldering?
7. How do you "tin" a soldering iron?
8. What should be the appearance of a proper solder connection?
9. What is bridging and what should you do about it?
10. Describe dip and wave soldering.

1.5 DESOLDERING

Sometimes it is necessary to remove solder from a connection. For example, a circuit board may have a bad part. The solder has to be removed before the part can be changed.

Desoldering may also be needed while troubleshooting wired boards. A mistake may have been made in wiring. If so, rewiring may be needed. Before you can reconnect parts, solder must be removed.

There are three basic techniques for desoldering:

- Solder wick
- Solder sucker
- Desoldering iron.

Solder Wick. This technique uses a material that absorbs hot solder. The material is braided wire. The connection is heated with an iron or pencil. Then the braided wire is placed in contact with the connection. The hot solder is absorbed by the wick. When the wick is withdrawn, the solder is removed, Figure 1-27.

Figure 1-25. Solder pot.

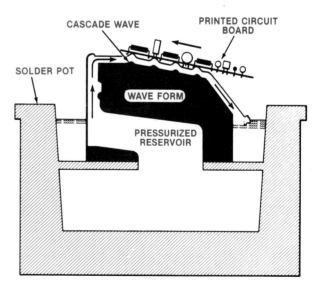

Figure 1-26. Drawing showing wave soldering.

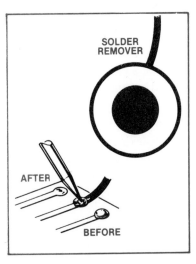

Figure 1-27. Desoldering wick.

Solder Sucker. A simple, inexpensive technique for removing solder is with a squeeze-type bulb. The connection is heated. This is done in the same way as for use of solder wick. Instead of using a wick, a bulb is squeezed, then released. The squeezing causes a suction. The hot solder is sucked into the device, Figure 1-28.

Desoldering Iron. This is a specialized tool for desoldering. It includes both a suction device and a heating iron. The tip of the desoldering iron is placed at the connection. The solder is heated and drawn off as it melts, Figure 1-29.

1.6 NONSOLDERED CONNECTIONS

On many electronic devices, connections are being made without the use of solder. Connections are made through use of wrap-around, or wire-wrap, techniques.

In wire wrapping of connections, a special device called a **wire-wrap gun** is used, Figure 1-30. Connections are made to tinned posts premounted in printed-circuit boards. The gun wraps the wire conductor around the peg, Figure 1-31. The connection is secured mechanically and electrically. It is not necessary to wait for solder to cool or for the connection to warm up.

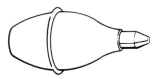

Figure 1-28. Solder sucker.

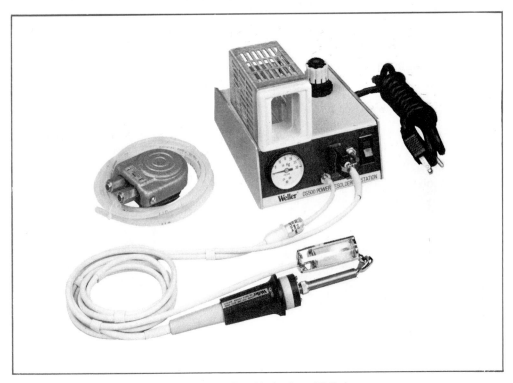

Figure 1-29. Desoldering iron. (Weller)

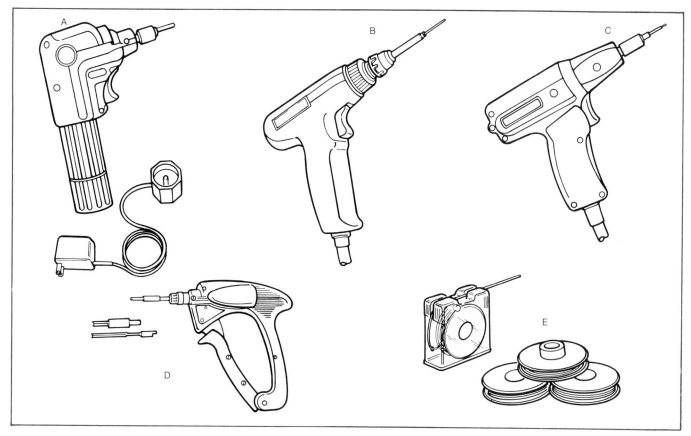

Figure 1-30. Wire-wrapping tools.
A. Battery-powered wire-wrapping tool for field repair or low-production applications where a power source is not available.
B. Lightweight and slim design for service and light production work. **C.** Heavy-duty design, easy to operate for production, installation, and service. **D.** Manual, hand-operated gun using interchangeable bits and sleeves. **E.** Wire dispenser contains three rolls of 30 AWG wire in one dispenser, usually one red, one white, and one blue. It has a built-in plunger that cuts wire to desired length. Refillable dispenser also has a built-in stripper capable of stripping 1 inch of insulation.

Figure 1-31. Wire-wrap connection.

This method is used in high-speed production lines. More people will be using wire-wrap methods in the future. They should be able to save money with this technique. The cost of solder has become so high that it can affect the selling price of electronic devices.

In some situations, these connections may also be soldered after they have been wrapped.

REVIEW QUESTIONS FOR SECTION 1.6
1. Describe wire wrapping.
2. What are some advantages of wire wrapping?
3. Why is wire wrapping less expensive than soldering?

1.7 PRINTED-CIRCUIT BOARDS

Printed circuits are defined as electrical circuits in which solid copper conductive patterns are bonded to an insulating base material.

Development of Printed Circuits

There are many methods used for putting a conductive pattern onto an insulating base. Nearly all of this development has taken place in a very few years, as very little work was done with printed circuits in the United States before World War II.

The proximity fuse, a printed-circuit device that exploded shells from anti-aircraft guns, required large numbers of very compact electronic devices. Techniques used to mass-produce these devices were developed into what is referred to as the first printed circuit.

In 1941, Dr. Paul Eisler, a research worker for Henderson and Spaulding, a printing firm in London, attempted to apply the techniques of the printing industry to the production of conductive circuits. The methods of the

15

photoengraver were used to apply a photosensitive, acid-resisting coating to the metal surface, which was then exposed through a photographic negative, developed, and etched in the manner used to produce printing plates. The etching process removed the unwanted foil from the insulated base and left a metal-foil layer in the form of the desired circuit, Figure 1-32.

Dr. Eisler worked on a small scale in his experimentation with printed circuits. His activities attracted little attention until 1946, when the U.S. Army Signal Corps took note of the work and extended his experimentation to develop uses for the new electronic circuits. At about the same time, and also for military purposes, the National Bureau of Standards, in conjunction with the Centralab Company, developed a ceramic-based conducting circuit utilizing mass-production methods. Manufacturers of commercial radio and television equipment took over where they left off and developed the printed circuit for peacetime uses.

Rockets, missiles, computers, transistorized radios, and television sets have utilized the printed circuit to such an extent as to make this method of production the largest in the field of electronics today. Automobiles are now using printed circuits to eliminate the maze of wires once found behind the automobile instrument panel. Electrical meters with printed circuits are not damaged by momentary overloads of as much as 10,000 times their normal capacity.

Methods of Production

There are a number of ways to produce a printed-circuit board. They can be etched (most are still done this way), or can involve stamping, embossing, inking with conductive inks, plating, use of pressed powder, and use of a laser to evaporate the copper foil.

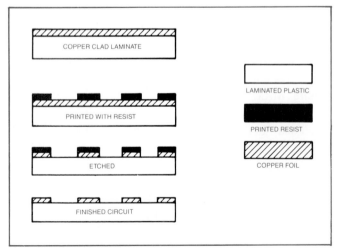

Figure 1-32. Etching a circuit.

Metal Foils

Electrolytic copper foil is used almost exclusively for all etched circuit applications because of its low cost, good solderability, and high conductivity.

Tin, aluminum, steel, silver, or alloys of various metals may be used in some special cases. Copper, however, is a standard material. Copper-clad boards may warp during the bonding operation, but they usually straighten out after etching when some of the copper has been removed.

One of the more inexpensive types of board for printed circuits is made of a paper base (100% rag or cotton-content paper) impregnated with a thermosetting product known as a phenolic resin binder. The paper and its binder are compressed under high temperature and pressure and come out of the process as a hard, solid sheet of high mechanical strength. Boards are available commercially in thicknesses of 1/64, 1/32, 3/64, 1/16, 3/32, 1/8, 3/16 and 1/4 inch. The electrical qualities of the material are rated according to grades that are designated *X*, *XX*, and *XXX*. A *P* indicates good punching qualities for the board. The X grade is the poorest quality and the XXXP grade is the best quality. The cost of materials runs highest for grade XXXP and lowest for grade X. Grade XXXP has high insulation resistance and low dielectric losses under high-humidity conditions. Fiberglass is used in some instances for high mechanical strength and high resistance to change because of humidity variations. Such materials as melamine, epoxy, and silicone may be used for their special insulating qualities.

An insulating board coated on both sides may be used for its natural capacitances. The board makes an excellent dielectric and the copper foil on each side serves as a plate for the capacitor. This can also become a problem when extremely high frequencies are used and the extra capacitance between conductors has to be taken into circuit-design consideration.

Arc-Over

Over 1500 volts is required for arc-over on 1/32″ line spacing between conductors on phenolic material.

Even small line widths can handle high currents. For instance, copper 0.0015″ thick and 1/64″ wide can handle three amperes of current. If the copper is 0.0067″ thick and 1/4″ wide, it will handle 48 amperes of current. These examples illustrate the ability of small "lines" of copper to carry large currents.

Planning the Etched Circuit

One of the most important parts of constructing a project is the planning phase. Proper planning eliminates mistakes. When working with printed circuits, a great deal of planning must be done before a project or completed circuit is made functional. The following steps are presented in an attempt to eliminate mistakes that beginners often make.

A schematic drawing of the desired circuit is a good starting point. Redraw the circuit so that the portion to be etched onto a copper-clad board has input and output terminals located for easy connection to other components, Figure 1-33.

Figure 1-34 shows the layout of the pc board. Figure 1-35 shows how the holes are lettered on top of the board where the components are inserted to match those on the

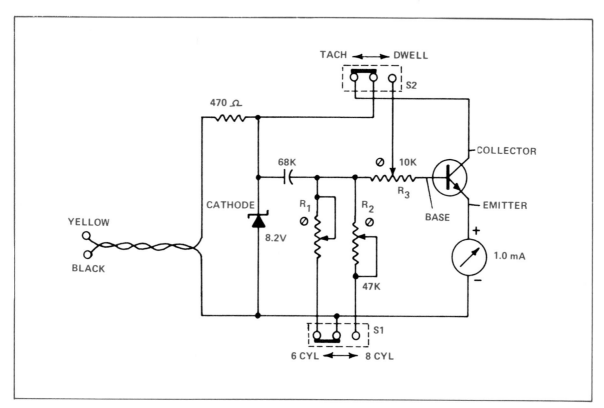

Figure 1-33. Schematic of a dwell tach for checking auto-engine timing.

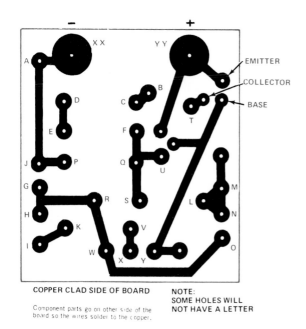

COPPER CLAD SIDE OF BOARD   NOTE:
                             SOME HOLES WILL
Component parts go on other side of the   NOT HAVE A LETTER
board so the wires solder to the copper.

Figure 1-34. Copper-clad side of board with circuit etched.

pc board copper side. Note that some holes will not be used. C to D indicates that a jumper wire is used; the pc board strips cannot touch or jump over one another without making electrical contact.

Figure 1-36 shows how the components are placed into the holes and bent to touch the foil. Figure 1-37 shows how all the components are mounted on top of the board and are ready for soldering. After these components have been hand-soldered or flow-soldered, the zip cord, Figure 1-38, is added and soldered. This is the input of the circuit. XX and YY are the output of the circuit, and snap onto the meter movement with clips soldered to the large dots.

Preparing for Circuit Etching

There are a couple of ways you can produce the pc board for experimental work. Use a pressure-sensitive tape for a resist. You can even use the black plastic electrician's

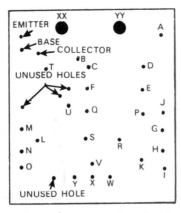

Figure 1-35A. Top of pc board with hole locations.

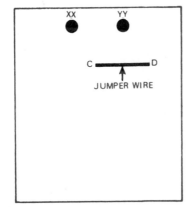

Figure 1-35B. Note how jumper is placed.

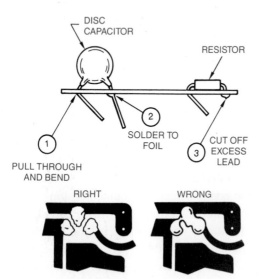

Figure 1-36. Mount components on top of pc board. Solder without making bridges.

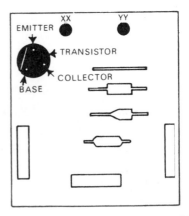

Figure 1-37. All parts are located on top of the board. The transistor needs special attention.

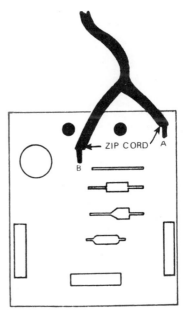

Figure 1-38. Zip cord represents the input to the circuitry in holes A and B.

Before applying either tape or paint, it is a good practice to clean the board thoroughly. Make sure there are no fingerprints on the copper. Use steel wool or a household cleaner to make sure the board is free of contamination. This also increases the life of the etchant solution.

Etching the Circuit

After the board has been prepared with acid-resist paint or tape, it is ready for the removal of surplus copper. The removal is a chemical process.

To prepare the etchant, use 16 ounces (1 lb.) of ferric chloride (powdered) to 1 gallon of lukewarm water. Dissolve slowly in a *glass container*. The solution will turn brown. Store the solution in a glass container. *After it has been exhausted, dispose of it properly according to EPA*

tape to mask the area you want saved. Or, you can use asphaltum-based paints to cover the strips you want to use as circuit paths. Tape-dispensing machines and dots are available commercially.

regulations. Do not discharge it down the sink into the sewage system.

Ferric chloride may be obtained from the chemistry lab in your school or from a local supplier.

Safety Tips

The ferric chloride solution is a relatively mild reagent, but it will leave stains on clothes and skin. Open cuts and sensitive skin should be covered with rubber gloves. People who react to photographic chemicals may experience ill effects other than stained fingers.

When mixing chemicals, note the increased temperature of the solution as water is taken on by the ferric chloride. The process is called *exothermic hydration.*

The board with the circuit protected by resist is placed in a glass or hard rubber tray (photographic darkroom trays work well) and covered with at least 1/4″ of etchant. The etching process can be speeded by rocking the tray gently to move the etchant across the board. It usually takes 15 to 20 minutes to remove the excess copper. This time increases as the solution becomes used. Progress may be checked by using photographic tongs to pick up the board and inspect the surface. If the board is left in the solution too long, the copper will be undercut and it will be difficult to solder the components in place without the copper lifting from the board.

Remove the resist from the board and inspect the circuit. In the case of tape, the resist may be removed by simply pulling it off the board. Paint may be dissolved by lacquer thinner, turpentine, or other appropriate solution.

Be sure the board is clean so that a good solder joint may be made later. Very fine steel wool will clean the board nicely before the component parts of the circuit are mounted.

Mounting Components

The last step is the easiest, and usually the most interesting, as all parts fit into their planned locations.

Dots and terminal locations must be drilled to allow the component leads to pass through the board to be soldered to the copper circuit on the other side.

Transformers and large components may be mounted to the board with machine screws or pop rivets. Jumpers, if required, may be made with insulated wire, or with uninsulated wire if used on top of the board.

Soldering

Soldering is done with a small iron. A 35-watt soldering iron is usually sufficient. The pc board can withstand about 450°F for only 5 to 7 seconds before the copper lifts from the board. Just make sure the component lead and the copper strip are hot enough to melt solder for a good solder joint. Use rosin-core solder.

REVIEW QUESTIONS FOR SECTION 1.7

1. What is a printed-circuit board?
2. When was the printed-circuit board first developed?
3. Where is the printed-circuit board used today?
4. What does *XXXP* mean?
5. What is meant by *arc-over*?
6. What is used to etch a printed circuit board?
7. What is exothermic hydration?
8. How much heat can a printed-circuit board stand during soldering?

SUMMARY

Life without electricity means life without calculators, computers, radios, radar, microwave ovens, electric lights, or home temperature-comfort control.

In 1600, William Gilbert began scientific inquiry into the nature of magnetism and started a process that has led ultimately toward today's electrification of most aspects of life.

Edison followed up with the light bulb and many other inventions. By the end of the 1800s, laws and formulas were established, and the science of electricity moved forward rapidly.

People were relieved of back-breaking labor by electrical machines. More people were needed to operate and repair the new machines.

Radio developed rapidly after the invention of the vacuum tube. World War II caused the development of radar, sonar, and infrared detectors, as well as communications systems.

The future of electricity and electronics is limited only by people's imagination. All the technologies are interdependent and rely upon electricity and electronics for advancement.

A schematic is a diagram of a circuit to be built. Schematics are drawn with standard symbols. These symbols tell you what components are needed for a project.

To work with electricity or electronics, you need certain special tools. Each of these tools does one or more specific jobs.

The most common technique for joining wires, lugs, and terminals in electrical and electronics circuits is soldering. Solder is an alloy metal. It is made from tin and lead. Dip soldering is used for mass production.

It is possible to solder a number of connections in one operation. This is done to secure components onto printed circuit boards. Wave soldering is done by a machine that heats the solder so it flows when pumped into waves. As the printed-circuit board comes in contact with the solder wave, it solders the components in place on the board.

A solder wick is used to absorb hot solder and clean up a joint or remove the solder so the component can be removed. A solder sucker sucks the hot solder from a joint so the component can be removed.

Wire wrap is a technique used in many television sets today. It wraps a wire around a peg without use of solder. A special gun is used for the job.

Printed-circuit boards were developed to get rid of some of the wiring that was originally part of every electronics device. Metal foils are attached to phenolic or fiberglass boards and then etched to make circuit paths. Various types of boards are available for different types of circuit requirements. Ferric chloride is used most frequently for etching purposes.

USING YOUR KNOWLEDGE

1. Draw a schematic of a project using solid-state components to produce a dimmer for a lamp.
2. Draw a schematic of a project for the stepping up of 12-V DC current to operate a low-wattage 120-V AC device in a car.
3. Draw a schematic of a project for the building of a crystal-set radio receiver.
4. Lay out the circuit for a neon lamp and resistor in series to provide a test lamp for 120 volts AC and DC.
5. Etch the circuit on a printed-circuit board.
6. Solder the neon lamp and resistor to the copper-clad side of the printed-circuit board.
7. Test your circuit by inserting it into a 120-volt AC receptacle. *Don't touch the exposed printed-circuit copper strips. ALSO, do not allow the printed-circuit-board side to touch a metal surface when the circuit is plugged into a power-live source.*

KEY TERMS

solder
flux
soldering irons
soldering pencils
diagonal-cutting pliers
side-cutting pliers
long-nose pliers
combination pliers
nut drivers
Allen wrenches
Phillips screwdriver

scratch awl
chassis punch
solder wick
solder sucker
desoldering iron
wire-wrap gun
rosin-core solder
acid-core flux
schematic
printed circuits

Chapter 2

SAFETY

After studying this chapter, you will

- *understand that electricity is a force that can kill or maim.*
- *understand the proper procedures to be followed when working around electricity.*
- *realize the importance of working safely and with high regard for the power of electricity.*
- *know that you must never work on a "live" circuit.*
- *know how to use and store tools properly.*

INTRODUCTION

You can work safely around electricity if you observe some common sense rules. It is also possible to become injured by the electrical energy that can do so many good things when properly harnessed.

Keep in mind that it is better to be a *safe* user of electricity than a victim of it.

Electricity can be a valuable tool and power source. However, it is necessary to keep it in its place. A study of some simple rules will aid you in utilizing this power source for your own good.

2.1 IMPORTANCE OF SAFETY

Electricity is valuable to you. From electricity, you get convenience. You use electricity as a regular part of your everyday life.

Many people use electricity without really thinking about their actions. But there is something about electricity you can never afford to forget: Electricity can be dangerous. You can be injured, even killed, by electricity.

In your classwork—as well as in your everyday life—you should be aware of the safety hazards you face. A little care can go a long way in protecting your safety—and the safety of others.

2.2 SAFETY RULES

Below are some simple rules for electrical safety. Follow them whenever you work with electricity. They will help you form safe habits. Of course, no set of rules will guarantee your safety in every situation. You still have

to understand what you are doing. Most importantly, you have to think about what can happen every time you use electricity or electrical devices. There is no substitute for common sense. Read and remember these rules.

1. Any time you handle electric wires or circuits, assume they are dangerous. Beware of a false sense of security.
2. Use only the right tool for every job. Make sure you use the protective devices provided.
3. Before you work on electrical equipment, remove all watches or jewelry.
4. Never tamper with safety devices. They are there for your protection.
5. Make sure you have a fire extinguisher available. Be sure it is the correct kind of fire extinguisher for the work you are doing.
6. Follow directions on the use of chemicals, including cleaning solvents. Improper use of chemicals can cause explosions, burns, or fires.
7. Use the proper protective clothing or safety equipment for each job.
8. Don't work on "live" circuits. Unplug devices before you begin working on them. If you are working on wiring, turn off the circuit breaker or fuse. If necessary, turn off the master switch.
9. If you can't turn off the power for any reason, follow the one-hand rule. Keep one hand in your pocket as you work. This will make it difficult for you to establish a grounding connection. You avoid shocks and other accidents in this way. Follow the same rule whenever you work with capacitors or other devices that store electricity.
10. Never work on electrical equipment when you are standing on a wet surface.
11. Never remove equipment grounds.
12. Inspect all electrical equipment from time to time. Look for loose wires or frayed insulation on cords. If you find any safety hazards, repair the equipment before you use it.

Throughout the book you will find Safety Tips highlighted when they are important to the discussion at hand and deal with the possibility of shock or personal injury.

Safety Tip

You might save a life with mouth-to-mouth resuscitation, Figure 2-1. Here's how to do it:

Clear victim's mouth of foreign matter, if any.

Lay victim on back, put folded coat or blanket under his shoulders to tilt his head back.

Pinch his nostrils shut, then take a deep breath, seal your mouth over his and breathe out until you see his chest rise. Then remove your mouth and listen for outflow of air.

Continue this every 5 seconds for an adult, every 3 seconds if victim is a child.

Figure 2-1. (Courtesy of National Safety Council)

2.3 TOOL SAFETY

The electrician and electronics technician have to observe certain rules and use common sense to work safely. Each uses a number of different types of tools. Some of them can be more dangerous than others. However, there are some rules for tool use that may be of aid to those who work with electrical equipment at all levels.

Screwdrivers

The professional requires a number of screwdrivers in a variety of sizes and types. The right driver is necessary for the fast, efficient driving and removal of screws in any kind of material. The wrong-size driver—too short or too long—or a driver with a point that doesn't fit the screw properly can waste time and cause trouble.

The basic rule is to fit the tool to the work. The size of the screw and the type of opening it has determine which driver you use. Slotted screws, Phillips screws, Robertson screws, and others come in many sizes and lengths. There are screwdrivers to fit all of them. But there are a few tips on how to use a driver from which we can all benefit, because screwdrivers are the most often misused and abused hand tools of all, Figure 2-2.

Attempts to repair most types of drivers are *not recommended*. The tip of a slotted screwdriver can be dressed

on a bench grinder, but care must be taken not to let the tip get too hot. If it gets hot to the touch, the temper or hardness has been drawn and the tip will no longer stand up well. Drivers with cracked handles, bent or twisted shafts, or worn tips should be discarded and replaced.

Wrenches

Whatever the job, only a wrench of the proper type and size will give you the kind of results you want. That one right wrench will do the job correctly, with less effort and more safety than any other wrench, Figure 2-3.

This is a basic and obvious fact. It is stated here only to reinforce the professional approach to tool selection. The person who makes a living with tools never needs to improvise, because the right tool is always available.

Attempts to repair box, open-end, or combination wrenches are not recommended. If any of these wrenches have bent handles, spread, nicked, or battered jaws, or rounded or damaged box points, the wrenches should be discarded and replaced.

Socket and adjustable wrenches can be repaired by the replacement of damaged parts. Periodic inspecting, cleaning, and light lubrication serve to maintain these wrenches and reverse any damage. An adjustable wrench with a spread or damaged fixed jaw or a bent handle should be discarded and replaced. Bent socket-wrench handles and

1. **Never use a driver to do another tool's job.** Using a driver as a pry bar, or a scraper, or a chisel can ruin the tool and spoil the work, not to mention the time you might lose or injury you might sustain. Limit screwdrivers to screws.

2. **Never push a driver beyond its capacity.** Make a pilot hole for the screw and your work will go more easily. Use a square-shank driver for heavy work requiring the use of a wrench to help do the turning.

3. **Never expose a driver to excessive heat.** Direct flame can draw the temper from the metal, weakening and possibly warping it, making it unsafe and inefficient to use.

4. **Never use a driver at an angle to the screw.** Always keep the shank perpendicular to the screw head. This is as important as using a driver with a point that fills the screw opening. Driving at an angle to the screw or using a too-small point can spoil the screw and also slip and damage the work.

5. **Never depend on a driver's handle or covered blade to insulate you from electricity.** Plastic and cushioned grip handles are intended only to provide a firm, comfortable grip. Insulated blades are intended only as a protective measure against shorting out components.

Figure 2-2. (Courtesy of Klein Tools, Inc.)

1. **Never use a wrench to do another tool's job.** You won't do the job as well and you might damage or even break the wrench. And using a wrench as a hammer or a pry bar or anything else can be dangerous. Take the time to get the right tool.

2. **Never use a wrench opening too large for the fastener.** Using a wrench opening too large for the nut or bolt can spread the jaws of an open-end wrench and batter the points of a box or socket wrench. A too-large wrench opening can also spoil the points of the nut or bolt head. And when selecting a wrench for proper fit, take special care to use inch wrenches on inch fasteners and metric wrenches on metric fasteners.

3. **Never push a wrench beyond its capacity.** Quality wrenches are designed and sized to keep leverage and intended load (torque) in safe balance. The use of an artificial extension on the handle of any wrench can break the wrench, spoil the work, and hurt the user. Instead, get a larger wrench or a different kind of wrench to do the job. The safest wrench is a box or socket type. (To free a "frozen" nut or bolt, use a striking-face box wrench or a heavy-duty box or socket wrench; never use an open-end wrench. And apply penetrating oil beforehand.)

4. **Never expose a wrench to excessive heat.** Direct flame can draw the temper from the metal, weakening and possibly warping it, making it unsafe to use.

5. **Never push on a wrench unless absolutely necessary.** There may be situations in which you can only push a wrench handle to loosen or tighten a nut or bolt. But *you should always pull on a wrench* to exert even pressure and avoid injury if the wrench slips or the nut breaks loose unexpectedly. (If you must push the wrench, do it with the palm of your hand and hold your palm open.)

6. **Never cock or tilt an open-end wrench.** Always be sure the nut or bolt head is fully seated in the jaw opening for both safety and efficiency. A box or socket wrench should be used on hard-to-reach fasteners. Adjustable wrenches should be tightly adjusted to the work and pulled so that the force is applied to the fixed jaw.

7. **Never depend on plastic-dipped handles to insulate you from electricity.** Plastic-dipped handles are for comfort and a firmer grip. *They are not intended for protection against electric shock.* (Special high-dielectric handle insulation is available, but it should only be used as a secondary precaution.)

Figure 2-3. (Courtesy of Klein Tools, Inc.)

extensions, and cracked or battered sockets should be discarded and replaced.

Caution

Plastic-dipped handles are NOT intended for protection against electrical shock. This includes high-dielectric plastic-dipped handles.

Pliers

Each type of plier does its own particular job better than any other type can. Professionals take the job—and the tools—seriously. They know all the rules about using pliers. They have heard them or read them or learned them through experience, Figure 2-4.

Tools safety can be summed up with an excerpt from *Safe Worker*, a publication of the National Safety Council.

1. Take care of your tools, so they don't "take care" of you.
2. Keep them clean. Check their condition before using them. If the heads of striking tools become mush-

1. **Never use pliers to do another tool's job.** A plier is not a hammer or a pry tool or a wrench. Using a plier instead of the proper tool risks damaging the plier, damaging the work, damaging yourself, and losing time. It's never worth it.

2. **Never push pliers beyond their capacity.** Bending stiff wire with light pliers or the tip of needle nose pliers can spring them or break them. Use a stronger, blunt nose plier. When you need greater leverage, use a plier with greater leverage. Don't extend the length of the plier handles. Bolts should be cut with a bolt cutter, large cable with a cable cutter. To each its own.

3. **Never expose pliers to excessive heat.** Direct flame on metal can draw the temper and ruin the tool. Cutting pliers are especially vulnerable to high, direct heat.

4. **Never cut hardened wire with ordinary pliers.** Pliers should not be used for cutting hardened wire unless they are specifically recommended for this use.

5. **Never rock pliers from side to side when cutting wire...and never bend the wire back and forth against the cutting knives.** Either practice can dull or nick the cutting edges. Cut wire at a right angle only. If it won't cut through readily, the knives may need sharpening. Or you may need a plier with greater leverage.

6. **Never cut any wire or metal unless your eyes are protected.** Safety goggles or other protective devices are an absolute must. It's easy to forget to wear them. It's a big bother to put them on for "just one cut." You've heard all the reasons and excuses. But none of them make any sense. They're all part of the lazy man's way, not the professional's way...the safe way.

7. **Never cut any wire or metal unless your fellow workers' eyes are also protected.** Now maybe you never heard that one before. But it makes sense. The wire that doesn't get you may get somebody else. So think about the "other guy" as well as yourself.

8. **Never depend on plastic-dipped handles to insulate you from electricity.** Plastic-dipped handles are for comfort and a firmer grip. *They are not intended for protection against electric shock.*

9. **Always wear protective goggles, Figure 2-5.**

Figure 2-4. (Courtesy of Klein Tools, Inc.)

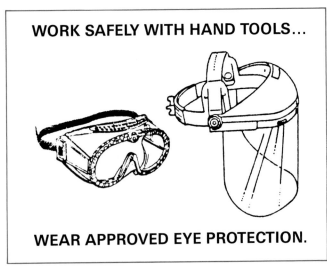

WORK SAFELY WITH HAND TOOLS...

WEAR APPROVED EYE PROTECTION.

Figure 2-5. Always wear eye protection! (Courtesy of Klein Tools, Inc.)

roomed or burred, have them replaced. If any handles are splintered, broken or loose, have them replaced.

3. Each tool should have its own storage place. Tools should be returned to their proper places, and not left lying where they could fall onto or trip you or someone else.

4. Carrying tools in clothing pockets is dangerous, especially if the tools are sharp or pointed. Use a tool belt and tool pouch.

5. Never use excessive pressure or force on any hand tool.

2.4 ELECTRICAL SAFETY

In order to work safely in any lab or shop, you should make sure that equipment wiring, switches, and motors are properly grounded. Most shops require dustproof switches and devices. Paint spray booths require explosion-proof wiring and lighting. *National Electrical Code®* and local ordinances should be followed in all school-shop electrical wiring. All portable power tools and fixed machines should be inspected for shorts on motors and wiring. All switches should be located on the front of machines.

OSHA (Occupational Safety and Health Act of 1970)

Enactment in the U.S. of the Occupational Safety and Health Act of 1970 (OSHA) has given a new force to an already extremely important aspect of on-the-job safety. Now, not only linemen, tree-trimmers, and others who

must work at elevated job sites, but all employees must be protected from falling by appropriate equipment. This regulation applies at locations such as towers, poles, aerial lifts, catwalks, and other elevated sites not adequately protected by enclosures, guardrails, or nets. The new federal standards cover all employees in general industry, construction, and utilities.

OSHA also has a checklist to make working in the shop safer.

1. Storage areas should be kept free of debris.
2. Aisles and passageways should be kept clear.
3. Every opening, floor, or platform 4 or more feet above ground level should be guarded by standard rail and toe boards.
4. All stairs with four or more risers should have standard hand rails.
5. Slippery conditions on floors are to be eliminated as soon as possible.
6. Combustible waste material and residue in building or operating area are to be kept in a covered metal receptacle.
7. Personal protective equipment for eyes, face, head, and extremities, and protective clothing, shields, and barriers are to be provided, used, and maintained.
8. All pieces of protective clothing and equipment are to be properly stored for ready use.
9. Suitable fire-extinguishing equipment is to be immediately available in the work area and is to be maintained in a state of readiness for instant use.
10. Extinguisher tops are to be not more than 5 feet from the floor. Those weighing over 40 pounds are to be not more than 3½ feet from the floor.
11. Fire extinguishers are to be inspected annually by a competent person and are to be operable.
12. Access to extinguishers and exits is not to be hindered in any way.
13. All portable electrical tools are to be equipped with hand-operated switches that must be manually held in a closed position.
14. All hand and portable power tools are to be in good operating condition, with no defects in wiring, and to be equipped with ground wires.
15. Extension cords used with portable electrical tools and appliances are to be of three-wire type unless they are of the U.L.-approved double-insulated type.
16. All potential sources of fire and/or explosion from gases, vapors, fumes, dust, and mists are to be inspected for correctable hazards.
17. All hazardous gases, liquids, and other materials are to be properly labeled and stored.
18. Lighting in work areas is to be adequate for jobs performed.

19. Each electrical box is to be provided with a cover that effectively protects against the hazard of accidental contact.
20. A trained person or persons are to be available to render first aid. First-aid supplies approved by the health department are to be readily available.
21. Safety meetings are to be scheduled and held at regular intervals.

(Information and illustrations on tool safety are provided by Klein Tools, Inc.)

SUMMARY

Safety is very important when you work with electricity. There are a number of safety rules that will aid you in this area. A number of them are enumerated in this chapter.

Mouth-to-mouth resuscitation is described and tool safety is emphasized. OSHA and its implications are discussed. A checklist provided by OSHA is also included.

Chapter 3

UNDERSTANDING ELECTRICITY

OBJECTIVES

After studying this chapter, you will know

- *that electricity is a form of energy that results when electrons flow through a conductor.*
- *that the word* current *describes the movement of electricity.*
- *that a conductor is a material that carries electric current, an insulator is a material that resists the flow of current, and a semiconductor is neither a good conductor nor a good insulator.*
- *that a circuit is a complete path for the flow of electricity.*
- *the basic structure of atoms and understand the relationships of atomic particles, including protons, electrons, and neutrons.*
- *that the binding force in atomic particles comes from attraction of negative and positive charges to each other.*
- *that atoms form elements, elements are basic forms of matter, and each element has its own atomic structure.*
- *that the force that drives electricity is called an electromotive force (emf) and is measured in volts.*
- *that current is measured in amperes, that resistance is measured in ohms, and that electric power is measured in watts.*
- *that Ohm's law describes the relationship of voltage, current, and resistance in circuits.*

INTRODUCTION

People depend upon electricity. You might say that electricity is a necessity for modern life.

Because electricity is so vital, its production and delivery are major activities, Figure 3-1. Billions of dollars are spent in providing electricity to businesses and homes. Billions more go into building and buying equipment that depends on electricity. We couldn't have modern skyscrapers without electricity to drive elevators. At home, people use electricity for everything from providing light, to opening cans of food, to keeping themselves warm at night with electric blankets.

This book is a tool to help you build the knowledge that can help you master the use of electricity. The knowledge you gain through use of this book will be of value personally: You will understand the way electricity is put to work in your own life and experiences. You will be in a better position to reach decisions about the electrical equipment you buy and use in your work and your life. At very minimum, the knowledge you gain will sharpen your skills as a consumer.

In addition, your growing knowledge of electricity may help in your choice of a career. Because of its important role, electricity offers many interesting, challenging, and rewarding jobs. Many millions of people are employed in producing electricity and electrical products. Many millions more work at bringing electricity to buildings. In addition, millions of others work at servicing and repairing equipment that depends upon electricity.

To begin the building of your knowledge, this chapter introduces you to the basics of electricity. The information you master will give you an understanding of what electricity is. You will learn where electricity comes from. And (in general terms) you will discover electricity is delivered for use by people.

Figure 3-2 illustrates an important source of electricity. This photo shows Norris Dam, in Tennessee. You can see the electric transmission lines fed by the dam's generators.

Figure 3-1. Electricity at work. Broadway, New York City.

Figure 3-2. Electrical source. Norris Dam, Tennessee. (Tennessee Valley Authority)

3.1 SOME BASIC ELECTRICAL TERMS

It is easy to understand, from your everyday life, that electricity is a vital tool. But it is more difficult to understand just what electricity is and how it works.

Most tools you use can be seen. You can understand a lot about how they work because you can see and hear what happens as they operate. With electricity, understanding can be more difficult. You can't see electricity. And you don't want to try to feel electricity—this can be dangerous. So, to understand electricity, the best starting point is to learn some basics.

Electricity

Energy is a basic need of modern life. **Electricity** *is a form of energy.* Electricity can be produced from a number of energy sources. To be useful, electricity must be controlled. This means that there must be controls to deliver electricity where it is needed, in useful amounts.

Current

The term used to describe the *movement of electricity* is **current**. Current flows from the source of electricity—

the place where it is created—to the point of use and back to its source. To be useful, current must be controlled and directed.

Conductor

A material that carries electrical current is called a **conductor**. Different materials behave in a variety of ways when they receive electrical currents. Some materials are better conductors than others. Some materials resist the flow of electricity. The degrees to which a substance conducts and resists the flow of electricity are known as its *electrical properties.*

Electric Circuits and Charges

To be controlled, the energy of an electric current must be directed. This direction comes from the nature of electricity. An electric current must start from someplace —from a source. The current must also have someplace to go—a destination.

The path followed by an electric current is known as a *circuit.* A circuit starts at the source. Within the circuit, there can be devices or equipment that use electricity. The current within the circuit completes its flow back at the other side of the source.

The points where current starts and completes its flow are called *terminals*. There are two kinds of terminals. A *negative terminal* (excess side) is the source of electricity. Another way of saying this is to say that **electrons** flow from (out of) the negative terminal. The *positive terminal* is the deficiency side. Electrons, then, flow from a negative point to a positive point.

A simple, typical circuit is illustrated in Figure 3-3. This shows current flowing along a wire from the negative terminal of a battery into a bell. The path of the current then moves to a button that keeps the circuit open until it is pressed. When the button is pressed, it closes a gap in the wire, permitting current to flow. The flowing current then moves on to the positive terminal of the battery. Devices used to complete or break circuits to control the flow of electric current are called *switches*. The button on a doorbell is a switch. The circuit is completed by the switch. This permits the flow of current when the button is pushed.

___ **REVIEW QUESTIONS FOR SECTION 3.1** ___

1. What is electricity?
2. What is current?
3. What is a conductor?
4. What is a circuit?
5. What is a terminal?

3.2 STRUCTURES OF MATTER

The energy that is electricity is released from matter. *Matter* is the term used to include all substances—solids, liquids, and gases—that make up the entire universe. All matter is formed in certain uniform patterns, or structures. Electricity results from the way these structures react with each other or with applied energy.

Elements

There are a limited number of structures that form all known types of matter. These basic structural forms are known as *elements*.

An element is a form of matter that has specific, basic structures and properties wherever it is found. All solids, liquids, and gases that make up our earth are formed from 92 natural elements. (Other elements can be formed through nuclear reactions.)

You probably know about many natural elements already. You may well have heard of such elements as hydrogen, helium, oxygen, carbon, aluminum, and copper. These are examples of some of the natural elements that comprise matter.

Atoms

For each element, the smallest unit of matter that still retains the structure of that element is called an **atom**. Atoms are so small that they can't be isolated and seen

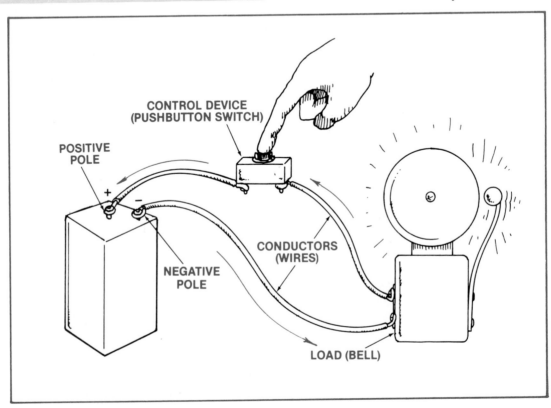

Figure 3-3. An electric circuit—a doorbell.

even with the most powerful optical microscopes. Scientists know about atoms and their structures by studying the way they behave.

The atoms of different elements are structurally different from one another. However, all atoms are composed of parts that are interchangeable. These parts, or particles, of atoms can move between atoms. As they do, energy may be released. Atomic particles can also exist on their own, outside of atomic structures.

Many of the separate parts of atoms are kept within their structures by principles of electrical energy. To understand how these principles apply, consider the illustration in Figure 3-4. This is a diagram of an atom of the simplest (and lightest) element, hydrogen. In the form shown, hydrogen has just two parts, or particles.

At the center of every atom is a core structure known as a *nucleus*. Within the nucleus of a hydrogen atom, there is a single proton. A *proton* is positively charged. Atoms other than hydrogen all have more than one proton in each nucleus. The number of protons within the nucleus determines the element formed by the atom.

Around the outside of the hydrogen atom in Figure 3-4 is a dotted circle. This represents a path of travel for another particle, known as an electron. An *electron* is negatively charged and is much smaller than a proton. (A proton weighs 1837 *times* as much as an electron.)

Thus, the nucleus of an atom, containing one or more protons, has a positive electrical charge. The electrons, which move in *shells* around the nuclei, are negatively charged. The positive force of the protons attracts the negative electrons, keeping the atomic particles in position in relation to each other. When an atom has the same number of protons and electrons, a balance exists between these opposite charges. This is known as a *binding force*, or *binding energy*.

Binding energy is, in effect, the force that holds our world together. The attraction of positive and negative charges keeps electrons in their shells around the positive nuclei of atoms. Thus, although materials you handle may seem solid, they are actually composed mostly of empty spaces between atomic particles. In its structure and behavior, an atom can be likened to our solar system. For example, if the nucleus of an atom were the size of the earth, the closest electron would be as far away as the moon. The force of binding energy is strong enough to hold atoms together to form the elements from which all matter is built.

There is also another particle that is present in the nucleus of atoms other than hydrogen. This is the neutron. A *neutron*, which is roughly the same size and weight as a proton, has no electric charge at all. It is said to be electrically neutral. Neutrons add weight to atomic structures.

The simplest example of an atom that always includes neutrons in its nucleus is helium, as shown in Figure 3-5. Helium is a gas. Its atomic structure includes two orbiting electrons and a nucleus with two protons and two neutrons. The presence of two protons determines that a given atom forms the element helium.

As elements become heavier, their atomic structures become more complex. For example, the drawing in Figure 3-6 shows an aluminum atom. The nucleus of this atom contains 13 protons and 14 neutrons. There are also 13 electrons. Note that the electrons **orbit** in three distinct shells. The first shell has two electrons, the second eight, and the third three. The forces of binding energy limit the number of electrons that can exist in any shell. At the same time, the positive charge of the protons in the nucleus seeks to attract electrons and result in a balance of charge. The structures of these electron shells and the principles of electrical balancing of charges within the atom are keys to the formation of electric currents.

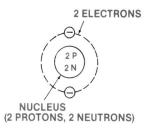

Figure 3-5. A helium atom.

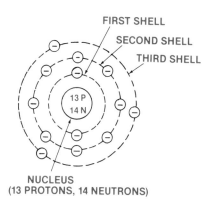

Figure 3-6. An aluminum atom.

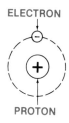

Figure 3-4. The simplest atomic structure, hydrogen.

Movement of Electrons

As mentioned earlier, the particles that form atoms have identities of their own. Particles can be separated from their atoms and move freely. The ability to separate from atomic structures is especially typical for electrons within certain atomic elements. Outside sources of energy (such as heat, light, or magnetism) can cause electrons to break loose from their atomic structures. When this happens, as illustrated in Figure 3-7, they are known as *free electrons*.

Free electrons retain their negative charge. Free electrons can be drawn into orbits of other atoms. These atoms contain a positive charge because they have lost electrons. They can also move with relative freedom within some types of materials. These materials are known as conductors. Within conductors, free electrons are drawn toward the positive terminal of a circuit. The number of electrons flowing through the conductor determines the current. See Figure 3-8.

REVIEW QUESTIONS FOR SECTION 3.2

1. Identify matter and some common types of matter.
2. What is an atom?
3. How are atoms identified?
4. What are electrons and what charge do they possess?
5. What are protons and what charge do they possess?

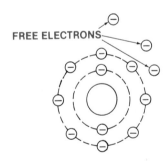

Figure 3-7. Free electrons leaving an atomic structure.

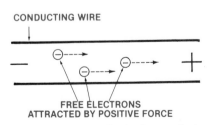

Figure 3-8. Electrons flowing in a conductor.

3.3 CONDUCTORS, INSULATORS, AND SEMICONDUCTORS

Elements are classified according to their ability to conduct electricity—or to resist the flow of current. The capacity to conduct electricity depends upon the number of free electrons present in a substance.

Conductors

A material that has a large number of free electrons is called a **conductor**. Conducting materials allow the flow of electrical current.

All metals are conductors of electricity to some degree. However, some metals are better conductors than others. Silver, copper, and aluminum are good conductors. This means that metals composed of these elements have a large number of free electrons.

In choosing conductors to be used for electrical and electronic functions, all of the properties of a substance are considered. For example, silver conducts electricity better than copper does. But silver is much more expensive. So, it is not practical for most situations. Copper is used most often as an electric conductor.

Aluminum doesn't conduct electricity as well as copper does. But it is light in weight and less expensive than copper. So, there are certain applications for which aluminum is the best choice. For example, aluminum is often used for heavy-duty, cross-country power lines. The saving in costs for the wire is important on long lines.

One of the factors that determine electrical conduction is the amount of material used. Conductors are shaped into tubes, bars, sheets, rods, and wires. Wires are used most commonly. But even wires come in many different weights and sizes. The more current, the heavier the wire. Figure 3-9 shows some types of wire used to conduct electricity.

Superconductors

Superconductors are materials that allow electrons to flow with no **resistance**. Since there is no resistance, there is no energy loss associated with current flow. Superconductive materials will make it possible to do things not presently possible in electronics.

The first superconductor was discovered in 1911 by H. K. Onnes while working with mercury. He found that this substance lost electrical resistivity when cooled to about absolute zero ($-479°$ Fahrenheit). It took over 75 years before researchers found additional low-temperature superconductors. There were no major breakthroughs until 1986, when IBM scientists announced a superconducting compound. Other discoveries were announced very quickly after that. Within months, a ceramic that worked

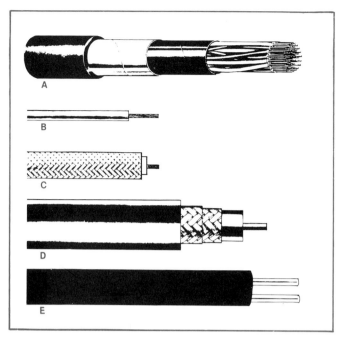

Figure 3-9. Types of wire conductors. (Plastoid)
A. Solid wire for telephone cable.
B. Stranded cable.
C. Stranded, braided cable.
D. Coaxial cable.
E. Twin-lead conductor.

Figure 3-10. This small piece of superconducting material has been cooled to − 284° Fahrenheit. At that temperature, it becomes superconducting and floats in the air above a magnet. (Courtesy Bellcore & NYNEX: photo © Bellcore 1987)

at minus 288° Fahrenheit was produced. This made it possible to chill the ceramic with liquid nitrogen instead of expensive liquid helium, see Figure 3-10. Today the record is held by copper oxide ceramic, which superconducts at minus 234° Fahrenheit.

Copper, barium, oxygen, yttrium, and lanthanum have been used to make superconductors. However, each superconducting material presents its own set of problems that have to be solved before the material can be used practically at room temperatures. Japan and the United States are putting millions of dollars into research trying to adapt the idea to practical use.

There are some devices that already utilize superconductors. Hospitals use a superconducting magnet for magnetic-resonance imaging that can scan the human body without use of x-rays. It can reveal bone and tissue structure for cancer diagnosis.

Superconducting chips enable computers to work faster and occupy a smaller space. However, keeping the temperature down is another problem yet to be solved. The ability to make a motor without core or copper wire is promising and could create trains that levitate and speed along at 300 miles per hour. Superconductive materials could also make it possible to launch rockets without all the present-day apparatus. Much work remains to be done with this amazing concept.

Insulators

A material with few free electrons is called an *insulator*. It takes large amounts of energy to cause bound electrons to flow as a current. Thus, insulators hold up, or impede, the flow of electric current. Insulators can be used for jobs on which it is necessary to limit current flow.

The wire that carries electricity from a wall plug to your toaster conducts electrical energy. This energy is used to heat coils that toast your bread. The electricity could be dangerous if you touched the wire. However, the wire is covered with an insulating material. This insulation provides protection.

Insulators, basically, are poor conductors. Some insulating materials are dry wood, rubber, mica, glass, and plastics. A number of typical insulators used in electronic systems are shown in Figure 3-11.

Semiconductors

Materials that are neither good conductors nor good insulators can be called *semiconductors*. In electronics, semiconductors are used to direct or control the flow of electricity. Among these are diodes and transistors. You will learn more about uses of semiconductors in a later chapter.

The atomic structures of semiconductor materials have some free electrons. But energy is needed to move these electrons. The two materials most often used to make semiconductors are germanium and silicon. For use in devices such as diodes and transistors, impurities must be

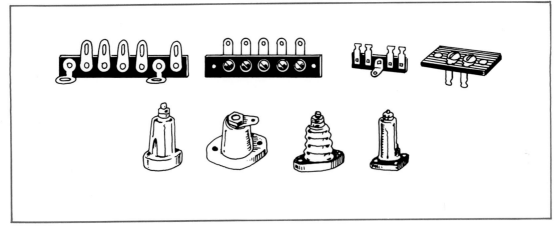

Figure 3-11. Insulators for electronics equipment.

added to germanium and silicon. These impurities, when controlled, provide the ability to control electron flow. A number of semiconductor devices are shown in Figure 3-12.

The chart in Figure 3-13 shows how different materials conduct or resist current.

REVIEW QUESTIONS FOR SECTION 3.3

1. What are conductors?
2. What are insulators?
3. What are semiconductors?
4. What are superconductors?

3.4 VOLTAGE (ELECTROMOTIVE FORCE)

The *external force* that drives electrons through a conductor is known as *electromotive force (emf)*. Emf is, in effect, electrical pressure, or voltage. The unit for measuring emf is the **volt**, abbreviated *V. Voltage* is the force applied to drive electrons through a conductor.

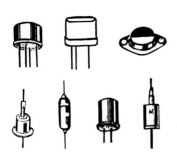

Figure 3-12. A sampling of semiconductors.

Sources of Emf

At present, there are six known, practical sources of emf.

- Magnetism
- Chemical action
- Friction
- Pressure
- Heat
- Light

A brief description of these sources follows.

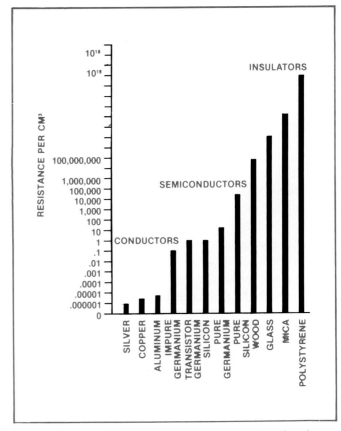

Figure 3-13. Properties of conductors, insulators, and semiconductors.

Magnetism. Magnetism is a force that attracts materials containing iron. Magnetism occurs in nature. It is also possible to create an electromagnet. This is done by passing an electric current through a coil of wire surrounding a structure containing iron. A *magnetic field* is the area around a magnet in which *magnetic force* is active.

Safety Tips

1. Be careful when you plug in an electric cord. Hold the plug at the back. Be sure your fingers are not in contact with the metal prongs. (Prongs are the metal pieces that fit into the wall outlet.)
2. Protect yourself before you work with electricity. Don't stand on a wet floor. If possible, wear rubber-soled shoes. (Rubber is a better insulator than leather.) Wear gloves any time you do electrical work.

Emf is generated when a wire is passed through a magnetic field. Emf also results if the magnetic field is moved near a stationary wire. As long as either the wire or the magnetic field is in motion, emf results.

Magnetism is the most economic means of generating electric power known today. Electricity produced through magnetism is used for lights, motors, and many other applications. One example is the alternator that provides electric power for an automobile, Figure 3-14.

Chemical Action. Emf can result from chemical reactions of elements with each other. These reactions power the flow of electrons. The reactions are contained in special structures known as *cells*. Two or more of these cells are connected to provide a source of emf. The device that results needs no introduction. It is called a *battery*.

Diagrams of cells appear in Figures 3-15 and 3-16. Chemically produced electricity powers transistor radios. Cars use batteries for starting, lighting, and other purposes.

Friction. Electricity is produced when certain materials are rubbed together. You have certainly experienced this source of emf. It is common if you wear clothing made of certain artificial fibers, such as nylon. As the cloth rubs together, emf builds up. You feel—and often see—sparks when you then touch metal objects. The emf built up in this way is known as *static electricity*. Your body conducts the static electricity to the metal object.

Static electricity is used in a number of practical ways. One example, the coating of sandpaper, is illustrated in Figure 3-17.

Pressure. Certain materials produce emf when they are put under pressure. These materials are crystals such as Rochelle salts or quartz. The pressure can be created when electricity is introduced into a crystal. Signals of some radio transmitters come from pressure-generated emf. Crystals are also used to generate electricity in some microprocessors, Figure 3-18. This effect is called the piezoelectrical effect.

Piezo alarms and trimmers are used in many circuits today instead of bells and buzzers, Figure 3-19 (page 35).

Crystals are used in some record players. They are located in the pickup head. The emf produced by the

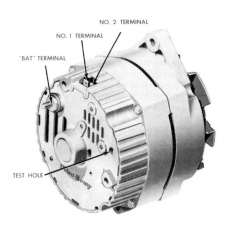

Figure 3-14. Automotive alternator.

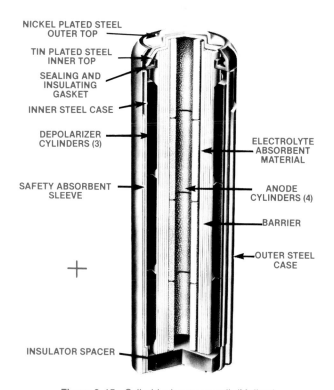

Figure 3-15. Cylindrical mercury cell. (Mallory)

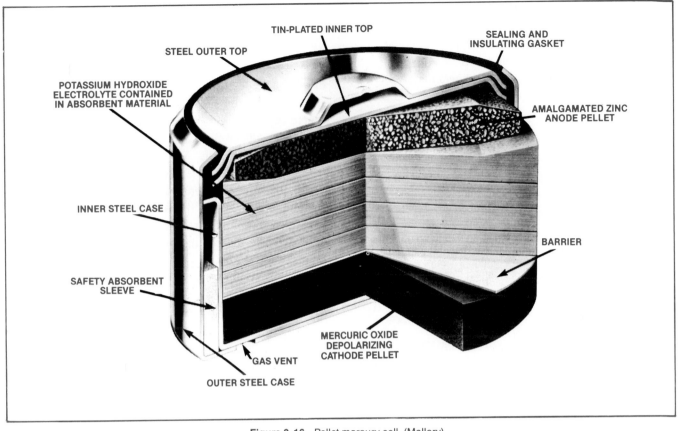

STEEL OUTER TOP

TIN-PLATED INNER TOP

SEALING AND
INSULATING GASKET

POTASSIUM HYDROXIDE
ELECTROLYTE CONTAINED
IN ABSORBENT MATERIAL

AMALGAMATED ZINC
ANODE PELLET

INNER STEEL CASE

BARRIER

SAFETY ABSORBENT
SLEEVE

MERCURIC OXIDE
DEPOLARIZING
CATHODE PELLET

GAS VENT

OUTER STEEL CASE

Figure 3-16. Pellet mercury cell. (Mallory)

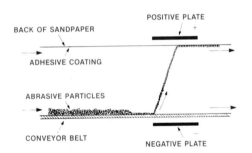

BACK OF SANDPAPER

POSITIVE PLATE
+

ADHESIVE COATING

ABRASIVE PARTICLES

CONVEYOR BELT

NEGATIVE PLATE

Figure 3-17. Using static electricity to coat sandpaper.

Figure 3-18. Quartz crystals used in receivers, transmitters, and microprocessors.

crystal produces the current that is amplified and fed to the speaker.

Heat. Emf results when heat is applied to certain combinations of materials. These are called *thermocouples*. *Thermo* comes from the Greek word for heat. *Couple* means "to join." In a thermocouple, two different pieces of metal with different properties are joined. When heat is applied, a flow of electrons is started, Figure 3-20.

Thermocouples are widely used to sense heat. A common example is the pyrometer. This device is used to measure temperatures of ceramic kilns.

Light. An emf is produced when light strikes certain materials. These substances are said to be *photosensitive*. (*Photo* means "light.")

There are many uses for light-sensitive materials. One is for control of exposures on cameras. A photosensitive device reads the light and generates electricity to control the camera. The electricity can change the opening of the lens. Or the emf can automatically change the lens opening or speed of the shutter.

Solar cells are now in use that convert sunlight directly to electricity, Figure 3-21.

Photosensitive devices can be used to turn lights on automatically when it gets dark, Figure 3-22.

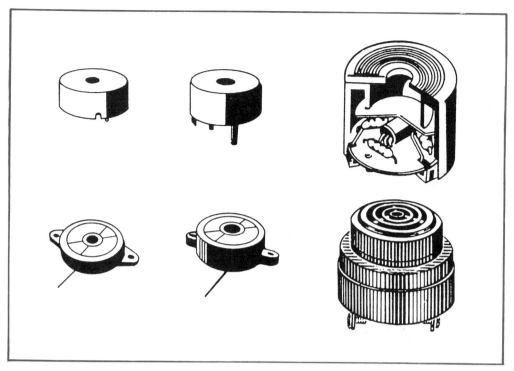

Figure 3-19. Piezo alarms and trimmers.

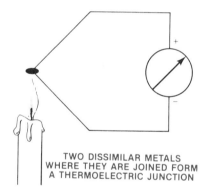

TWO DISSIMILAR METALS
WHERE THEY ARE JOINED FORM
A THERMOELECTRIC JUNCTION

Figure 3-20. A thermocouple.

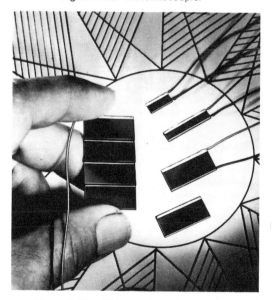

Figure 3-21. Silicon solar cells. (International)

Figure 3-22. Photoelectric cell to turn lights on and off. (Fisher-Pierce)

REVIEW QUESTIONS FOR SECTION 3.4

1. What is electromotive force, and how is the term abbreviated?
2. How is electromotive force measured?
3. How is magnetism used to create electricity?
4. Name a device that generates electricity through chemical action.
5. What generates electricity in a thermocouple?
6. What is the piezoelectrical effect?

Activity: GENERATING ELECTRICITY

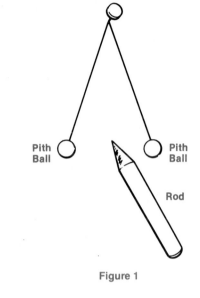

Figure 1

_____ OBJECTIVES _____

1. *To investigate several representative methods of generating electricity.*
2. *To determine an efficient method of generating electricity.*

MATERIALS NEEDED

1 Meter, 1-0-1 mA DC
1 Pair test leads, with alligator clips on both ends
2 Wires for connections, No. 18 or No. 20, flexible
1 Copper strip, 5″ × 3/4″
1 Friction rod and wool cloth
1 Zinc strip, 5″ × 3/4″
1 Wire, copper, bare, 5″ long

1 Jar, for salt-water solution
1 Coil of wire, 300 turns on 3/4″ bobbin
1 Magnet, permanent, bar type, at least 5″ long
1 Wire, iron, 5″ long
1 Solar cell
2 Pith balls on threads
1 Candle, 1/2″ diam. × 5″ long

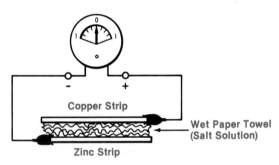

Figure 2

PROCEDURE

1. Straighten the threads attached to the pith balls and place them on the bench so that they can be picked up easily. The threads should be held at least 6 inches from the balls and the balls should be touching one another when suspended. Hold the friction rod in the left hand and the piece of wool cloth in the right hand. Rub the rod briskly with the wool cloth for several seconds.

2. Still holding the rod with the left hand, place the wool cloth on the bench. Pick up the pith balls, holding them suspended from the threads. Bring the charged friction rod close to the pith balls. See Figure 1. What happens to the pith balls? After the rod has touched the pith balls and is withdrawn, what happens to the pith balls? Why does this happen?

3. Using alligator clip leads, connect the copper strip to the positive terminal of the meter movement of a 1-0-1 mA meter. Connect the zinc strip to the negative terminal of the meter movement. See Figure 2. Moisten a piece of paper towel (or other absorbent paper) with the salt solution and place it between the copper and zinc strips. What indication is there of an emf generation?

4. Repeat Step 3, using a piece of paper moistened with saliva. What indication is there of an emf being generated? You may have experienced the sensation of pain when a piece of metal (aluminum foil from a piece of chewing gum) has touched a fresh filling in a tooth. In view of the findings in generating an emf with saliva, what is the probable cause of this pain?

5. Cross the copper and iron wires about 2 inches from one end. Twist the wires together for the remainder of their lengths. In order to compress the length of the twisted section, wrap it around a pencil. Remove the pencil. See Figure 3. Connect the copper wire to the positive [+] terminal of the meter (1-0-1 mA) and the iron wire to the negative [−] terminal of the meter. Light the candle and heat the twisted section of the wire. (Another source of heat may be used if preferred.) Is this method of emf generation as efficient as the chemical method?

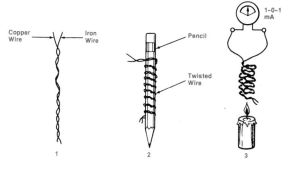

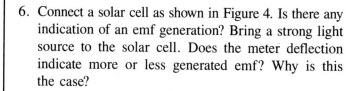

Figure 3

Figure 5A

6. Connect a solar cell as shown in Figure 4. Is there any indication of an emf generation? Bring a strong light source to the solar cell. Does the meter deflection indicate more or less generated emf? Why is this the case?

Figure 4

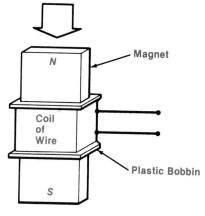

Figure 5B

7. Connect the meter movement to a coil with 300 turns of wire. See Figures 5A and 5B.
8. Push the north pole of a bar magnet into the coil. Is there any indication of emf generation? With the north pole inserted into the coil first, what is the direction of the meter deflection (left or right)? When the magnet is pulled out of the coil, which way does the meter needle move? Now, take the other end (south pole) of the magnet and insert it into the coil. What happens? Which way did the meter movement's needle move when the south pole was inserted first? When the south pole was removed from the coil, in what direction did the meter needle move? If the magnet is inserted into the coil and left there without moving,

what is the meter reading? If the magnet is held steady, and the coil is moved up and down, what happens to the meter movement needle?
9. Disconnect all circuits used. Clean all equipment and materials and return them to their proper storage places.

SUMMARY

1. List the different methods of emf generation used in this activity.
2. From your observations, what was the most efficient method for generating electricity?

3.5 CURRENT

The movement of electrons along a conductor, as you already know, is called *current*. The term *current* is also used to describe the rate at which numbers of electrons move. A *volt* is a measure of electrical pressure. An *ampere* is a measure of the amount of electron movement per unit of time.

One ampere is described as the flow of 6,250,000,000,000,000,000 electrons past a single point in one second. This number is 6.25 quintillion. In scientific shorthand, this amount could be shown as 6.25×10^{18}. Abbreviations used for ampere are *amp* or *A*.

Current is often described in terms of parts of an ampere. One milliampere (mA) is one-thousandth of an

37

ampere. It is written as 0.001 A or 1 mA. One micro-ampere (μA) is one-millionth of an ampere. It can be written as 0.000001 A or 1 μA. The device used to measure electrical current is the *ammeter*.

If 6.25×10^{18} electrons are stored and not moving, this is known as one **coulomb**.

Current Flow

The direction of current flow is from negative to positive. This is because electrons are negatively charged. Thus, a negative condition represents a surplus of electrons. These excess electrons are attracted by a deficiency of electrons at a positive terminal.

REVIEW QUESTIONS FOR SECTION 3.5

1. What is electric current?
2. What is the standard measure of current?
3. What is a milliampere? A microampere?
4. What is a coulomb?
5. What is the abbreviation for ampere?

3.6 RESISTANCE

Holding back, or impeding, the movement of electrons along a conductor is known as **resistance**. The amount of resistance in a material is determined by the number of free electrons in their orbits. With fewer free electrons, it is more difficult for current to flow.

In circuit design, resistance is an important tool. Materials become hotter and tend to give off extra heat when they resist current. Resistance also makes it possible to control electric power. The photo in Figure 3-23 shows the interaction between high voltage and resistive materials.

The resistance of an electric conductor depends on four factors:

Material
Length of conductor
Cross section of conductor
Temperature

Material. Materials differ in their ability to conduct or resist electric currents. Refer to Figure 3-23.

Length. Most conductors are wires. The longer the wire through which current must travel, the higher is the resistance.

Cross Section. The cross section of a conductor is its thickness *diameter*. With wire conductors, the larger the

Figure 3-23. Insulators resist electron flow. High voltage is necessary to force movement of electrons in insulation materials. (Lapp Insulator Company)

diameter, the lower is the resistance for each foot of length.

Temperature. For most materials, resistance increases as temperature rises. The higher the temperature, the greater is the resistance. Some electrical devices are built around this factor. They increase their resistance as they are heated by the passage of current.

There are also some devices that work on the opposite principle. That is, their resistance is lowered as the temperature is increased. These are devices called *thermistors* and *surgistors*. Thermistors are used to control current and are also used as temperature sensors. Several thermistors are shown in Figure 3-24. Surgistors are used to prevent a large current in a device when it is first turned on. Surgistors are often used as protective devices in television receivers.

Measuring Resistance

The unit of measure used to describe resistance is the **ohm**. One ohm of resistance is present when one ampere of current flows under one volt of pressure. Thus, volts, amperes, and ohms are closely related. The abbreviation for ohm is the Greek *omega* (Ω).

38

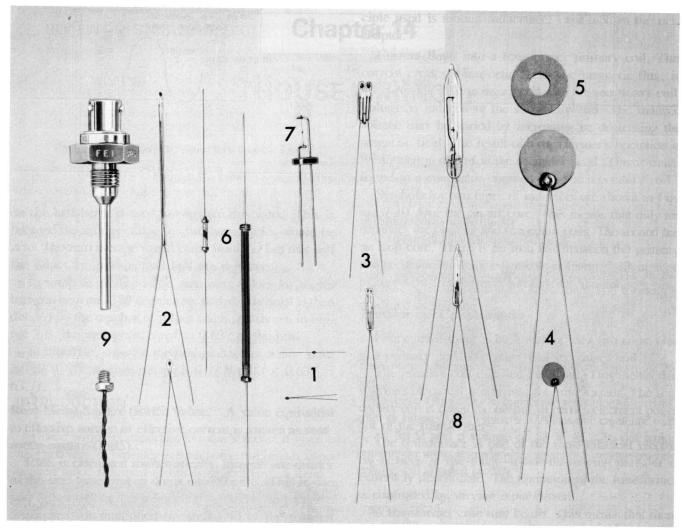

Figure 3-24. Thermistors. (Fenwal)
1. Glass-coated beads.   2. Glass probes.   3. Interchangeable ISO-Curve.   4. Discs.   5. Washers.   6. Rods.
7. Mounted beads.   8. Vacuum and gas-filled assemblies.   9. Probe assemblies.

REVIEW QUESTIONS FOR SECTION 3.6

1. What is electrical resistance?
2. How does the material used in a conductor affect resistance?
3. How does conductor length affect resistance?

3.7 OHM'S LAW

The relationships of voltage, current, and resistance to each other are covered by a set of basic electrical principles. These are known as **Ohm's law**. The law is named after its discoverer, Georg Simon Ohm. It was first published in 1827.

To understand and use electricity, you must be able to apply the principles of Ohm's law. This law can be stated in three ways. The different statements tell you how to find voltage, current, and resistance.

To find voltage: Voltage needed to force a certain amount of current through a circuit is equal to the current multiplied by the resistance.

To find current: The current flowing through a circuit is equal to the voltage divided by the resistance.

To find resistance: The resistance within a circuit is equal to its voltage divided by the amount of current.

Notice that these three statements cover one single law of electricity. If you know any two of the electrical properties of a circuit, you can use Ohm's law to find the third.

Voltage, current, and resistance are all stated in numbers. A circuit, for example, may have a voltage of 120 V, a current of 20 A, and a resistance of 6 ohms. This means that you may use simple math and Ohm's law to find voltage, current, or resistance.

Activity: *CIRCUITS, INSULATORS, AND CONDUCTORS*

OBJECTIVES

1. *To investigate the properties and characteristics of the complete electrical circuit.*
2. *To investigate common substances to determine whether they are insulators or conductors.*

MATERIALS NEEDED

1 6-volt power supply
3 Wires for connections
1 Switch
1 Lamp, 6-volt
1 Lamp socket
2 Clips for holding test strips
1 Iron wire, 3.5" long
1 Copper wire, bare, 3.5" long

1 Aluminum strip, 5" × 3/4"
1 Carbon strip, 5" × 3/4", or pencil lead
1 Plastic rod, 5" × 1/4"
1 Dowel rod, 5" × 1/4"
1 Piece of paper, 5" × 1/4"
1 String, 4" long

PROCEDURE

1. Connect the circuit shown in Figure 1. (**CAUTION:** *Be sure the power supply is off.* Always turn off the power supply before working on a circuit.)

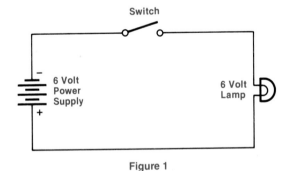

Figure 1

2. With the switch on the circuit board in the *off* position, *turn on the power supply* and adjust the output to 6 volts DC.

3. Turn the switch *on*. What happens to the light? Does this indicate that the circuit is now complete? Explain.

4. Remove the lead on one side of the lamp. Does the lamp continue to glow? Explain.

5. Turn off the power supply. Remove the connections to the switch.

6. Connect the circuit shown in Figure 2. You will note that the test clips have been substituted for the switch in this circuit. When a conductor is placed in the test clips, a complete circuit will exist. However, when an insulator is placed in the clips, the circuit will not be complete and there will be no path for electron flow.

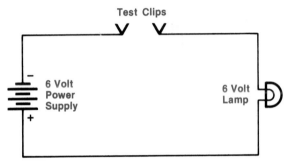

Figure 2

7. The chart below lists a number of materials that you are to place between the test clips. You will then record whether the light is *on* or *off* in each case. From this, you will conclude whether the material is a conductor or an insulator. Turn on the power supply (with the output set at 6 volts DC). Test each material.

1. Iron wire
2. Copper wire
3. Aluminum strip
4. Carbon strip
5. Plastic rod
6. Wood dowel
7. Paper
8. String

8. *Turn off the power supply.* Disconnect the circuit. Return all equipment and materials to their proper storage places.

SUMMARY

1. What is necessary to have a complete circuit?
2. What causes some materials to be conductors, whereas others are insulators?

Three statements of Ohm's law can be given as equations. An equation, basically, is a statement. It is given in mathematics rather than in words. It says that the total value of items on one side of the equal sign (=) is the same as the total on the other side. Ohm's law can be stated in the equations that follow. In these equations, *E* represents emf or voltage. *I* stands for current, given in **amperes**. *R* is for resistance, given in ohms. To illustrate how the formulas work, sample electrical values are used.

For voltage (emf): $E = I \times R$ $120\text{ V} = 20\text{ A} \times 6\Omega$

For current: $I = \dfrac{E}{R}$ $20\text{ A} = \dfrac{120\text{ V}}{6\ \Omega}$

For resistance: $R = \dfrac{E}{I}$ $6\ \Omega = \dfrac{120\text{ V}}{20\text{ A}}$

Using Ohm's Law

For an example of how Ohm's law works, look at the diagram of a simple circuit in Figure 3-25. When circuits are diagrammed, certain symbols are used to represent voltage, current, and resistance.

The power source for the circuit in Figure 3-25 is represented by short and long lines next to each other. This is a direct current (DC) power source. Positive and negative polarities are shown. The short line always represents the negative side of a DC power source. The long line is the positive side.

Current moving through the circuit is shown in a circle. A zigzag line represents resistance.

Remember these symbols. You will have to recognize and use them often.

In Figure 3-25, you know that the circuit has 2 A of current and 5 Ω of resistance. Your job is to find voltage (emf), represented by the letter E. To do this, you multiply current by resistance. With 2 amperes and 5 ohms, your product is 10. This is a 10-V circuit.

For an example of how current is calculated, look at Figure 3-26. This is the same circuit as in Figure 3-25, except that voltage and resistance are given. To figure current, you divide resistance into voltage. Dividing 5 Ω into 10 V gives a current of 2 A.

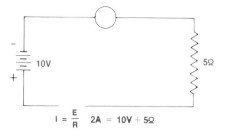

$$I = \frac{E}{R} \quad 2\text{A} = 10\text{V} \div 5\Omega$$

Figure 3-26. Circuit with E and R given.

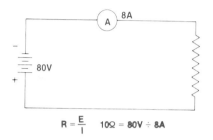

$$R = \frac{E}{I} \quad 10\Omega = 80\text{V} \div 8\text{A}$$

Figure 3-27. Circuit with E and I given.

In Figure 3-27, you have basically the same diagram. But the numbers have been changed. This circuit shows 80 V of power and 8 A of current. To figure resistance, you divide current into emf. Dividing 8 A into 80 V gives you 10 Ω of resistance.

Remembering Ohm's Law

A simple technique to help you remember and use Ohm's law is shown in Figure 3-28. This is a circle divided into three parts. One part, labeled *E*, takes up the top half of the circle. The bottom half is divided into two equal parts, labeled *I* and *R*.

Think of the parts of the circle as Ohm's law. Cover one part and the relationship of the other two parts shows you the formula to use.

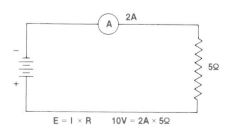

$$E = I \times R \quad 10\text{V} = 2\text{A} \times 5\Omega$$

Figure 3-25. A simple circuit.

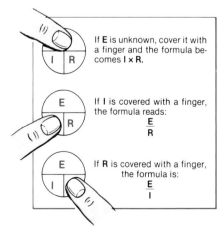

Figure 3-28. A key to remembering Ohm's law.

Activity: OHM'S LAW

_____ OBJECTIVES _____

1. *To gain familiarity with Ohm's law in its several forms.*
2. *To demonstrate and prove the mathematical relationships of Ohm's law in operating circuits.*

MATERIALS NEEDED

1 Meter, 1-0-1 mA DC
1 Resistor, 10,000 ohms, ±5%, 1 watt
1 Resistor, 15,000 ohms, ±5%, 1 watt
1 Voltmeter, DC, 0–10 volts

1 Power source, capable of delivering up to 7.5 V DC
3 Wires for connections

PROCEDURE

1. Connect the circuit shown in Figure 1. **CAUTION:** Do not turn on the power supply until instructed to do so in the procedure for each activity.

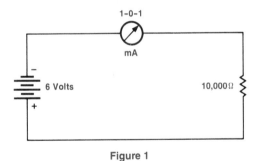

Figure 1

2. Set the voltage control for the power supply at the minimum output. *Turn on the power supply.* Slowly advance the voltage control to an output of 6 volts. (If you have a fixed 6-volt power supply, it will not be necessary to bring it up to 6 volts slowly.)
3. The circuit that you have just connected contains a single 10,000-ohm resistor connected to a source of 6 volts. By Ohm's law, what should the current through the resistor be? Change this value to milliamperes (1 ampere = 1000 mA).

4. What current is flowing in the circuit, as indicated by the reading of the milliammeter? Is there a difference between this reading and the answer in Step 3? If there is a difference what would account for it?
5. *Turn off the power supply.* Disconnect the circuit you have been using and connect the circuit shown in Figure 2.

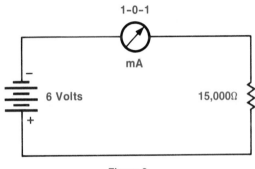

Figure 2

6. In this circuit, a 15,000-ohm resistor has been substituted for the 10,000-ohm resistor in the preceding circuit. *Turn on the power supply.* Adjust the voltage output to 7.5 volts. What is the current through the resistor as determined by the reading of the milliammeter?
7. Reduce the voltage to 6 volts. What is the current through the resistor?
8. *Turn off the power supply.* What happens to the current in a circuit when the voltage remains the same and the resistance of the circuit is increased?
9. What happens to the current in a circuit when the resistance remains the same and the source voltage is increased?
10. Disconnect all circuits used. Return all equipment and materials to their proper storage places.

SUMMARY

1. What does Ohm's law state?
2. How do your findings in this activity prove or disprove Ohm's law?

REVIEW QUESTIONS FOR SECTION 3.7

1. What is Ohm's law?
2. Give the formula for finding power in a circuit.
3. Give the formula for finding current in a circuit.
4. Give the formula for finding resistance in a circuit.
5. How are emf, current, and resistance measured?

3.8 POWER

The *rate* of doing work is called *power*. Power can be either electrical or mechanical. When a mechanical force is used to lift or move a weight, work is done. The rate at which the weight is moved identifies the mechanical power, or *horsepower*, applied.

Horsepower

The energy needed to lift a 550-pound weight one foot for one second is one *horsepower*. This is a unit of mechanical energy. However, there is an electrical equivalent. This is used to describe the power of electric motors. The rule is: The electrical energy required to equal one horsepower of mechanical energy is 746 watts. (The watt is a basic unit of electric power.)

Watt

Electric power (P) is measured in **watts**. The power within a circuit is found by multiplying its voltage by the current. Thus, 1 watt (W) of power is produced by 1 V of force at 1 A of current.

The basic formula for determining power in watts is as follows:

$$P = E \times I \qquad 80 \text{ V} \times 8 \text{ A} = 640\text{W}$$

This means that if you know the voltage (E) and the current (I) of a circuit, you can figure the power. Power is equal to E times I. Refer back to Figure 3-27. This shows a circuit with 80 V of pressure and 8 A of current. This means that the power is 640 W.

Can you find electric power if you don't know both E and I?

Yes. You can if you know two of the three quantities used in Ohm's law. Suppose you know voltage and resistance, but not current. You know that current is equal to voltage divided by resistance. So, you substitute these values into the basic formula:

Formula:	$P = E \times I$	$640 \text{ W} = 80 \text{ V} \times 8 \text{ A}$
Substituting:	$P = E \times \dfrac{E}{R}$	$640 \text{ W} = \dfrac{80 \text{ V} \times 80 \text{ V}}{10 \text{ } \Omega}$
Consolidating:	$P = \dfrac{E^2}{R}$	$640 \text{ W} = \dfrac{6400}{10}$

In consolidating this version of the formula, you multiply the two items on the right of the equation by each other. This gives you $E \times E$ over R. E times E can also be written as *E to the second power*, or *E-squared*. This is written E^2.

If you know current and resistance, but not voltage, you figure power this way:

Formula:	$P = E \times I$	$640 \text{ W} = 80 \text{ V} \times 8 \text{ A}$
Substituting:	$P = I \times R \times I$	$640 \text{ W} = 8 \text{ A} \times 10 \text{ } \Omega \times 8 \text{ A}$
Consolidating:	$P = I^2 R$	$640 \text{ W} = 64 \times 10 = 640$

Other Units of Power

Power is often described in units of 1000 W, or *kilowatts*. The abbreviation of kilowatt is *kW*. You will often encounter measurements that include the prefix *kilo*. *Kilo* means 1000.

The use of power is measured in **kilowatthours**, abbreviated *kWh*.

Power is also measured in parts of watts. A *milliwatt* is one-thousandth of a watt. The term is abbreviated *mW*.

A *microwatt* is one-millionth of a watt. The abbreviation is *μW*.

Another term used to measure the use of energy is the **joule**. A joule is equal to usage of one watt per second. This unit is widely used in countries employing the metric system of measurement.

REVIEW QUESTIONS FOR SECTION 3.8

1. What is the definition of electrical power?
2. What is the definition of a watt?
3. If a circuit has 10 V and 3 A, how many watts are there?
4. How many watts are in a kilowatt?
5. How many watts are in a milliwatt?
6. What is a microwatt?
7. What is a joule?

SUMMARY

Electrical terms such as *electricity, current, conductor, terminal,* and *circuit* are defined and discussed. The structure of matter in terms of its elements, atoms, and charged particles are examined in detail. Like charges repel; unlike charges attract.

A coulomb is a unit of charge. It is 6.25×10^{18} electrons stored on a surface. When put in motion, one coulomb gives rise to an ampere when it passes a given point in one second. Voltage and potential difference are the same.

Conductors have free electrons; insulators do not. Superconductors are the basis of some interesting research at this time, with many future possibilities. Semiconductors are still being developed and put to use in all forms of electronic devices.

Sources of emf are ways of generating electricity. They are magnetism, chemical action, friction, pressure, heat, and light. Piezo alarms and trimmers are very useful electronic devices made today.

Resistance is the opposition to electron flow or current flow. Resistors are devices made to produce specific amounts of resistance. They come in fixed and variable forms. Thermistors are the opposite of resistors in terms of reaction to temperature changes.

Ohm's law states that current in any circuit is equal to the voltage divided by the resistance.

One horsepower is equal to 746 watts of electrical energy. The watt is the unit of measure for electrical power. The joule is equal to usage of one watt per second.

USING YOUR KNOWLEDGE

1. Draw a schematic diagram of an atom of hydrogen.
2. Draw a schematic diagram of an atom of aluminum.
3. A circuit has a 25-W bulb and 120 V of emf. How much current, in amperes, flows through the circuit?
4. A circuit has a 100-W light bulb and 120 V of emf. What is the resistance of the bulb?
5. What is the resistance of a 300-W light bulb rated at 120 V?

KEY TERMS

electricity	resistance
current	ohm
electron	volt
atom	ampere
nucleus	watt
proton	Ohm's law

neutron	joule
orbit	coulomb
conductor	kilowatthour
semiconductor	superconductor
insulator	

PROBLEMS

1. An ammeter is connected in a direct-current (DC) circuit and indicates a current of 4 amperes. How many electrons will flow through the meter in 10 seconds?
2. If a circuit is connected using a battery of 10 volts as the source and has a total circuit resistance of 20 ohms, how much current exists in the circuit?
3. What voltage is necessary to force 4 amperes through 8 ohms of resistance?
4. What resistance would limit the current in a circuit to 4 amperes when the source voltage is 12 volts?
5. If a circuit has 240 ohms resistance and has a 120-volt power source, what would be the current through the resistance?
6. A circuit has 100 ohms resistance and 100 volts applied. What is the current in the circuit?
7. A circuit has 1000 ohms resistance and 100 volts applied. What is the current in the circuit?
8. A circuit has 10,000 ohms resistance and 100 volts applied. What is the current in the circuit?
9. A circuit has 200 volts applied and a current of 5 amperes. What is the resistance?
10. A circuit has 240 volts applied and a current of 6 amperes. What is the resistance?
11. A circuit has 10 volts applied and a current of 5 amperes. What is the resistance?
12. A circuit with 100 ohms has a current of 5 amperes. What voltage was needed to push this amount of current through the circuit?
13. A circuit with 200 ohms resistance has a current of 5 amperes. What is the voltage needed to push this current through the circuit?
14. A circuit with 5000 ohms resistance has a current of 2 amperes. What is the voltage needed to push this current through the circuit?
15. A circuit has three resistors in series. The resistors are 111 ohms, 1234 ohms, and 567 ohms. What is the voltage needed to push one ampere through the circuit?

Chapter 4

PUTTING ELECTRICITY TO WORK

OBJECTIVES

After studying this chapter, you will know

■ *that resistance is the basis for converting electrical energy to heat, and is also a means for controlling the flow of electrons within a circuit.*

■ *that a load is any device that consumes electrical energy.*

■ *that a resistor is a device that limits, or impedes, the flow of electrons through a conductor.*

■ *how to identify different types of resistors by size and color bands.*

■ *the functions and uses for variable resistors, including potentiometers and rheostats.*

■ *the principles of series, parallel, and series–parallel circuits.*

■ *how to use Ohm's law to determine voltage, current, and resistance for various circuits.*

■ *how to draw various circuit configurations.*

INTRODUCTION

You now know about the basic principles of electricity. You should understand the relationships of electromotive force, current, resistance, and electrical power. In this chapter, you will learn how circuits are constructed to put electricity to work on a *controlled* basis.

One of the main tools used in controlling electricity is the **resistor**. So, the first portion of this chapter deals with resistors and their functions.

Then, the chapter covers some basic types of circuits— series and parallel.

You also learn how to use Ohm's law on circuits. This law helps you find out what to expect from a circuit.

4.1 RESISTANCE AND RESISTORS

There are three roles for resistance in circuits:

1. Resistance converts electrical energy into heat. So, resistance is used in electric heaters, stoves, or toasters.
2. Resistance can be used to limit the flow of electrons within a circuit. This ability makes it possible to assure that the proper voltage or current reaches a certain point in an electrical circuit.

3. Resistance provides a load within circuits. A **load** is any device that uses electric current or power. Loads may be used to control current or voltage within a circuit.

Where there are special reasons to add resistance to a circuit, resistors are used. A *resistor* is a device that limits, or controls, the flow of electrons through a circuit. A resistor adds resistance to a circuit. The amount of resistance needed in any circuit depends upon the natural resistance already within the **conductor** itself. All conductors present some resistance to current flow. This resistance is determined by:

* length of the conductor
* size of the conductor, including diameter
* temperature of the conductor
* type of material used as a conductor.

The resistor symbol is a zigzag line, Figure 4-1.

REVIEW QUESTIONS FOR SECTION 4.1

1. What is electrical resistance?
2. What is a resistor?
3. What are the three functions of resistors in circuits?
4. What are the factors that determine resistance in a circuit?

4.2 TYPES OF RESISTORS

Resistors may be fixed or variable. A *fixed* resistor has one specific resistance value. This is built into the unit.

A *variable* resistor can be set for different resistance values. A control is provided to vary the resistance settings. Use of variable resistors is described later in this chapter.

Figure 4-1. Symbol for a resistor.

Resistors are also identified by the type of construction used in building them. The main types of construction used are wire-wound and carbon-composition. These and other resistor structures are described below.

Wire-Wound Resistors

Resistant wire is wound on an insulated core to form *wire-wound resistors*. These units are used to hold back the flow of electrical power of 3 watts or more. There are several types of wire-wound resistors. Some of these are described below.

Power-type resistors are used to dissipate, or carry off, large amounts of heat. These devices are physically large. The resistors must be made of materials that can stand temperatures as high as 300 degrees Celsius.

Most high-power, wire-wound resistors consist of a round (cylindrical) tube. A single layer of resistant wire is wound around this ceramic tube. The entire unit is enclosed in a ceramic material. (Ceramics are made from fired clay. Such materials are excellent insulators.)

There are terminals at each end of a power-type resistor. These terminals are used to connect the resistor into the circuit. An illustration of a power-type resistor is shown in Figure 4-2.

Low-power and precision-type resistors are also wire-wound. Because of their low-power rating, they do not generally give off large amounts of heat. However, these resistors are extremely accurate.

Accuracy of electrical devices is measured by performance. Each device has a rating. For example, a resistor may be rated at 10 Ω. Precision, then, is the accuracy with which this rated service is delivered. Precision-type resistors are often accurate to plus or minus ($\pm$) 1 percent. Thus, a 10-Ω precision resistor with 1-percent accuracy would deliver between 9.9 Ω and 10.1 Ω. (Since 1% of 10 Ω = 0.1 Ω).

The greatest use for low-power, precision-type resistors is in meters and electronic devices, Figure 4-3.

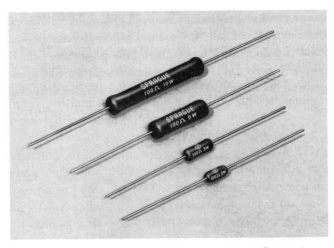

Figure 4-3. Low-power, precision-type resistors. (Sprague)

Flat wire-wound resistors are constructed with a single layer of bare or insulated wire wound on a flat, insulated card. This flat card is usually enclosed in an insulated cover.

Carbon-Composition Resistors

The word **composition** means that these devices are made from a combination of materials. In carbon-composition resistors, this combination includes a conductor, an insulator, and a bonding material. The materials blended into a resistor determine its resistance.

Three types of resistant materials are used in carbon-composition resistors. These are:

1. Lacquer, used for volume controls
2. **Resin-bonded** substances, used for general-purpose resistors
3. **Ceramic-bonded** materials, used for heavy-duty resistors.

In manufacturing, the combined materials are pressed into the required resistor shape. After that, the resistors are fired in an oven. This fuses, or bonds, the materials together.

The wires used to connect resistors into a circuit may be pressed into the composition material. They may also be connected with cement. Cementing is done after the resistor body is formed and before further heat is applied.

Radial terminals or caps may also be used. When this is done, the ends of the resistor are metalized. The leads are either pressed or soldered into the metalization.

Carbon-composition resistors come in either insulated or noninsulated types. Resistance levels may be either fixed or variable.

Carbon-composition resistors are usually small. The size of the units determines their wattage. See Figure 4-4. These resistors are made in standard sizes, including 1/8 W, 1/4 W, 1/2 W, 1 W, and 2 W.

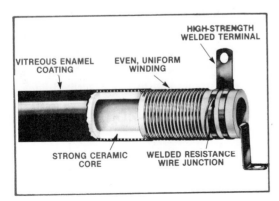

Figure 4-2. Cutaway view of a wire-wound resistor. (Ohmite)

Most carbon-composition resistors are used in circuits requiring low heat dissipation. Dissipation means loss of heat. Resistors above 2 W are usually wire-wound.

Other Types of Resistors

There are also other types of resistors with which you should be familiar.

Deposited-carbon resistors use a film of resistant material. This film is deposited on a ceramic surface. The resistant material is crystalline carbon. This carbon, when heated to high temperatures, becomes a gas. The gas is spread uniformly over the ceramic.

Wire leads are attached with either graphite or a special metallic paint. Metal caps are then forced over these coatings.

Carbon-deposited resistors are accurate and stable. They are used where accuracy is important. Test instruments require this type of resistor.

Metal-film resistors use alloys coated onto glass tubes or rods. An alloy is a combination of metals formed to do a special job. In this case, the alloy provides certain resistance properties. The resistant alloy in a metal-film unit is evaporated through use of heat. Then, a film of this material is coated onto the glass. An insulating material is then added for protection. The ends of the resistors are coated with silver. The coatings are used to attach wire leads to the metal film. Metal-film resistors add less noise to the circuit than do carbon-composition resistors.

REVIEW QUESTIONS FOR SECTION 4.2

1. Describe the construction of wire-wound resistors.
2. What are the main features of power-type resistors?
3. What are the main features of low-power, precision-type resistors?
4. How are carbon-composition resistors made?
5. How are film-type resistors constructed?

4.3 RESISTOR COLOR CODING

Wire-wound resistors are usually marked with numbers or letters. These markings indicate the power and resistance ratings for the devices.

Carbon-composition resistors are too small to be marked with numbers and letters. Instead, there are two different ways to tell their ohmic value.

The **power**, or wattage, **rating** is indicated by physical size. The larger the carbon-composition resistor, the more

power it can handle. As you come across carbon resistors in your work, you will have to learn to judge their **wattage rating**. This comes easily with experience. Note that Figure 4-4 is marked to indicate the relationship of the sizes of carbon-composition resistors to their wattage ratings.

The resistance of a carbon-composition resistor is marked with color bands. This color-band coding system is shown in Figure 4-5. Take a moment to study the colors and their meanings.

These number values are applied to markings on all carbon-composition resistors. Slightly different systems are used to mark resistors rated at more than 10 Ω and those rated at less than 10 Ω. You will learn about markings of resistors rated at more than 10 Ω first.

The color bands start at the left end of the resistor. You can identify the starting end because the color bands are closer to that end of the resistor. Read the number values from the starting end. The first-band is the first digit of the number for the resistance rating. The second band is the second digit. The third band is a multiplier. The fourth

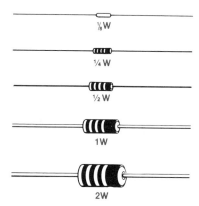

Figure 4-4. Carbon-composition resistors. Size determines resistance ratings. (Ohmite)

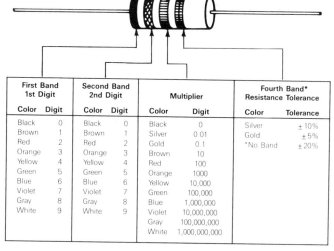

First Band 1st Digit		Second Band 2nd Digit		Multiplier		Fourth Band* Resistance Tolerance	
Color	Digit	Color	Digit	Color	Digit	Color	Tolerance
Black	0	Black	0	Black	0	Silver	± 10%
Brown	1	Brown	1	Silver	0.01	Gold	± 5%
Red	2	Red	2	Gold	0.1	*No Band	± 20%
Orange	3	Orange	3	Brown	10		
Yellow	4	Yellow	4	Red	100		
Green	5	Green	5	Orange	1000		
Blue	6	Blue	6	Yellow	10,000		
Violet	7	Violet	7	Green	100,000		
Gray	8	Gray	8	Blue	1,000,000		
White	9	White	9	Violet	10,000,000		
				Gray	100,000,000		
				White	1,000,000,000		

Figure 4-5. Resistor color-code chart.

band indicates the tolerance of the resistor. This fourth band is either silver or gold in color.

For practice in using this system, look at the color coding on the resistor in Figure 4-6. The first band is red. This color has a value of 2. The second band is violet, representing a 7. The third band is yellow. This means the multiplier is 4.

The first two numbers, then, are 27. A multiplier of 4 means that you add four zeros. So, the resistance rating in ohms is 270,000.

Now, check the fourth band in Figure 4-6. This is silver. A silver band in the fourth position means that the tolerance of the resistor is ±10 percent.

This resistance rating means that the actual resistance will be between 10 percent more or 10 percent less than the rated resistance. This is known as *resistance range*.

Now, figure the resistance range for the coding in Figure 4-6. The rating is 270,000 Ω. Ten percent of this is 27,000 Ω (270,000 × 0.10 = 27,000). So, the resistance may be as low as 243,000 Ω (270,000 − 27,000 = 243,000). The resistance may also be as high as 297,000 Ω (270,000 + 27,000 = 297,000). Any resistance between 243,000 Ω and 297,000 Ω is within the range for this resistor.

As indicated above, the fourth band may also be gold. A gold band indicates the tolerance is ±5 percent. To interpret the coding of a resistor with a gold band, look at Figure 4-7. The first band is yellow, for a value of 4.

The second is violet, representing 7. The third band is orange, meaning the multiplier is 3. So, the rating is 47,000 Ω.

The gold band in the fourth position means the tolerance is ±5 percent. This means the resistance may vary by as much as 2,350 Ω (47,000 × 0.05 = 2,350). Thus, resistance for this unit may be as low as 44,650 Ω (47,000 − 2,350 = 44,650). It may also be as high as 49,350 Ω (47,000 + 2,350 = 49,350).

The lower the tolerance, the more resistors cost. Thus, resistors with gold bands cost more than those with silver bands.

Occasionally, you may find a resistor that has no fourth band. This means the tolerance is ±20 percent. Resistors of this type are not made any more. But you may still find them in older electronic equipment.

Now you are ready to review markings of carbon-composition resistors with ratings of less than 10 Ω.

On these low-power resistors, the first two bands have the same meaning as described above. That is, they use the colors you already know to represent number values. The third band on a low-power resistor is silver or gold. A gold band means that you divide the first two numbers by 10. A silver band means that the first two numbers are divided by 100.

The fourth band still has the same meaning. It indicates tolerance. A silver band means a tolerance of ±10 percent. A gold band means a tolerance of ±5 percent.

Resistors come in standard sizes, Table 4-1. The Electronics Industry Association (EIA) sets standards for the manufacture of resistors. This helps to standardize the number of sizes of resistors made. It also makes it easy to find spare parts to use for repair jobs.

Precision Wire-Wound Resistors

Precision wire-wound resistors that resemble carbon-composition resistors in appearance have the first band of four with a wider-than-usual thickness, Figure 4-8. If one of the bands is twice the width of the others, it is not

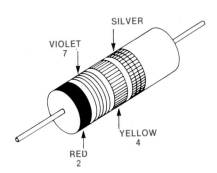

Figure 4-6. Color-coded resistor.

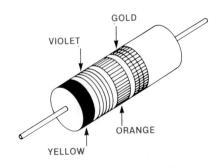

Figure 4-7. Color-coded resistor.

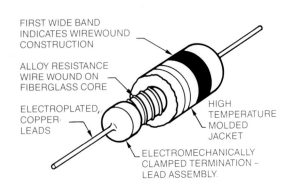

Figure 4-8. Precision wire-wound resistor.

10.0	19.1	36.5	69.8	133	255	487	931	1.78K	3.40K	6.49K	12.4K	23.7K	45.3K	84.5K	158K	294K	549K
10.2	19.6	37.4	71.5	137	261	499	953	1.82K	3.48K	6.65K	12.7K	24.3K	46.4K	86.6K	162K	301K	562K
10.5	20.0	38.3	73.2	140	267	511	976	1.87K	3.57K	6.81K	13.0K	24.9K	47.5K	88.7K	165K	309K	576K
10.7	20.5	39.2	75.0	143	274	523	1.00K	1.91K	3.65K	6.98K	13.3K	25.5K	48.7K	90.9K	169K	316K	590K
11.0	21.0	40.2	76.8	147	280	536	1.02K	1.96K	3.74K	7.15K	13.7K	26.1K	49.9K	93.1K	174K	324K	604K
11.3	21.5	41.2	78.7	150	287	549	1.05K	2.00K	3.83K	7.32K	14.0K	26.7K	51.1K	95.3K	178K	332K	619K
11.5	22.1	42.2	80.6	154	294	562	1.07K	2.05K	3.92K	7.50K	14.3K	27.4K	52.3K	97.6K	182K	340K	634K
11.8	22.6	43.2	82.5	158	301	576	1.10K	2.10K	4.02K	7.68K	14.7K	28.0K	53.6K	100K	187K	348K	649K
12.1	23.2	44.2	84.5	162	309	590	1.13K	2.15K	4.12K	7.87K	15.0K	28.7K	54.9K	102K	191K	357K	665K
12.4	23.7	45.3	86.6	165	316	604	1.15K	2.21K	4.22K	8.06K	15.4K	29.4K	56.2K	105K	196K	365K	681K
12.7	24.3	46.4	88.7	169	324	619	1.18K	2.26K	4.32K	8.25K	15.8K	30.1K	57.6K	107K	200K	374K	698K
13.0	24.9	47.5	90.9	174	332	634	1.21K	2.32K	4.42K	8.45K	16.2K	30.9K	59.0K	110K	205K	383K	715K
13.3	25.5	48.7	93.1	178	340	649	1.24K	2.37K	4.53K	8.66K	16.5K	31.6K	60.4K	113K	210K	392K	732K
13.7	26.1	49.9	95.3	182	348	665	1.27K	2.43K	4.64K	8.87K	16.9K	32.4K	61.9K	115K	215K	402K	750K
14.0	26.7	51.1	97.6	187	357	681	1.30K	2.49K	4.75K	9.09K	17.4K	33.2K	63.4K	118K	221K	412K	768K
14.3	27.4	52.3	100	191	365	698	1.33K	2.55K	4.87K	9.31K	17.8K	34.0K	64.9K	121K	226K	422K	787K
14.7	28.0	53.6	102	196	374	715	1.37K	2.61K	4.99K	9.53K	18.2K	34.8K	66.5K	124K	232K	432K	806K
15.0	28.7	54.9	105	200	383	732	1.40K	2.67K	5.11K	9.76K	18.7K	35.7K	68.1K	127K	237K	442K	825K
15.4	29.4	56.2	107	205	392	750	1.43K	2.74K	5.23K	10.0K	19.1K	36.5K	69.8K	130K	243K	453K	845K
15.8	30.1	57.6	110	210	402	768	1.47K	2.80K	5.36K	10.2K	19.6K	37.4K	71.5K	133K	249K	464K	866K
16.2	30.9	59.0	113	215	412	787	1.50K	2.87K	5.49K	10.5K	20.0K	38.3K	73.2K	137K	255K	475K	887K
16.5	31.6	60.4	115	221	422	806	1.54K	2.94K	5.62K	10.7K	20.5K	39.2K	75.0K	140K	261K	487K	909K
16.9	32.4	61.9	118	226	432	825	1.58K	3.01K	5.76K	11.0K	21.0K	40.2K	76.8K	143K	267K	499K	931K
17.4	33.2	63.4	121	232	442	845	1.62K	3.09K	5.90K	11.3K	21.5K	41.2K	78.7K	147K	274K	511K	953K
17.8	34.0	64.9	124	237	453	866	1.65K	3.16K	6.04K	11.5K	22.1K	42.2K	80.6K	150K	280K	523K	976K
18.2	34.8	66.5	127	243	464	887	1.69K	3.24K	6.19K	11.8K	22.6K	43.2K	82.5K	154K	287K	536K	1.00M
18.7	35.7	68.1	130	249	475	909	1.74K	3.32K	6.34K	12.1K	23.2K	44.2K					

Table 4-1. Standard Resistor Values

the tolerance band, but is read as the first band as it normally follows the color code for composition resistors. The wider band on a four-band resistor indicates that it is wire-wound.

In some instances, the resistor information is coded or digitized on the resistor. If it reads 3273, it means the resistance is 327,000 ohms, with the last digit indicating the multiplier or number of zeros. Keep in mind, though, that 1000 would mean 100 ohms, since the last zero has no meaning as a multiplier.

Wire-wound resistors with 0.1% precision can be obtained in metal-film form also. Wire-wound resistors pose a problem in terms of inductance when used in circuits with high frequencies (above 5 MHz), whereas metal film resistors can be used in circuits with frequencies up to 50 MHz.

Many 1% precision metal-film resistors are used in stereo amplifiers to produce good quality sound with little inherent noise. These resistors do have other designations that should be kept in mind when a circuit is being designed. Their temperature coefficient of resistance, called TCR, is specified in terms of resistance change in ppm/°C (parts per million per degree Celsius). This rating is not too important on Earth, but is very important when the resistors are used in space communications and under other temperature extremes.

Five-Band Resistors

There may be some confusion associated with five-band resistors inasmuch as the fifth band can represent either the reliability factor (used in Military Specifications, or MIL-R-39008-type resistors) or the tolerance.

Reliability factor has become a more important concept now that equipment is expected to last and resistors are not expected to change their values outside their tolerance specifications. A fifth band has been added to resistors to indicate this reliability factor, Figure 4-9. The metal-film five-band resistor has its last band separated from its first four. In MIL SPEC resistors this band indicates the reliability factor. The reliability factor gives the percentage of failure rate per 1000 hours of use. A 1% failure rate means 1 out of every 100 will fail to fall within the tolerance range after 1000 hours of use. Colors of the fifth band represent the reliability percentage. Brown, red, orange, and yellow are used for this type of resistor. This

Figure 4-9. Five-band resistor, with fifth band indicating the reliability factor.

allows easy identification in terms of MIL SPEC 39008. Some of these resistors are declared surplus for one reason or another and find their way onto the civilian market. Keep the following in mind.

Brown   1%
Red   0.1%
Orange   0.01%
Yellow   0.001%

As the accuracy of the resistors improve with improved methods of manufacture, the fifth band becomes more important. It can be used to indicate the *tolerance* of the resistor instead of the failure rate or reliability factor. In some cases, the five-band resistor is utilized in better-quality electronics equipment. It can be specified to a greater degree of accuracy than the four-band. Figure 4-10 indicates a five-band resistor where the first, second, and third bands are significant. The fourth band reveals the

multiplier and the fifth band reveals the tolerance. The fifth band is easily recognized as the tolerance band since it has a gold or silver color rather than the brown, red, orange, or yellow of the MIL SPEC resistors.

_____ REVIEW QUESTIONS FOR SECTION 4.3 _____

Refer to the resistor in Figure 4-11 to answer the following questions.

1. What is the resistance?
2. What is the tolerance of the resistor?
3. What are the upper and lower values of the resistor?

Refer to the resistor in Figure 4-12 to answer the following questions.

4. What is the resistance?
5. What is the tolerance of the resistor?
6. What are the upper and lower values of the resistor?

4.4 VARIABLE RESISTORS

Some resistors are variable. This means that the amount of resistance can be changed. Variable resistors may be either carbon-composition or wire-wound.

These devices are used for special circuits. On these circuits, the amount of voltage or current that is delivered must be varied. A common example is the volume control on your radio or television set, Figure 4-13.

You can recognize variable resistors easily because they have three connection points for leads. The center lead is usually the variable contact. A variable resistor that is connected into a circuit at all three points is called a **potentiometer**, Figure 4-14. A schematic diagram for a potentiometer is shown in Figure 4-15. A potentiometer is often called a *pot* for short.

Usually, a potentiometer is used to vary **voltage**. The potentiometer is connected for maximum resistance across a voltage source. The variable-setting device is then used to change the voltage that is available from the variable resistor.

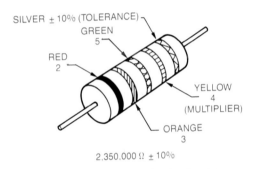

Figure 4-10. Metal-film five-band resistor.

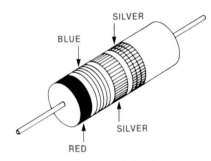

Figure 4-11. Color-coded resistor.

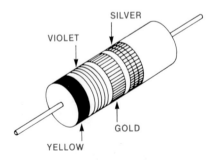

Figure 4-12. Color-coded resistor.

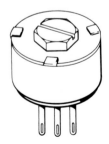

Figure 4-13. Variable resistor.

Figure 4-14. Potentiometer, a cutaway view. (Dale)

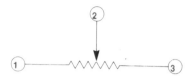

Figure 4-15. Symbol for a potentiometer.

Another type of variable resistor is called a *rheostat*. Rheostats, in general, are designed to handle higher currents than are potentiometers. Very few rheostats are used today because the same jobs are being done by semiconductors. (Semiconductors are covered in Chapter 16.) Usually, a rheostat is connected to a circuit at only two points. A diagram of the symbol for a rheostat is shown in Figure 4-16. A rheostat unit is shown in Figure 4-17. Rheostats are inserted in a circuit in series so they drop voltage.

Variable resistors can have a wide range of adjustment. For example, volume controls typically use carbon resistors. Resistance ratings can be adjusted from 0 to 10,000,000 Ω—that is, from 0 to 10 megohms (*Mega* means "million.")

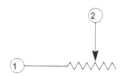

Figure 4-16. Symbol for a rheostat.

Figure 4-17. A rheostat to control voltage. (Ohmite)

Many potentiometers have what is called a *nonlinear resistance element*. This simply means that resistance does not change at a fixed, or uniform, rate as adjustments are made. Usually, there are small, or fine, changes at the low end. At the high end, settings lead to large resistance changes. This *nonuniform* resistance leads to what is called a *tapered control*. Such devices are usually used to adjust sound volumes. So, they are called *audio taper* units.

There are also **linear-taper** variable resistors. These have a uniform change of resistance as settings are adjusted.

Wattage ratings for variable resistors are established by manufacturers.

REVIEW QUESTIONS FOR SECTION 4.4

1. What is one way to identify a variable resistor on sight?
2. How many leads will you find on most variable resistors?
3. What is the most commonly used type of variable resistor?
4. What is the meaning of linear taper in a variable resistor?
5. What is the meaning of the prefix mega?

4.5 THE ELECTRICAL CIRCUIT

A complete electrical circuit needs: a source of electrical energy, a conductor, and a consuming device. The consuming device is the user of the electricity. It produces heat or work. In addition, there is usually a control device, such as a switch. The control device opens or closes the circuit. This starts or stops the flow of electrons to the consuming device. A diagram of a typical circuit designed to deliver electricity to a consuming device is shown in Figure 4-18.

When you wire a circuit, always remember three simple rules:

1. Connect one side of the power source to one side of the consuming device. (In the diagram, start by connecting point F to point E.)

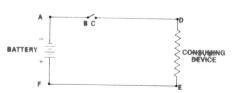

Figure 4-18. A complete circuit, including electrical source, conductor, and load. Switch opens and closes the circuit.

2. Connect the other side of the power source to the control device (switch). (Wire point A to point B.)
3. Connect the other side of the control device to the consuming device. (Wire point C to D.)

This produces a complete circuit. When a circuit is complete, you must have a complete path for electrons. Electrons must be able to leave a power source and return to the original source.

REVIEW QUESTIONS FOR SECTION 4.5

1. What are the requirements for a complete circuit?
2. What is the function of the load?
3. What is the function of a switch?
4. What are the procedures for wiring a circuit?

4.6 SERIES RESISTANCE CIRCUITS

Electrical circuits have voltage (emf), current, and resistance. To include these in a circuit, you need a source of emf, a conductor for the current, and consuming devices (resistance).

One way to set up an electric circuit is to connect the conductor and consuming devices in series. In a **series circuit**, one wire connects the power source and consuming devices. Current follows a path called a circuit. Electrons start from the negative battery terminal and travel through closed switches and consuming devices to the positive battery terminal.

The diagram in Figure 4-18 shows a series circuit that includes a control device and a consuming device. The source of electricity in this circuit is a battery. Note the symbol for battery. The consuming device is indicated by a symbol for a resistor. All consuming devices resist current to some degree.

Often, a circuit must provide voltage and current to more than one consuming device. When this is done in a series circuit, the resistors are connected to each other by a single conductor. A schematic diagram showing this type of resistor connection is illustrated in Figure 4-19.

Notice the way the resistors are identified in Figure 4-20. There are three resistors in this circuit. They are identified through use of subscripts. A *subscript* is a number or letter placed below the normal reading line. A resistor is identified with a capital R. When there is more than one resistor in a circuit, subscript numbers are used. The first resistor is R_1. The second resistor is R_2.

The third resistor is R_3. If more resistors are included, additional subscript numbers are added. The letter R always identifies resistors. For the circuit as a whole, total resistance is identified as R_T. In reading these notations, you say "R sub T" or "R sub 1," and so on.

Determining Resistance

In a **series circuit**, *total resistance is equal to the sum of individual resistance values.* You find total resistance by adding the resistance ratings for all resistors in a circuit. Mathematically, the value for resistance of the circuit in Figure 4-19 is expressed this way:

$$R_T = R_1 + R_2 + R_3$$
$$R_T = 10 \ \Omega + 20 \ \Omega + 5 \ \Omega = 35 \ \Omega$$

The total resistance of the three resistors in this circuit is 35 Ω.

Current Is Constant

Within a series circuit, **current** *is the same at each resistor.* There is only one path for the current to follow, as shown in Figure 4-20. The flow of electrons remains constant all along this path.

This can be proved if a circuit is broken and ammeters are inserted. Ammeters are devices that measure current. Figure 4-21 is a diagram showing how this is done. The symbol for a meter is a circle. An ammeter is indicated by a circle with an A inside. As indicated, the readings through all the ammeters are the same. For this circuit, the current is 2 A all along the conductor. The total current for this circuit is 2 A.

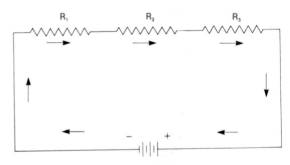

Figure 4-20. Schematic showing current flow in series circuit.

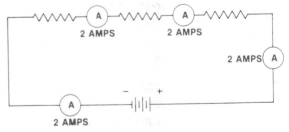

Figure 4-21. Schematic demonstrating uniform current level in series circuit.

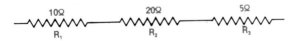

Figure 4-19. Schematic of resistors in series.

Determining Voltage

Within a series circuit, **voltage drop** *varies with the individual resistors*. If the voltages across the individual resistors are added, they equal the total, or *applied voltage*, of the circuit. For an example, look at Figure 4-22.

To calculate the voltage across any resistor, use Ohm's law.

Look at the first resistor in Figure 4-22. *I*, or current, is 2 A. *R*, or **resistance**, is 10 Ω. Multiply 2 × 10. *E* equals 20 V. Using the same technique, you get 40 V for the second resistor and 10 V for the third. Adding 20 + 40 + 10 gives a total emf for this circuit of 70 V.

$$E = I \times R$$

$$20 \text{ V} = 2 \text{ A} \times 10 \text{ }\Omega$$

Laws of Series Circuits

There are three laws of series circuits. They apply for any number of resistances in the series.

1. **Voltage.** Applied voltage (E_A) is equal to the sum (total) of voltage drops across all resistors. Voltage across an individual resistor is also called *voltage drop*.

$$E_A = E_{R_1} + E_{R_2} + E_{R_3} + \dots$$

2. **Current.** Total current is the same in any part of the circuit.

$$I_T = I_{R_1} = I_{R_2} = I_{R_3} = \dots$$

3. **Resistance.** Total resistance is equal to the sum of the individual resistances.

$$R_T = R_1 + R_2 + R_3 \dots$$

Advantages of Series Circuits

Series circuits are economical because only one wire is needed. Another advantage is that current is uniform at all points in the circuit.

The greatest use of series circuits is for voltage distribution.

Disadvantages of a Series Circuit

If there is a single break in the flow of electricity at any point, the series circuit fails to function. This can happen if any single resistor or consuming device burns out.

4.7 PARALLEL RESISTANCE CIRCUITS

In a **parallel circuit**, resistors are connected differently than in series. Think of the resistors as being side by side, or parallel, as shown in Figure 4-23. A schematic diagram for the same circuit is shown in Figure 4-24.

Determining Resistance

Resistance values are not added in a parallel circuit, the way they are in a series circuit. When there are two or more resistors in a parallel circuit, it is necessary to calculate the value mathematically.

One method is used if there are only two resistors in parallel. In this case, you first multiply the resistance values by each other. Then, you divide that product (total of the multiplication) by the sum of the resistance values.

Suppose the two resistors in Figure 4-24 have ratings of 10 Ω each. You multiply 10 × 10. The product is 100. Then you divide by the sum (10 + 10 = 20). Dividing

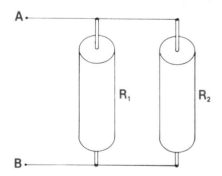

Figure 4-23. Pictorial view, resistors in parallel.

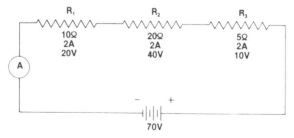

Figure 4-22. Schematic of resistance in series. Ratings of resistors are given.

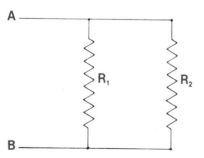

Figure 4-24. Schematic of resistors in parallel.

Activity: RESISTANCE IN SERIES

OBJECTIVES

1. *To investigate circuits with more than one resistor, and to study the relationship between voltage, current, and total resistance of a series circuit.*
2. *To obtain supporting evidence for the mathematical relationships that hold for circuits containing more than one resistor.*

MATERIALS NEEDED

1 Power supply, 6 volts DC

1 Meter, 1-0-1 mA DC

1 Resistor, 5000 ohms, ±5%, 1 watt

1 Voltmeter, DC, 0–10 volts

1 Resistor, 10,000 ohms, ±5%, 1 watt

4 Wires for connections

PROCEDURE

1. Connect the circuit shown in Figure 1.

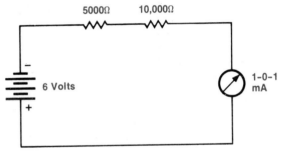

Figure 1

2. Set the voltage control for the power supply at the minimum output. *Turn on the power supply.* Slowly advance the voltage control to an output of 6 volts.

3. The circuit that you have just connected is a series circuit. It contains a 10,000-ohm resistor and a 5000-ohm resistor. By use of the formula for series resistance, compute the equivalent resistance of these two resistors in series.
4. Using Ohm's law, what should the current through this series circuit be? What current is indicated by the meter? Is there any difference between these two values? If so, what would account for the difference?
5. Set the voltmeter to read in the 0–10-volt range. Be careful of polarity (unless you have a digital voltmeter, which will indicate incorrect polarity). Read the voltage across the 5000-ohm resistor.
6. Read the voltage across the 10,000-ohm resistor.
7. *Turn off the power supply.* Disconnect the circuit you have been using. Return all equipment and materials to their proper storage places.
8. Using Ohm's law and the current you obtained in Step 4, what should the voltage across the 10,000-ohm resistor be?

SUMMARY

1. What happens to the voltage in a series circuit?
2. What happens to the current in a series circuit?
3. What happens to total resistance of a series circuit when another resistor is added?
4. Do all resistors in a series circuit have the same current flowing through them? Why or why not?
5. What would happen if one of the connections in the circuit came loose?

---

20 into 100, you get a result of 5. This means that if you have two 10-Ω resistors in parallel, the resistance rating for the circuit is 5 Ω.

$R_1 \times R_2 = 10 \times 10 = 100$

$R_1 + R_2 = 10 + 10 = 20$

$100 \div 20 = 5$

In this example, the ratings of both resistors are equal. A rule for this situation is: *If a circuit has two equal resistors in parallel, the total resistance is one-half the value of the resistor.*

$10 \times 0.5 = 5$

Also: *If there are three equal parallel resistors, divide the value of one of the resistors by three.* For four, five, or more equal resistors, divide the value of one resistor by the total number of resistors in the parallel circuit.

Now, suppose a circuit has two unequal resistors. Assume the resistors have ratings of 10 Ω and 40 Ω. Multiplying these ratings by each other produces a total of 400. Now, adding the values together, you get 50. Dividing 50 into 400 gives you an answer of 8. The effective resistance of this circuit is 8 Ω.

$R_1 \times R_2 = 10 \times 40 = 400$

$R_1 + R_2 = 10 + 40 = 50$

$400 \div 50 = 8$

You can also have more than two resistors with unequal values. This is the case in Figure 4-25. In this illustration, the resistors have values of 5 Ω, 10 Ω, 20 Ω, and 20 Ω. Resistance value for the circuit can be calculated in either of two ways. One is a shortcut you can work in your head. The other uses a mathematical formula.

To use the shortcut, figure the combined value of two resistors at a time and keep combining your results. For example, take the two resistors rated at 20 Ω. You divide 20 by 2 (see the rule above for equal resistors) and get a value of 10 Ω. (In effect, you now have a circuit like the one in Figure 4-26.) Now, you take the value of 10 you have just calculated and combine it with the resistor rated at 10 Ω. Dividing by 2 again, you get a value of 5 Ω. This leaves you with two values of 5 Ω. (See Figure 4-27.) Once again, divide by 2. The total resistance of this parallel circuit is 2.5 Ω.

To make the same calculations mathematically, here's the formula you use:

$$R_T = \cfrac{1}{\cfrac{1}{R_1} + \cfrac{1}{R_2} + \cfrac{1}{R_3} + \cfrac{1}{R_4}}$$

This formula can also be stated as follows:

$$\frac{1}{R_T} = \frac{1}{R_1} + \frac{1}{R_2} + \frac{1}{R_3} + \frac{1}{R_4}$$

Use the values from Figure 4-25 in this formula:

$$\frac{1}{R_T} = \frac{1}{5} + \frac{1}{10} + \frac{1}{20} + \frac{1}{20}$$

Now convert all fractions to equal values.

$$\frac{1}{R_T} = \frac{4}{20} + \frac{2}{20} + \frac{1}{20} + \frac{1}{20} = \frac{8}{20}$$

$$\frac{1}{R_T} = \frac{8}{20}$$

Now invert to remove the reciprocal. This gives R_T the value of a whole number.

$$\frac{R_T}{1} = \frac{20}{8}$$

$$20 \div 8 = 2.5$$

The circuit has a resistance of 2.5 Ω.

Determining Voltage

Within parallel circuits, *the same voltage is present across all resistors*. This is illustrated in Figure 4-28. This diagram shows that the same voltage is present across all three resistors.

This condition results because the resistors are connected to wires that are in parallel. Thus, the wire connected to point A runs to the leads on that end of all three resistors. Similarly, the wire leading to point B is connected to the other leads of all three resistors.

Remember: The same voltage is present across all resistors of a parallel circuit.

$$E_T = E_1 = E_2 = E_3 = \ldots$$

Determining Current

In a series circuit, current is equal through all resistors. However, in a parallel circuit, *current varies according to resistance*. For an example, see Figure 4-29.

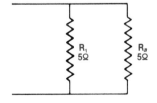

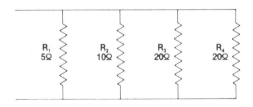

Figure 4-25. Schematic showing four resistors in parallel.

Figure 4-27. Schematic of second step in shortcut method for determining resistance in parallel circuit.

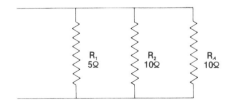

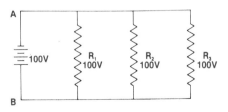

Figure 4-26. Schematic of first step in shortcut method for determining resistance in parallel circuit.

Figure 4-28. Schematic showing voltages of parallel resistors.

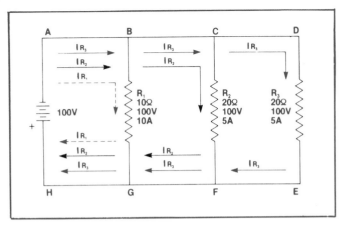

Figure 4-29. Schematic showing current flow in parallel circuit.

To determine current in a parallel circuit, use Ohm's law:

$$I = \frac{E}{R}$$

This means that, to find current, you divide voltage by resistance. You do this for each resistor in a circuit.

For example, in Figure 4-29, 100 V is delivered to the entire circuit. The first resistor has a rating of 10 Ω. Dividing 10 into 100, you get a current of 10 A. Resistors 2 and 3 are rated at 20 Ω each. Dividing 20 into the voltage of 100 gives you a current of 5 A at each of these resistors.

Resistor 1:   100 V ÷ 10 Ω = 10 A
Resistors 2 and 3:   100 V ÷ 20 Ω = 5 A

The total current for the circuit is then equal to the sum of the current through the three resistors. The parts of the circuit connected to individual resistors are called *branches*. The three branches in the circuit in Figure 4-29 have currents of 10 A, 5 A, and 5 A. So, total current of this circuit is 20 A.

$$I_T = 10\ A + 5\ A + 5\ A = 20\ A$$

Advantages of Parallel Circuits

If one resistor opens, the circuit keeps working. Current flows through other consuming devices.

Total resistance of a circuit can be varied with the addition or removal of resistors.

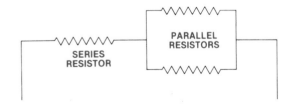

Figure 4-30. Schematic of series–parallel circuit.

Disadvantage of Parallel Circuits

Two wires are needed to connect each consuming device to the power source.

REVIEW QUESTIONS FOR SECTION 4.7

1. What is a parallel circuit?
2. How are resistors connected in a parallel circuit?
3. How is resistance determined in a parallel circuit?

4.8 SERIES-PARALLEL RESISTANCE CIRCUITS

It is possible to have circuits that combine series and parallel connections. Within these circuits, some resistors or consuming devices are connected in series. Others are connected in parallel. An example of such a circuit is diagrammed in Figure 4-30.

Another technique for wiring a series-parallel circuit is diagrammed in Figure 4-31.

One of the features of a series–parallel circuit is obvious from these diagrams: There must be at least three resistors or consuming devices.

Determining Resistance

To determine resistance for a **series–parallel circuit**, follow a simple method: *Calculate the resistance for the parallel units first. Then treat the entire circuit as a series circuit.* This means that, after you have reduced the values of the resistors that are in parallel to a series equivalent, you add all resistance values together.

For example, look at the diagram in Figure 4-32. Both parallel resistors have values of 20 Ω. This means the parallel circuit has a total resistance of 10 Ω. Remember the rule: For two resistors of equal value in parallel, divide the value of one resistor by two.

$$20\ \Omega \div 2 = 10\ \Omega$$

In effect, you wind up with a series circuit like the one in Figure 4-33. (You can actually redraw diagrams to help figure values for series–parallel circuits.) Figure 4-33 shows two resistors in series. Each has a value of 10 Ω. So the total resistance for the circuit is 20 Ω. Remember

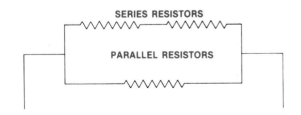

Figure 4-31. Schematic of alternative series–parallel circuit.

Activity: *RESISTANCE IN PARALLEL*

_____ OBJECTIVES _____

1. *To become familiar with circuits with resistances connected in parallel.*
2. *To connect parallel circuits.*
3. *To work parallel-resistance problems.*
4. *To obtain total current in a parallel circuit.*
5. *To obtain voltages for resistors in parallel.*

MATERIALS NEEDED

1 Power supply, 6 volts DC
1 Meter, 1-0-1 mA DC
1 Resistor, 10,000 ohms, ±5%, 1 watt

1 Voltmeter, DC, 0–10 volts
1 Resistor, 15,000 ohms, ±5%, 1 watt
5 Wires for connections

PROCEDURE

1. Connect the circuit shown in Figure 1.

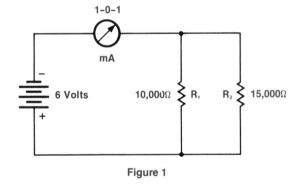

Figure 1

2. In this circuit, a 15,000-ohm resistor and a 10,000-ohm resistor are connected in parallel. They form a parallel resistance circuit. *Turn on the power supply.* Adjust the output to 6 volts DC. What is the current in the circuit indicated by the milliammeter?
3. Use the formula for parallel resistance. What is the equivalent resistance of a 10,000-ohm and a 15,000-ohm resistor connected in parallel? Keep in mind that only two resistors are connected here.

$$R_T = \frac{R_1 \times R_2}{R_1 + R_2}$$

4. Using Ohm's law, the total resistance just found, and the current just measured, what is the voltage necessary to cause this current to flow through the circuit?
5. Use a 0–10 voltmeter to check the voltage across the resistors R_1 and R_2.
6. Does this measured voltage read the same as the voltage figured in Step 4?
7. Does the voltage across both resistors read the same? What does this tell you about the voltage across resistors in parallel?
8. Turn off the power supply. Disconnect all circuits used. Return all equipment and materials to their proper storage places.

SUMMARY

1. What happens to the current in a parallel circuit?
2. What happens to the voltage in a parallel circuit?
3. What are the two formulas for finding total resistance in a parallel circuit?

the rule: To determine total resistance in a series circuit, add the value of all resistors.

To review these steps for a more complex circuit, look at Figure 4-34. The diagram shows a circuit with five

Figure 4-33. Schematic showing series equivalent of series–parallel circuit.

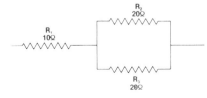

Figure 4-32. Schematic of series–parallel circuit, showing resistance values.

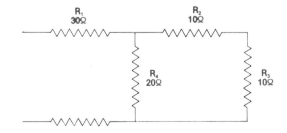

Figure 4-34. Schematic of complex series–parallel circuit.

resistors. To figure the resistance for this circuit, you can follow the steps shown in Figure 4-35.

In drawing A of Figure 4-35, the values of two 10-Ω resistors in series are added together. The total value for these two resistors is 20 Ω. Drawing B then shows this value as part of a parallel circuit. This parallel circuit includes the combined values of the two series resistors. These have been redrawn in parallel with another 20 Ω resistor. The effect is to combine the series resistors within the parallel portion of the circuit from the original drawing. The two 20-Ω parallel resistors have a total resistance of 10 Ω. In diagram C, this value is included as part of a series. The three resistors in the final drawing have values of 30 Ω, 10 Ω, and 25 Ω. So, adding these, the total resistance for the circuit is 65 Ω.

The rules for determining resistance can be summed up as follows:

1. When necessary for clarity, you can redraw the series–parallel circuit.
2. Solve all branches with series resistors first. Simply add their values.
3. Find the equivalent, or total, resistance of those resistors connected in parallel.
4. Add all the resistance values. Do this after you have series values for all resistors.

Determining Current

To determine current in a series–parallel circuit, follow the current path through the entire circuit. Remember your rules of current: In a series circuit, current is the same through all resistors. In a parallel circuit, you divide resistance into voltage to determine current.

Current flow in a series–parallel circuit is shown in Figure 4-36. The calculations for amperage in this circuit are included in the following section. They are part of the discussion and calculations of voltage values for the same circuit.

Determining Voltage

To find voltage drop across resistors in series–parallel circuits, follow the rules you already know: When resis-

tors are in series, add voltages. When resistors are in parallel, voltage is the same across all resistors.

As an example for the finding of voltage (and current) in a series–parallel circuit, look at the diagram in Figure 4-37. This diagram provides total voltage at the source and resistance values for all resistors. To find values for voltage and current, here are the simple steps to follow:

1. Redraw the circuit for clarity, if necessary. Follow the steps illustrated in Figures 4-38, 4-39, and 4-40.
2. Locate all branches with series resistors. This has been done in drawings (1) and (2) of Figure 4-38.
3. Locate all branches with parallel resistors. This has been done in drawing (3) of Figure 4-38.
4. Locate all branches that have series resistors in series with other resistors. Look back at Figure 4-37. Resistors 1, 2, and 5 are in series. These are shown in series as Loop 1 in Figure 4-39. Resistors 1, 3, 4, and 5 are also in series. These are shown as Loop 2 in Figure 4-40.
5. Apply the rules for series or parallel circuits to calculate voltage and current.

To figure current, divide total resistance into total voltage. Total voltage is given as 100 V. Resistance values are shown in Figure 4-37. Look at the resistance for the parallel portion of this circuit first. Resistors 3 and 4 are in series. These values of 10 Ω each are added for a total of 20 Ω. This gives the effect of two 20-Ω resistors in parallel, for a total value of 10 Ω. This value is then added in series to the values of resistors 1 and 5. The values added are 5 Ω, 10 Ω, and 10 Ω, for a total of 25 Ω. Dividing 25 Ω into 100 V produces a quotient, or answer, of 4 A.

$$R_3 + R_4 = 10\ \Omega + 10\ \Omega = 20\ \Omega$$

$$20\ \Omega \div 2 = 10\ \Omega$$

Resistance of parallel circuit $= 10\ \Omega$

$$10\ \Omega + R_1 + R_5 = 10\ \Omega + 5\ \Omega + 10\ \Omega = 25\ \Omega$$

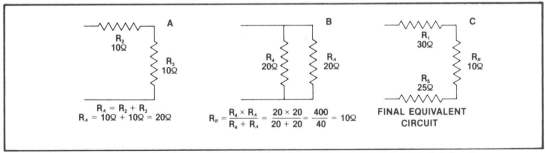

Figure 4-35. Set of schematics redrawn for determining resistance in series–parallel circuit.

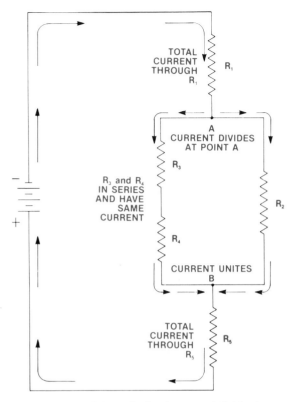

Figure 4-36. Schematic showing current division in series–parallel circuit.

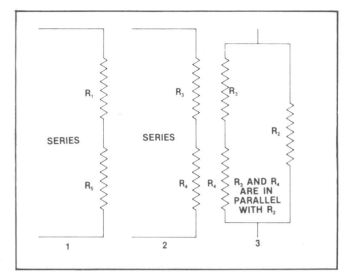

Figure 4-38. Schematic of complex series–parallel circuit redrawn in three branches.

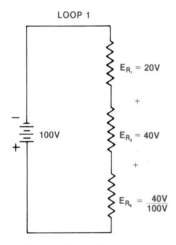

Figure 4-39. Schematic showing voltages in Loop 1 of series–parallel circuit.

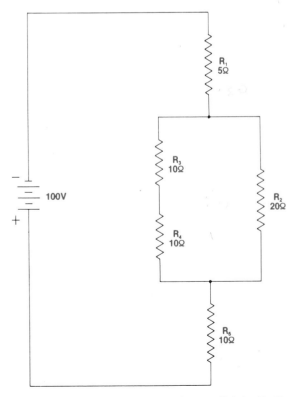

Figure 4-37. Schematic of complex series–parallel circuit with resistance values given.

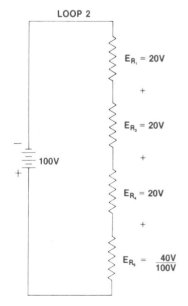

Figure 4-40. Schematic showing voltages in Loop 2 of series–parallel circuit.

Total resistance for the circuit is 25 Ω. Now, to determine current, divide voltage (100 V) by resistance.

$$100 \text{ V} \div 25 \text{ }\Omega = 4 \text{ A}$$

The current through the circuit is 4 A. This current divides through the parallel portion of the circuit. So, current at resistors 2, 3, and 4 is 2 A. At resistors 1 and 5, the current is 4 A.

Note that resistors 1 and 5 are repeated in the series drawings in Figures 4-39 and 4-40. This is important to remember: In determining voltage in a series–parallel circuit, reduce the branches to an equivalent series value. In doing so, consider all resistors in each series.

To determine voltage for each resistor in a series branch, multiply current by resistance. Resistor 1 has a rating of 5 Ω and a current of 4 A. Its voltage is 20. Resistor 2 has 20 Ω of resistance and 2 A. Its voltage is 40. Resistor 5 has a rating of 10 Ω and 4 A of current, for 40 V. Resistors 3 and 4 are rated at 10 Ω each. These resistors have 2 A of current. Voltage across these resistors is 20 each.

Here's a simple way to test the accuracy of your voltage calculations: *In a series–parallel circuit, the combined voltage in each series always equals total voltage for the circuit.* Both loops have combined voltages of 100 V. The circuit has a total of 100 V.

_____ **REVIEW QUESTIONS FOR SECTION 4.8** _____

1. What is a series–parallel circuit?
2. What is the minimum number of resistors needed in a series–parallel circuit?
3. How do you determine resistance in a series–parallel circuit?
4. How do you determine current in a series–parallel circuit?

4.9 KIRCHHOFF'S LAWS

The rules you have learned for figuring voltage and current within circuits are known as **Kirchhoff's laws**. Kirchhoff's laws are used to help find values of current or voltage when more than one of these items is not known. (When you know two of the three values, you can find the third by using Ohm's law. Kirchhoff's laws apply when two or more of the values are missing.)

Kirchhoff's First Law

This law applies to the determination of current: *Current entering a junction is equal to the current leaving the*

junction. A junction is a point where two or more conductors join. Principles of this law are illustrated in Figure 4-41. Three circuits are diagrammed. In each, the principle is the same: The current leaving a junction is equal to the sum of the currents entering the junction.

Kirchhoff's Second Law

This law states that *the sum of the voltage drops around a complete circuit loop is equal to the applied voltage.* This principle is illustrated in Figure 4-42, which shows a series–parallel circuit with two loops. Note that the combinations of voltage drops add up to the same total, 60 V, for both loops of this circuit.

Kirchhoff's second law can also be applied to solve circuits with two sources of emf. A circuit with two power sources is diagrammed in Figure 4-43. Two batteries are connected in a series with a load resistor. Battery E_1 has 20 V of power and 2 Ω of internal resistance. Battery E_2 has 5 V of power and 2 Ω of resistance.

Note that Battery E_2 is connected so that it opposes the flow of current from Battery E_1. That is, the flow of current from the two batteries moves in opposite directions around the circuit. Therefore, the current from Battery E_2 has a minus $(-)$ effect on total current carried through the circuit. Therefore, the power rating of Battery E_2

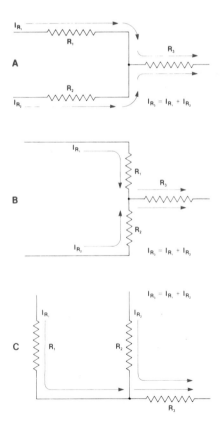

Figure 4-41. Schematic illustrating Kirchhoff's First Law.

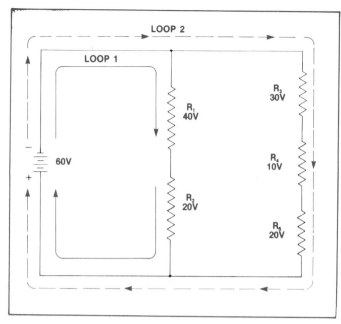

Figure 4-42. Schematic of series–parallel circuit with two loops, or branches.

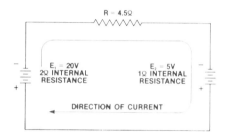

Figure 4-43. Schematic of circuit with two power sources.

should be subtracted from the voltage of Battery E_1. This is done through algebraic addition: Adding $+20$ V and -5 V produces a sum of $+15$ V.

Total resistance should also be added algebraically: Internal resistance of the batteries is $+2\ \Omega$ and $+1\ \Omega$, for a total of $+3\ \Omega$. The series load resistor has a resistance of $+4.5\ \Omega$. Adding $+4.5\ \Omega$ and $+3.0\ \Omega$ produces a sum of $+7.5\ \Omega$.

Current across this circuit can be figured through use of Ohm's law:

$$I = \frac{E}{R} \qquad I = \frac{15\ V}{7.5\ \Omega} \qquad I = 2\ A$$

Thus, I, or current, is 2 A throughout the circuit. To determine voltage drops across individual resistors, you also use Ohm's law: $E = I \times R$. For Battery E_1, the current is 2 A and the resistance is 2 Ω. So, the voltage drop is 4 V (2 A $\times$ 2 Ω). For the series load resistor, the voltage drop is 2 A $\times$ 4.5 Ω, or 9 V. For Battery E_2, the voltage drop is 2 A $\times$ 1 Ω, or 2 V.

Now, to test this work, you can add the voltage drops algebraically: There is the drop of 5 V from the opposing current of Battery E_2 added to the drops of 9 V, 4 V, and 2 V calculated above. Adding $+5$ V, $+9$ V, $+4$ V, and $+2$ V produces a sum of $+20$ V. This is equal to the voltage rating of Battery E_1. Thus, *the sum of voltage drops around a complete circuit equals the applied voltage.*

Circuits with Two Power Sources

Many applications for circuits with two power sources exist. In dealing with them, keep in mind the two laws that Kirchhoff developed and proved many years ago. They deal with the circuit current and the circuit voltage.

Current entering a junction is equal to the current leaving the junction. In order to use this law, it is necessary to establish the direction of current flow in a circuit and through each resistor. Alternatively, if the direction is assumed incorrectly, a negative answer results.

The sum of the voltage drops around a complete circuit loop equals the applied voltage. This law also comes in handy when trying to determine the voltage drops in a series–parallel circuit.

In order better to understand Kirchhoff's laws for circuits with two power sources, let us take an example and follow it through to see whether our answers agree with these two laws.

EXAMPLE

Determine the direction and magnitude of the current through resistor R_3 in Figure 4-44. It is assumed that the current through R_3 is as the arrows indicate. However, if the resulting answer is negative, this will indicate that the basic assumption as to current direction was incorrect.

1. Kirchhoff's first law states that the current entering a junction is equal to the current leaving the junction.

Activity: *SERIES–PARALLEL RESISTANCE CIRCUIT*

OBJECTIVES

1. *To study the relations that exist between voltage, current, and values of resistance in circuits containing resistors connected in series–parallel configurations.*
2. *To prove by measurement the mathematical results obtained in solving series–parallel problems.*

MATERIALS NEEDED

1 Power supply, 6 volts DC

1 Voltmeter, 0–10 volts DC

1 Meter, 1-0-1 mA DC

1 Resistor, 5000 ohms, 5%, 1 watt

5 Wires for connections

1 Resistor, 10,000 ohms, ±5%, 1 watt

1 Resistor, 15,000 ohms, ±5%, 1 watt

PROCEDURE

1. Connect the circuit shown in Figure 1.

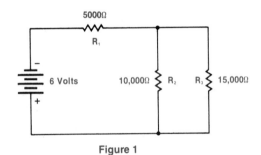

5000Ω

R_1

6 Volts

10,000Ω R_2 R_3 15,000Ω

Figure 1

2. The circuit you have just connected is a series–parallel resistance circuit. Resistors R_2 and R_3 are connected in parallel. This network of two resistors is then connected in series with R_1. What is the equivalent resistance of R_2 and R_3?
3. To determine the resistance of the total circuit, combine the equivalent resistance of the network that you found in Step 2 with the resistance of R_1. What is the total circuit resistance?
4. Use the voltmeter to measure the source voltage. The meter should be set to read from 0 to 10 volts DC. Measure the source voltage. What is the source voltage?

5. Determine the total circuit current by using Ohm's law:

$$I_T = \frac{E_A}{R_T}$$

I_T = total circuit current
E_A = applied voltage or source voltage
R_T = total circuit resistance

6. Since the total circuit current passes through R_1, to find the voltage drop across R_1, multiply the value found in Step 5 by the resistance of R_1.
7. Use the multimeter to measure the voltage drop across R_1. How does this value of voltage compare with the value you found in Step 6? What would account for any difference?
8. Use the voltmeter to measure the voltage across R_2 and R_3. Are the readings the same for each resistor? Why? What is the voltage across the parallel resistors?
9. Subtract the value of voltage found in Step 6 from the source voltage to determine the voltage across R_2 and R_3. How does this answer compare with the voltage drop measured in Step 8?
10. The current through R_2 and R_3 can be determined by Ohm's law:

I_2 = current through R_2
E_2 = voltage across R_2 (Step 9)
R_2 = value of resistance
I_3 = current through R_3
E_3 = voltage across R_3 (Step 9)
R_3 = value of resistance

11. Do the values of I_2 and I_3 combine to give the value of the total circuit current? Why would this be so?
12. Disconnect the circuit you have been using. Return all equipment and materials to their proper storage places.

SUMMARY

1. In what ways did the values of current and voltage in the experiment demonstrate the rules of current and voltage in the series and parallel circuits?
2. List in the proper order the mathematical procedure for determining the values of resistance, voltage, and current in the series–parallel circuit used in this activity.

2. Establish the current loops by inspection: I_1 goes from F to E, through R_3 to B, through R_1 to A, and through E_A to F; I_2 goes from D to E, through R_3 to B, through R_2 to C, and through E_B to D.

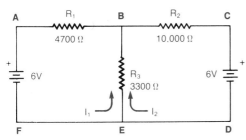

Figure 4-44. Finding current through a resistor.

3. Substitute terms in the formulas:

$$E_A = (R_3)(I_1 + I_2) + (R_1)(I_1)$$
$$E_B = (R_3)(I_1 + I_2) + (R_2)(I_2)$$

4. Substitute values for one power source (E_A) and simplify:

(a) $6 = 3300(I_1 + I_2) + 4700 I_1$
(b) $6 = 3300 I_1 + 3300 I_2 + 4700 I_1$
(c) $6 = 8000 I_1 + 3300 I_2$

5. Do the same for the other power source (E_B):

(a) $6 = 3300(I_1 + I_2) + 10,000 I_2$
(b) $6 = 3300 I_1 + 3300 I_2 + 10,000 I_2$
(c) $6 = 3300 I_1 + 13,300 I_2$

6. Solve I_2 by simultaneous equations:

E_A: $6 = 8000 I_1 + 3300 I_2$
(multiply this equation by 33)
E_B: $6 = 3300 I_1 + 13,300 I_2$
(multiply this equation by 80)

This will result in both equations having the same values for I_1.

7. Multiply both sides of the equal sign in both equations:

E_A: $198 = 264,000 I_1 +   108,800 I_2$
E_B: $480 = 264,000 I_1 + 1,064,000 I_2$
$\overline{-282 =   -955,200 I_2}$

Both negative values so we have
a + answer

Subtract E_B from E_A.

$$I_2 = \frac{282}{955,200} = 0.000295 \text{ A or } .295 \text{ mA}$$

8. To find I_1, substitute the value of I_2 in the E_A formula:

$I_2 = 0.000295$ A
E_A: (a) $6 = 8000 I_1 + 3300 I_2$
(b) $6 = 8000 I_1 + (3300)(0.000295)$
(c) $8000 I_1 + 0.9735$
(d) $5.0265 = 8000 I_1$
(e) $I_1 = \frac{5.0265}{8000} = 0.000628$ A or .628 mA

9. Determine the current through R_3:

current through $R_1 = I_1$, which is 0.628 mA
current through $R_2 = I_2$, which is 0.295 mA
current through $R_3 = I_1 + I_2$, or 0.923 mA.

Since the value of R_3 is positive, the assumed direction of current is correct. If the value were negative, the absolute value would be correct, but the direction would be wrong. Since both current values are positive, the assumed direction of current is correct. The current flows upward through R_3.

10. Voltage drop across $R_1 = (R_1)(I_1)$

$E_1 = 4700 \times 0.000628$
$E_1 = 2.95$ volts

11. Voltage drop across $R_2 = (R_2)(I_2)$

$E_2 = 10,000 \times 0.000295$
$E_2 = 2.95$ volts

12. Voltage drop across $R_3 = (R_3)(I_3)$

$E_3 = 3300 \times 0.000923$
$E_3 = 3.05$ volts

13. Kirchhoff's second law states that the sum of the voltage drops around a complete loop equals the applied voltage. Therefore:

$$\begin{array}{r} 2.95 \\ +3.05 \\ \hline 6.00 \end{array}$$

REVIEW QUESTIONS FOR SECTION 4.9

1. What are Kirchhoff's laws used for?
2. What is Kirchhoff's first law?
3. What is Kirchhoff's second law?
4. How are Kirchhoff's laws used to determine current?
5. How are Kirchhoff's laws used to determine voltage?

SUMMARY

Resistance plays three *important* roles in circuits. Resistors may be fixed or variable, carbon-composition or wire-wound. Carbon-composition resistors have color bands that designate their ohmic resistance and tolerance. Variable resistors include rheostats and potentiometers.

A complete circuit needs a source of energy, a conductor, and a consuming device. In a series circuit, the current is the same in all parts of the circuit. The total resistance in series is found by addition. The main advantage of a series circuit is that the current is the same in all parts. The disadvantage of a series circuit is its complete failure when any one component opens or the wiring is interrupted. The parallel circuit has the same voltage across every component in the circuit. The current divides according to the resistance.

A series–parallel circuit obeys the laws derived for both series and parallel circuits. To find total resistance of a series–parallel circuit, the parallel branches have to be reduced to an equivalent resistance. Once the whole circuit has been reduced to a series equivalent, the resistances are added to arrive at the total.

Kirchhoff's first law applies to the determination of current. Kirchhoff's second law applies to the distribution of voltage drops around a complete circuit loop.

Circuits with two power sources have many applications. Kirchhoff's two laws of voltage and current are needed in order to solve the problems associated with circuits that have two power sources.

USING YOUR KNOWLEDGE

1. Draw a circuit with three resistors in series.
2. Draw a circuit with three resistors in parallel.
3. Draw the symbol for a potentiometer.
4. A 3300-Ω resistor has a ± 10 percent tolerance. Figure the maximum and minimum levels of its resistance range.
5. The resistors in a series circuit have voltage drops of 10 V, 20 V, and 35 V. What is the applied voltage of the circuit?

KEY TERMS

resistor
conductor
composition
potentiometer
linear taper
series circuit
resin-bonded
ceramic-bonded
deposited carbon resistor
metal-film resistor
wattage rating
voltage
power rating
parallel circuit
voltage drop
current
resistance
series–parallel circuit
load
Kirchhoff's laws

PROBLEMS

1. What is the total resistance of four resistors in series if their individual resistances are 1000, 10,000, 5000, and 2000 ohms?

2. What is the total resistance of five resistors with 2200 ohms each?

3. If there are three resistors in series and one of them has a current of 10 milliamperes, what is the current through the rest of the resistors?

4. If you had a 22,000-ohm resistor in series with a 2,200-ohm resistor, and the connection between the two resistors opened, what would the total resistance of the series then be?

5. What is the current through a series circuit if the total resistance is 2000 ohms and the applied voltage is 20 volts?

6. Three resistors of 6000 ohms each are connected in parallel. What is the total resistance?

7. A 200-ohm resistor is connected across a 2-megaohm resistor. What is the resistance if the two are connected in parallel?

8. What is the total resistance of the circuit shown in Figure 4-45?

9. What is the applied voltage for the circuit shown in Figure 4-45?

10. What is the voltage drop across R_1 in Figure 4-45?

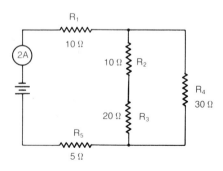

Figure 4-45. A circuit containing five resistors.

Project: *NEON VOLTMETER*

A simple voltmeter can be constructed with a minimum of time, cost, and effort, by means of a neon lamp. Gas in such a lamp ionizes and glows at about 55 volts.

This meter operates on the extinction principle. When the lamp goes out, the voltage indicated by the pointer is the voltage present in the circuit being checked.

Two rows of numbers will appear on the face of the meter if it is calibrated for use on AC and DC. Since the reading for 100 V DC will appear at a different location than that of 100 V AC, it is suggested that the two scales be marked in different colors and labeled.

An AC voltmeter of known accuracy is needed, along with a variable AC power supply, to calibrate the meter. Set the knob completely counterclockwise and apply the highest voltage available. Do not adjust over 600 volts. Adjust the knob until the neon lamp goes out. Mark the face plate at this point with the voltage read from the voltmeter being used to calibrate. Repeat this operation, reducing the voltage each time, until a voltage of 60 is reached. The lamp fails to ionize at voltages less than 52. Since each neon lamp ionizes at a slightly different voltage, two neon voltmeters made at the same time may vary slightly in their lowest voltage reading.

To calibrate the face plate for DC, obtain a DC voltmeter of known accuracy and a variable DC power supply. Proceed as described for calibrating the AC scale, Figure 1.

You will need the following materials (see Exploded View, Figure 2).

1 Face plate—fiber, tin plate, or aluminum—to fit box
1 5-terminal tie strip
1 Box, Bakelite, approximately $1.5'' \times 3'' \times 4''$
1 Knob
1 Rubber grommet
5 Machine screws
1 Neon lamp—NE-2, 1/25 watt
1 Resistor, 150,000 ohms, 0.5 watt, 10%
1 Potentiometer, 500,000 ohms, 0.5 watt, carbon, linear taper
1 Nut for potentiometer
1 #4-40 steel, hex
2 Probe tips
2 Probe leads
2 Test leads

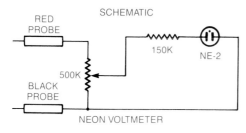

SCHEMATIC

Figure 1

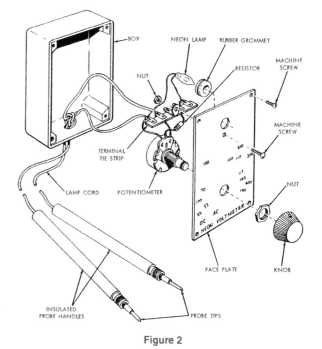

Figure 2

Chapter 5

MAGNETISM

OBJECTIVES

After studying this chapter, you will know

- *that permanent magnets may be made through magnetic induction.*
- *that low-carbon steel forms temporary magnets.*
- *that the ability of a material to become magnetized is called its permeability.*
- *that magnets have poles, that like poles repel, and that unlike poles attract.*
- *that magnetism decreases with distance from the poles.*
- *the meaning of the term* magnetic domain.
- *that electromagnetism depends upon electrical current.*
- *that the direction of current flow in an electromagnet determines its polarity.*
- *that when a coil of wire receives current, the degree of magnetism depends on the wrapping of the wire.*
- *that magnetic fields seek the path of least reluctance.*
- *the operation of solenoids and doorbells.*

INTRODUCTION

You have seen magnets at work since you were a child. At some time, you probably played with a toy driven by an electric motor. Many of these motors have permanent magnets that control the toy.

Every time you press a doorbell, an electromagnet causes it to ring. When you finish your work in this chapter, you will know how a doorbell works. You will also know about magnets and electricity.

Magnetism is the force that draws some materials to a magnet. Some magnets exist in nature. One kind of natural magnet is called a **lodestone**. Lodestone is a type of iron ore known as **magnetite**. The chemical formula for magnetite is Fe_3O_4. Fe_3O_4 is not molecular. Therefore the term **basic unit** is used instead of molecule in order to be technically correct. This means each basic unit of magnetite contains three parts of iron and four parts of oxygen.

The relationship of magnetism to electricity is easily demonstrated. This can be done with a compass. A compass has a magnetic needle that always points north. The needle in a compass lines up with the magnetic field that surrounds the earth. If a compass is placed within the field of an electric circuit, the needle is affected, Figure 5-1.

5.1 NATURE OF MAGNETISM

Some magnets are permanent. That is, they keep their ability to attract iron and to react to electrical current. They do not need electricity to function.

Other magnets are temporary. They attract iron under certain conditions. When these conditions are removed, the magnetism is gone as well.

Some magnets require electricity to function. These are *electromagnets*. When the electric current is removed, there is no magnetism.

Both permanent magnets and electromagnets have important roles in electricity.

Permanent Magnets

One type of permanent magnet is the natural magnet, the lodestone. It is also possible to make permanent magnets. For example, if you stroke a piece of high-carbon steel with a lodestone, the steel becomes magnetized.

Normally, the steel stroked with the lodestone keeps its magnetism permanently. When this happens, the basic units that form the steel bar line up in the same direction. This is shown in Figure 5-2. The basic units that line themselves up in this way point toward the ends, or *poles*, of the magnet. Each magnet has two poles, which are identified as north and south.

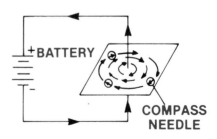

Figure 5-1. Electricity affects magnetism of a compass.

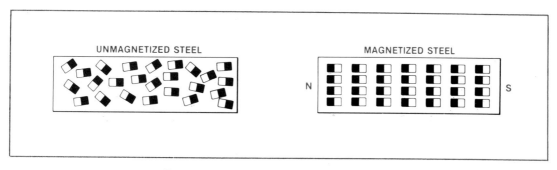

Figure 5-2. Magnetism aligns the iron in the steel.

A permanent magnet can be used to create other magnets. Figure 5-3 shows a permanent magnet used to stroke a bar of high-carbon steel. This creates a second magnet.

Magnets made of high-carbon steel sometimes lose their magnetism. This happens if the magnetized steel bar receives a strong shock. Such a shock may result if the bar is dropped or if it is struck with a hammer. When this happens, the basic units lose their alignment. This leads to a loss of magnetism. Magnetism can also be lost if a magnet is heated to high temperatures.

A magnet is formed by rubbing a piece of steel with another magnet. This is called **magnetic induction**. This was the only known way of making magnets until electro-magnetism was discovered.

Temporary Magnets

A bar of low-carbon steel attracts iron particles when it is in direct contact with a magnet. For example, you can place a lodestone against one pole of a bar of low-carbon steel as shown in Figure 5-4. When this is done, the bar attracts iron particles. Removing the lodestone removes the magnetism. Then the iron particles fall away. The same thing happens if you place a permanent magnet against one end of a bar of low-carbon steel.

Low-carbon steel does not retain, or keep, magnetism.

This material can serve as a *temporary magnet*. But it cannot be used for a permanent magnet.

Electromagnets

Magnetism also results when electric currents are passed through iron materials. **Electromagnetism** is important to electricity. A discussion of techniques and uses of electromagnetism is presented later in this chapter.

REVIEW QUESTIONS FOR SECTION 5.1

1. What is a natural permanent magnet called?
2. How can one permanent magnet be used to produce another?
3. What kind of magnet can be made from high-carbon steel?
4. What kind of magnet results when low-carbon steel is used?
5. How is electricity used to produce a magnet?

5.2 MAGNETIC THEORY

To understand and use magnetism, it is necessary to know some basics about how magnetism behaves.

Scientists still do not know exactly how magnets behave. There are few laws or fixed rules for magnetism. Instead, there are a number of theories. These theories explain the behavior patterns that can be observed.

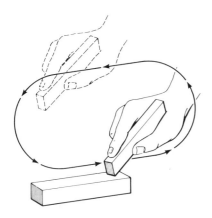

Figure 5-3. Permanent magnets can create other magnets.

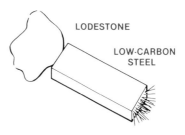

LODESTONE

LOW-CARBON STEEL

Figure 5-4. Lodestone and bar of low-carbon steel form a temporary magnet.

Magnetic Permeability

The ability of a material to become magnetized is called **permeability**. The magnetic force that attracts iron materials is called **magnetic flux**. The greater the permeability of a material, the higher is its magnetic flux.

Numbers are used to indicate the permeability of materials. The higher the number, the more easily a material is magnetized. The relative permeabilities of a number of materials are listed in Figure 5-5.

At lower levels, materials are considered to be non-magnetic. This applies to aluminum and all materials listed above it in Figure 5-5.

Materials with higher ratings—including nickel, cobalt, iron, and permalloy—have high permeability. Permalloy, a combination of metals, has an extremely high permeability. When a bar of permalloy is held in a north–south position, it becomes a magnet. When the bar is turned to point east and west, it loses its magnetism.

Shapes of Magnets

Magnets can be found or made in a variety of shapes. Some of these are shown in Figure 5-6. One common shape for magnets is the rod or bar. Another is a shape similar to the letter "U." A magnet with this shape is sometimes called a *horseshoe magnet*.

Magnets made in a horseshoe shape have an advantage: Their ends, or *poles*, are close together. This means that horseshoe magnets have a strong field of magnetic attraction, or **flux field**, Figure 5-7. By comparison, a bar magnet of the same strength has a flux field with less attraction. This is demonstrated in Figure 5-8.

Poles of Magnets

Dip a bar magnet into a pile of iron filings. When it is withdrawn, the filings will gather at the ends of the magnet, Figure 5-9 (page 70). This demonstrates that magnetism is strongest at the ends. These are the *poles* of a magnet.

Each magnet has two poles. However, there is a difference between the poles. If you suspend a magnet from a string, it turns until one pole points north, the other south.

This is the principle behind the operation of a compass: One pole of a magnet is north-seeking. A bar magnet is suspended on a bearing to reduce friction. The **north-**

Bismuth	0.999833
Quartz	0.999985
Water	0.999991
Copper	0.999995
Liquid Oxygen	1.00346
Oxygen (S.T.P.)	1.0000018
Aluminum	1.0000214
Air (S.T.P.)	1.0000004
Nickel	40.
Cobalt	50.
Iron	7,000.
Permalloy	74,000.

Figure 5-5. Relative permeabilities of selected materials.

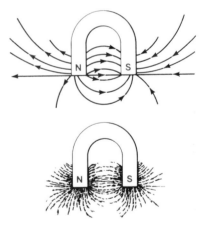

Figure 5-7. Horseshoe magnet flux field.

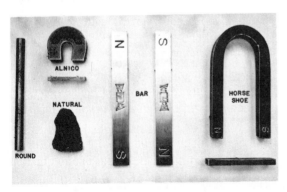

Figure 5-6. Some shapes of magnets.

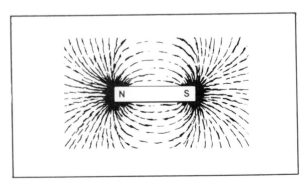

Figure 5-8. Bar magnet flux field.

Activity: BASIC MAGNETISM

OBJECTIVES

1. *To study the process of magnetic induction.*
2. *To test common materials to determine whether they are magnetic or nonmagnetic.*

MATERIALS NEEDED

1 Bar magnet	1 Nickel strip
1 Carbon strip	1 Tin strip
1 Zinc strip	1 Plastic rod
1 Copper strip	1 Wood rod
1 Lead strip	1 Hard steel rod
1 Aluminum strip	50 Washers, steel, #8
1 Iron strip	

PROCEDURE

1. Spread a few iron washers on the top of a table. Holding a bar magnet in your hand, approach one of the iron washers with the end of the magnet, Figure 1. Is there any attraction between the magnet and the iron washer? Bring more iron washers to the end of the magnet. Are they also attracted? How many washers can the bar magnet hold at one time?

Figure 1

2. Reverse the end of the magnet and determine whether the opposite end of the magnet will attract the same number of iron washers. Is there any difference in the strength of the magnetism at the two poles? From this observation, could you say that the magnetic flux is the same at each pole of the magnet?
3. Can the center of a magnet attract washers? Why should this be true?
4. Try to pick up an iron washer with the hard steel rod. Can it attract any washers? Hold the end of the bar magnet to the end of the hard steel rod, Figure 2. With the opposite end of the hard steel rod, try to pick up iron washers. Is there any attraction of the iron washers? Why would this happen?

Figure 2

5. Rub the hard steel rod with the bar magnet, as shown in Figure 3. Will the steel rod now attract an iron washer? Has the steel rod now become magnetized? Will either end of the steel rod attract a washer?

Figure 3

6. Magnetism can be destroyed in a magnet if the magnet is subjected to a physical jarring sufficient to cause a movement of the domains of the magnet. Throw the hard steel rod forcibly to the floor several times. Try again to pick up a washer with the hard steel rod. Does it now attract the washer as readily as it did before? Can magnetism be destroyed in a magnet by physical abuse?
7. Use a bar magnet to determine whether the following materials are attracted by the magnet.

Carbon	Iron
Zinc	Nickel
Copper	Tin
Lead	Plastic
Aluminum	Wood

Record your observations in the handout provided by your teacher.
8. Clean all equipment and materials and return them to their proper storage places.

SUMMARY

1. What is magnetic induction? What is the procedure for making a magnet by magnetic induction?
2. What characteristics must a material have to be magnetic?
3. What is the relationship between the pole of a magnet and the strength of the magnetic field?

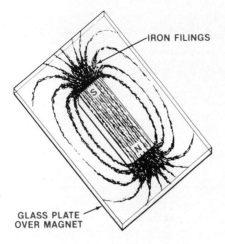

Figure 5-9. Magnetic flux is strongest at poles.

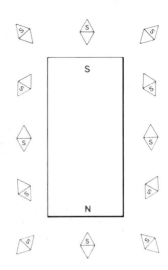

Figure 5-11. Pattern of attraction around bar magnet.

seeking pole points north and the other pole is directed south (**south-seeking pole**). Therefore, the poles of a magnet are referred to as the north pole and the south pole.

If you point the north poles of two bar magnets at each other, they are not attracted. Actually, there is a reverse force pushing the magnets apart. This force repels them from each other. This is a rule of magnetism: _Like poles repel._

If you place the north pole of one magnet near the south pole of another, there is a strong attraction. This illustrates another rule of magnetism: _Unlike, or opposite, poles attract._ This means that the north-seeking pole of a magnet is actually the south pole, Figure 5-10.

The pattern of attraction around a bar magnet is shown in Figure 5-11.

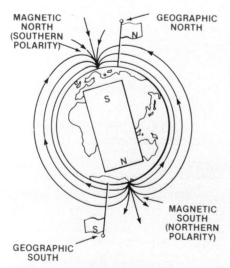

Figure 5-10. Polar attraction of magnets.

Magnetic Flux

You already know that the force of attraction of a magnet is called _magnetic flux_. One explanation of magnetic flux came from Michael Faraday. He was a pioneering scientist in magnetism. Faraday felt that a magnetic field consists of _lines of force_. This helped build an understanding of magnetic flux.

Actually, however, these specific lines do not actually exist. It is now known that magnetism has _continuous force fields_. That is, magnetism is strongest around a pole, then drops off with distance. This happens in a continuous pattern, rather than in lines. However, you may still hear the term _lines of flux_ used to describe magnetic force.

Force fields are frequently demonstrated with the method shown in Figure 5-12. A bar magnet is placed beneath a piece of paper. Iron filings are sprinkled onto the paper. The small particles of iron line up and form continuous chains. These chains follow the flux field. They loop around the magnet, joining the fields at the north and south poles. These patterns indicate continuous flux fields around the magnet.

For another demonstration, two bar magnets can be placed under a sheet of paper. The magnets are positioned so that the north poles of both magnets are close to each other, Figure 5-13. Clear boundaries are formed in the patterns of the iron particles. These indicate the points at which the force fields of the like poles of the two magnets repel each other.

If the positions of the two magnets are reversed, the results are different. In Figure 5-14, two magnets are placed so that the north and south poles are close to each other. The fields of flux attract the iron filings in a con-

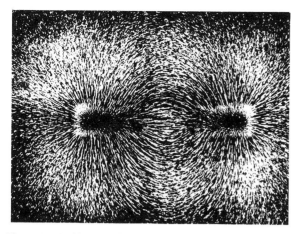

Figure 5-12. Magnetic force fields are strongest around poles.

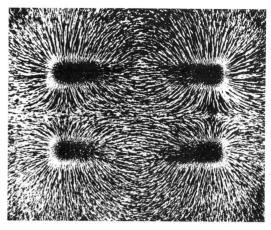

Figure 5-13. Force fields of like poles repel.

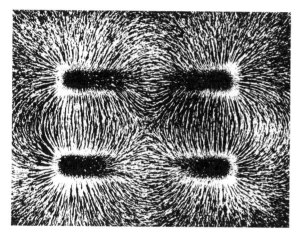

Figure 5-14. Force fields of unlike poles attract.

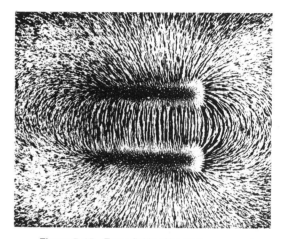

Figure 5-15. Force fields of horseshoe magnet.

tinuous pattern. The same kind of pattern results around the two fields of a horseshoe magnet, Figure 5-15.

The rule, once again, is: *Like poles repel; opposite poles attract.*

The *Domain Theory* of Magnetism

In recent years, scientists have found that some magnetic materials have a glasslike, or crystalline, structure. These crystals can be lined up so that their individual magnetic properties add to, or build upon, each other. This causes the entire piece of material to become a magnet.

These crystal structures are called *domains*. In a nonmagnetized sample of such material, the domains are random, or in no special positions in relation to one another. When the crystals are placed in a strong magnetic field, the domains change position. They actually show a movement within the crystal structure. The north pole

of one domain lines up next to the south pole of another. In this way, the many domains add their magnetism to one another. An example is seen in the coating of magnetic tape. The principle is illustrated in Figure 5-2 (page 67).

REVIEW QUESTIONS FOR SECTION 5.2

1. What does *magnetic permeability* mean?
2. What is a horseshoe magnet?
3. Why does a horseshoe magnet have a stronger attraction than a bar magnet?
4. What are magnetic poles and how are they identified?
5. What is magnetic flux?
6. What are magnetic lines of force?
7. What is meant by *magnetic domain*?

Activity: *MAGNETIC ATTRACTION AND REPULSION*

_____ OBJECTIVES _____

1. *To study the attraction and repulsion of magnetic poles.*
2. *To observe the flux patterns of two magnets when their poles attract and when their poles repel.*

MATERIALS NEEDED

2 Bar magnets   1 Piece of paper,
1 Vial of iron filings   8.5″ × 11″

PROCEDURE

1. Place two bar magnets as shown in Figure 1. Cover the magnet ends that are close together with a piece of paper about 8.5″ × 11″. Sprinkle iron filings onto the top of the paper. Observe the pattern that the iron filings make. Since the two poles are the same, what does the flux pattern revealed by the iron filings indicate?
2. Remove the paper and iron filings. Carefully return the filings to their container. Reverse the two magnets, so that the two south poles are facing one another but do not quite touch. Cover the ends of the magnets with a sheet of paper as before, and again sprinkle iron filings onto the top of the paper. What does the pattern of iron filings in the flux field indicate?

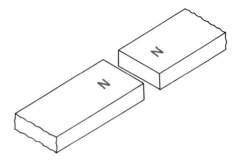

Figure 1

3. Remove the paper as before, and return the filings to their container. Reverse *one* of the two magnets, so that a north pole and a south pole are facing one another. Cover the ends of the magnets with a sheet of paper as before and again sprinkle iron filings onto the paper. Since these two poles are *unlike*, what does the pattern of the iron filings in the flux field indicate?
4. Return the filings to their container. Return all equipment and materials to their proper storage places.

SUMMARY

1. What is the basic law of magnetism?
2. In view of the basic law of magnetism, how would you explain attraction and repulsion?

5.3 ELECTROMAGNETISM

The relationship between electric current and magnetism was discovered by Hans Christian Oersted in 1820.

The connections between current and magnetism are direct. When current flows through a conductor, there is a magnetic field around the conductor.

The direction of the current flow determines the force field of an electromagnet. Currents flowing in the same direction set up connecting fields of force, Figure 5-16. Currents flowing in opposite directions set up repelling force fields. See Figure 5-17.

The polarity of an electromagnet is determined by the direction of current flow, Figure 5-18.

The strength of an electromagnetic field is proportional to the current flowing through the conductor. The more current, the greater is the magnetism.

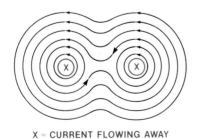

X = CURRENT FLOWING AWAY
AWAY FROM YOU.

Figure 5-16. Electromagnetic fields follow direction of current flow.

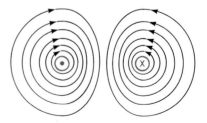

(•) CURRENT FLOWING TOWARD YOU.
(X) CURRENT FLOWING AWAY FROM YOU.

Figure 5-17. Opposite-flowing currents create repelling force fields.

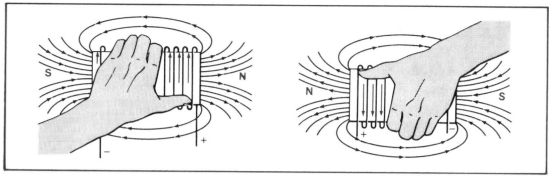

Figure 5-18. Polarity of electromagnet is determined by direction of current flow.

Magnetism in a Coil of Wire

The magnetic field that surrounds a conductor depends on the form into which the wire is shaped. See Figure 5-19.

The magnetic field surrounding a single loop of wire is shown in Figure 5-19A.

Additional loops are wound in the wire shown in Figure 5-19B. These loops are in the form of a spiral, or **helix**. This shape is also known as a *helical coil*. Figure 5-19B shows magnetic flux around loops of wire wound next to one another. The more loops in the wire, the larger the magnetic field becomes. This is shown in Figure 5-19C.

The magnetic strength of a coiled conductor depends upon two factors—the amount of current flowing in the coil and the number of turns of wire.

Remember also that the direction of current flow also determines magnetic polarity. This is shown in Figures 5-17 and 5-18.

Electromagnets

The usual form of an electromagnet is a coil of wire wound around a soft iron core, Figure 5-20. The iron core provides an easy path for the magnetic field created by the coil. An electromagnet with an iron core forms a stronger magnet than the same coil of wire without the iron core.

The size of the iron core can help determine the strength of an electromagnet. For full strength, the core should be large enough to absorb all of the magnetism from the coil. Creation by the coil of more magnetism than the core can absorb is called saturation. If the core is too small, there is too much magnetism to be absorbed. All of the magnetism of the coil can be used when the core has a capacity slightly larger than the magnetic flux created.

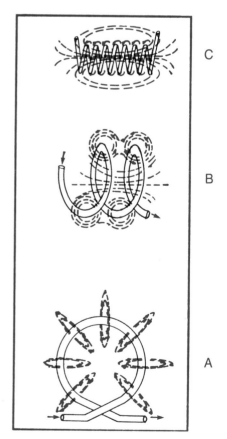

Figure 5-19. Magnetic fields form around shapes of magnetic coils.

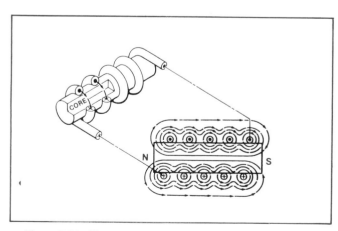

Figure 5-20 Electromagnets are made by winding wire around a soft iron core.

REVIEW QUESTIONS FOR SECTION 5.3

1. Where does magnetism build up when current flows through a conductor?
2. What determines the field of force of an electromagnet?
3. What determines the polarity of an electromagnet?
4. How does the amount of current affect an electromagnet?
5. How does the shape of the conductor affect an electromagnet?
6. What is the role of the core of an electromagnet?

5.4 USES OF ELECTROMAGNETISM

Electromagnets play an essential role in the generation and use of electricity. They have many important uses. Just a few of these are illustrated in the remainder of this chapter.

The Solenoid

A magnetic field seeks the path of minimum *reluctance*. Reluctance in magnetism is much the same as resistance in electrical circuits. The lower the reluctance, the greater is the attraction of materials to a magnetic field.

A **solenoid** is a device that uses these principles. A helical coil of wire produces a magnetic field. An iron core fits loosely within the coil of wire. When the current is off, the core rests outside the area of the coil. When current is applied, the core is sucked into the coil.

A diagram showing the operation of a solenoid is shown in Figure 5-21.

This magnetic action is frequently used in devices that require a small physical movement. A solenoid device is pictured in Figure 5-22.

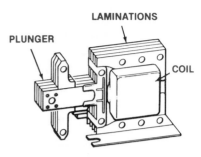

Figure 5-22. A solenoid.

Solenoid actions are used for many jobs. One common use of solenoids is in devices called solenoid valves. See Figure 5-23. Basically, a valve is a device that opens and closes to permit liquids or gases to flow. In a common gas valve, for example, you turn a handle or knob for opening or closing action. This starts or stops the flow of gas. Solenoid valves are used widely as safety devices. They are found in gas lines, air lines, and water lines.

The solenoid in the valve in Figure 5-23 is closed when there is no current through its coil. The core is held in a closed position by a spring that applies light pressure. When the valve is closed, gas cannot flow.

Current is applied when a person or thermostat turns on a heater. The current draws the movable core, or **armature**, into the coil. This opens the gate to the flow of gas. When the current is turned off, the spring moves the core back into closed position.

The same principle is used widely in solenoid-type relays. A solenoid relay is like an electrically operated switch. The coil in this type of relay pulls a core piece that has a number of electrical contacts. The contacts are designed so that they may either open or close electrical circuits. The relay contacts themselves can be designed to handle large amounts of current. But the coil of the con-

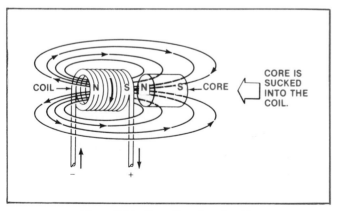

Figure 5-21. Operation of a solenoid.

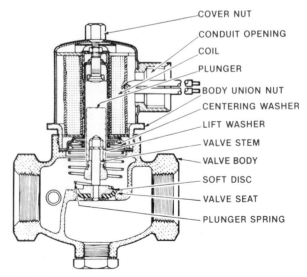

Figure 5-23. Cutaway view of solenoid valve. (Honeywell)

trol solenoid may operate on only a fraction of an ampere. The effect is that electricity is used to control the flow of electricity, Figures 5-24 and 5-25.

A heavy-duty power relay that operates from solenoid action is shown in Figure 5-26. In this device, a spring pulls the solenoid core, or armature, away from the electrical contact when the current is off. When current flows, the armature is pulled toward the coil. An electrical contact connected to the armature is either closed or opened by this action. Note that electromagnetism can be used either to open or close the relay. The action taken depends on the design of the relay.

Doorbells and Buzzers

Electromagnetism is used commonly in doorbells and buzzers, Figure 5-27.

When the doorbell is not in use, the contact strip is held against the contact point by a spring. This contact strip is connected to the armature. The pressing of the button causes current to flow through the coil. At this time, the armature is attracted to the coil. The contact strip then

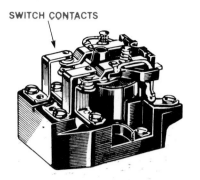

Figure 5-26. Heavy-duty power relay.

moves to strike the bell. At the same time, the movement of the contact strip causes the circuit to open. Current is removed from the coil. The armature drops out of the coil. The contact strip closes again. The process is repeated as long as the button is pressed.

Without the hammer and bell, the same basic mechanism operates as a buzzer.

Electromagnets for Handling or Removing Iron

An obvious use for an electromagnet is for lifting iron. Modern electromagnets can lift up to 100 times their own weight in iron. Many lifting magnets are used in the steel industry. A common job is the lifting and moving of scrap iron and steel that are being used to make new steel.

Portable electromagnets are used to remove loose pieces of iron and steel from airport runways. This eliminates the chance that the pieces of metal will be sucked into jet engines.

Another typical value can be seen in the use of magnetic separators. In some industries, it is important to remove bits of iron or steel from materials being processed. This is done by placing electromagnetic fields at points along the pathway of flowing materials.

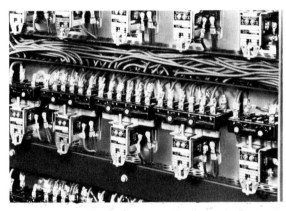

Figure 5-24. Installation of solenoid relays.

Figure 5-25. Photo showing working parts of a solenoid relay.

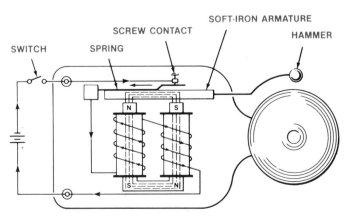

Figure 5-27. Schematic of a doorbell.

Figure 5-28. Permanent-magnet chuck on grinder. (Brown & Sharpe)

Magnetic Chucks

A chuck is a device that holds a piece of material in place in a machine such as a surface grinder or milling unit. The workpiece must be held tightly. Movement or adjustment of the workpiece must also be convenient. Figure 5-28 shows a grinding machine with a permanent-magnet chuck.

Other Uses of Electromagnetism

Electromagnets are used in many electrical and electronic circuits. You use electromagnets regularly. Electromagnets make electric motors and generators possible. Telephones use electromagnets.

REVIEW QUESTIONS FOR SECTION 5.4

1. What is reluctance?
2. How is electromagnetism used in a solenoid?
3. Describe the action of a solenoid valve.
4. Describe the action of a solenoid relay.
5. Describe the operation of a doorbell.

SUMMARY

Magnets called lodestones are found in nature. The process of induction may be used to make permanent magnets. Low-carbon steel forms temporary magnets. Low-carbon steel is used in most electromagnets. The ability of a material to become magnetized is called permeability. Like poles repel; unlike poles attract. The domain theory of magnetism says that magnetic materials have a glasslike, or crystalline, structure. These crystals can be lined up so that their individual magnetic properties add to, or build upon, each other to become a magnet.

The magnetic field that surrounds a conductor depends on the form into which the wire is shaped. The usual form of an electromagnet is a coil of wire wound around a soft iron core. The direction of current flow in an electromagnet determines its polarity. When a coil of wire receives current, the degree of magnetism depends on the wrapping of the wire. The more turns in the wrapping, the stronger is the magnetism. The magnetic fields seek the path of less reluctance. Reluctance in magnetism is similar to resistance in electricity.

A solenoid is a coil of wire. It is a device that uses the principles of an electromagnet. Solenoid actions are used for many jobs. The most common use is in electrical switching and in valves.

USING YOUR KNOWLEDGE

1. Take a clean sheet of paper and place it over a bar magnet. Sprinkle iron filings over the piece of paper. Observe the pattern formed. Draw the pattern on another sheet of paper.
2. Take a clean sheet of paper and place it over a horseshoe magnet. Sprinkle it with iron filings. Observe the pattern formed. Draw the pattern on another sheet of paper.
3. Take the sheet of paper with the iron filings on it and very carefully dip it into a container of paraffin. Remove the dipped paper with the filings and allow it to cool. You will have a permanent record of the pattern.
4. Take a coil of wire with an appropriate low-power source. Place a large nail part way into the coil. Turn on the power. Observe the sucking effect of the coil. What happened to the nail?
5. Take a coil of wire and place a soft iron core into its center. Turn on the power. Move the end of the coil close to a pile of iron filings. What happens? Record your observations.

KEY TERMS

lodestone	force field
magnetite	electromagnetism
magnetic inductance	solenoid
magnetic flux	armature
permeability	helix
flux field	domain theory
north-seeking pole	magnetic chuck
south-seeking pole	

Activity: ELECTROMAGNETISM

OBJECTIVES

1. *To study the physical structure of an electromagnet.*
2. *To demonstrate the effect of the number of ampere turns on electromagnet strength.*

MATERIALS NEEDED

1 Power supply, 0–6 volts DC	1 Bar magnet
1 3/4″ square laminated core	2 Coils, 300 turns each
	50 Washers, iron, #8

PROCEDURE

1. Place the 3/4″ laminated core in one coil of wire, Figure 1. Hold the coil and bar so that one end is pointing upward. Connect the top lead of the coil to the positive terminal of the power supply. The other lead goes to the negative terminal of the power supply. Turn the voltage control of the power supply to *minimum*. Turn on the power supply. Adjust the output of the power supply to 3 volts DC.

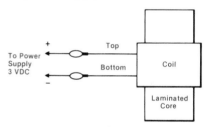

Figure 1

2. Use the bar magnet and the law of magnetic attraction and repulsion to determine whether the top end of the coil is a north pole or a south pole. Which end of the bar magnet is attracted to the top end of the electromagnet? (NOTE: The bar magnet will be attracted to the laminated core with or without the power being on. In order to test the setup, you need to bring the bar magnet close to the core so that it is not pulled to it quickly. Then turn on the power supply and see whether it is attracted across the short distance to the core or is repelled as you try to move it nearer.) Is the top end of the electromagnet a north or south pole?

3. *Turn off the power supply.* Reverse the connections to the coil of the electromagnet, so that the top lead is connected to the negative terminal of the power supply and the bottom is connected to the positive terminal of the power supply. Turn on the power supply with the output set at 3 volts DC. Repeat the same pro-

cedure as before and determine whether the top of the coil is a north or south pole. Which pole of the bar magnet does the electromagnet now attract? Which magnetic pole is the top end of the electromagnet now?

4. Place a number of iron washers on the top of a table. *Adjust the power supply to 3 volts.* How many washers does the electromagnet pick up?

5. *Turn on the power supply.* Adjust the output voltage to 6 volts. Again test to see how many washers the electromagnet picks up. Is the magnet stronger with the higher voltage? Why does (or doesn't) the increase in voltage cause a difference?

6. Turn off the power supply. Disconnect the electromagnet and place a second coil, with the same number of windings, on top of the first. Connect the coils so that the top lead of one fits the bottom lead of the other. This puts them in series. Make sure the two coils are wound in the same direction and that the two top leads are both properly located according to the winding direction. Connect the coils as shown in Figure 2.

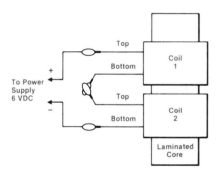

Figure 2

7. *Turn on the power supply.* Adjust the output voltage to 6 volts. This is the same voltage used in Step 5. How many washers does the two-coil electromagnet attract? This electromagnet has twice as many windings as the one in Step 5. How does the number of windings affect the strength of an electromagnet?

8. *Turn off the power supply.* Disconnect the circuit and return all equipment and materials to the proper storage places.

SUMMARY

1. On what does the strength of an electromagnet depend?
2. On what does the polarity of an electromagnet depend?

Project: METAL-MARKING PEN

The metal-marking pen is a handy tool to have around the house. It can be used to mark metal tools permanently.

You will need the following materials.

Handle, conduit, 6″ long
Bobbin
Core, 1/4″ steel rod, 3″ long
Tip, No. 18 copper wire, insulation removed, 1″ long
Flexible arm, with one end rounded and a 1/8″ hole in each end (made from transformer lamination)
Battery clips, 10-ampere size
Cord, lamp, 30″ long
Pop rivet gun and 1 pop rivet, 1/8″
Screw, machine, plated, 4-40, 1/4″ long
Nut, machine, hex, plated, 4-40
Electrician's tape strip, 6″ long
Wire, magnet, No. 18, copper, 7 feet long
Solder, 60/40 rosin core, 6″ long
Solder lug, No. 6
Duco cement

Operation of the Pen

When the tip of the metal-marking pen touches a metal, it completes an electrical circuit. This energizes a coil of wire. When energized, the coil creates a magnetic field and its core pulls the flexible arm up. This breaks the circuit, creating a small arc. Once the contact with the other part of the circuit is broken, the coil is de-energized and releases the tip. Tension on the flexible arm causes the tip to make contact again, and the entire cycle is repeated. The intense heat of the arc produces a permanent mark on the metal object.

Construction Hints

Drill holes in each end of the bobbin to allow the wire to pass through.

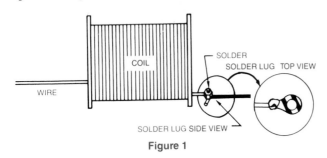

Figure 1

Push about 4 inches of wire through the hole in the bobbin next to the core location. Start winding three layers

of wire and push the end of the wire through the hole in the bobbin that is located near the outside of the bobbin, Figure 1.

Cut the wire to length. Remove the insulation and attach the solder lug as shown.

Scrape off about 1/2 inch of insulation from the other coil wire and attach one end of the flexible leads and solder. Tape with electrician's tape.

Attach alligator clips to the other end of the flexible lead. Attach alligator clips to both ends of the other piece of flexible wire you make from lamp cord that is ripped apart, Figure 2.

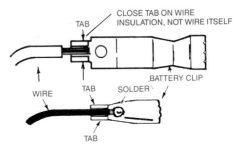

Figure 2

Follow Figures 3, 4, and 5 to make the rest of the parts. Assemble the parts inside the conduit handle and attach them with a screw or pop rivet.

Keep in mind that the power source (**CAUTION: NO MORE THAN 12 VOLTS AC or DC**) should have one lead going to the tool to be marked and the other lead clipped to the metal-marking-pen lead that comes out of the handle, Figure 6.

A light pressure on the pen produces the best results. The surface of the object being marked should be free of paint, oil, or rust.

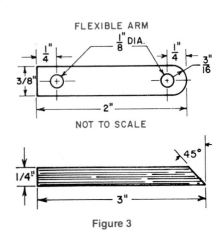

Figure 3

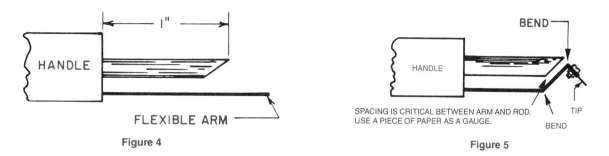

1"
HANDLE
FLEXIBLE ARM

Figure 4

BEND
HANDLE
SPACING IS CRITICAL BETWEEN ARM AND ROD.
USE A PIECE OF PAPER AS A GAUGE.
TIP
BEND

Figure 5

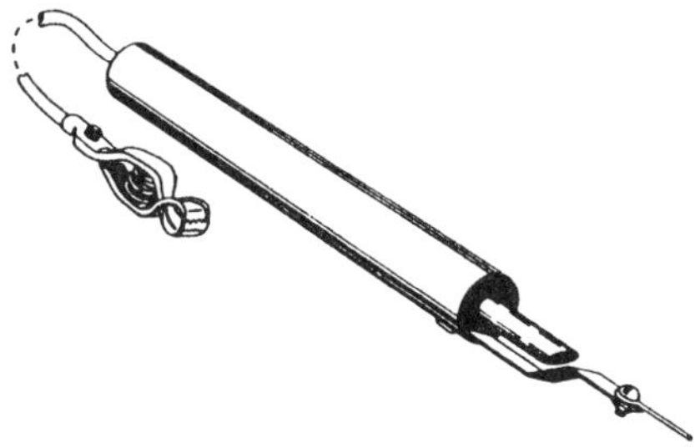

Figure 6

Chapter 6

ELECTRICAL AND ELECTRONIC METERS

OBJECTIVES

After studying this chapter, you will know

- *the parts and operation of the d'Arsonval meter movement.*
- *the basic types of scales provided on d'Arsonval meters.*
- *how to use an ammeter.*
- *how to use a voltmeter.*
- *how to use both DC and AC meters.*
- *how to use ohmmeters.*
- *how to use multimeters.*
- *how to use digital meters.*

INTRODUCTION

To meter means *to measure*. To use electricity safely, you must be able to measure voltage, current, and resistance. Through your work in this chapter, you will understand and learn to use devices that measure electricity. You will learn when, where, and how they are used best in electrical and electronic jobs.

Measuring electricity presents special challenges. That's because you can't see, hear, smell, or taste electricity. You can feel electric current. But don't try. You could get hurt.

It takes electrical instruments—meters—to measure electricity. Meters measure the three elements of electricity about which you have learned—voltage, current, and resistance. Figure 6-1 shows a multimeter. This is a single instrument that can measure voltage, current, and resistance.

6.1 HOW METERS WORK

There are many types of meters. One design for the working parts of a meter is shown in Figure 6-2. This is known as the **d'Arsonval meter movement**. It is named after its inventor, Arsene d'Arsonval. The movement is adapted from d'Arsonval's *galvanometer*. A galvanometer is a device with a rectangular coil. It was the first instrument used to sense and measure electric current. The design shown in Figure 6-2 also includes improvements made by Edward Weston.

In this meter movement, there is a small, rectangular coil of wire. The coil is suspended in a magnetic field.

The magnetic field is created by a permanent magnet. When current is applied to the coil, an electromagnet is produced. The coil then lines up with the poles of the permanent magnet. The amount of current applied to the coil controls its movement. This is d'Arsonval's basic principle: *An electromagnet that is free to move will align its axis with the magnetic axis of a fixed magnet.* The axis of an electromagnet is a straight line between its poles.

To enable the coil to align itself with the magnetic axis, it must be free to rotate. The coil in the meter is mounted on pivots that permit easy rotation. Two small springs are mounted on top and bottom. These offer slight resistance to the rotation of the coil. The springs control the position of the coil when there is no current flowing. Keep in mind that like poles repel and unlike poles attract.

When current flows, the magnetic flux overcomes the force of the springs and moves the coil. A pointer on top of the coil rotates to mark the amount of movement. The greater the current through the coil, the more it turns—and the farther the pointer moves. The pointer then stops in front of a marked scale on the face of the meter. The position of the pointer indicates the meter measurement.

To help clarify what happens, see Figure 6-3. This is a drawing showing an overhead view of a meter movement.

A typical meter scale is shown in Figure 6-4 (page 82).

Figure 6-1. Multimeter measures voltage, current, and resistance.

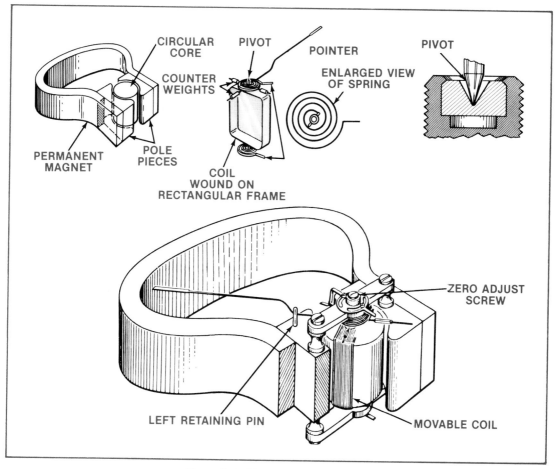

Figure 6-2. D'Arsonval meter movement.

Meters like the one illustrated can be extremely sensitive. For example, a relatively inexpensive meter can give a full-scale deflection of 10 microamperes. *Full-scale deflection* is the total range of a meter scale. A **microampere** is one-millionth of an ampere.

The amount of deflection of a meter pointer on a scale depends on three factors:

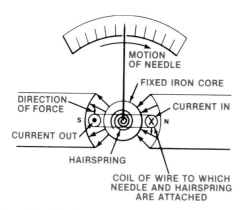

Figure 6-3. Overhead view of d'Arsonval meter movement.

- Magnetic force energizes the coil. This force is determined by the number of turns of wire in the coil and the amount of current flowing in the coil.
- The strength of the permanent magnet affects the positioning of the coil.
- The tension of the springs and the friction of the bearings help determine the sensitivity of the meter.

Of these factors, only the amount of current flowing through the meter is changeable. The other factors are built into the meter. This means that movement of the pointer depends upon the amount of current flowing through the coil. The scale of the meter is *calibrated* to show the type of reading it gives and its sensitivity. Calibrations are the markings on the meter scale. For example, the scale in Figure 6-4 is calibrated for reading DC (direct current) voltage. (Direct and alternating currents are explained in later chapters.)

Before you use a meter, first read the instruction manual. There are many types of meters. Operations differ. To get best results, always read and follow instructions.

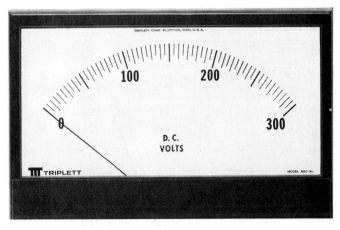

Figure 6-4. Meter scale.

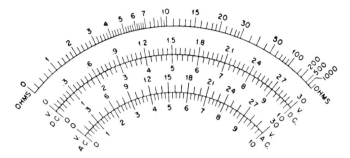

Figure 6-5. D'Arsonval multimeter scale.

6.2 METER MOVEMENT CHARACTERISTICS

A scale for one model of a meter using a d'Arsonval movement is shown in Figure 6-5. Be aware that meter scales vary. They show different measurements and sensitivities.

Meters are usually described in terms of *full-scale deflection*. Full-scale deflection refers to the action of the pointer. It is fully deflected when it moves to the highest position on the scale.

Full deflection for the scale shown in Figure 6-5 is one **milliampere**. This is one-thousandth of an ampere. A meter with this type of scale would be called a *one-milliampere meter.*

NOTE: Many meters show their readings by displaying numbers. These displays are called *digital readouts*. The meters are called *digital meters*. These meters operate electronically. Thus, they have no coils or magnets. Their number displays replace the scales. Digital meters are gaining popularity. However, analog-type units will be used for many years. Analog meters have a needle to indicate, on a scale, a given reading. Further, analog-type units make it

easier to show how meters work. That's why this chapter reviews analog-type meters first. Digital meters are introduced later in the chapter.

The cost of any meter depends upon the quality of its movement. Quality, in turn, is judged mainly by the method used to suspend the coil. The pivot may rest on metallic bearings. Or jeweled bearings may be used. Bearing jewels are often made from synthetic ruby or sapphire. (Refer to Figure 6-2 for a diagram about bearing function.)

The accuracy of the coil winding also affects the cost and quality of a meter. The copper wire used in the coil has some resistance to the flow of electricity. So, each meter has some internal resistance. On typical one-milliampere meters, resistance may range from 30 to 300 Ω.

Also important is the accuracy to which the scale is prepared. In general, the larger a meter scale, the greater is its accuracy.

6.3 READING AN ANALOG METER

Use of an analog meter (scale display) requires you to understand and be able to read the scale. A reading is indicated by the position of the pointer on the scale. The skill needed lies in finding the meanings and values of these readings.

A scale with four separate readings is shown in Figure 6-6. This shows what is known as a **linear scale**. A linear scale has readings divided into equal portions over

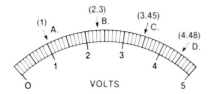

Figure 6-6. Meter scale with four readings.

its entire area. The full-scale deflection is 5. This could mean 5 A, 5 V, 5 mA (milliamperes), or 5 of any other value of voltage or current. The information on what is being measured usually appears on the face below the scale.

There are also nonlinear scales. One type of nonlinear scale is illustrated in Figure 6-7. This is known as a square-law scale. The lines in this scale increase in squares. That is, each larger scale marking indicates that the measured value is multiplied by itself (squared). Spacing of nonlinear scales such as the square-law scale is uneven.

A multimeter usually has a number of scales. Each scale on a multimeter is defined individually, Figure 6-5.

As an example of meter reading, assume the scale in Figure 6-6 is calibrated in volts. The range is 0 through 5 V. With these values, the pointer in the position indicated as A shows a reading of 1 V.

At point B, the reading is between 2 and 3 V. To be more accurate, you refer to the smaller marks on the face of the meter. Referring to these lines, the pointer at B rests on the third line above the 2-V mark. Therefore, the reading at B is 2.3 V.

At position C, the pointer is between two of the small lines. For this reading, it is necessary to estimate. If the pointer were at the small division line to the left of C, the reading would be 3.4 V. At the line to the right of the pointer, the reading would be 3.5 V. But the pointer appears to be at a midway point between these lines. The position can be read as 3.45 V.

At position D, the pointer is just below 4.5 V. This calls for a closer, more careful estimate. In this case, the reading could be given as 4.48 V.

It is not a good idea to attempt to read a meter scale any closer than is done in these examples. Specifically, do not try to estimate readings of a scale to a third decimal position. One reason is that few meters are

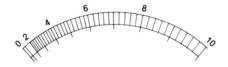

Figure 6-7. Nonlinear scale.

Figure 6-8. Meter with mirror in scale.

accurate enough to justify this kind of estimate. Even a relatively accurate, sensitive meter will have a readout tolerance of ± 1 percent. The size of the meter scale also affects the accuracy of the readings that can be made. The larger the scale, the more closely you can estimate readings when the pointer is between lines.

In reading a meter, be careful to avoid what is known as a parallax error. A **parallax error** results from the fact that there is a distance between the pointer and the scale behind it. If you look at the pointer from an angle, you will get an incorrect reading. From the side, the reading will be different than it is when taken from directly in front. It is also possible to have a parallax error if you use both eyes. Each eye is likely to be to one side of the pointer. The best practice is to use one eye. Make sure your view is from directly in front of the pointer.

Many meters have a small mirror attached to the scale to reduce parallax error, Figure 6-8. To minimize parallax error, line up the pointer with its reflection in the mirror.

Remember this: Meters are most accurate in the middle 80 percent of their range. Accuracy falls off in the lowest and highest 10 percent of the scale.

REVIEW QUESTIONS FOR SECTION 6.3

1. What is a linear scale?
2. What is a nonlinear scale?
3. What kind of scale is used on a multimeter?
4. What is a square-law scale?

6.4 THE AMMETER

It is important to fit the meter to the job. For example, a one-milliampere meter is ideal for measuring direct currents of up to one milliampere. But such a meter does not have sufficient sensitivity to measure currents of less than 100 microamperes (0.1 mA).

The Ammeter Shunt

A one-milliampere meter can be used to measure direct currents that are larger than this level. To do this, a resistor is connected in parallel with the meter movement. This resistor is called a **shunt**. To measure 2 milliamperes, the shunt requires a resistor that can pass 1 mA.

Use of a shunt is diagrammed in Figure 6-9. Notice that the symbol for a meter is a circle with a pointer inside. In the illustration, the resistance of the shunt is equal to the resistance of the meter. This splits the current equally between the two parallel resistances, enabling the meter to handle 2 mA—1 mA in each branch.

If current through the circuit is 2.0 mA, one half of this current goes to the meter and one half to the shunt. The current in both the shunt and the meter is 1.0 mA. In effect, the meter has been modified by the shunt. Therefore, the current in the meter is within its normal measurement range. To use the meter under these conditions, you double the reading you observe on the scale. This offsets the fact that half the current is taken up by the shunt.

Thus, a reading of 0.5 mA on the scale would be doubled to 1.0 mA.

Figure 6-10 shows how a shunt is connected across an ammeter. In ammeter circuits, individual shunts may have resistance values as low as 0.001 Ω. In some meters the shunt appears to be made of a bar of metal. This is shown in Figure 6-10.

Now, assume you have the same 1-mA meter. You want to measure current with a full-scale deflection of 1 A. This means the shunt must carry 999 mA of the current.

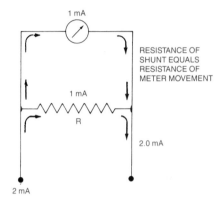

Figure 6-9. Schematic of ammeter and shunt.

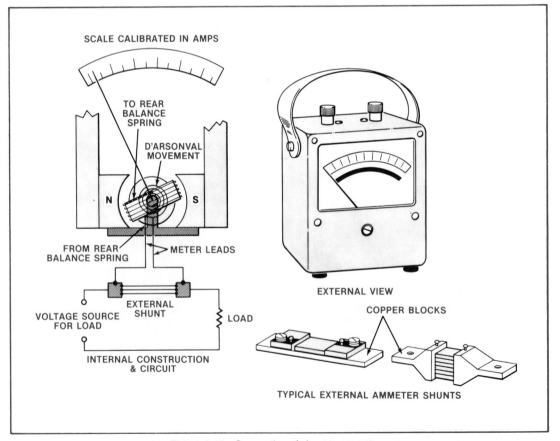

Figure 6-10. Connection of shunt to ammeter.

rent. Putting it another way, the shunt must carry 999 times as much current as the meter. Therefore, resistance of the shunt must be 1/999 as great as for the meter movement. This situation is diagrammed in Figure 6-11.

Assume that the meter movement has 100 Ω of resistance. The value of the shunt resistance must be 100 Ω divided by 999. This is the amount of current the shunt must be able to conduct. Therefore, the amount of *R* needed in the shunt is approximately 0.1001 Ω.

The Multirange Ammeter

In electronics work, it is necessary to measure a wide range of currents. To meet this need, many persons use multirange ammeters like the one illustrated in Figure 6-1.

In effect, a multirange ammeter has a number of built-in shunts that can be selected through the use of a switch. There are also multiple scales. You read the scale that corresponds with the setting of the selecting switch.

A circuit diagram of a range switch for an ammeter is shown in Figure 6-12. The range switch must be built so that it completes the circuit of one shunt before it breaks the circuit of another. Otherwise, the meter movement is unshunted when the switch is open between ranges. If this happened, the meter would be ruined.

Connecting an Ammeter

The rule for connecting an ammeter is vital: *An ammeter is always inserted in a circuit in series with the load.*

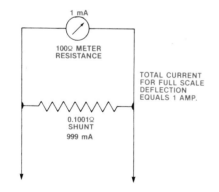

Figure 6-11. Schematic showing values for ammeter and shunt.

6.5 THE VOLTMETER

A 1-mA meter can be used as a voltmeter. To illustrate, suppose a meter movement is connected across a potential of 0.05 V. The meter movement has an internal resistance of 100 Ω. Using Ohm's law, you determine that the current flowing through the meter is 0.0005 A, or 0.5 mA.

In this situation, a difference in potential of 0.05 V gives a half-scale reading on the meter. The meter is thus calibrated to indicate a full-scale deflection of 0.1 V. The scale would be read accordingly.

Note that this is an extremely small voltage. It is more likely that you would require readings in a higher voltage range.

To use the same meter movement for a voltmeter with a higher range. you have to add resistance. This resistance is added in series with the meter movement. The resistor used is called a voltmeter **multiplier**.

Voltmeter Multipliers

A voltmeter with a 100-V full-scale deflection is diagrammed in Figure 6-13. The resistor indicated in the illustration is the meter multiplier. The multiplier's resistance rating is 99.9 kΩ. The notation *k* represents 1000. So, the resistance is 99,900 Ω. Let's review how this resistance requirement was calculated.

To start, it was determined that a full-scale deflection of 100 V was desired on the meter. The meter movement measures current with a full-scale deflection of 1 mA. The

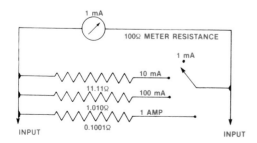

Figure 6-12. Schematic of multirange ammeter.

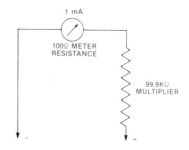

Figure 6-13. Schematic showing voltmeter multiplier.

Activity: THE AMMETER

OBJECTIVES

1. *To investigate the operational characteristics of an ammeter.*
2. *To be able to figure meter shunts of an ammeter.*

MATERIALS NEEDED

1 D cell, 1.5 volts
1 Potentiometer,
  1500 ohms, linear taper
3 Wires for connections

1 Meter movement, 1-0-1
  mA, 43 ohms internal
  resistance (If the meter
  movement is not
  43 ohms, adjust the
  shunt accordingly.)

PROCEDURE

1. CAUTION: This meter can be easily damaged by any overload that may occur in the performance of this activity. Follow the directions precisely. Turn the zero-adjust control fully clockwise. Then, move it fully counterclockwise so it is pointing to the left. This is the potentiometer shown in Figure 1. Connect the circuit as shown in this figure.

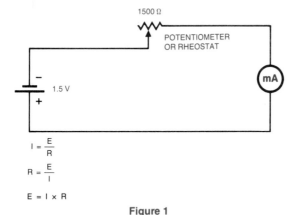

$$I = \frac{E}{R}$$
$$R = \frac{E}{I}$$
$$E = I \times R$$

Figure 1

2. The zero-adjust control has a maximum resistance of 1500 ohms. With the connections shown in Figure 1, there is a complete circuit. The power supply is 1.5 volts, furnished by the D cell. The resistance in the circuit is 1500 ohms. Use Ohm's law to calculate the current in this circuit. (Resistance of the meter movement is 43 Ω, so the potentiometer is adjusted to 1457

ohms.) Keep in mind that Ohm's law provides an answer in amperes. You have to convert to milliamperes or parts of a milliampere to use this meter. Record the meter reading.

3. There may be some variations between the amount of current that should have been read and the amount of current actually indicated by the meter. This is due to the tolerance of the zero-adjust control. These variable resistances are usually made so that the tolerance is in a positive direction—that is, with resistance greater than 1500 ohms. For this reason, it may be necessary to adjust the zero-adjust control to give full-scale deflection. Very slowly turn the zero-adjust knob in a clockwise direction until the meter pointer rests exactly on the 1-mA indication. Was it necessary to move the zero-adjust control very far? What should the resistance of the variable resistor now be with full-scale deflection?

4. The circuit now connected can indicate a current flow of only one milliampere (1 mA or 0.001 A). In order to be able to determine values of current in excess of the full-scale deflection of the meter, it is necessary to add a shunt.

5. Connect the shunt across the meter terminals as shown in Figure 2. What happens to the deflection of the meter? Why? (NOTE: The voltage source and zero-adjust resistor used in this experiment are not part of the ammeter circuit. Only the shunt and the meter movement form an ammeter.)

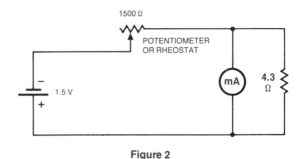

Figure 2

6. The meter movement used in this activity has an internal resistance (made up of the moving coil wire) of 43 ohms. The shunt has a resistance of 4.3 ohms.

Note that this shunt is one-tenth of the meter movement. Considered as resistors, the shunt and the meter movement are a part of the resistance circuit. What is the resistance of the combination of the resistances in parallel? How does this value of equivalent resistance for the two resistors compare with the resistance of the meter movement? (In other words, what is the ratio of resistances?)

7. Disconnect the circuit and return the equipment to its proper storage space.

SUMMARY

1. What components are necessary in every ammeter?
2. What is the purpose of the shunt in an ammeter?
3. What does a milliammeter measure?
4. How does a milliammeter compare with an ammeter?

internal resistance of the meter is 100 Ω. The values to be used in this equation are: $E = 100$ and $I = 0.001$. You divide 0.001 into 100. The answer is that $R = 100,000$ Ω. This is the *total resistance* of the circuit.

Now, you want to find the needed *multiplier resistance*. To do this, you subtract the meter movement resistance from this total resistance of the series circuit (100,000 Ω − 100 Ω). The difference, or answer, is 99,900 Ω. This is the value of your multiplier.

Compare the internal resistance of the meter movement with the value of the multiplier. At 100 Ω, this value is only 0.1 percent of the required multiplier. In most situations, a tolerance of ± 1 percent would be allowed. In this situation, a resistor rated at 100 kΩ would probably be used.

Meter multipliers are usually precision-type resistors. Their tolerance is typically ± 1 percent. Such a resistor automatically gives the circuit an additional limit of accuracy of ± 1 percent.

Connecting a Voltmeter

The rule for connecting a voltmeter should always be followed to avoid an incorrect reading of the meter: **A voltmeter is always connected in parallel**. It is placed across a voltage source or across a resistor, as shown in Figure 6-13.

The Multirange Voltmeter

A **multirange** voltmeter uses a switch to select among multipliers. Scales are arranged so that voltage is measured near the center of the meter scale. The center portion of a meter scale is always the most accurate part of the deflection range. This is the same principle that you learned in the discussion of ammeters.

A diagram of a multirange voltmeter is shown in Figure 6-14. Notice that the range switch makes connections at various points along a series of resistors. The range

switch is used to select the multiplier to be used. A separate, single resistance is used for each range. The resistors are then connected into the circuit in series to give the necessary resistance for the desired deflection reading of the meter.

Safety Tips

1. Don't handle the probes of a meter if the voltage is more than 100 V. Attach the probes to the source while the power is off, and then turn on the circuit to be measured.
2. Make sure the meter movement is properly selected for use with AC or DC.
3. Do not use an ohmmeter on a circuit with power applied.

REVIEW QUESTIONS FOR SECTION 6.5

1. What kind of meter may be used to measure voltage?
2. What is a voltmeter multiplier?
3. How is a voltmeter used to measure resistor voltage?
4. How is a voltmeter connected in a circuit?
5. What is the function of a range switch on a multirange voltmeter?

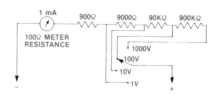

Figure 6-14. Schematic of multirange voltmeter.

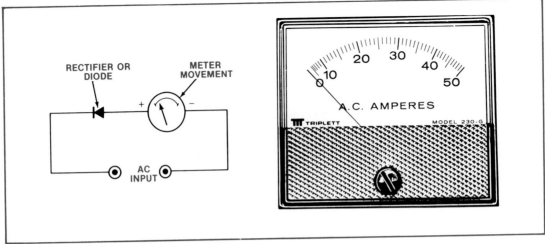

Figure 6-15. Schematic and scale for AC ammeter.

6.6 THE AC AMMETER

AC is the abbreviation for alternating current. Most household electricity in the United States is AC.

The meters discussed so far in this chapter have been direct-current (DC) devices. If these are used with AC, the needle vibrates very rapidly about the zero point on the scale. A DC meter could be damaged by AC current.

DC meters can be modified to measure AC current. This is done by connecting a **rectifier** in series with the meter movement. A rectifier is a device that changes the AC to DC. An AC scale and rectifier circuit for an AC meter are shown in Figure 6-15. Today, most rectifiers are semiconductor devices. In the past, vacuum tubes were often used for this purpose. Rectifiers and their uses are discussed further in another chapter.

Within AC meters, rectifiers are usually placed inside the meter case. In effect, the rectifier functions for the AC–DC conversion. The proper resistance value for the AC shunt is determined in the same way as for a DC shunt. Resistance of the meter is the sum of the resistances of the rectifier and the meter coil.

REVIEW QUESTIONS FOR SECTION 6.6

1. What is the function of a rectifier or diode in an AC voltmeter?
2. What type of device is added to create a multi-range AC voltmeter?
3. How is resistance determined for an AC shunt?

6.7 THE AC VOLTMETER

An AC voltmeter is made with a rectifier. One type of circuit that can be used for this purpose is illustrated in Figure 6-16. The rectifier used in this circuit is a diode. This is a semiconductor device. The rectifier is connected in series with the meter movement. Its resistance is added in determining the total resistance for full-scale deflection. A multirange AC voltmeter could be created by adding a selector switch to the circuit.

Another diode arrangement for an AC voltmeter movement is shown in Figure 6-17. This is known as a bridge rectifier. The meter movement is inserted in the middle of the diode arrangement.

The d'Arsonval meter movement can be used for AC and DC voltmeters and ammeters. This is done by adding resistors or rectifiers to the circuits. With this approach, a single type of meter movement may be used for a wide range of purposes. The d'Arsonval meter has many other uses. One of the most common is for direct measurement of resistance. This function is covered next.

REVIEW QUESTIONS FOR SECTION 6.7

1. What is the function of a rectifier or diode in an AC voltmeter?
2. What type of device is added to create a multi-range AC voltmeter?

6.8 THE SERIES OHMMETER

Although there are other methods for measuring resistance of a circuit, the ohmmeter is most convenient. The ohmmeter is a single instrument. It has two probes. These probes can be placed across the unknown resistance. The value of the resistance is displayed on the meter scale.

Inside the case of an ohmmeter, there is a voltage source. This is usually one or more dry cells. Fixed resistors and a variable resistance element are also built into

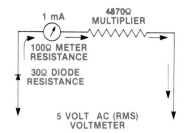

Figure 6-16. Schematic of AC voltmeter.

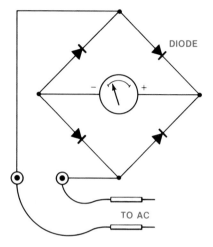

Figure 6-17. Schematic for AC voltmeter with bridge rectifier.

the meter. The variable resistor is called a **zero adjust**.

An ohmmeter circuit is diagrammed in Figure 6-18. The movement of this meter gives a full-scale deflection at 1 mA. Internal resistance of this meter is 100 Ω. The fixed resistor has a value of 3900 Ω. The adjustable resistance element is a potentiometer with a maximum resistance of 1000 Ω. The power source consists of three dry cells connected in series. They provide 4.5 V.

When the two meter terminals are shorted together (placed in contact with each other), the meter movement gives approximately full deflection. If the dry cells are new, the voltage is a little greater than 4.5 V. If the cells are old, the voltage is usually less than 4.5 V. This variation is offset by adjusting the 1000-Ω variable resistance element. For the adjustment, the potentiometer is set at the point where the meter delivers full-scale deflection.

Full-scale deflection of the meter indicates 0 Ω between the two terminals. If there is no connection between the two terminals, there is no deflection of the meter pointer. Normally, the zero point on the meter scale indicates infinite ohms. The term *infinite* means that the amount is too large to measure.

To illustrate, suppose a 4500-Ω resistor is connected between the terminals of the ohmmeter. The circuit now

has an approximate resistance of 9000 Ω. Using Ohm's law, it can be determined that current through the meter is 0.5 mA, or 0.0005 A. This produces a mid-scale deflection. Therefore, the mid-scale reading of this ohmmeter is 4500 ohms.

The meter may be calibrated (readings determined and marked on the scale) by use of resistors of known value. As resistors are placed between the terminals, the deflection points are observed. The scale is then marked at the different deflection points. A scale for an ohmmeter is illustrated in Figure 6-19. This particular instrument is calibrated to measure resistances between 50 and 100,000 Ω.

REVIEW QUESTIONS FOR SECTION 6.8

1. What is the function of an ohmmeter?
2. Why does an ohmmeter have its own voltage source?
3. What test has to be performed before an ohmmeter is used?

6.9 THE SHUNT OHMMETER

Series ohmmeters usually do not measure resistance below 300 Ω. For lower levels of resistance, another type of ohmmeter is used. This is the **shunt ohmmeter**. It is possible for a single instrument to serve both as a series ohmmeter and a shunt ohmmeter. When this is done, both circuits usually have the same values in their components, or parts.

In a shunt ohmmeter circuit, current is applied to the meter continuously. This occurs whether or not the meter

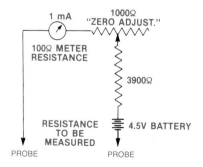

Figure 6-18. Schematic of ohmmeter circuit.

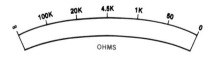

Figure 6-19. Ohmmeter scale.

is being used for measurements. Thus, a disadvantage of the shunt ohmmeter circuit is that prolonged operation of the meter movement will discharge the battery. Therefore, commercial shunt ohmmeters usually have an on–off switch. This makes it possible to conserve the battery.

A shunt ohmmeter circuit is shown in Figure 6-20. This meter movement has an internal resistance of 100 Ω. Assume a 100-Ω resistor is placed across the probes. This means that one-half the current will flow through the meter movement. The other one-half flows through the resistance being measured.

If the probes of the ohmmeter are shorted together, resistance is zero. There is no deflection of the pointer. This indicates a condition of zero ohms of resistance.

Half-scale deflection is achieved with a resistance equal to the meter movement, or 100 Ω. The measured resistance at the midpoint of the scale is 100 Ω. If no resistance is connected to the ohmmeter circuit, this results in full-scale deflection (200 Ω). Therefore, full-scale deflection on a shunt ohmmeter indicates infinite ohms of resistance. A scale for the ohmmeter described above is shown in Figure 6-21.

The scales of a shunt ohmmeter and a series ohmmeter are exactly opposite. Use care in reading the scale to be certain that the correct scale is being used. The *ohms-adjust* setting on a shunt ohmmeter is used to adjust full-scale deflection of the meter. This represents infinity.

A photograph of a meter that can be used as a shunt ohmmeter is shown in Figure 6-22.

Connecting an Ohmmeter

In using an ohmmeter, remember that it contains its own power source. Therefore, *the circuit or device being tested should never have power applied*.

REVIEW QUESTIONS FOR SECTION 6.9

1. What is the function of a shunt ohmmeter?
2. How is a shunt ohmmeter adjusted?

6.10 THE MULTIMETER

Voltmeters, ammeters, and ohmmeters are often combined into a single instrument. Such an instrument is called a **multimeter**, Figure 6-23.

Figure 6-22. Voltmeter usable as shunt ohmmeter.

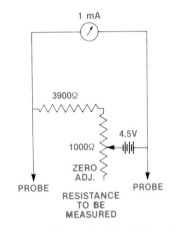

Figure 6-20. Schematic of shunt ohmmeter.

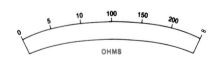

Figure 6-21. Shunt-ohmmeter scale.

Figure 6-23. Multimeter.

All multimeters, from all manufacturers, have similar controls. One such control is a range switch that selects the scale. Another control is a function switch to control its use as an ohmmeter, an ammeter, or a voltmeter. Multimeters also have a *zero-adjust* knob for the ohmmeter circuits. There are also jacks for the probes.

The functions and ranges of a multimeter are determined by the selector switches.

REVIEW QUESTIONS FOR SECTION 6.10

1. What functions are performed by a multimeter?
2. What is the function of a multimeter range switch?
3. What is the function of a multimeter zero adjust knob?

6.11 THE WATTMETER

A **wattmeter** is used to measure power. It indicates power used in watts per second. It does this by measuring and computing readings for voltage multiplied by current. This means there are two types of coils in the meter, Figure 6-24.

A current coil is inserted in series with the load. In most instances, there are two current coils in series. These coils are made of heavy wire so that they can handle all of the current through the circuit.

The voltage coil has many turns of fine wire. This makes it possible to place the coil across a line without using so much current that it overheats.

Positioning of the coils in the meter is important. The voltage coil is allowed to move on pivot points. The current coils are stationary. The electromagnets produce magnetic fields. The relationship between these magnetic fields determines the repulsion of the voltage coil.

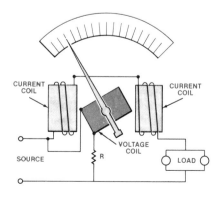

Figure 6-24. Functional design of wattmeter.

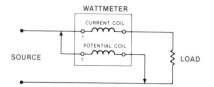

Figure 6-25. Schematic of wattmeter.

The voltage coil has a needle attached. A scale is marked with the amount of power being consumed at any given point. A spring returns the coil to its resting place when power is removed.

An electrical diagram of a wattmeter and its connection into a circuit is shown in Figure 6-25. Note the location of the voltage (potential) coil. It is placed across the line or source. The current coil is in series with the load.

REVIEW QUESTIONS FOR SECTION 6.11

1. How does a wattmeter operate?
2. What does the reading of a wattmeter show?

6.12 THE MEGGER

The **megger** is a device for measuring very high resistances. It is used to check for insulation breakdown in motors and other devices. A megger is diagrammed in Figure 6-26.

On a megger, a hand crank is attached to a coil. The coil is placed inside a permanent magnetic field. A high voltage is produced when the crank is rotated. This high voltage is required to drive small amounts of current through high resistances. The meter movement has to be extremely sensitive.

This type of meter can produce shocking voltages. You can receive an uncomfortable shock from a small turn of the crank. So be careful when using a megger. Also be careful of your electrical and electronic equipment: Do not attach the leads to anything that will be easily damaged by high voltage.

REVIEW QUESTIONS FOR SECTION 6.12

1. What is a megger?
2. How does a megger operate?

6.13 THE DIGITAL METER

Digital meters are entirely electronic. There are no coils or magnets.

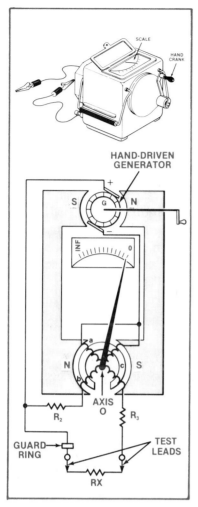

Figure 6-26. Functional diagram of megger.

Figure 6-27. Digital multimeter (DMM). (Triplett)

The meter shown in Figure 6-27 is a digital type. It is a bipolar, 3½-digit, portable, self-contained **digital multimeter**. The abbreviation for this type of meter is *DMM*. This particular meter is designed for laboratory and field use. It is lightweight, shock-resistant, and convenient to use.

The meter illustrated presents its reading in numbers on a liquid-crystal display. This is the type of digital readout found on many instruments, clocks, and watches.

This meter has five AC and DC voltage ranges. There are five AC and DC current ranges, as well as six resistance ranges. The displaying of 26 ranges is accomplished through use of an autorange feature on functions for volts, milliamperes, amperes, **kilohms**, and **megohms**. Some digital meters automatically select the range. These have what is called *autoranging*.

Specifications

The meter illustrated in Figure 6-27 has excellent accuracy. It can measure up to 1000 volts, 10 amperes,

and 20 megohms. Ampere ranges are provided through a separate jack. Two disposable 9-V batteries provide power for operation up to 200 hours. The half-inch-high numbers on the display can be read in ambient light (ordinary room or outside light).

Resistance is extremely high. Because of this, the meter hardly becomes a part of any circuit it measures. It has an input impedance of 10 megohms. Accuracy is within ± 0.1 percent.

A digital meter set up to measure voltage is shown in Figure 6-28. In this case, the measurement would be for DC voltage. You can tell this because the dial selector is at V. A switch above the dial selector is at the DC setting. Full-scale voltage is 1000 V in this instance. Autoranging occurs automatically, covering ranges from 1.999 to 1000 V.

Using the Digital Meter

The digital meter is easy to use. Note the features on the meter in Figure 6-28. It has a function switch. A DC–AC Ohms switch selects the mode of operation. The selection of a setting for V, mV, A/mA, ohms, k ohms, or M ohms automatically activates the autoranging. Leads go into the common and V-ohms-mA holes in the side of the meter. If you use the wrong polarity, a minus sign shows in the display. One lead has to be moved to measure 10 amperes.

For use, simply turn the meter to the needed function. Then select Ohms and use the probes across the resistor to measure resistance. If a circuit is being measured for resistance, make sure there is no power turned on when the meter is connected. If you use this unit as a milli-ammeter, you have to insert it properly into the circuit being measured.

To extend battery life, turn the meter off whenever it is not in use.

Prices of digital meters are dropping rapidly. This is the same pattern followed with electronic calculators. In time, the digital meter can be expected to replace all other types. However, for some purposes, the digital meter can be outperformed by the d'Arsonval movement. For example, some meter indicators simply need a deflection of the needle to show proper operation. In the digital, you have to wait for the numbers to be counted up or down. It takes concentration on the part of the user. The d'Arsonval movement simply shows a deflection of the needle.

Selection of meter type is up to the user. Selection tends to be based both on the job to be done and the training or experience of the user.

REVIEW QUESTIONS FOR SECTION 6.13

1. How do digital meters measure electricity?
2. How are readings displayed on digital meters?
3. What is the role of a function switch on a DMM?

6.14 CLAMP-ON METERS

A clamp-on meter comes in handy whenever you want to see whether a piece of equipment is drawing too much current, Figure 6-29. Thus it is very handy to have when troubleshooting AC electric motors. In that instance, you can check the current by clamping the end of the meter around *one* of the wires leading to the consuming device. The transformer action causes a small current to be induced in a coil in the meter. The small current is measured with respect to the quantity needed to induce that amount in the coil, and then the scale is calibrated accordingly. This type of meter can also be used as a volt–ohmmeter and milliammeter when a battery is used to supply power for the ohmmeter operation and test leads are used.

The main advantage of this type of meter is in testing motors, relays, transformers, and control circuits, as well as ordinary power and lighting circuits. Load balance tests and AC current measurements are made without equipment shutdown or power-line interruption. The meter can measure up to 300 amperes AC without attachments.

Figure 6-29. Clamp-on AC ammeter. (Simpson)

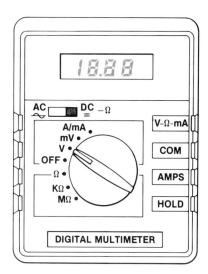

Figure 6-28. Digital meter set for voltage measurement.

The jaw shape allows it to be inserted into wire clusters. The finger-operated lever locks the scale pointer and allows reading of measurements at some convenient point away from the measuring position, if desired.

SUMMARY

To meter means *to measure*. Meters are used to measure electrical current, resistance, and voltage. Different types of meter movements are utilized to measure AC and DC, as well as ohms and voltage and current. Some meter movements are made up of a fixed magnet and a moving magnet. The electromagnet may or may not move. However, one or the other magnet moves in order to indicate current through the coil of the meter.

Multimeters are used to measure a number of functions of a circuit. An analog meter has a number of scales on the face of the meter. The digital meter displays the voltage, resistance, or current in digits.

The clamp-on meter is used to check current flow in a circuit without breaking the wires. It comes in handy in troubleshooting electric motors while they are operating.

The AC voltmeter uses a diode to change AC to DC so it can be measured with a DC meter movement. The series meter movement measures high resistance; the shunt ohmmeter measures low resistance. An ohmmeter has its own power source and should never be connected to or in a live circuit.

The megger is used to measure extremely high resistances. The wattmeter is used to measure power consumed in a circuit.

_____ USING YOUR KNOWLEDGE _____

1. Draw a circuit diagram showing the connection of a DC ammeter. The ammeter should measure 10 A of current through a 1-Ω resistor.
2. Use an ohmmeter to test at least five resistors with different ratings. Zero-adjust the ohmmeter. Then compare the values of your readings with the rated values of the resistors. Look at the tolerance values of the resistors. See whether your readings match the tolerance ratings for the resistors.
3. Draw a scale for a DC voltmeter. The scale should be four inches wide. Markings should be evenly spaced for readings of 10 V in units of 0.1 V.

_____ KEY TERMS _____

d'Arsonval meter movement	zero adjust
linear scale	voltage (potential) coil
parallax error	megger
microamperes	digital multimeter
milliamperes	kilohms
shunt	megohms
multiplier	autoranging
multirange	multimeter
shunt ohmmeter	wattmeter
series ohmmeter	rectifier

_____ PROBLEMS _____

1. What size shunt is needed for a 1-mA meter movement with 50 ohms internal resistance, to measure 100 mA?
2. What size shunt is needed for a 10-mA meter movement with 500 ohms internal resistance, to measure 1 ampere?
3. What size shunt is needed for a 10-microampere meter movement to measure 100 mA, if the internal resistance is 5000 ohms?
4. What size shunt is needed for a 1-mA meter movement to measure 10 amperes, if the internal resistance is 50 ohms?
5. What size shunt is needed for a 1-mA meter movement with 500 ohms internal resistance, to measure 100 amperes?
6. What size resistor would be needed as a multiplier for a voltmeter circuit if the meter movement was 1 mA at 50 ohms and you wanted to measure 500 volts?
7. What size resistor would be needed as a multiplier for a voltmeter if the meter movement was 10 mA at 5000 ohms and you wanted to measure 50 volts?
8. What size multiplier is needed for a 50-ohm, 1-mA meter movement to make it capable of measuring 1000 volts?
9. If a 1500-ohm resistor is placed between the probes of a series ohmmeter and the meter movement reads half scale, what is the internal resistance of the meter?
10. Where, on the scale of a shunt ohmmeter, does the infinite reading appear? (left or right).

Activity: THE VOLTMETER

_____ OBJECTIVE _____

To learn how to recognize and use a voltmeter.

MATERIALS NEEDED

1 D-cell, 1.5 volts	1 Resistor, 100K, 5%,
1 Power supply,	0.5 W
0–10 V DC	1 Meter movement,
1 Resistor, 10K, 5%, 1 W	1-0-1 mA, 43 ohms
1 Resistor, 5K, 5%, 1 W	3 Wires for connections

PROCEDURE

1. Set up the circuit as shown in Figure 1.

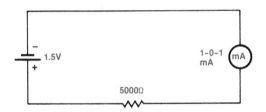

Figure 1

2. Use Ohm's law to determine the amount of current that should flow in the circuit. How does this compare with the reading of the milliammeter? How much voltage would be necessary for full-scale deflection with 1 mA? (Since this is the amount of voltage necessary for full-scale deflection, the meter is a 0–5 volt DC voltmeter.)
3. Since the meter is a 0–5 volt DC voltmeter, how much voltage does the voltmeter indicate for the source voltage of this circuit?
4. Change the circuit to that shown in Figure 2.

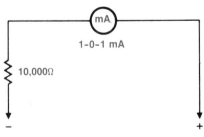

Figure 2

5. The circuit you have connected is a 0–10 volt DC voltmeter. Adjust the voltage control on the power supply to its *minimum* output. Then, turn on the power supply. Slowly increase the voltage control of the power supply until it is 5 volts DC. Connect your voltmeter to the output terminals of the power supply. What is the reading of the voltmeter you have connected?
6. With the meter still connected to the power supply, *slowly* increase the output of the power supply to 10 volts. Compare the meter indicating the output of the power supply and the meter you have connected. Do you read the same value?
7. *Turn off the power supply.* Disconnect from the power supply the meter you have made, and change the circuit to the one shown in Figure 3. This is a 0–100 volt DC voltmeter. *Turn on the power supply.* Adjust the output of the power supply to 10 volts DC. Connect your meter to the power supply as in Step 6. What is the voltage reading of your meter now?

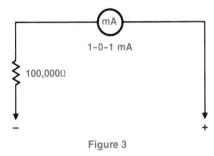

Figure 3

8. If you wanted to make a 0–1000 volt DC voltmeter, what would the value of the multiplier resistor be?
9. *Turn off the power supply.* Disconnect all circuits and return the equipment to its proper storage places.

SUMMARY

1. What components are necessary in a voltmeter?
2. What is the purpose of the multiplier resistor?

Project: MAKING AN OHMMETER

This ohmmeter utilizes a standard d'Arsonval meter movement that is used in most electronic instrumentation. When used in various special circuits, this type of meter movement can accurately measure most of the common electrical units. This particular project places the meter movement in a circuit that measures resistance in ohms. This ohmmeter is a very useful instrument.

You will need the following materials.

Project case, 2 7/8" × 4"
Cover, XP phenolic
Meter, 0–1 mA, with ohmmeter scale to be added
3 Solder lugs, #4
Potentiometer, 2000 ohms, 0.5 W
SPDT slide switch
Resistor, 1200 ohms, 0.5 W, 10%
Resistor, 9760 ohms, 0.5 W, 1%
24 ga. brass sheet, 5/8" × 6"
Test lead, 36", red
Test lead, 36", black
Grommet, 3/8" O.D.
4 Machine screws, 3/8" #4-40
4 Machine screws, 1/4" #4-40
4 Nuts, #4-40, hex
Battery connector
Knob
#22 hook-up wire, 6"

Planning

Study the list of materials and identify each component part. Study the schematic diagram and illustrations to see how the ohmmeter is constructed.

Construction Notes

The layout of this project must be done accurately in order to assure that all components will fit properly. For this reason, a front panel template locating all holes has been shown in Figure 1.

It is much safer and easier to drill or punch the three holes in the brass clamps before cutting the brass strip, Figure 2. The four meter mounting studs are machine screws set in plexiglas. Excessive tightening of the hex nuts for mounting the meter will crack the plexiglas. The plastic meter face is also made of plexiglas. It can be scratched quite easily. Therefore it is advisable to temporarily tape a piece of protective paper over the meter face while wiring the ohmmeter.

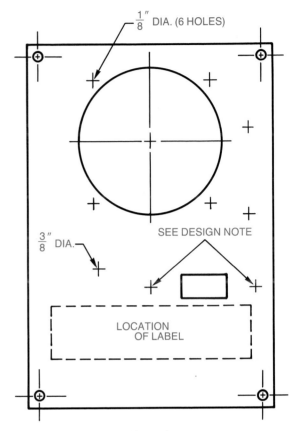

Figure 1

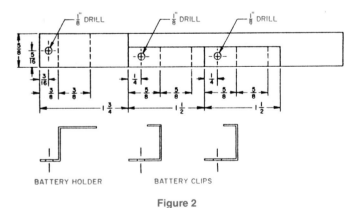

Figure 2

Design Notes

Slide the switch to either of its positions, Figure 3. Note that it comes very close to the screw heads. If the hex nuts were on the inside, the slide switch would not be able to travel fully, possibly making poor electri-

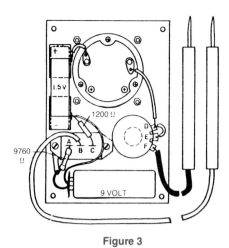

Figure 3

cal contact. The ohmmeter cover template is shown in Figure 1.

The 9760-ohm resistor is a deposited film 1% tolerance precision type. Be careful not to chip or scratch it, as this may alter the resistance. The accuracy of the meter is dependent upon the tolerance of this resistor.

Testing

Place the *High–Low* switch in the *Low* ohms position. Short the two probes together. The needle should move to the right, toward the zero marking. Touch the probes together and turn the zero-adjust knob until the needle is directly over the zero marking on the scale. The ohmmeter is now calibrated to measure resistance up to 20,000 ohms. Place the switch in the *High* ohms position and again short the probes together. Readjust the zero-adjust knob until the needle indicates zero. The ohmmeter is now calibrated to measure up to 500,000 ohms.

Using the Ohmmeter

Measure resistance by first determining which scale to use. If you think the resistance to be measured is less than 20,000 ohms, use the low scale. Use the high scale for all other values. If you are not sure which scale to use, try the high one first. If the reading is below 20,000 ohms, switch to the low scale for a more accurate reading.

Remember: The meter must be zeroed each time you change scales. When the ohmmeter is not in use, be sure the probes are not touching. Continued shorting of the test leads will cause excessive battery drain. When the meter is not in use (actually measuring resistance), no current is being drawn from the batteries.

If the ohmmeter is not used for a period of time, (several weeks or more), be sure to remove the batteries, since they may leak or corrode, thus causing damage inside the ohmmeter case. Figure 4 shows the schematic of the ohmmeter for easy reference in case there are problems with construction or operation.

Safety Note

Never test a live circuit with an ohmmeter. To do so will result in serious damage to the meter. If there are any capacitors in the circuit you are measuring, be sure to discharge them before attempting any resistance measurements. They can be discharged by shorting across the terminals with a screwdriver blade or a piece of insulated wire with bare tips that make contact with the capacitor leads or terminals.

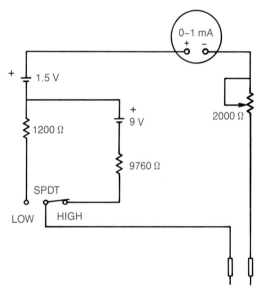

Figure 4

Chapter 7

ALTERNATING CURRENT

OBJECTIVES

After studying this chapter, you will know

- *that alternating current varies periodically.*
- *the term* hertz *and its meaning.*
- *the term* frequency *and its time-related meaning.*
- *how to describe the operation of an alternator.*
- *how a sine wave is generated.*
- *how a basic transformer operates.*
- *how transformer losses are overcome.*
- *the relationship between voltage in the primary and secondary of various transformers.*

INTRODUCTION

AC meters are described in Chapter 6. Otherwise, almost all discussions so far have been about direct current (DC). In this chapter, you look closer at alternating current (AC). Your study includes analysis of the current you use in your home. You will learn just what AC is and how it is used and measured. You will also learn about several AC devices, including transformers. In addition, this chapter covers frequency and phase.

Some Background

To alternate means *to change at regular intervals*. **Alternating current** changes directions as it flows through a wire. The change of directions occurs many times each second.

AC, alternating current, was developed by Nicholas Tesla. Its use was introduced around 1900. The word *polyphase* means *many phases*. In the world of commercial electrical power, *polyphase AC* refers to three-phase current. This is current delivered by three conductors. Today, three-phase current is generated at almost all commercial facilities.

Alternating current has an important advantage over direct current: AC is a more efficient form of electricity for transmission over power lines, Figure 7-1. The device that helps make electric transmission possible is the **transformer**. A transformer is used to step up or step down AC voltage. As the voltage is stepped up, or increased, the current is stepped down.

This ability takes advantage of the nature of conductors. Wire can transmit more power at high voltages. Current is the critical factor in wire capacity. With transformers, smaller wires are used to transfer large voltages with smaller currents over great distances. This is done with less power loss from resistance.

Alternating current is usually generated at 13,800 V. This is then stepped up for transmission. The most commonly used voltages for long-distance transmission are 138,000 V, 250,000 V, and 750,000 V. Once the AC power reaches its destination, it is stepped down to 240 V for commercial use.

Long-distance transmission and delivery of AC power are shown in Figure 7-2. A substation that steps down the power for commercial use is shown in Figure 7-3.

7.1 THE NATURE OF ALTERNATING CURRENT

Alternating current changes its direction of flow as it moves along a wire. It flows in one direction first, then in the other. This is different from DC, which moves in one direction only.

Another important item of information about electricity is the speed at which it travels. Electrical current travels at the speed of light. The rate of travel is 186,000 miles

Figure 7-1. AC is used for electric-power transmission.

LONG LINE LOSSES

GENERATION

CONSUMED

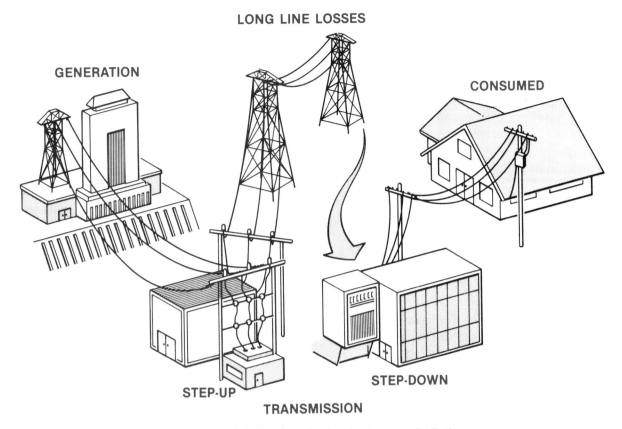

STEP-UP

TRANSMISSION

STEP-DOWN

Figure 7-2. Drawings showing electric-power distribution.

[299 274 kilometers] *per second*. This speed is hard to imagine. In effect, it means that electricity moves *instantly* from the generator to your home. This is true even if the generator is several hundred miles away. For some idea of the speed involved, consider this: Electrical current

Figure 7-3. Power-distribution substation.

can move the equivalent of almost eight times around the world in one second.

Frequency

The term used to indicate the changes in direction of AC is **frequency**. The AC you use changes directions 120 times per second. This means it moves *back* and *forth* 60 times per second. This current is described as having a frequency of 60 *hertz*. It is said to be *60-hertz AC* current. *Hertz* is abbreviated *Hz*. The unit of time related to Hz is always considered to be one second. Each hertz has two alternations, Figure 7-4.

The operation of an AC generator is diagrammed in Figure 7-5 and explained further below. One-half of one hertz of power generation is referred to as an **alternation**. Two alternations make one hertz, Figure 7-4.

AC frequency is generally described in terms of hertz. Thousands of hertz are **kilohertz** (*kHz*). Millions of hertz are **megahertz** (*MHz*). If AC changes directions 8 million times per second, it has a frequency of 4 MHz. If the change of directions occurs 6000 times per second, the frequency is 3 kHz.

You may also hear the term *cycle* used to describe AC current. This is an older term. But you may still come across it on some equipment you find in use. A cycle is equivalent to 1 Hz, or two alternations.

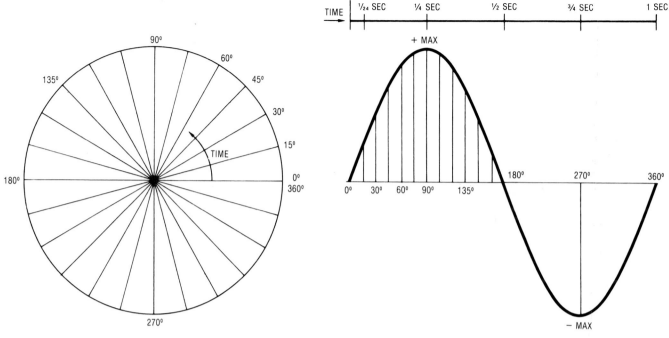

Figure 7-4. Graph showing AC being generated.

7.2 THE AC GENERATOR, OR ALTERNATOR

Generators that produce alternating current are called **alternators**. You will probably hear the term *alternator* frequently. One use of alternators is in modern automobiles. Your car's engine operates an alternator.

Generator Operation

Parts of a simple generator are labeled in Figure 7-5. This shows an armature loop (coil) suspended in a magnetic field. The brushes in the diagram serve to direct current into the load resistor as the coil rotates in the magnetic field.

Note the position of the armature loop in Figure 7-5A. In this position, the top and bottom portions of the loop are parallel with the magnetic lines of force. No lines of force are cut by the armature loop when it is in the position shown in Figure 7-5A. Unless lines of force are cut by the rotating loop, no current is applied to the load resistor.

When the armature loop moves to the position shown in Figure 7-5B, current is generated. The movement of the wire through the magnetic field produces electricity. The loop position in Figure 7-5B represents a movement of

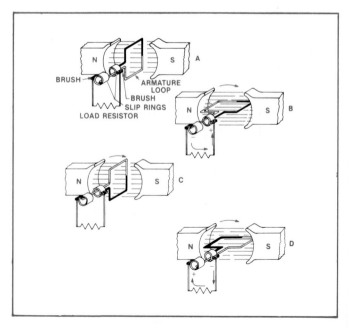

Figure 7-5. Simplified illustration of AC generation.

90 degrees. That is, the loop has turned one quarter of a full circle from the position shown in Figure 7-5A.

As the loop rotates to the 90-degree position, current hits a peak of output. This is illustrated in Figure 7-6, which shows a **sine wave**. A sine wave is a graph of the electric output of a generator. The production of current moves to a positive high point first. Then it alternates as the armature turns.

Now look at Figure 7-5C. This drawing shows the armature in an open position again. This is a 180-degree rotation, or a half-circle from the starting point. In this position, there is zero output of current. This is reflected also in the notation on the sine wave in Figure 7-6.

Figure 7-5D shows the loop once again breaking the magnetic field. At this point, there has been 270 degrees of rotation. This is three-quarters of a full circle, Figure 7-6. This shows that the sine wave is at its maximum negative position when the coil has rotated 270 degrees.

When the loop rotates back to the starting position, current generation stops again. The loop has traveled a full circle in the generator. It has completed one cycle of operation, or one Hz of AC output.

REVIEW QUESTIONS FOR SECTION 7.2

1. Why are AC generators called alternators?
2. What is the role of magnetism in AC generation?
3. What is an armature loop within an AC generator?

7.3 AC WAVEFORMS

Alternating current is constantly changing. These changes are in *magnitude* and *direction*. This AC pattern is called a **waveform**. When graphed, an AC cycle is shown as a *sine wave*. By comparison, DC does not change magnitude or direction. It can be represented by a straight line. Figure 7-7 shows this contrast in patterns for DC and AC.

An AC voltage causes an alternating current to flow. AC at 240 V is delivered by power companies to homes in the United States and Canada. Once inside the home, the voltage is divided. This serves to deliver 120 V AC to electrical devices. This is a standard power requirement. The standard AC frequency in the United States and Canada is 60 Hz.

An AC generator with only one loop operates at 3600 revolutions per minute, or rpm, to produce 60 Hz. It is possible, through generator design, to vary this pattern. One objective is to keep the frequency the same—60 Hz. Another objective is to slow down the rotation of the generator. Lower generator speeds reduce equipment wear. Costs are also reduced.

Adding a second loop makes it possible to cut generator speed in half. With six loops in the magnetic field, 60-Hz current can be produced at 600 rpm. However, the frequency of the waveforms can change. This is illustrated in Figure 7-8. The illustration relates Hz to the passage of time. In one diagram, there are 4 Hz. In the other, there is a single Hz.

The time base of one second is standard in the electrical field. This is the time in which 60 Hz of AC current is delivered. So, whenever you see a reference to 60 Hz, understand this meaning: AC current at 60 Hz means that 60 sine waves are generated each second.

REVIEW QUESTIONS FOR SECTION 7.3

1. What is the standard voltage supplied by power companies in the U.S.?
2. What is the standard hertz rating of AC power in the U.S.?

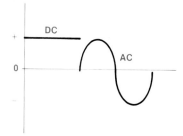

Figure 7-7. Graphic comparison of DC and AC.

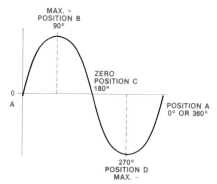

Figure 7-6. Sine-wave graph.

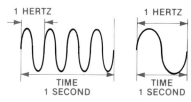

Figure 7-8. Graph showing time–hertz relationship in AC generation.

Activity: THE OSCILLOSCOPE AND THE AC WAVEFORM

OBJECTIVES

1. *To study the AC waveform's characteristics.*
2. *To be able to adjust an oscilloscope to display an AC waveform.*

MATERIALS NEEDED

1 Oscilloscope with probes

1 Step-down filament transformer (120 V to 6.3 V or 12.6 V)

PROCEDURE

1. Plug in the oscilloscope and turn it on. Allow time for the tube to warm up and show a trace or line.
2. Attach the oscilloscope probes to the transformer leads. (Polarity is not important: either lead of the scope can go to either lead of the transformer secondary).
3. Plug in the step-down transformer.
4. A trace of some sort should appear on the face of the scope.
5. Have your instructor explain the purpose of each of the controls on the front of the oscilloscope.
6. Answer the following questions on a separate sheet of paper.
   A. What does the *Intensity* control do?
   B. What does the *Focus* control do?
   C. What does the *Horizontal Position* control do?
   D. What does the *Vertical Position* control do?
   E. What control allowed you to see only one sine wave?
7. Does your sine wave look like the one shown in Figure 1?
8. Unplug the transformer from its 120-volt power source.

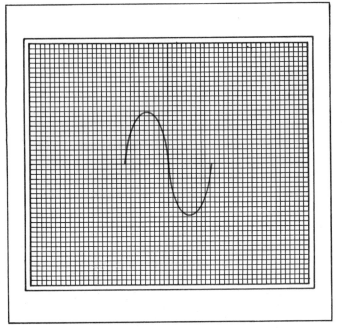

Figure 1

9. Is there still a trace on the scope? Draw it on the same piece of paper you used for answering the questions in Step 6.
10. Turn off the oscilloscope. Remove its power plug from the socket.
11. Return all the equipment to its proper storage place.

SUMMARY

1. What caused the sine wave to settle down and show only one waveform?
2. How can the scope be used to check other waveforms?
3. How does the AC waveform relate to the rotation of the generator that produces it?

7.4 MAXIMUM AND PEAK VALUES OF AC

In describing AC, three values are used as references:

- Peak
- Average
- Root-mean-square (RMS)

Peak Value. The maximum point on a sine wave is its **peak value**. Both peaks of a single hertz may be included in a reference. If so, it becomes a *peak-to-peak* value. A peak value of 100 V means that the peak-to-peak value is 200 V. This is illustrated in Figure 7-9.

Average Value. The average of all instantaneous values of a generator is measured at regular intervals. Instantaneous values are taken at selected points in the generating process. The average of these is the **average value** of AC current.

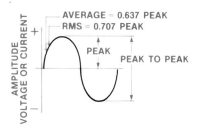

Figure 7-9. Graph of AC values for sine wave.

The half hertz is used to compute this value. This is because the average value for full-hertz cycles would be zero. The zero average would come from the fact that half the values are positive and half are negative.

To compute average value, sine wave values for angles between zero and 180 degrees are added. The total is then divided by the number of values taken. As shown in Figure 7-9, the average is equal to 0.637 of the peak.

To illustrate, suppose the peak voltage of a sine wave is 100 V. The average voltage is 63.7 V (100 × 0.637 = 63.7).

Root-Mean-Square (RMS) Value. A value equivalent to effective voltage or effective current is known as **root-mean-square** (*RMS*).

RMS is calculated mathematically. Imagine one-quarter of the sine wave broken down into 90 parts. This is one part for each degree of a hertz. The value of each degree is squared, or multiplied by itself. All of the squared values are added together. Then the total is divided by 90. This produces the average of the squares. A square root is then figured for this average. This is the RMS value.

The RMS value is the reading you get on an AC voltmeter or ammeter. It is the amount of AC that effectively produces the same amount of heat as would DC of equal value.

REVIEW QUESTIONS FOR SECTION 7.4

1. What is the meaning of *peak value of AC current*?
2. What is the meaning of *average value of AC current*?

7.5 TRANSFORMERS

A device that transfers electrical power from one coil to another is called a *transformer*. Transformers may also change the value of voltage during the transfer.

A transformer functions with no physical connection between the source and receiving conductors. The prin-

ciple used is mutual inductance, explained in the next chapter.

Current flows into a transformer primary coil. This current creates a magnetic flux. The magnetic flux, in effect, couples the **primary coil** with the **secondary coil**. Voltage is induced in the secondary coil. The induced voltage may be varied by increasing or decreasing the magnetic field. The result of a transformer's operation is the induction of emf in the secondary coil. This is transferred to a conductor connected to the secondary coil.

Symbols for two types of inductors are shown in Figure 7-10. One has an air core. This means that only air separates the primary and secondary coils. The second has an iron core. There is an iron bar between the primary and secondary coils. The symbols in Figure 7-10 are used to represent transformers in electrical diagrams.

Transformer Construction

Every transformer is built with at least two coils. One is a primary coil. The other is a secondary coil.

The primary coil is called the **input**. This means that it brings electrical power into the transformer. The secondary coil is called the **output**. It carries electrical power out of the transformer.

The transformer is one of the most efficient electrical devices in use today. It has no moving parts. It is extremely dependable. The operation of the transformer is controlled by varying input current.

A transformer core may be air. This means that there is no solid material between the coils. Power transformers and **audio frequency** transformers use an iron core. The iron core helps to concentrate the lines of force. Efficiency is increased.

Voltage Transfer

The transformer is a *power transfer device*. Power into the primary is transferred to the secondary. Also, power required by the secondary is reflected back as a power

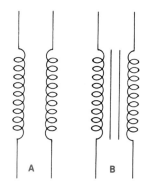

Figure 7-10. Symbols for air-core (A) and iron-core (B) transformers.

required by the secondary is reflected back as a power requirement to the primary. The rule is simple: Power in is equal to power out, less losses.

Thus, if there is no load connected to the secondary coil, the current flowing into the primary is virtually zero. The only current required would be the amount necessary to support part of the losses. This is almost zero. Current flows in the primary only when the secondary is connected and there is **output current**.

A moderate overload causes the temperature of the transformer to be increased. Therefore, the current available is actually a function of the amount of heating permissible in the transformer itself.

Step-Up and Step-Down Transformers

Transformers are either *step-up* or *step-down*. If input voltage is higher than output voltage, the device is a step-down transformer. In a step-up transformer, the output voltage is higher than the input.

Turns Ratio

The input and output voltage relationship of a transformer depends upon the **turns ratio**. Turns ratio describes the relationships of windings of the primary and secondary coils. The first number in a given ratio is for the secondary. The second number is for the primary.

Safety Tips

1. Be careful when you work around a transformer. Transformers can step voltages up high enough to kill people.
2. Make sure all transformers used in your home have metal covers.
3. Remember that autotransformers do not isolate the ground in home or industrial circuits from the primary. It is possible to obtain the voltage you need with this type of transformer. But you may get a shock if you touch the water pipe or ground at the same time you are in contact with the "hot" side of the transformer. Keep one hand in your pocket when working around an energized transformer.

To find the turns ratio of a transformer, divide the number of turns of the secondary by the number of turns of the primary. For example, if there are 1000 turns in the secondary and 100 turns in the primary, divide 1000 by 100. This produces a ratio of 10 to 1. This is written 10:1. The formula used to find turns ratio is:

$$\text{Turns ratio} = \frac{N_s}{N_p} = \frac{1000}{100} = \frac{10}{1} \text{ or } 10:1$$

N_p is equal to the number of turns in the primary. N_s is equal to the number of turns in the secondary.

The voltage ratio of a transformer can be found in the same manner. The formula is:

$$\frac{E_s}{E_p} = \frac{N_s}{N_p}$$

E_p is the voltage of the primary. E_s is the voltage of the secondary.

If the transformer has a turns ratio of 10:1, the voltage ratio is 10:1. In the example given, if the transformer has 100 V applied to its primary, the secondary would have 1000 V measured on its output.

Isolation Transformers

Transformers that have the primary and secondary separately wound and not physically connected are called **isolation transformers**. That means the ground connection on the power coming from the power company is isolated from those items plugged into the secondary. This is advantageous when using equipment that has a chassis that is grounded and can become "hot" if the plug is incorrectly inserted into the wall socket. It can also be of some advantage when a computer is plugged into the outlet. The transformer has a tendency to reduce the fluctuations produced on the power line.

Power Transformers

Power transformers use multiple secondary coils. These can deliver a number of secondary voltages. Examples of power transformers are illustrated in Figures 7-11 and 7-12. These devices may be used for a wide variety of jobs. They may provide power for factories. They may power broadcast stations. Or they may be built into special equipment as a power supply.

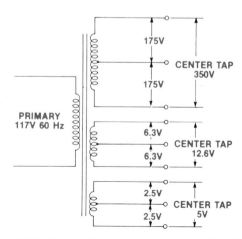

Figure 7-11. Schematic of power transformer with multiple secondary coils.

Flat Compact Printed-Circuit Power Transformers

The flat compact transformer, Figure 7-12B, has been created for use in any low-voltage application in which space is critical. Good examples of its use are in single- or dual-output DC supplies, isolated control circuits, and reference supplies. It incorporates nonconcentric windings, in which the secondary is located next to the primary instead of being wound directly over it. The 2-volt-ampere-size flat compact transformer permits 3/4-inch card spacing and is only 0.650 inch high. The 12-volt-ampere size has 1 1/4-inch card spacing and is 1.065 inches high. Plug-in pins of wire are used in order to handle properly the weights and current required for the 24-VA and 48-VA units. A dual primary is capable of handling 115 or 230 volts. It also diminishes the effects of a radiated magnetic field.

Mini-ISO-mite Transformers

The mini-ISO-mite transformer is an isolation (iso) transformer for low-voltage, plug-in use. It is designed for any low-voltage power supply application in which tight space restrictions are a consideration. It uses eight-pin dual 115/230-V primaries. Primaries and secondaries are wound nonconcentrically—that is, side-by-side. This construction results in smaller units and eliminates the need for costly electrostatic shielding. (Electrostatic shielding prevents radiation of the magnetic field to adjacent components.) It is tested for 2500 volts HIPOT and is built to military specifications. (*HIPOT* means it is tested for high potentials between the windings and the laminations and from coil-to-coil or primary-to-secondary.) It is available in sizes from 1.1 VA to 36 VA. This transformer can also be ordered with a single 115-volt primary, Figure 7-12C.

Newer-Type Power Transformers

Newer-type transformers have been designed for use with the latest electronic equipment that requires the arrangement of the open-lead-style, clamp-fixing transformer, Figure 7-12D. The budget-range transformers are ideal for "slimline" electronic equipment. They are 50% lighter and have 50% lower volume than do traditional stacked-laminated types. They have no air gap and less magnetic resistance. They also have an 8.1 reduction of electrically induced hum. They save open-circuit power and offer a stacking-factor capability of 95% of theoretical weight. These high-efficiency, high-flux-density transformers have low iron losses. The audible noise caused

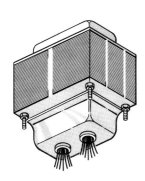

A. Power transformer for electronics equipment.

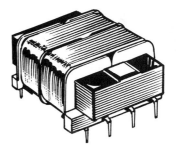

B. Flat PC power transformer.

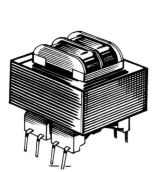

C. Mini-ISO-mite low-voltage plug-in power transformer.

D. Open-lead style, clamp-fixing transformer.

Figure 7-12.

by magnetostriction is considerably reduced. Secondary windings can be connected in series or parallel for dual-voltage operation. They will operate on 47 to 400 Hz. Secondary voltage tolerance is within 3% at nominal input and full load. The transformers are PVC-insulated and will take 2.5 kV peak. They will operate at 105°C above ambient temperature.

Audio and Radio Transformers

Audio transformers change voltages for use in the audio range of frequencies. Audio frequencies are from 16 Hz to 16,000 Hz. An audio transformer is shown in Figure 7-13.

To deliver voltage at higher frequencies, no core is used. This is because an iron core consumes too much power above 16,000 Hz. Air core transformers, therefore, are usually used in radio-frequency circuits.

For electronics equipment, special miniature transformers are used. A series of these devices is shown in Figure 7-14.

Autotransformers

Autotransformers have only one coil. Input and output come from this single coil. The amount of voltage output is determined by the position of the secondary tap (connection point). Placement of the secondary tap changes the turns ratio.

When an autotransformer is used to step up the voltage, part of the single winding is used as the primary. The entire winding acts as the secondary, Figure 7-15.

When an autotransformer is used to step down voltage, the entire winding acts as the primary. Part of the winding acts as the secondary.

Power is transferred from the primary to the secondary by the changing magnetic field. The turns ratio determines the voltage output.

An autotransformer lacks isolation. It will not serve to remove the ground found in most home and industrial wiring systems. This is because the primary and secondary use the same turns.

This type of unit can be made into a variable transformer. A variable transformer can produce differing output voltages. A slider arm is placed over the windings. It makes contact with points where the insulation on the wire has been removed. This type of variable transformer is called a variac.

Iron-Core Transformers

There are three basic types of iron-core transformers:

- *Open core.* This type is used in power transformers, Figures 7-16 and 7-17.
- *Closed core.* This type is more efficient than the open core. As shown in Figure 7-18, the flux path is contained within the core. This increases the strength of the magnetic field. Transfer of energy is improved.
- *Shell core.* This is the most efficient type of transformer. As shown in Figure 7-19, this design makes for extremely efficient magnetic-flux patterns.

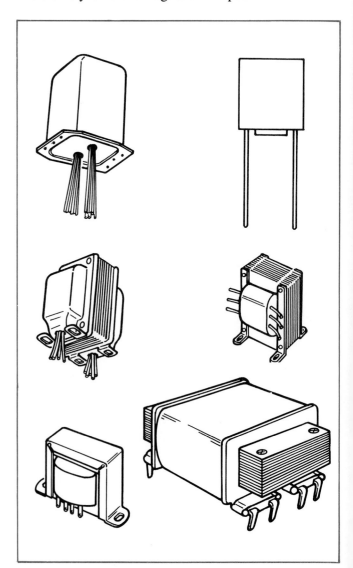

Figure 7-14. Miniature transformers for electronic circuits.

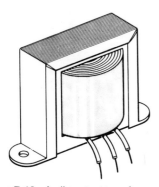

Figure 7-13. Audio output transformer.

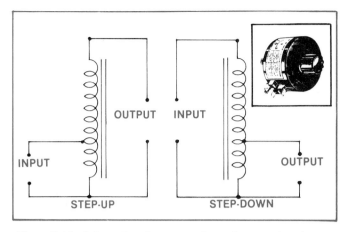

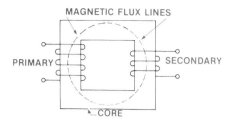

Figure 7-15. Schematics of step-up and step-down autotransformers.

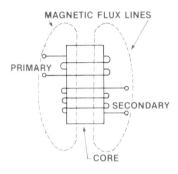

Figure 7-16. Functional drawing of open-core transformer.

Figure 7-17. Open-core transformer, usually called an ignition coil.

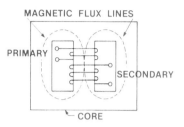

Figure 7-18. Functional drawing of a closed-core transformer.

Figure 7-19. Functional drawing of a shell-core transformer.

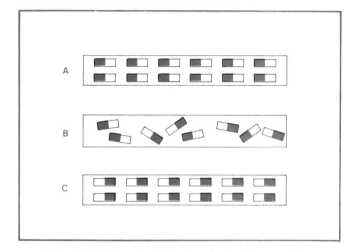

Figure 7-20. Functional drawings showing hysteresis loss.

Transformer Losses

Transformers are not 100-percent efficient. Three types of losses take place within transformers:

- *Copper losses*. These are due to the resistance of the wire in the primary and secondary coils. Large-size wire helps to minimize these losses.
- *Hysteresis losses*. These are due to the properties of the iron core. Iron is slow to change polarity with changes in current and magnetic field polarity. The delay is known as *hysteresis*, or a slowness to change properties, Figure 7-20. In Figure 7-20A, the magnetic domains of an iron core are facing in one direction. In Figure 7-20B, the domains are in a mixed state. This is in response to a change in magnetic fields. In Fig-

ure 7-20C, the domains are lined up in the opposite direction. This happens when the magnetizing force of AC is at its peak. Hysteresis losses in transformers are minimized by use of silicon steel. This type of metal offers small opposition to changing polarity.

- *Eddy currents*. These are small currents created in the core when magnetic fields are changed. See Figure 7-21. Eddy currents generate heat. They flow in the direction opposite to the current that induced them. The effect is to resist the flow of current in the core. Eddy currents can be minimized by use of lamination techniques. Laminating is the building of an object through use of several layers of material. When this method is used, each lamination is varnished. Varnishing insulates the layers from each other. This increases resistance to eddy currents.

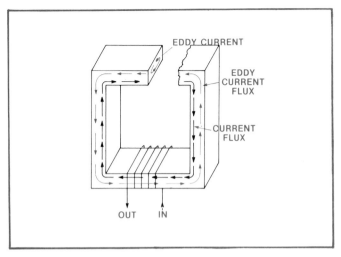

EDDY CURRENT

EDDY
CURRENT
FLUX

CURRENT
FLUX

OUT   IN

Figure 7-21. Diagram of eddy currents in an open-core transformer.

iron-core transformers are used where there are audio frequencies. Turns ratio between the primary and secondary of a transformer determines the voltage output from the device. An isolation transformer is used to remove the ground from the most-often-used type of power supplied to homes and businesses in the United States. Usually, isolation transformers have a 1:1 ratio.

Autotransformers have only one coil; the input and output come from that single coil. Voltage output is determined by the position of the secondary connection point. They can be used for step-up or step-down of voltages.

The iron-core transformer may be either an open core, closed core, or shell core. Transformer losses are due to copper losses, hysteresis losses, or eddy currents.

_____ USING YOUR KNOWLEDGE _____

1. Draw the symbol for an air-core transformer.
2. Draw the symbol for an iron-core transformer.
3. Draw the symbol for an autotransformer.
4. Draw the symbol for a power transformer.
5. Draw the AC waveform for 60 Hz.
6. Sketch outputs for DC and AC generators.

_____ REVIEW QUESTIONS FOR SECTION 7.5 _____

1. How is voltage transferred through a transformer?
2. What are the primary and secondary coils, and what do they do?
3. What is the turns ratio of a transformer?
4. What is the voltage ratio of a transformer?
5. What is meant by "power in is equal to power out less losses"?
6. What is the main construction feature of a power transformer?
7. What is the frequency range of audio and radio transformers?
8. How does an autotransformer operate?
9. How do transformers lose power?

_____ KEY TERMS _____

alternator	output current
alternating current	audio frequency
frequency	sine wave
megahertz	turns ratio
kilohertz	autotransformer
waveform	alternation
peak value	transformer
RMS	input
average value	output
primary coil	isolation transformers
secondary coil	

SUMMARY

Alternate means *to change at regular intervals*. Alternating current varies at different rates, depending upon its frequency. The hertz (Hz) is the unit for measuring frequency. A transformer is used to step up or step down alternating current. Alternating current is ideal for transmission over wires for long distances.

The *alternator* is another name for the AC generator. The alternator generates a sine-wave type of output power. This AC type of output pattern is called a waveform. Alternating current has three values: peak, average, and rms (root-mean-square). RMS is the most often used and most commonly referred to in AC circuits.

Transformers are made with air cores or iron cores. Air-core transformers are used for radio frequencies and

_____ PROBLEMS _____

1. What is the peak value of 100 volts RMS?
2. What is the RMS value of 100 volts peak?
3. What is the peak-to-peak value of 100 volts RMS?
4. What is the RMS value of 100 amperes peak?
5. What is the RMS value of 500 volts p-p?
6. A step-up transformer is constructed, using 575 turns on the primary and 1725 turns on the secondary. If a voltage of 120 volts is connected to the primary, what is the voltage across the secondary?
7. A step-down transformer connected to a source of 240 volts AC delivers 6 volts at the secondary. If the current in the secondary is 10 amperes, what is the current in the primary?

8. A step-down transformer has 100 turns in the secondary and 1000 in the primary. If 100 volts is applied to the primary, what is the secondary voltage?
9. A step-up transformer has 100 turns in the primary and 1000 turns in the secondary. If 100 volts is applied to the primary, what is the secondary voltage?
10. A step-up transformer has 120 volts at 1 ampere applied to the primary. If the output is 240 volts, how much current can be drawn from the secondary when it operates at 100% efficiency?

Project: TRANSFORMER

The transformer can easily be wound, using commonly available tools, provided a plastic bobbin of the proper size is available. If not, a bobbin may be constructed of stiff cardboard. It serves to hold the wire in position while the laminations are placed inside the core.

You will need the following materials.

Cord, 36″, with AC cap
42 Laminations, E and I, 3/4″
Bobbin, Nylon, with 3/4″ core
Wire, on plastic bobbin, #30, enough for 1050 turns
Wire, on plastic bobbin, #26, enough for 105 turns
Insulation tubing, 15″
2 Transformer mounting brackets
Tape, electrician's, black plastic, 1 roll

Construction

Begin construction by winding the transformer with two separate coils. First you have to decide what voltage you need for your purposes. In our case, we have chosen a transformer that will step down 120 volts AC to 12 volts AC. Your ratio may vary according to the output voltage needed for your purposes. The transformer consists of two coils of wire wound around a plastic bobbin that is placed on a laminated steel core. The first winding, called the primary, consists of 1050 turns of #30 AWG (American Wire Gage) magnet wire. Magnet wire is made of cop-

per and is covered with a thin, tough coating of varnish/plastic insulation. The second winding is called the secondary. It is wound on top of the primary and consists of 10 turns of #26 AWG magnet wire. The wire used for the secondary is similar to that used for the primary, except that the former is of larger diameter. This means it will carry more current. The two windings are separated by insulation tape that ensures complete isolation (lack of electrical continuity) between them. This is needed for safe operation.

Use a small block of wood to fit into the core of the bobbin. Its size may vary according to the bobbin you are going to use. Figure 1 shows how the bobbin is held in place for winding by use of the wood block. Insert the 1/4″-×-2″ machine screw into the hole drilled through the wood block and tighten it in place with a nut. The protruding part of the machine screw can be fit into the hand-drill chuck, Figure 1.

Examine Figure 2 and see how the wire is doubled for strength before starting to wind the coil. Since #30 wire is quite small in diameter and is made of soft copper, it will break if abused, kinked, twisted, or bent excessively. Work with care. Figure 3 shows how to slip a piece of

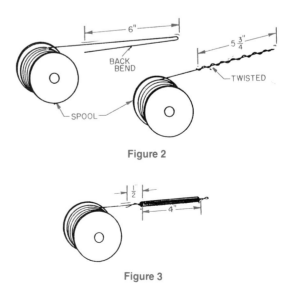

Figure 2

Figure 3

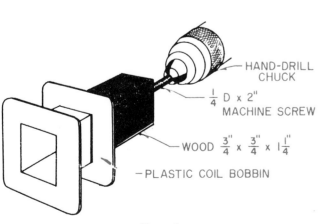

HAND-DRILL CHUCK

$\frac{1}{4}$ D x 2″ MACHINE SCREW

WOOD $\frac{3}{4}$″ x $\frac{3}{4}$″ x $1\frac{1}{4}$″

PLASTIC COIL BOBBIN

Figure 1

"spaghetti" insulation over the twisted wire. Then, in Figure 4 you can see how tape is used to secure the insulated wire in place before starting to wind the coil.

Next, wind the 1050 turns onto the coil very carefully. Don't lose count or you'll have to start over. It is not necessary to "layer wind" the primary coil evenly. However, taking a little time to keep the winding level neat will yield a much-higher-quality transformer.

Construction Hints

If you use the hand drill for winding the coils, be sure to account for the gear ratio between the crank and the chuck. One turn of the crank may give four or five turns at the chuck.

Check Figure 5 for the finish winding of the coil. Then Figure 6 shows how to put the insulation onto the twisted wire end to make it a finished primary. Figure 7 shows how tape is wound around the primary winding.

Next, wind the secondary, treating the start and ending windings the same as in the primary. Twist or double over the wire so it will be stronger and will take the necessary handling later, Figure 8.

Attach the line cord to the leads coming from the primary. Solder and tape. Use several turns of tape to tape the ends where connections were made to the body of the coil.

Place the laminations into the core, Figure 9. Note that the E laminations are placed in the core one at a time and then the I laminations are used to fill in the gaps after the E's have all been inserted tightly. The last lamination should fit quite snugly; otherwise, the transformer will produce a buzzing noise during operation.

Transformer Check

Set the ohmmeter on R × 1 or the lowest ohms range, and read the resistance of the primary winding. It should read about 15 to 20 ohms. Keep in mind that the ends of

the wires have had the insulation removed so that you can make electrical contact on the 1/4″ that has been treated. The secondary should read about 2 to 5 ohms. Then check the reading from one lead of the primary to one lead of the secondary. There should be an infinite reading.

Figure 6

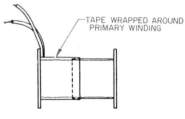

Figure 7

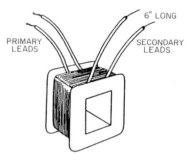

Figure 8

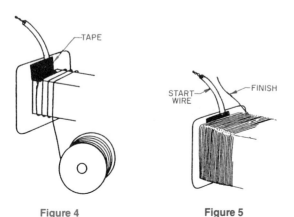

Figure 4

Figure 5

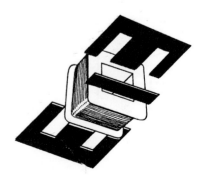

Figure 9

If there is a finite reading, there is a short somewhere between the primary and secondary coils. Also, check the reading from one lead of the primary coil to the laminated core. This reading should also be infinite. Touch one of the secondary leads and the laminated core. Again, this should read infinite. If not, one of the laminations has cut through the insulation on the coil and will, in some cases, cause the coil to "self-destruct" if plugged into 120 volts. Plug in, and test for 12 VAC.

Make a bracket for holding the transformer if one is not furnished with the covers for the transformer, Figure 10.

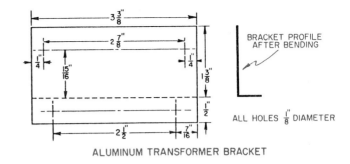

ALUMINUM TRANSFORMER BRACKET

Figure 10

INDUCTORS AND INDUCTANCE

INTRODUCTION

In the last chapter, you learned that transformers operate through mutual inductance. In this chapter, you are ready to learn about the nature and uses of inductance.

This chapter describes how inductors are manufactured, labelled, and used. Information is provided on the types of circuits in which inductors are included. Symbols for various types of inductors are presented.

There is also information about a number of devices that use the principles of inductance.

8.1 TYPES OF INDUCTORS

The property of a coil that *opposes* any change in circuit current is called **inductance**. Inductance is abbreviated *L*.

Coils are also known as **inductors** or **chokes**. The inductance of a coil is measured in **henrys** (*H*). One henry is the *amount of inductance present when a current variation of one ampere per second results in an induced electromotive force of one volt*. Small coils are measured in *millihenrys* (thousandths) or *microhenrys* (millionths), Figure 8-1.

Inductors or chokes are used in circuits to suppress the flow of AC without affecting the flow of DC. Inductors are diagrammed in circuits through use of the symbol

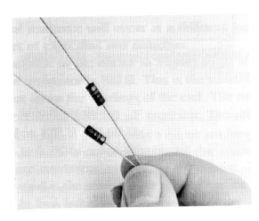

Figure 8-1. Small inductors use resistor color-code bands.

shown in Figure 8-2. The two lines drawn above the symbol for the coil represent an iron core. Actually, the core is usually inserted inside the coil. This increases inductance.

The symbol for a radio-frequency inductor is shown in Figure 8-3. Note that this inductor has no metal core. This means it uses air for its core. Most radio-frequency inductors use air cores. They are known as **radio-frequency chokes** (*RFC*).

There are also variations of these two types of inductors. One variation involves the use of a *powdered-iron core*. An inductor of this type is diagrammed in Figure 8-4.

Figure 8-2. Symbol for iron-core inductor.

Figure 8-3. Symbol for air-core choke.

Figure 8-4. Symbol for powdered-iron-core choke.

Many types of *variable inductors* are also made. Diagrams for several of these are shown in Figure 8-5. Figures 8-6, 8-7, and 8-8 show the actual appearance of a number of types of inductors.

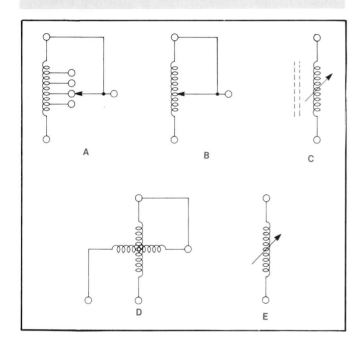

Figure 8-5. Symbols for variable inductance with
A. tapped coil, **B.** slider contact on coil,
C. adjustable powdered-iron slug, **D.** variometer,
E. variable inductance (usual symbol).

Figure 8-6. Radio-frequency choke for transmitter. (National)

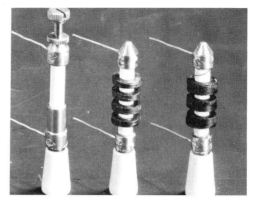

Figure 8-7. Radio-frequency chokes. (National)

Figure 8-8. Air-core or radio-frequency chokes and transformers. (Spaulding)

8.2 METHODS OF CHANGING INDUCTANCE

Inductance of a coil depends on four factors:

- Inductance changes with the *number of turns*. The more turns a coil has, the greater is its inductance. The relationship is that *inductance is proportional to the square of the number of turns.*
- Inductance varies with the *diameter of the coil*. The wider the cross section of a coil, the higher is the inductance.
- Inductance varies with the **permeability** *of the core* material. The more permeable the core, the better are its magnetic properties. Higher permeability leads directly (proportionally) to higher inductance.
- Inductance varies with the *length of the* **coil**. The shorter the coil, the higher is the inductance. Coil length and inductance are said to be inversely proportional. This means that there is a direct relationship: As length gets shorter, inductance increases. Even if a shorter coil had the same number of turns as a longer coil, its inductance would be higher.

Change any of these factors and you change the inductance. Note that the factor of turns varies by the square, whereas length is inversely proportional. These conditions can be important in designing a coil or transformer.

8.3 SELF-INDUCTANCE

The ability of a conductor to induce voltage in itself when the current changes is **self-inductance**. This ability can be valuable. Self-inductance is measured in henrys.

Self-inductance is produced by varying current in a coil. The current produces a magnetic field around each turn of the coil. The field around each turn then cuts across other turns. This makes for a moving magnetic field. Voltage is induced by this moving field.

The magnetic field and the voltage that creates it move in *opposite directions* (by Lenz's law). This generates what is called a **counter-electromotive force (cemf)**.

The cemf effect is diagrammed in Figure 8-9. In Figure 8-9A, a current increases rapidly from zero to maximum. This causes the magnetic field to expand. The expanding magnetic field produces a cemf that moves away from the input current. The cemf cuts across windings in the path of the current. The increasing cemf opposes the increase in current in the circuit.

In Figure 8-9B, the circuit is broken by a switch. When this happens, the magnetic field collapses. At this time, the current in the circuit changes from its maximum value to zero. The collapsing field induces voltage across the coil. The induced voltage opposes the decrease in current. The current is thus prevented from dropping quickly to zero. The gradual decline in current level due to self-inductance is illustrated in Figure 8-10.

8.4 INDUCTANCE-RESISTANCE TIME CONSTANT

The time required for a circuit to reach a level of 63.2 percent of its current level is called the **time constant**.

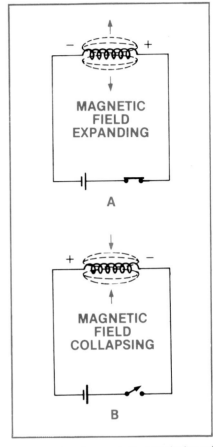

Figure 8-9. Diagram showing cemf effect producing self-inductance.

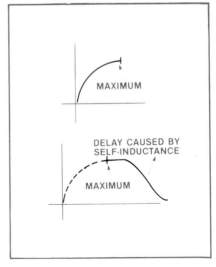

Figure 8-10. Diagrams showing time lag produced by an inductor.

The time constant for a circuit is calculated with the following formula:

$$\text{Time (seconds)} = \frac{\text{Inductance (henrys)}}{\text{Resistance (ohms)}} \qquad \text{Or: } T = \frac{L}{R}$$

In this equation, T indicates the time required for the current to reach 63.2 percent of its maximum. In each time constant thereafter, 63.2 percent of the remaining current is added. During each new time constant, the added current is equal to 63.2 percent of the difference between the existing and maximum levels.

To illustrate, look at Figure 8-11. This shows a circuit with 1 A of total current. Resistance is 100 Ω. Emf is 100 V. Inductance is 10 H. The time constant for this circuit is calculated as follows:

$$T = \frac{L}{R} \qquad T = \frac{10 \text{ H}}{100 \text{ }\Omega} \qquad T = 10 \div 100 \qquad T = 0.1 \text{ sec.}$$

The time constant is one-tenth (0.1) of a second. This is the time required for the circuit to realize 0.632 A of current, or 63.2 percent of total current.

Figure 8-12 relates time constants and current flow for this circuit: In 0.2 seconds, the current level would be 0.865 A. For each time constant thereafter, 63.2 percent of the remaining current is added. This is equal to 63.2 percent of the difference between the point reached in the previous period and the total current in the circuit. Thus, Figure 8-12 shows that in 0.3 seconds, there would be 0.95 A; in 0.4 seconds, 0.982 A; and in 0.5 seconds, 0.993 A. These time delays are created because the coil opposes passage of current.

As the values of inductance increase, the time constant becomes longer. If a shorter time is desired, a smaller value of inductance is placed in the circuit.

Resistance also helps to determine the time constant. If inductance of a circuit or coil is fixed, the time constant is increased by decreasing resistance. The time constant is decreased by increasing resistance.

Remember, a coil opposes increases or decreases in current. This is valuable in filter circuits. In this situa-

tion, inductance protects devices against sudden increases or drops in electrical current.

An inductor, in effect, creates a *lag* or *delay* between voltage and current. To illustrate, consider a purely inductive AC circuit. This is one with an inductor only. In such a circuit, **current lags** voltage by 90 degrees. This action is illustrated in Figure 8-13. This diagram assumes an inductive circuit without resistance. When the circuit is closed, voltage is present immediately across the inductor. The current takes time to build to its full value. The rate of lag is equal to one-quarter of an AC cycle, or hertz. When the conditions shown in Figure 8-13 are present, the voltage and level are said to be *out of phase*.

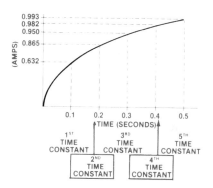

Figure 8-12. Graph of L/R time constant.

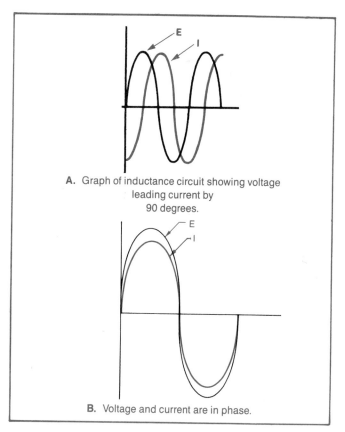

A. Graph of inductance circuit showing voltage leading current by 90 degrees.

B. Voltage and current are in phase.

Figure 8-13.

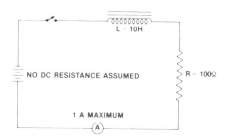

Figure 8-11. Diagram of inductance–resistance circuit.

REVIEW QUESTIONS FOR SECTION 8.4 ____

1. What is the time constant of an inductance circuit?
2. How is the time constant for an inductance circuit calculated?
3. What does voltage leading current mean?
4. What happens when current and voltage are out of phase?

8.5 MUTUAL INDUCTANCE

A condition in which two circuits share the energy of one circuit is called **mutual inductance**. The energy in one circuit is transferred to the other circuit. The two coils have a mutual inductance of 1 H if a current change of 1 A per second in one coil induces 1 V in the other coil. Magnetic lines of flux cause a coupling between the circuits.

A mutual-inductance circuit is diagrammed in Figure 8-14. The symbol at the left of this diagram indicates an alternating-current source. Coil L_2 is near L_1. The magnetic flux surrounds the wires making up L_2. This induces a voltage across the coil L_2.

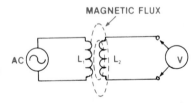

Figure 8-14. Diagram of mutual-inductance circuit.

As soon as the magnetic field is built up by the AC, the current flow is reversed. The magnetic field begins to collapse. As the magnetic field collapses, it also induces a voltage in L_2. This **induced voltage** moves in the opposite direction from when it built up to a maximum.

L_2 is part of a circuit with an electrical load. Current flows through the load. Current in the circuit of coil L_2 also produces a magnetic field. This is in opposition to the field that induced the current in L_2. If a heavier load is placed in the circuit with L_2, a greater opposition is presented to the induced current. That means L_1 must draw more current from its power source. It needs this additional current to meet the increased opposition provided by the increased load across L_2. There is no load

on L_1 until L_2 has a load connected across its terminals. This is typical transformer action.

Inductors used for mutual inductance may have two types of cores. These are illustrated in Figure 8-15. The types of cores are air and iron.

As you read earlier, air-core coils are used in circuits with radio frequencies. These frequencies cannot be heard by people. Iron-core coils are used in circuits containing audio frequencies. Audio frequencies can be heard.

Air cores are the basis for figuring coupling between coils. An iron core concentrates lines of force. This increases the inductance. On the other hand, copper, when used as a core, decreases inductance. Air cores are the constant used for comparison.

REVIEW QUESTIONS FOR SECTION 8.5 ____

1. What is mutual inductance?
2. How does mutual inductance affect circuit operation?

8.6 INDUCTIVE REACTANCE

An inductor reacts to alternating current. When alternating current passes through an inductor, a **phase shift** occurs. Voltage and current that were *in phase* become out of phase. Phase shifting occurs repeatedly. (Phase shifting causes the current to lag behind the circuit voltage in an

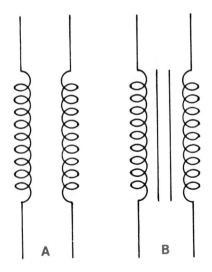

Figure 8-15. Symbols for coils with mutual inductance: **A.** air core, **B.** iron core.

inductor.) This results from the opposition to AC by an inductor. This opposition is referred to as *reactance*. Since the opposition takes place in an inductor, it is called **inductive reactance**.

This type of opposition, or reactance, does not affect DC current. The only opposition to DC comes from resistance of conductors.

Measurement of Inductive Reactance

Inductive reactance is measured in ohms. This is because it opposes alternating-current flow. The symbol for inductive reactance is X_L. The X represents reactance. The L represents inductance. When used together, they become X_L.

A number of factors determine X_L. One is the frequency of alternating current, which affects reactance. Another is the size of the inductor. The formula used to calculate X_L is:

$$X_L = 2\pi fL$$

In this equation, f is frequency, measured in hertz. L is inductance, measured in henrys. π is a standard mathematical term with a value of approximately 3.14 (rounded from 3.1415927). So, 2π equals 6.28. Thus, if either frequency (f) or inductance (L) is increased, inductive reactance (X_L) also increases. If either of these factors is reduced, inductive reactance also becomes smaller.

The effects of inductive reactance are illustrated in Figure 8-16. Note that a new symbol is used. This symbol is at the right of each drawing. It indicates a lamp in a circuit. Alternating current is presented in Figures 8-16A and B. Direct current is shown in Figures 8-16C and D.

Resistance of the light bulb is 100 Ω in all circuits. All circuits have 100 V applied. Therefore, use of Ohm's law indicates a current of 1 A for the circuits without inductance (A and C). Figure 8-16B has an inductor with a reactance of 10,000 Ω. The current in Figure 8-16B is limited to 0.01 A, or one-hundredth of its value without inductance. The low current means the light bulb in B is dimly lit. This means, in turn, that the AC is meeting opposition other than the resistance of the lamp. So, the low light level of the lamp results from inductive reactance.

In Figure 8-16D, the bulb is in a DC circuit with an inductor. The resistance of the coil to DC is low, approximately 1 Ω. So, the bulb glows brightly.

Figures 8-16B and D are identical except for power source. Thus, the difference in results comes from the effect of inductive reactance on AC.

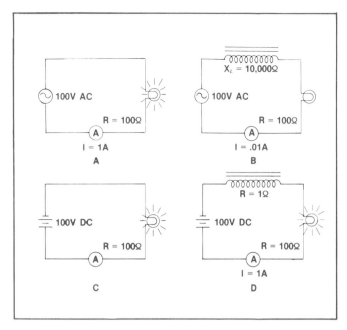

Figure 8-16. Schematics showing effects of inductive reactance.

Inductive Reactances in Series

Just as resistors can be placed in series, so can inductive reactances. The total value of X_L in a circuit with two or more inductive reactances is found by adding the X_L values. For example, Figure 8-17 shows two inductances, of 10 Ω and 20 Ω. So, the total inductive reactance of this circuit is 30 Ω. This assumes there is no coupling between the two inductors.

Mutual inductance may be a factor in series circuits. If so, the value of the mutual inductance must be added or subtracted. The formula for inductance in series with mutual inductance (L_M) included is $L_T = L_1 + L_2 \pm 2L_M$.

Inductive Reactances in Parallel

Inductive reactances in parallel are treated as resistors in parallel. That is, for two reactances, the sum of the reactances is divided into the product of the reactances.

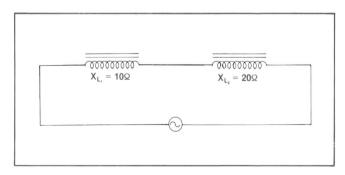

Figure 8-17. Schematic showing inductive reactances in series.

117

As an example of a two-reactance circuit, see Figure 8-18. The reactances in parallel are rated at 10 Ω and 20 Ω. The formula and solution are as follows:

$$X_{L_T} = \frac{X_{L_1} \times X_{L_2}}{X_{L_1} + X_{L_2}}$$

$$X_{L_2} = \frac{10 \times 20}{10 + 20} = \frac{200}{30} = 6.667 \ \Omega$$

The inductive reactance of the circuit in Figure 8-18 is 6.667 Ω.

A rule for resistors in parallel that also holds true for inductive reactance in parallel circuits is: *Total inductive reactance is always smaller than the rating of the smallest inductive reactance in a circuit.*

For circuits with three or more inductive reactances in parallel, the total opposition is figured by using the same formula applied for resistors:

$$\frac{1}{X_{L_T}} = \frac{1}{X_{L_1}} + \frac{1}{X_{L_2}} + \frac{1}{X_{L_3}}$$

In this equation, X_1 is the total inductance. X_{L_1} is the first inductive reactance, X_{L_2} is the second inductive reactance, and so on.

A circuit with three inductors in parallel is diagrammed in Figure 8-19. Inductive reactances are 10 Ω, 20 Ω, and 30 Ω. Total inductance is calculated as follows:

$$\frac{1}{X_{L_T}} = \frac{1}{10} + \frac{1}{20} + \frac{1}{30}$$

$$\frac{1}{X_{L_T}} = \frac{6}{60} + \frac{3}{60} + \frac{2}{60}$$

$$\frac{1}{X_{L_T}} = \frac{11}{60}$$

$$\frac{X_{L_T}}{1} = \frac{60}{11} = 5.4545 \ \Omega$$

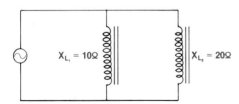

Figure 8-18. Schematic of circuit with inductive reactances in parallel.

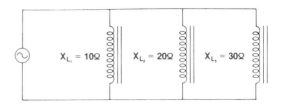

Figure 8-19. Schematic of circuit with three inductors in parallel.

Safety Tips

1. Do not complete a circuit, even with low voltage, if an inductor is in a path to a power source. It can shock you with the inductive kick that a coil produces.
2. Never touch any part of an inductor when it is energized.
3. Strong magnetic fields can damage watches and other delicate devices. Keep them away from coils and electromagnets.
4. Keep magnetic recording tape away from a coil with AC. It can erase the tape.

Uses of Inductive Reactance

The major use of inductance is to provide a minimum reactance for low frequencies. Inductors produce high opposition to higher frequencies.

One specific use of inductance is in filters. **Filters** are used when certain frequencies are desired and others have to be avoided. An inductor is used that has an X_L that passes certain frequencies and opposes others.

The main use for inductive reactance is in electronic circuits. Such circuits, along with capacitors, tune in certain frequencies and reject others. The inductor passes low frequencies and opposes high frequencies. An example is the tuner for a radio or TV set.

REVIEW QUESTIONS FOR SECTION 8.6

1. What is inductive reactance?
2. What is a phase shift?
3. How is inductive reactance measured?
4. How do you determine inductive reactance in a series circuit?
5. How do you determine inductive reactance in a parallel circuit?

Activity: *INDUCTIVE REACTANCE (X_L)*

_____ OBJECTIVES _____

1. *To study the property of inductance known as inductive reactance (X_L).*
2. *To become familiar with the way in which the addition of coils to an inductor affects inductance and inductive reactance.*

MATERIALS NEEDED

1 Power supply, 0–100 volts AC

1 Voltmeter, 0–150 volts AC

1 60-Watt lamp and socket

2 Coils, 300 turns each

1 3/4" laminated core and inductor

6 Wires for connections

PROCEDURE

1. Connect the circuit shown in Figure 1. CAUTION: Inductive circuits can be quite dangerous in some instances; consider the effect of the inductance when the circuit is suddenly interrupted.

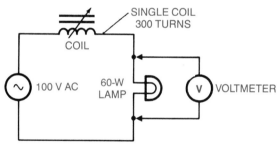

Figure 1

2. Set the voltmeter to read 0–100 volts AC. *Turn on the power supply.* Adjust the output of the power supply to 100 volts AC. With the laminated "I" bar completely covering the poles of the inductor, Figure 2, what is the voltage across the 60-watt lamp?
3. Slowly remove the "I" bar by sliding it off the poles of the inductor. What happens to the voltage across the lamp as the inductance of the inductor is reduced in this manner?
4. *Turn off the power supply.* Disconnect the circuit you have been using and connect the circuit shown in Figure 3.

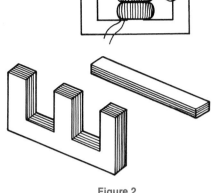

Figure 2

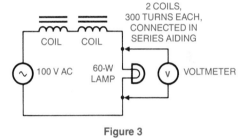

Figure 3

5. The circuit you have just connected has two coils (each with 300 turns) connected in series—in effect, doubling the number of turns on the coil of the inductor. The voltmeter should again be set to read 0–100 volts AC. *Turn on the power supply.* Adjust the output to 100 volts AC. With the laminated core "I" bar completely covering the poles of the inductor, what is the voltage across the 60-watt lamp? How does this compare with the voltage in Step 2 when only one coil was used? What would account for this difference?
6. *Turn off the power supply.* **DO NOT disconnect the circuit you have been using.** Connect the circuit shown in Figure 4. (*Simply pick up one coil and turn it over and place it on top of the other one. This causes one coil to be wound in one direction and the other in the opposite direction.*) NOTE: This is the same circuit as Step 4, except that the two coils are now connected in *series opposing* instead of *series aiding.*
7. The voltmeter should again be set to read 0–100 volts AC. *Turn on the power supply.* Adjust the output to 100 volts AC. With the laminated core "I" bar com-

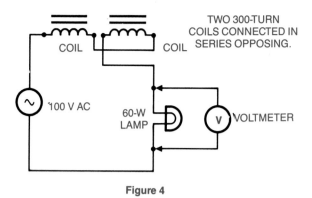

Figure 4

pletely covering the poles of the inductor, what is the voltage across the 60-watt lamp? How does the brilliance of the lamp compare with that in Step 5 and Step 2?

8. Slowly remove the core piece ("I" bar) by sliding it off the poles of the inductor. What happens to the voltage across the lamp as the inductance of the inductor is reduced in this manner? Is there any inductance effective when the two coils are connected in series aiding?

9. *Turn off the power supply.* Disconnect the circuit you have been using. Return all equipment and materials to their proper storage places.

SUMMARY

1. What is inductive reactance? In what way was inductive reactance evident in this activity?
2. Explain the terms *series aiding* and *series opposing.*

SUMMARY

The property of a coil that opposes any change in circuit current is called inductance. Inductance is abbreviated *L.* Coils are also known as chokes and inductors. Inductance is measured in henrys (H).

Inductors are used in circuits to slow down the flow of current. Inductors have air cores, iron cores, or powdered-iron cores. They operate on radio frequencies and audio frequencies. Four factors determine the inductance of an inductor. The ability of a conductor to induce voltage in itself when the current changes is self-inductance. Self-inductance is measured in henrys.

The abbreviation of counter-electromotive force is cemf. The inductive–resistance time constant is equal to the inductance divided by the resistance. The time constant is 63.2% of maximum current. An inductor creates a delay or lag in a circuit's current as compared to its voltage. It creates a phase angle between voltage and current.

Mutual inductance occurs when two circuits share the energy of one circuit and the energy in one circuit is transferred to the other.

Inductive reactance is equal to the frequency times the inductance times 2π. The result is measured in ohms, but an ohmmeter is not capable of measuring inductive reactance. Inductive reactances in series and in parallel are treated the same as if they were resistances in series and parallel.

USING YOUR KNOWLEDGE

1. Draw the symbol for an air-core inductor.
2. Draw the symbol for an iron-core inductor.
3. Draw the curve showing the functioning of an inductor when current is first turned on.
4. Draw the curve showing the functioning of an inductor when the power is removed or turned off.

KEY TERMS

inductor	henry
inductance	cemf
choke	current lag
coil	induced voltage
permeability	phase shift
self-inductance	inductive reactance
mutual inductance	filter

PROBLEMS

1. What is the inductive reactance of an inductor of 1 henry connected to a source of 60 Hz AC?
2. An inductor has an inductive reactance of 3769.911185 ohms in a 60-Hz circuit. What is the inductance?
3. Two inductive reactances of 15 ohms and 290 ohms are connected in series. What is their total reactance?
4. Two inductive reactances of 20 ohms and 30 ohms each are connected in parallel. What is the total reactance?

5. What is the inductive reactance of an inductor of 100 millihenrys (mH) at a frequency of 10 kHz?
6. What is the inductive reactance of a circuit that has 10 MHz with an inductance of 100 microhenrys?
7. What is the inductance of a circuit that has an inductive reactance of 1884.455925 ohms at 60 Hz?
8. What is the inductive reactance of a circuit that has three coils of 60 ohms each X_L? The coils are connected in parallel.
9. What is the time constant if the coil has 5 henrys and a 100-ohm internal resistance?
10. What is the current in a DC circuit at the end of the first time constant if it has an inductor of 10 henrys with 1000 ohms internal resistance and 100 volts applied?

Project: JUMPING RING

The jumping ring (Figure 1) demonstrates a number of electrical properties. It shows how magnetism repels and how electricity produces heat. This device can suspend a ring in the air with no visible means of support. If held down near the coil, the ring gets hot quickly.

You will need the following materials.

1 Switch, push-button
2 Fahnstock clips
2 Screws, self-tapping, No. 4
1 Core, 5/16″ × 8″ long, steel
1 Coil form, 5/16″ hole, 1″ diameter, 1.25″ long
1 Base, wood
2 Washers, aluminum, 1″ O.D., 5/8″ I.D.
20 feet Wire, magnet, Formvar, No. 20

Operation

The jumping ring shows how electromagnetism can cause movement. Slip one aluminum washer onto the steel core, and, using a ruler, measure how high the washer rises when the circuit is closed. USE A 6-VOLT AC POWER SUPPLY. Complete the circuit several times and record the average height. Measure the height at which the aluminum ring floats when you leave the circuit closed for several seconds. Are the results of the two measure-

ments the same? Now hold the switch closed and hold the washer down near the coil. How long does it take the washer to get too hot to hold on to?

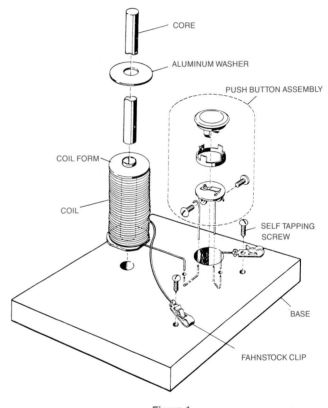

Figure 1

121

CAPACITORS AND CAPACITANCE

After studying this chapter, you will know

- *how a capacitor opposes any change in circuit voltage.*
- *how a capacitor operates and how it is constructed.*
- *that the capacitance of a capacitor is affected by three things.*
- *the unit of measurement for capacitance.*
- *the various types of capacitors and their symbols.*
- *what WVDC means and how it is used effectively in designing circuits.*
- *the facts associated with capacitors in series and parallel.*
- *how capacitors react to AC and DC circuits.*
- *how to calculate resistive-capacitive time constant for a circuit.*
- *the concept of capacitive reactance and be able to use formulas involving X_C.*
- *how Ohm's law is associated with circuits containing capacitance and capacitive reactance.*
- *why voltage lags current in a purely capacitive circuit.*

INTRODUCTION

Capacitors play important roles in building circuits. In this chapter, you will learn about capacitors and capacitance.

You will learn how to connect capacitors for use in circuits.

You will also master techniques for determining circuit needs. And you will learn to specify and select capacitors.

9.1 THE CAPACITOR

A **capacitor** is a device that opposes any change in circuit voltage. That property of a capacitor that opposes voltage change is called **capacitance**.

Capacitors make it possible to *store electric energy.* Electrons are held within a capacitor. This, in effect, is stored electricity. It is also known as **electrical potential** or an **electrostatic** field. Electrostatic fields hold electrons. When the buildup of electrons becomes great enough, the electrical potential is discharged. This process takes place in nature: Clouds build electrostatic fields. Their discharge is seen as lightning.

In summary, a capacitor performs two functions within a circuit:

- A capacitor opposes changes in circuit voltage.
- A capacitor stores electrostatic energy.

These properties have many uses in electricity.

A diagram of a capacitor is shown in Figure 9-1. Two plates of a conductor material are isolated from each other. Between the plates is a **dielectric**. Dielectric material does not conduct electrons easily. Electrons are stored on the plate surfaces. The larger the surface, the more area is available for stored electrons. Increasing the size of the plates, therefore, increases capacitance.

Action of a Capacitor

If a capacitor has no charge of electrons, it is uncharged. This happens when there is no voltage applied to the plates. An uncharged capacitor is shown in the drawing in Figure 9-2A. Note the symbol for a capacitor in this drawing. This is the preferred way to show a capacitor: a straight and a curved line facing each other. Note that the circuit has a DC current source and a three-position switch that is in the open position.

In Figure 9-2B, the switch has been closed to position 1. This causes current to be applied. A difference in

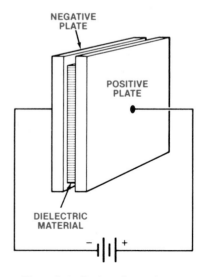

Figure 9-1. Design of capacitor.

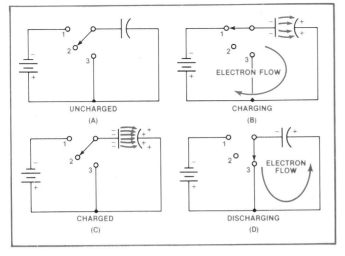

Figure 9-2. Schematic of capacitor circuits.

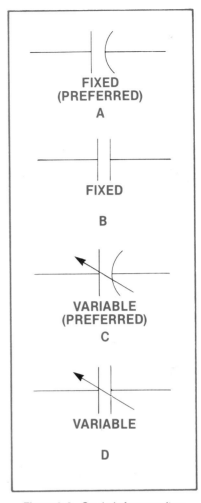

Figure 9-3. Symbols for capacitors.

potential is created by the voltage source. This causes electrons to be transferred from the positive plate to the negative plate. This transfer continues: 1) as long as the voltage source is connected to the two plates, and 2) until the accumulated charge becomes equal to the potential difference of the applied voltage. That is, charging takes place until the capacitor is charged.

In Figure 9-2C, the voltage has been removed. The switch is open. At this point, the potential difference, or charge, across the capacitor remains. That is, there is still a surplus of electrons on the negative plate of the capacitor. This charge remains in place until a path is provided for discharging the excess electrons.

In Figure 9-2D, the switch is moved to its third position. This opens the path for discharging the surplus electrons. Notice that the **discharge** path is in the opposite direction from the charge path. This demonstrates how a change in circuit voltage results in a change in the capacitor charge. Some electrons leave the excess (negative) plate. They do this to try to keep the voltage in the circuit constant.

To review: This ability to oppose a change in circuit voltage is called *capacitance*. The capacitor places stored electrons back into the circuit. This is an attempt to prevent a change in circuit voltage. Capacitance attempts to hold down increases in circuit voltage. When voltage in a circuit decreases, capacitance tries to hold it up. Capacitance, then, opposes voltage changes in either direction, up or down.

DC voltage varies only when current is turned on and off. Therefore, there is little capacitive effect in DC circuits. However, AC changes systematically. So, the capacitance effect is continuous in an AC circuit.

Symbols for capacitors are shown in Figure 9-3. Note that two opposite straight lines are acceptable. However,

a straight line opposed by a curved line is the preferred symbol.

REVIEW QUESTIONS FOR SECTION 9.1

1. What are the principal properties of a capacitor?
2. What is meant by *capacitor discharge*?
3. What is the function of the plates of a capacitor?
4. What is the function of a dielectric?
5. How does the size of plates affect capacitance?

9.2 CAPACITY OF A CAPACITOR

The two plates of a capacitor can be made from almost any material. The only requirement is that the material allow electrons to collect on its surface.

The dielectric between the plates of a capacitor is an insulating material. Some dielectrics are air, vacuum, wood, mica, plastic, rubber, bakelite, paper, and oil.

Capacitance may be fixed or variable. See Figure 9-4. Three factors determine capacitance:

- Area of the plates
- Distance between the plates
- Material used as a dielectric.

Area of Plates.   Area of the plates determines the ability of a capacitor to hold electrons. The larger the plate area, the greater is the capacitance.

Distance Between Plates.   Distance between plates determines the effect that the electrons on the plates have upon one another. There is an electrostatic charge. Electrons on the plates store energy when voltage is applied. That is, capacitance increases as the plates are brought closer together. Capacitance decreases as the plates are moved apart. This is demonstrated easily with a variable capacitor. Plates on these units can be moved out of mesh, or away from each other. When this is done, capacitance decreases.

Dielectric Material.   One of the effects of the dielectric material is determined by its thickness. The thinner the dielectric, the closer the plates will be. Thus, a thin dielectric helps to increase the capacitance.

In addition, some dielectrics have better insulating qualities than others do. This insulating property is referred to as a **dielectric constant**. Table 9-1 gives dielectric constants for a number of materials.

Breakdown Voltage

The voltage at which the dielectric of a capacitor no longer functions as an insulator is the *breakdown voltage*.

At this level, the capacitor permits the free flow of current. The result is known as a "shorted" capacitor. If the breakdown is partial, the result is a "leaky" capacitor.

Thus, the dielectric strength of the material used is important to the functioning of a capacitor. Dielectric strengths of a number of materials are shown in Table 9-2.

___ **REVIEW QUESTIONS FOR SECTION 9.2** ___

1. How does the area of the plates affect capacitance?
2. How does the distance between plates affect capacitance?
3. How does the dielectric material affect capacitance?
4. What is a dielectric constant?
5. What is breakdown voltage?
6. What is a shorted capacitor?
7. What is a leaky capacitor?

9.3 BASIC UNITS OF CAPACITANCE

A basic unit for capacitance is the **farad**. A farad is the size of a capacitor which has stored 1 coulomb of electrons. It has a potential difference of 1 V between its plates. The abbreviation for farad is *F*.

MATERIAL	DIELECTRIC CONSTANT (K)
Air or Vacuum	1
Rubber	2–4
Oil	2–5
Paper	2–6
Mica	3–8
Glass	8
Ceramics	80–1200

Table 9-1. Dielectric Constants

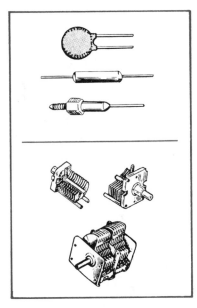

Figure 9-4. Fixed and variable capacitors.

MATERIAL	DIELECTRIC STRENGTH (VOLTS PER MIL)
Paper	1250
Mica	600–1500
Ceramics	600–1200
Rubber	450
Oil	375
Glass	200–250
Air or Vacuum	20

Table 9-2. Dielectric Strengths

use in circuits.
The most fre-
...nd **picofarads**.
...s one-millionth
...ro is the Greek
...crofarad is μF.
...symbol.)
One picofarad
...for *pico* is the
...s *pF*.
...marked *PF* or
...ed. The abbre-
...ame meaning.

...ION 9.3

...CITANCE

...e values. This
...markings on
...rself working
...nd drawings

...he farad. To
...farads, just

...ecimal point
...001 F. Also,

...mal point 12
...0000000001

...the decimal
...100,000 pF;

...iated 10 k.
...Usually, you
...ne capacitors.
...t means 0.01

...e left.
- μF to F   Move decimal six places to the left.
- F to μF   Move decimal six places to the right.
- μF to pF   Move decimal six places to the right.
- pF to F   Move decimal 12 places to the left.
- F to pF   Move decimal 12 places to the right.

1. How do you convert from farads to microfarads?
2. How do you convert from farads to picofarads?
3. How do you convert from microfarads to pico-farads?

9.5 TYPES OF CAPACITORS

Seven general types of capacitors most widely used are:

1. Air
2. Mica
3. Paper (oil filled)
4. Ceramic
5. Electrolytic
6. Tantalum
7. DIP.

The electrolytic capacitor is marked with a polarity, Figure 9-5.

A photo of one electrolytic capacitor is shown in Figure 9-6.

Other capacitors do not have polarity markings. However, the outside foil of each of these capacitors is marked with a band, as shown in Figure 9-7. This band is printed

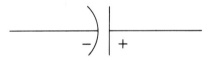

Figure 9-5. Symbol for an electrolytic capacitor. (Note − and + polarity.)

Figure 9-6. Electrolytic capacitor. (Sprague)

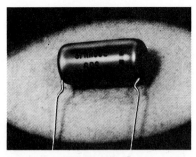

Figure 9-7. Nonelectrolytic capacitor with outside foil band. (Sprague)

on one end of the capacitor. The lead at this end is used in a circuit as the outside foil. That is, this lead is connected as though it had a negative value or to a ground connection on the chassis.

Mica Capacitors. Aluminum foil is used as the plate material in mica capacitors. Between the aluminum foil plates is a thin sheet of mica. Sometimes the mica is sprayed with a conducting paint. The paint then forms the plate on one side of the mica. Mica capacitors are usually sealed in bakelite, Figure 9-8. They have capacitances of 50 to 500 pF.

Paper Capacitors. Aluminum is also used as the plate material in paper capacitors, Figure 9-9. However, the plates are separated by a paper dielectric. The materials

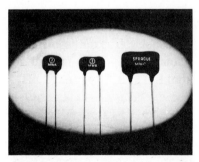

Figure 9-8. Mica capacitors. (Sprague)

Figure 9-9. Two paper capacitors in a metal can. (Sprague)

(paper and aluminum) are rolled into a cylindrical shape. A wire is connected to the alternate ends of the foil, Figure 9-10.

The cylinder is placed in a cardboard container. This is then sealed in wax. Sometimes the cylinder is placed in a mold and plastic is formed around it. Metallized paper capacitors are made of aluminum foil vaporized onto a paper dielectric. These are sealed in plastic.

Paper capacitors usually have a 0.001- to 1.0-μF capacitance.

Some capacitors use Teflon® or Mylar® as a dielectric. These capacitors have such features as high breakdown voltage and low losses. They operate without trouble over a longer period than do paper capacitors.

Oil-Filled Capacitors. Oil-filled capacitors are paper capacitors encased in oil. They are sometimes referred to as bathtub capacitors.

Values of oil-filled capacitors are usually not over 5 μF. The main advantages of these capacitors are sturdy construction and higher voltage-breakdown ratings. They are used in places where grease and oil are likely to be encountered.

Ceramic Capacitors. Ceramic dielectric materials make high-voltage capacitors. They undergo very little change in capacitance due to temperature changes. These small capacitors usually consist of a ceramic disc coated on both sides with silver. They are made in values from 1 pF up to 0.05 μF. Breakdown voltages of ceramic capacitors run as high as 10,000 V and more.

However, not all ceramic-disc-type capacitors are of the high-voltage type. In some transistor circuits, ceramic capacitors have low breakdown voltages. The maximum voltage is usually stamped below the capacitance marking. Miniature ceramic-disc capacitors are used on p.c. boards and can be adjusted or changed in value, Figure 9-11.

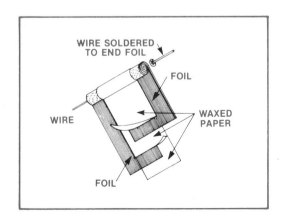

Figure 9-10. Construction of tubular paper capacitor.

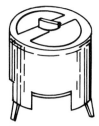

Figure 9-11. A miniature ceramic-disc capacitor.

Ceramic-Chip Capacitors

The newer *surface mount* printed circuit boards for computers and other electronic equipment utilize a differently configured capacitor, Figure 9-12. The inner electrode is multi-layered and the outside edges are easily soldered using a machine or other method to melt the solder on the surface of the board. These capacitors are used in TV tuners, video cameras, radio and tape recorders, electronic wristwatches, computers, and telecommunications equipment.

DIP Capacitors

It is very difficult to distinguish an integrated circuit chip (IC) from a DIP capacitor. They are both encased in *d*ual-*in*line *p*ackages (DIP), Figure 9-13. The capacitors are connected between the pins as shown. The DIP capacitors shown in Figure 9-13 are the two-pin type, which has only one capacitor, and the 16-pin, which contains eight capacitors. They are also available in other arrangements. The DIP monolythic ceramic capacitors have separate capacitor sections that are useful in electronics circuits for bypass and coupling applications.

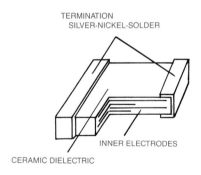

TERMINATION
SILVER-NICKEL-SOLDER

INNER ELECTRODES

CERAMIC DIELECTRIC

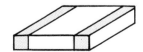

Figure 9-12. Layers in a ceramic-chip capacitor.

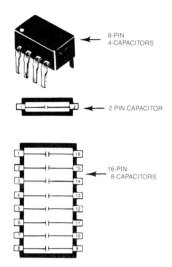

8-PIN
4-CAPACITORS

2 PIN CAPACITOR

16-PIN
8-CAPACITORS

Figure 9-13. DIP Monolythic ceramic capacitors.

REVIEW QUESTIONS FOR SECTION 9.5

1. Describe a mica capacitor.
2. Describe a paper capacitor.
3. Describe an oil-filled capacitor.
4. Describe a ceramic capacitor.
5. What is the capacitance range of mica capacitors?
6. What is the capacitance range of paper capacitors?
7. What is the capacitance range of ceramic capacitors?

9.6 ELECTROLYTIC CAPACITORS

There are two types of electrolytic capacitors: wet and dry.

The wet type uses a liquid **electrolytic**. It is rather large in size. But it has high capacitance. Wet electrolytics are generally used in transmitters and other large stationary electronic equipment.

The dry electrolytic is the one used in most electronic circuits, Figure 9-14.

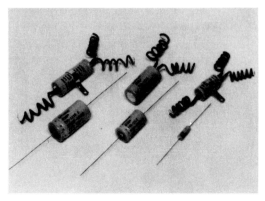

Figure 9-14. Dry electrolytic capacitors. (Sprague)

The electrolytic capacitor is used for capacitances of 1 μF up to 1 F. These units can be kept rather small in physical size for such large capacitances. This is done by using an oxide film as a dielectric, Figure 9-15. Larger sizes, in thousands of μF, are used in computer power supplies for filtering.

Computer Capacitor

Many people in electronics have long dreamed of having a source of power, other than a battery, to power devices that may require small currents for long periods of storage time. (Such is the case for the memory in computers or calculators.) This is now possible with the production of an electrolytic capacitor of sufficient size (1 farad). The 1-farad capacitor has long been thought of as a desirable item for many uses. It is now possible to package a capacitor in the 1-farad size in a 1.1-inch-diameter unit only 0.55 inch high, Figure 9-16.

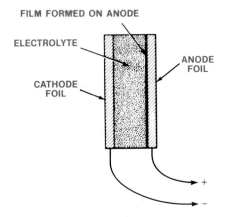

Figure 9-15. Electrolytic-capacitor construction.

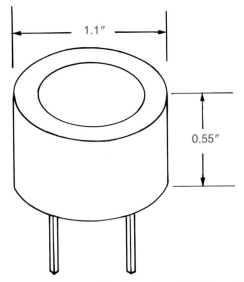

Figure 9-16. A 1-farad capacitor with dimensions of 0.55″ by 1.1″.

This energy source makes it possible to support digital-system backup applications without batteries. It has fast recharge time, easy interface, and virtually unlimited life. It is especially well suited for applications where the energy of a battery is not required, and reliability, long life, low cost, and simple design and implementation are of primary importance. The 64K-bit CMOS memory of a computer chip can retain data while dissipating typically minuscule power of 0.1 microwatt. This low power consumption can be supported by the 1-farad capacitor for several weeks.

Other applications for this type of capacitor are: relays, solenoids for starters, igniters, and actuators. Small motors and alarms for disc drives, coin metering devices, and security systems, as well as toys, can make use of a capacitor of this size. There are home appliances, such as TVs, microwave ovens, dishwashers, refrigerators, energy-management controls, personal computers, thermostats, vending machines, point-of-sale terminals, telephone autodialers, and programmable controllers, that can also use such a capacitor.

Forming an Electrolytic Capacitor

Consider the diagram in Figure 9-15. To produce this type of capacitor, special methods are used. DC voltage is applied to the electrolytic as part of the manufacturing process. This produces an electrolytic action. As a result, a molecular-thin layer of aluminum oxide with a thin film of gas is deposited. This gas is at the junction between the positive plate and the electrolyte.

The oxide film is a dielectric. There is capacitance between the positive plate (electrode) and the electrolyte through the film. The negative plate (electrode) provides a connection to the electrolyte. This film allows many layers of foil to be placed in a can or a cardboard cover. Capacitances produced in this way run into several microfarads.

Some electrolytic capacitors are made with more than one unit in a can. They may appear with a coded symbol stamped on the can or cardboard container. The symbol indicates the value. Figure 9-15 shows a diagram for an electrolytic capacitor with four units in a single can.

Usually, all of the capacitors in the same can will have their negative leads tied together. This connection is brought out of the can in a black lead. The negative lead may also be connected to the can, Figure 9-17.

Connecting Electrolytic Capacitors in a Circuit

Electrolytic capacitors have polarity. If these capacitors are not connected in a circuit properly, the oxide film is not formed. If connections are improper, no capacitance is available. Reverse electrolysis forms a gas when the

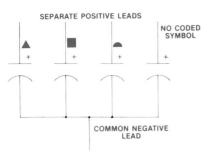

Figure 9-17. Symbol showing four coded electrolytic capacitors in one can.

Figure 9-18. Symbol showing AC electrolytic capacitors connected back to back.

capacitors are connected in opposite polarity. The capacitors become hot and may explode.

Electrolytic capacitors may "go bad" if they are left unused for too long.

Safety Tips

1. Do not pick up a charged capacitor.
2. Discharge a capacitor before you handle it. To do this, short the terminals with a screwdriver or shorting wire.
3. Never use a DC electrolytic on AC. It can **explode**.
4. Use safety goggles when working with electrolytics.

These units should be used in circuits in which approximately 75 percent of their working voltage is available. This keeps the capacitor formed.

Nonpolarized electrolytic capacitors can be used in AC circuits. The nonpolarized condition exists when two electrolytics are connected "back to back." This type of connection forms one AC electrolytic, as indicated in Figure 9-18. Connecting two electrolytics in series in this way reduces the capacitance.

REVIEW QUESTIONS FOR SECTION 9.6

1. What are the main uses for wet electrolytic capacitors?
2. What are the main advantages of dry electrolytic capacitors?
3. Describe the construction of a dry electrolytic capacitor.
4. What are the special requirements for connecting polarized electrolytic capacitors into circuits?
5. What can happen if a polarized capacitor is connected incorrectly?
6. How can you tell polarity of electrolytic capacitors?

9.7 TANTALUM CAPACITORS

Important improvements have been realized in recent years through use of new capacitor materials. One of these high-performance units is the **tantalum** capacitor.

There are two forms of tantalum capacitors. One is made directly from sintered-porous tantalum. It is known as the slug type. The other is the foil type, Figure 9-19.

The tantalum-foil capacitor consists of two tantalum-foil electrodes in a roll. The foil is thin, measuring 0.0005 to 0.001 inch [0.00127 cm to 0.00254 cm] in total thickness. Sometimes the foil is etched to increase the surface area. An oxide is formed on the surface of the positive foil during manufacture. This provides the capacitor dielectric. The thickness of the oxide is related directly to current applied to cause its formation.

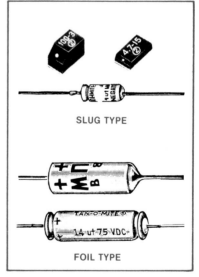

SLUG TYPE

FOIL TYPE

Figure 9-19. Tantalum capacitors.

REVIEW QUESTIONS FOR SECTION 9.7

1. What are the main advantages of tantalum capacitors?
2. What are the two forms of tantalum capacitors?

9.8 CAPACITOR TOLERANCES

General-purpose ceramic-disc capacitors usually have tolerances of $\pm$ 20 percent.

Paper capacitors usually have tolerances of ± 10 percent. Some have ± 5 percent.

Mica and ceramic tubular capacitors are used when closer tolerances are needed. Their tolerances range from ± 2 to ± 20 percent.

If very close tolerances are necessary, silver-plated mica units are used. These have tolerances of ± 1 percent.

Electrolytic capacitors usually have a wide tolerance. For instance a 40-μF electrolytic capacitor may have a tolerance of −10 percent and +75 percent. This means the range of capacitance can be between 36 and 70 μF. The lower minus tolerance helps assure that there is enough capacitance in a circuit to prevent damage.

Tantalum capacitors come in a range of tolerances, including ± 1, ± 5, ± 10, and ± 20 percent.

REVIEW QUESTIONS FOR SECTION 9.8

1. What are the usual tolerances for ceramic-disc capacitors?
2. What are the usual tolerances for paper capacitors?
3. What are the usual tolerances for mica capacitors?
4. What are the usual tolerances for ceramic tubular capacitors?
5. What are the available tolerances for silver-plated mica capacitors?

9.9 WORKING VOLTS, DIRECT CURRENT (WVDC)

The maximum safe operating (working) voltage of a capacitor in a DC circuit is identified as *working volts, direct current*, or *WVDC*. Above the WVDC level, a capacitor is expected to puncture and develop a short.

If the temperature in which a circuit operates reaches 60 degrees C or higher, the voltage rating is lowered.

Voltage ratings for mica, paper, and ceramic capacitors are usually 200, 400, and 600 V. Oil-filled capacitors have voltage ratings ranging from 600 to 7500 V.

As voltage ratings get high, the physical size of capacitors becomes larger. This is especially true for electrolytic capacitors. These units commonly have 25-V, 150-V, and 450-V ratings. At higher ratings, the units are much larger.

Electrolytic capacitors are also available for use in transistorized circuits. For these units, breakdown voltages are as low as 1.5 V. Thus, you should never use an ohmmeter with a 3-V power supply to check these capacitors.

Applied voltage across a capacitor should never exceed its WVDC rating. Electrolytic capacitors should be operated at about 75 percent of their rated voltage. This increases their useful life.

REVIEW QUESTIONS FOR SECTION 9.9

1. What does *WVDC* mean?
2. What is the WVDC rating of a capacitor?
3. What happens if applied voltage across a capacitor exceeds its WVDC rating?

9.10 CAPACITORS IN PARALLEL

To determine total capacitance for capacitors connected in parallel, add their values. This is the same formula as is used in finding total resistance in series circuits.

As an example, note the circuit in Figure 9-20. Two 100-μF capacitors are placed in parallel. So, total capacitance of the circuit is 200 μF.

In effect, adding a capacitor to another in a parallel circuit is the same as increasing the plate area, Figure 9-21.

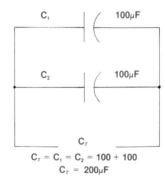

$$C_T = C_1 = C_2 = 100 + 100$$
$$C_T = 200\mu F$$

Figure 9-20. Schematic showing capacitors in parallel.

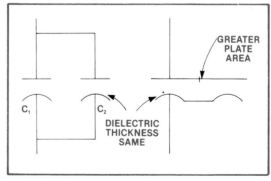

Figure 9-21. Schematic demonstrating that greater plate area increases capacitance.

___ REVIEW QUESTIONS FOR SECTION 9.10 ___

1. How do you determine total capacitance of a circuit with capacitors in parallel?
2. What happens to the effective plate area when capacitors are connected in parallel?

9.11 CAPACITORS IN SERIES

Connecting capacitors in series has the same effect as widening the separation between the plates of a single unit. This is illustrated in Figure 9-22.

If two equal capacitors are connected in series, the total capacitance of the circuit is equal to half of one of the units. For example, assume two 100-μF capacitors are connected in series. The total capacitance is equal to one-half the value of one of the capacitors. The capacitance of this circuit is 50 μF.

The formula for finding capacitance in series is the same as for resistance in parallel.

When there are only two capacitors in the circuit, the formula is:

$$C_T = \frac{C_1 \times C_2}{C_1 + C_2}$$

For two 100-μF capacitors in series, the formula would be applied as follows:

$$C_T = \frac{100 \times 100}{100 + 100} = \frac{10,000}{200} = 50$$

For three or more capacitors, the formula is:

$$\frac{1}{C_T} = \frac{1}{C_1} + \frac{1}{C_2} + \frac{1}{C_3} + \cdots$$

If a series circuit had capacitors at 25, 25, and 50 μF, the formula would be applied as follows:

$$\frac{1}{C_T} = \frac{1}{25} + \frac{1}{25} + \frac{1}{50}$$

$$\frac{1}{C_T} = \frac{4}{100} + \frac{4}{100} + \frac{2}{100} = \frac{10}{100}$$

$$C_T = \frac{100}{10} = 10$$

___ REVIEW QUESTIONS FOR SECTION 9.11 ___

1. How do you determine total capacitance of a circuit with capacitors in series?
2. What happens to the effective plate area when capacitors are connected in series?

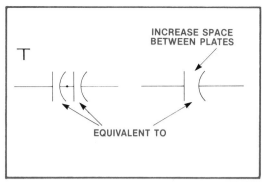

Figure 9-22. Schematic showing effects of capacitors in series.

9.12 RESISTANCE–CAPACITANCE TIME CONSTANT

When current is applied, a capacitor does not reach its full charge instantly. It takes time to reach full capacitance. This time **factor** is even greater if a resistor is placed in series with the capacitor. This situation is illustrated in Figure 9-23.

The time required for a capacitor to reach 63.2 percent of its full charge is its **time constant**. This is similar to the time constant for inductance. The time constant for a capacitor can be found through multiplication: Multiply the resistance of a circuit, in ohms, by its capacitance, in farads. You can also multiply if the resistance is in megohms and the capacitance is in microfarads. The product of this multiplication is time, in seconds.

This can be expressed as a formula:

T = RC

In this equation, T is time in seconds. R is resistance in ohms. C is capacitance in farads.

Under this formula, the greater the resistance, the larger is the time constant. Or, the greater the resistance, the longer it will take the capacitor to reach 63.2 percent of its full charge. This is because the resistance opposes the current required to charge the capacitor.

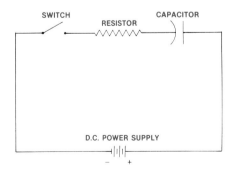

Figure 9-23. Schematic of series RC circuit.

131

As an example, look at the circuit in Figure 9-24. Resistance is 1,000,000 Ω. Capacitance is 10 μF. To find the time constant, you multiply 1,000,000 by 0.000010.

$$T = 1,000,000 \times 0.000010 = 10$$

The time constant is 10 seconds. That is, it will take 10 seconds for the capacitor to charge to 63.2 percent of its full charge. Remember that the full charge is the same as full voltage. The source voltage is 100 V. So, in 10 seconds the capacitor will be charged to 63.2 V.

The time constant for a full charge of this capacitor is shown in Figure 9-25. This indicates that it would take 50 seconds for a full charge.

The same time constant (RC) is used for the discharging of a capacitor. A discharge of 63.2 percent of the full charge occurs in the first time constant. It takes approximately five time constants for a capacitor to discharge fully.

A capacitor discharges whenever its voltage is greater than the applied voltage. The discharge continues at the time constant (RC) rate until the voltage equals the applied voltage. If voltage is increased, the discharge stops when applied and capacitor voltages are equal.

RC circuits are used as timing circuits, for filtering, and for wave-shaping.

REVIEW QUESTIONS FOR SECTION 9.12

1. How is the time constant for a capacitor determined?
2. If a circuit has a 1000-Ω resistor and a 25-μF capacitor, what is its time constant?
3. If a circuit has a 100,000-Ω resistor and a 100-μF capacitor, what is its time constant?

9.13 THE CAPACITOR IN A DC CIRCUIT

A capacitor reacts differently to AC than to DC. When DC is applied, a capacitor charges. The charge builds to the level of the source voltage. No current flow takes place once a capacitor is charged. A charged capacitor offers infinite (unlimited) opposition to DC. In DC circuits, it is assumed that a charged capacitor has infinite resistance.

If resistance is present in a DC circuit, it can permit a capacitor to discharge. Figure 9-26 shows the voltage source removed and a resistance placed across the capacitor. The capacitor discharges through the resistor. This discharge takes place in a direction opposite to the flow of the current that charged the capacitor.

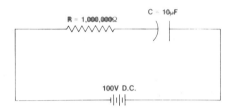

Figure 9-24. Schematic of series RC circuit.

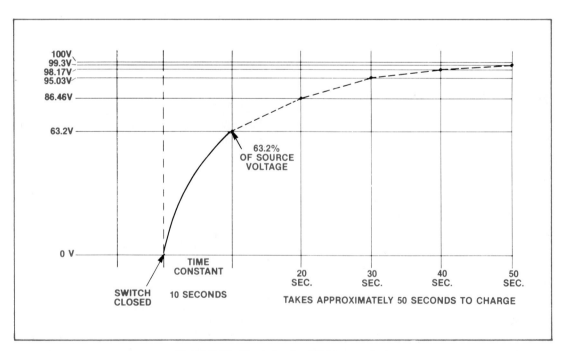

Figure 9-25. Chart showing RC time constant.

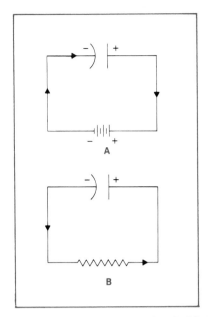

Figure 9-26. Schematic showing capacitors in DC circuits.

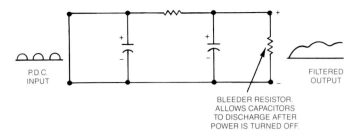

Figure 9-27. Pi-type filter.

BLEEDER RESISTOR
ALLOWS CAPACITORS
TO DISCHARGE AFTER
POWER IS TURNED OFF.

In a power-supply circuit the capacitor is used as part of a filter to smooth out pulsating DC and make it resemble the pure DC output of a battery, Figure 9-27.

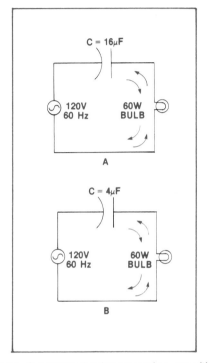

Figure 9-28. Schematic showing current in a capacitive circuit.

REVIEW QUESTIONS FOR SECTION 9.13

1. What is the effect of a capacitor upon a DC circuit?
2. How does a capacitor react to the flow of DC current?
3. What happens when a DC circuit has both a resistor and a capacitor?

9.14 THE CAPACITOR IN AN AC CIRCUIT

In an AC circuit, **capacitive reactance** is determined by capacitance and frequency. The behavior of a capacitor in an AC circuit is illustrated in Figure 9-28. Figure 9-28A shows an AC source at 60 Hz, a capacitor at 16 μF, and a light bulb in a single circuit. As the current changes, it moves in first one direction. Then, it returns to zero and moves to the maximum level in the opposite direction. As this happens, the bulb keeps glowing as the capacitor charges and discharges. At 60 Hz, the bulb is turned off and on at a rate of 120 times per second. This on–off cycle is too fast for the human eye to see. So, the bulb appears to be lit continuously.

In Figure 9-28B, there is a smaller capacitor. It has a rating of 4 μF. With this smaller capacitor in the circuit,

the bulb glows dimly. This indicates that there is less current in the circuit. The smaller capacitor is the only difference between Figures 9-28A and 9-28B.

Compare the performance of the two circuits. The current in the circuit with the 4-μF capacitor is less. The circuit with the 16-μF capacitor carries more current. Yet the voltage source is the same for both circuits. The difference lies in capacitor performance. The current remains at a high value for a longer time in the larger capacitor. Thus, since frequency is the same, higher capacitance means larger current flow.

Figure 9-29 demonstrates that a high frequency allows only a short time between each half-hertz for capacitor discharge. Figure 9-30 demonstrates that a lower frequency allows more time for capacitor discharge.

In an amplifier circuit, the capacitor is used to block DC and to allow the AC signal to pass.

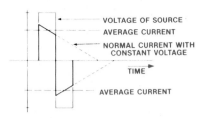

Figure 9-29. Diagram showing frequency–time relationship for high frequency in AC capacitor circuit.

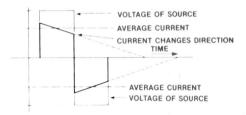

Figure 9-30. Diagram showing frequency–time relationship for low frequency in AC capacitor circuit. Lower frequency requires a longer time for the capacitor to charge.

In summary: The higher the capacitance in an AC circuit, the greater is the current flow. Also: The higher the frequency in an AC circuit, the greater the current flow will be.

REVIEW QUESTIONS FOR SECTION 9.14

1. What is the effect of a capacitor upon an AC circuit?
2. What is the frequency–time relationship at high frequency in an AC circuit with a capacitor?
3. What is the frequency–time relationship at low frequency in an AC circuit with a capacitor?

9.15 CAPACITIVE REACTANCE

The degree to which a capacitor opposes the flow of current is **capacitive reactance**. The capacitive reactance of a circuit is determined by capacitance and frequency. These two factors are used in a formula to determine capacitive reactance of any given circuit. The formula is:

$$X_C = \frac{1}{2\pi fC}$$

In this formula, X_C is the symbol for capacitive reactance in ohms. C is the symbol for capacitance in farads. And f is frequency in hertz.

For an example, assume the following values.

C = 10 μF or 0.00001 F.
f = 100 kHz or 100,000 Hz.
2π = 6.28.

$$X_C = \frac{1}{6.28 \times 100,000 \times 0.00001}$$

$$X_C = \frac{1}{6.28} = 0.16 \ \Omega$$

If a value of 60 Hz is substituted for f and all the other parts of the formula remain the same, capacitive reactance will be 265 Ω.

If a value of 60 Hz is used for f and 5 μF is used for C, capacitive reactance is 530 Ω.

Thus, these conclusions can be drawn:

• If capacitance decreases, the capacitive reactance increases for the same frequency.
• If capacitance increases, the capacitive reactance decreases for the same frequency.
• If frequency decreases, the capacitive reactance increases.
• If frequency increases, the capacitive reactance decreases, provided the capacitance remains the same.

Capacitive Reactance in Circuits

In either an AC or DC circuit, capacitive reactance has the same effect as resistance. That is, capacitors oppose the flow of current as well as opposing any change in circuit voltage.

Current opposition of capacitors is illustrated in Figure 9-31. Figure 9-31A shows a series circuit. In series circuits, X_C is calculated in the same way as resistance: The X_C ratings of all capacitors are added together. Thus, with X_C values of 50 Ω and 100 Ω, the total capacitive reactance for the circuit is 150 Ω.

Figure 9-31B shows a circuit with two capacitors connected in parallel. Again, the formula is the same as for resistance:

$$X_{CT} = \frac{X_{C_1} \times X_{C_2}}{X_{C_1} + X_{C_2}}$$

$$X_{CT} = \frac{100 \times 100}{100 + 100}$$

$$X_{CT} = \frac{10,000}{200} = 50 \ \Omega$$

Figure 9-31C shows a circuit with three capacitors connected in parallel. The formula for this calculation is also the same as for resistance:

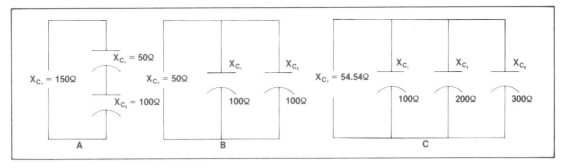

Figure 9-31. Schematic showing capacitive reactance in series and parallel circuits.

$$\frac{1}{X_{CT}} = \frac{1}{X_{C_1}} + \frac{1}{X_{C_1}} + \frac{1}{X_{C_2}}$$

$$\frac{1}{X_{CT}} = \frac{1}{100} + \frac{1}{200} + \frac{1}{300}$$

$$\frac{1}{X_{CT}} = \frac{6}{600} + \frac{3}{600} + \frac{2}{600}$$

$$\frac{1}{X_{CT}} = \frac{11}{600}$$

$$\frac{X_{CT}}{1} = \frac{600}{11}$$

$$X_{CT} = 54.54 \ \Omega$$

9.16 OHM'S LAW FOR CAPACITIVE REACTANCE

To apply Ohm's law for capacitive reactance in AC circuits, X_C is substituted for resistance. The formula is modified as follows:

$$E_C = I_C \times X_C$$

The other equations are also modified to substitute X_C for R.

To apply Ohm's law to capacitive reactance of circuits, refer to Figure 9-32. Figure 9-32A has one capacitor with X_C of 50 Ω and a voltage of 50 V AC. To determine current in amperes, divide capacitive reactance into voltage (50 divided by 50).

$$I_C = \frac{E_C}{X_C} = \frac{50}{50} = 1$$

The quotient, or answer, is 1. Figure 9-32A has 1 A of current.

Figure 9-32C has two capacitors connected in parallel. To find capacitive reactance for this circuit, you use the same formula as for resistors in parallel:

$$X_{CT} = \frac{X_{C_1} \times X_{C_2}}{X_{C_1} + X_{C_2}}$$

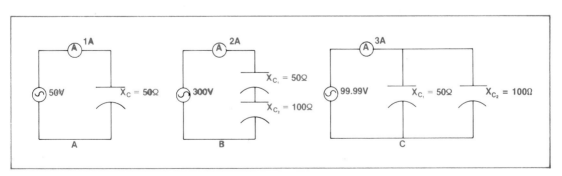

Figure 9-32. Schematic showing circuits with capacitive reactance.

Activity: CAPACITIVE REACTANCE

OBJECTIVES

1. *To investigate the operation of a capacitor placed in a circuit with AC and in a circuit with DC.*
2. *To study the capacitive reactance of a capacitor in an AC circuit.*

MATERIALS NEEDED

1 Voltmeter, 0–120 volts AC and 0–120 volts DC
1 Power supply, 120 volts AC and 120 volts DC
2 Capacitors, 1.0 μF at 400 WVDC (not electrolytic)
1 Lamp, 10-watt, 120 volts, and socket
1 Lamp, 25-watt, 120 volts, and socket
1 Lamp, 60-watt, 120 volts, and socket
5 Wires for connections

PROCEDURE

1. CAUTION: When you remove the circuit in this activity, make sure you discharge the capacitors. Connect the circuit shown in Figure 1.

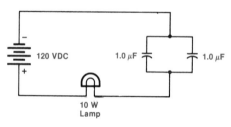

Figure 1

2. Set the voltage control of the power supply to the minimum output position. *Turn on the power supply.* Adjust the power supply for 120 volts DC. The two 1.0-μF capacitors have been connected in parallel to give 2.0 μF of total capacitance for this activity.

3. When the power supply is turned on, what happens to the lamp? Can DC flow through a capacitor? (If you can't see the lamp glow, substitute a NE–34 lamp and note how one plate glows on the charge cycle and other plate glows on the discharge cycle.)

4. Turn off the power supply. Discharge all capacitors by shorting across them. Disconnect the circuit you have been using and connect the circuit shown in Figure 2.

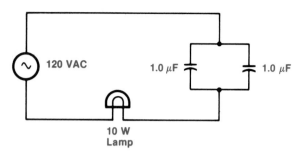

Figure 2

5. *Turn on the power supply.* Adjust the output voltage to 120 volts AC. What happens to the lamp? Can alternating current appear to flow through a capacitor?

6. Set the voltmeter to read 0–100 volts AC. Connect the probes across the lamp. What is the voltage across the lamp?

7. Remove the 10-watt lamp and place the 25-watt lamp in the socket. Does it glow as brightly as the 10-watt lamp? Measure the voltage across the 25-watt bulb.

8. Remove the 25-watt lamp and replace it with the 60-watt lamp. Does it glow as brightly as with the 10-watt lamp? Measure the voltage across the 60-watt lamp with the voltmeter. How does this compare with the reading across the 10-watt lamp? How does this compare with the reading across the 25-watt lamp?

9. The higher the wattage rating of an incandescent lamp, the more current it requires for proper operation. How does this fact, together with the effect of the reactance of the capacitor, affect the operation of the lamps?

10. *Turn off the power supply.* Discharge all capacitors by shorting across them. Disconnect the circuit. Place all equipment and materials in their proper storage places.

SUMMARY

1. How does the operation of a capacitor in DC circuits differ from the operation of a capacitor in AC circuits?
2. How does the capacitive reactance affect the flow of current in an AC circuit?

Apply this formula for the circuit in Figure 9-32C.

$$X_{CT} = \frac{50 \times 100}{50 + 100} = \frac{5,000}{150} = 33.33 \ \Omega$$

The answer is 33.33 Ω.

To verify the current in the circuit, you can then apply Ohm's law. To find current, you divide total capacitive reactance (33.33) into the applied voltage (99.99). The answer is 3 A.

REVIEW QUESTIONS FOR SECTION 9.16

1. In applying Ohm's law to capacitive circuits, what factor is substituted for resistance?
2. What is the formula for finding capacitive reactance?
3. What is the formula for determining total capacitive reactance in a circuit with two capacitors in parallel?

9.17 USES OF CAPACITORS AND CAPACITIVE REACTANCE

Capacitors and capacitive reactance are useful in electronic circuits with both AC and DC. X_C offers low opposition to AC. In a DC circuit, capacitance serves to block the flow of current. It is possible to use a capacitor to permit AC to flow while preventing the flow of DC. Thus, capacitors are valuable tools for circuits that use both AC and DC. This capability is used in coupling AC frequencies from one amplifier stage to another.

REVIEW QUESTIONS FOR SECTION 9.17

1. What type of opposition does capacitive reactance offer to AC current?
2. What type of opposition does a capacitor offer to DC current?
3. What is the effect of a capacitor on circuits with both AC and DC?

9.18 VOLTAGE LAGS CURRENT

In a circuit with a capacitor only, the voltage *lags*, or follows, the current by 90 degrees. This is because a purely capacitive circuit has no resistance.

This phase difference in a capacitive circuit is opposite from that of an inductive circuit. You will remember that voltage *leads* current in an inductive circuit.

Assume you have a purely capacitive circuit. There is no resistance. The voltage across the capacitor is present only after electron flow charges its plates. The power source is removed. At this point, assume voltage is at its maximum. The capacitor is charged. If a discharge path is provided, the current is maximum at the point where the voltage is zero. Zero voltage occurs when the capacitor is fully discharged.

A charging capacitor requires electron movement. This movement is from one plate to the other. Electron movement is greatest when the capacitor is charging or discharging. Electron movement reaches its minimum at the moment the capacitor is completely discharged.

Maximum voltage of a capacitor exists when it is fully charged. Charging a capacitor creates a deficiency of electrons on the positive plate. An excess of electrons builds on the negative plate. This difference in charge on the plates produces the potential difference. When voltage reaches its maximum, current flow stops. Thus, maximum voltage is matched by zero current. The reverse is also true: Maximum current is matched by zero voltage.

In a purely capacitive circuit, current and voltage are said to have a *phase relationship*. The phase difference is 90 degrees. Figure 9-33 illustrates this relationship in wave form.

Current and voltage relationships of capacitance and inductance can be combined in circuits. This combination of electrical properties makes possible radio and TV.

Determining Capacitive Phase Angle

Variations between voltage and current in a capacitive circuit depend upon the resistance placed in the circuit. The resistance in the circuit is due to the wire and other resistive elements.

The *ratio of capacitance to resistance* in a circuit determines the *phase angle*. The phase angle (or relationship) varies from zero degrees to 90 degrees. In a resistive–capacitive circuit, the phase angle indicates that voltage lags current. In a resistive–inductive circuit, the phase angle indicates that voltage leads current.

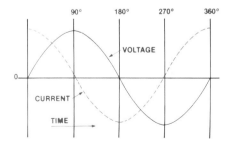

Figure 9-33. Diagram showing voltage–current waveform in purely capacitive circuit.

1. What is the relationship of voltage and current in a purely capacitive circuit?
2. How does the relationship of voltage and current in a capacitive circuit compare with the relationship of voltage and current in an inductive circuit?

SUMMARY

Capacitors are an important part of electrical and electronic circuits. Capacitors are made up of two plates with a dielectric in between. They indicate how much voltage they can stand without breaking down.

Capacitors in series behave like resistors in parallel when it comes to determining their total capacitance. When in parallel, the total capacitance is found by adding the individual values.

Capacitive reactance (X_C) is measured in ohms, but cannot be measured by an ohmmeter. It is treated as resistance when connections are in series and in parallel.

Capacitors are made in a variety of sizes and shapes, and with different working voltages as well as various cases. They may be either fixed in value or variable.

The basic unit of measurement of capacitance is the farad (F), with smaller units called microfarads (0.000001 F) and picofarads (0.000000000001 F).

Electrolytic capacitors have values over 1 microfarad and are usually polarized. Some nonpolarized electrolytics are used on AC circuits that require large capacitances.

Resistive–capacitive time constants are the product of resistance in ohms and capacitance in farads: $T = R \times C$. The time constant is 63.2% of the maximum voltage. Capacitors oppose any change in circuit voltage, therefore producing a difference in the phase angle between voltage and current in a capacitive circuit. There is a 90-degree lag in voltage in a capacitive circuit.

USING YOUR KNOWLEDGE

1. Find the total capacitance of two capacitors when you place them in parallel. They are 0.05 μF and 10 k pF.
2. Find the total capacitance of two capacitors connected in series. They are marked *0.05 μF* and *1.0 μF.*
3. Draw the symbol for a variable capacitor.
4. Draw the symbol for a fixed capacitor.

5. Sketch the inside of a typical capacitor. Show and label its parts.

KEY TERMS

discharge	microfarad
dielectric	picofarad
time constant	electrolytic
factor	tantalum
electrostatic	electrical potential
capacitance	capacitive reactance
capacitor	dielectric constant
farad	

PROBLEMS

1. How long does a capacitor of 1 μF connected in series with 10,000 ohms take to obtain a full charge?
2. How long does it take a capacitor that is fully charged to discharge if the capacitor is 10 μF and the resistor through which it is discharging is 10 ohms?
3. What is the voltage across a 2.0-μF capacitor in a series circuit composed of a capacitor and a resistor of 2 megohms when the circuit has been connected to a source of 100 volts for 8 seconds?
4. A capacitor of 1 μF and a resistor of 10 megs are connected in series. How long would it take a capacitor of this size to charge fully?
5. What is the time constant for a capacitor–resistor combination when the size of the resistor is 3.3K and the capacitor is 0.47 μF?
6. What is the capacitive reactance of a capacitor of 5.0 μF in a circuit with 60 Hz?
7. Two capacitive reactances of 20 ohms and 30 ohms are connected in series. What is the total capacitive reactance of the circuit?
8. What is the capacitance of a capacitor having a capacitive reactance of 2653.8 ohms when operated on 60 Hz?
9. What is the total capacitance of three capacitors in series if they are 20 μF, 30 μF, and 12 μF?
10. What is the working voltage of two capacitors connected in series if they are 1.0 μF at 15 volts DC, and 10 μF at 100 volts DC?
11. What is the working voltage of two capacitors connected in parallel if they are 1.0 μF at 1.5 volts DC, and 10 μF at 100 volts DC?
12. What is the total capacitance of two capacitors connected in parallel if they have capacitances of 20 μF and 0.05 μF?

Chapter 10

CIRCUITS WITH RESISTANCE, CAPACITANCE, AND INDUCTANCE

After studying this chapter, you will know

- *that impedance is the total opposition to current flow in a circuit.*
- *that impedance is the vector sum of resistance and reactance.*
- *how to solve problems involving RCL circuits in series and in parallel.*
- *the properties of a circuit with series and parallel resistances and reactances.*
- *how to utilize the cosine of* theta *to solve for the power factor.*
- *how phase angle is arrived at and used.*
- *the flywheel effect of a tank circuit.*
- *how to draw a series and a parallel resonant circuit using given values of each, and how to solve for voltage, current, impedance, and phase angle.*

INTRODUCTION

So far, you have covered resistance, inductance, and capacitance. These are three important components of circuits. Resistors have been combined with these three components only twice. This was done in the discussions of time constants in the chapters on inductance and capacitance.

In this chapter, you learn about circuits containing all three components—resistors, inductors, and capacitors. You learn how these components are used—primarily in tuning and filter circuits.

In short, this chapter will help you to build a foundation in the basics of electronics.

Resistance, Capacitance, and Inductance

Resistance, capacitance, and inductance affect both alternating and direct current. Any or all of these elements may be present in any circuit. Each causes changes in voltage and current. The relationship of current and voltage within circuits is vital. An understanding of resistance, capacitance, and inductance, therefore, is vital. With this understanding, you can determine relationships of current and voltage. This is the basis for understanding and using electronic circuits.

10.1 IMPEDANCE

The total opposition to the flow of current within a circuit is **impedance**. The symbol for impedance is Z.

In DC circuits, the only opposition is resistance. In AC circuits, opposition to voltage and current comes from a combination of factors. These include **resistance**, **inductance**, and **capacitance**.

The effect of combining these elements in a circuit is best introduced in steps. Start with resistance. Then add inductance and/or capacitance. When combined in a circuit, a resistor and an inductor and/or capacitor produces impedance. Impedance is a combination of effects. It consists of resistance and the reactance provided by an inductor and/or a capacitor.

It might seem reasonable to add the resistance to the reactance to obtain total opposition. But this is not the way a circuit functions. The resistance opposes both AC and DC in the circuit. Inductance and/or capacitance produces reactance. Reactance opposes AC only.

Remember, voltage leads current across an inductor and lags current across a capacitor. This is by a phase angle of 90 degrees. The effect on current is of special interest in this situation. The reactance effect is 90 degrees out of phase with the resistive effect. To combine these, it is necessary to obtain the vector sum of the two quantities.

Impedance is the *vector sum* of resistance and reactance. A **vector** is a line segment used to represent a quantity that has both direction and magnitude. Vectors are used to represent current. A single vector line, or curve, can show two dimensions of current. It can show direction of flow. It can also show the magnitude, or amount, of current flowing. A vector sum is a line representing the total of two or more vectors. Impedance is stated in terms of a vector sum.

Finding A Vector Sum Graphically

Inductive reactance causes the current in an inductor to lag 90 degrees behind the voltage. Therefore, a graphic representation of impedance for current can be drawn as shown in Figure 10-1. In this drawing, resistance is plotted on the horizontal line (AD). The length of line AD is proportional to the amount of resistance in the circuit. *Proportional* means that the quantity of resistance within the circuit is represented by line AD. Zero resistance is indicated at point A. The value of resistance in the circuit is indicated at point D. Using the same scale, the amount of inductive reactance is plotted on a line 90 degrees from the resistance line.

The vertical line AB represents inductive reactance. This is also proportional. Zero inductive reactance is represented by point A. The value of inductive reactance is indicated at point B.

Impedance (Z) is the vector sum of the two lines. It is represented by line AC. To find the value of C, begin by constructing a parallelogram. A parallelogram is a four-sided figure whose opposite sides are parallel and equal. In Figure 10-1, the dotted line CD is parallel to AB. And BC is parallel to AD.

C is the point where the parallelogram is completed. The value of Z is found by drawing a straight line between C and A.

The greatest value of this graphical method is that it helps you to visualize the procedure. In practice, there are faster methods for calculating impedance. This is a simple operation on most electronic calculators. Using a calculator becomes even easier once you can visualize and understand the values involved.

Impedance in a Series RL Circuit— Graphical Method

Figure 10-2 diagrams an AC circuit with a resistor and an inductor connected in *series*. The resistor has a value of 3 Ω. The inductor has a reactance of 4 Ω.

These values are indicated by the solid horizontal and vertical lines in Figure 10-3. In this illustration, a parallelogram has been formed and the value of Z has been plotted. The length of line BD is five units. This is

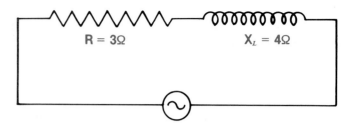

Figure 10-2. Schematic of RL circuit.

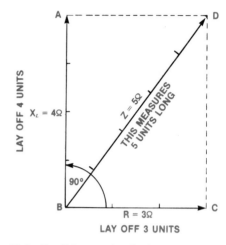

Figure 10-3. Parallelogram showing impedance of RL circuit.

proportional with the three units in BC and the four units in BA. Therefore, the impedance (Z) of the circuit is 5 Ω.

A Mathematical Solution

There is also a mathematical method for finding the value of Z. The equation used is known as the **Pythagorean Theorem**: *The square of the hypotenuse of a right triangle is equal to the sum of the squares of the other two sides.*

Refer to Figure 10-3. The lines representing resistance and inductance form a 90-degree angle, or right angle. The hypotenuse is the line in a right triangle that is opposite the right angle. Line BD is the hypotenuse of triangle ABD. The formula for the Pythagorean Theorem is:

$$Z^2 = X_L^2 + R^2$$

A simplified form of the same equation is:

$$Z = \sqrt{X_L^2 + R^2}$$

The equation can then be used as shown below to find the value of Z for the circuit in Figure 10-2.

$$Z = \sqrt{(4)^2 + (3)^2}$$
$$Z = \sqrt{16 + 9}$$
$$Z = \sqrt{25}$$
$$Z = 5 \ \Omega$$

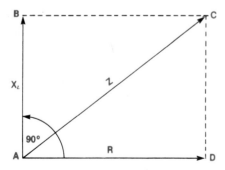

Figure 10-1. Diagram of impedance.

These equations illustrate how the theorem is used. If you have an electronic calculator available, simply use the square and square-root keys.

Impedance in a Series RC Circuit

A similar method can be used to find Z in circuits that have a resistor and capacitor in *series*. Such a circuit is diagrammed in Figure 10-4.

In solving this circuit for Z, the main difference is in the structure of the parallelogram. In a capacitive circuit, the current leads the voltage by 90 degrees. Expressed another way, the voltage lags the current by 90 degrees. Because the voltage lags the current, reactance is considered to be at −90 degrees with respect to resistance. Because of this, the vertical line for the reactance is drawn down below the line for resistance. This is demonstrated in Figure 10-5. The line AD represents capacitive reactance. This is below the resistance line, AB. Otherwise, the solution is the same as in the diagram in Figure 10-3.

The Pythagorean Theorem is also used in the same way as demonstrated above. In solving for capacitance, X_C is substituted for X_L in the sample equation.

In either of the impedance circuits shown, the **phase angle** is determined by the size relationship between the impedance line and the resistance line. The phase angle of any circuit can, therefore, be between +90 and −90

degrees. The angle depends upon whether the reactive part of the impedance is inductive or capacitive. The angle also depends upon the relative values of the reactance and the resistance.

There is only one other circuit parameter (factor) that affects the phase angle. Besides reactance and resistance, phase angle can be affected by the frequency of the applied voltage. Frequency affects reactance in both capacitive and inductive circuits. By affecting reactance, the frequency can also affect the amount of impedance in the circuit.

--- **REVIEW QUESTIONS FOR SECTION 10.1** ---

1. Define and give the symbol for impedance.
2. What is a vector sum? How is it used in representing current?
3. How is impedance measured?

10.2 SERIES RCL CIRCUITS WITH RESISTANCE, CAPACITANCE, AND INDUCTANCE

A resistor, a capacitor, and an inductor are connected in *series* in Figure 10-6. The resistor has a resistance of 6 Ω. The capacitor has a reactance of 2 Ω. And the inductor has a reactance of 10 Ω.

A vector diagram of these values is shown in Figure 10-7. Note that the vertical line in this diagram is no longer than the horizontal line. The horizontal line represents resistance. The vertical line shows reactance.

The vertical line represents the fact that the two reactances are in opposition to each other. So, a *resultant* of the vectors must be developed for X_L and X_C. This is done by subtraction. Subtraction can be used because of the differences in phase angles. Angles of +90 and −90 degrees, in effect, cancel each other.

The vertical line is positioned so that the value of X_C is below the resistance line. The 2 Ω of capacitive reactance is thus subtracted from the 10 Ω of inductive reactance.

The resulting reactance is still larger than the total resistance. So, part of the vertical line is above the level

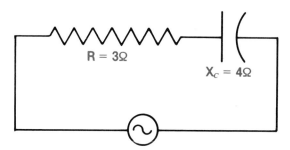

Figure 10-4. Schematic of RC circuit.

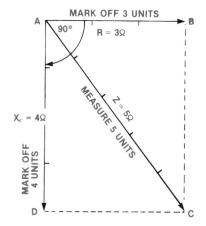

Figure 10-5. Parallelogram showing impedance in RC circuit.

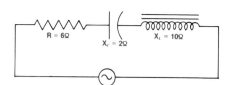

Figure 10-6. Schematic of RCL series circuit.

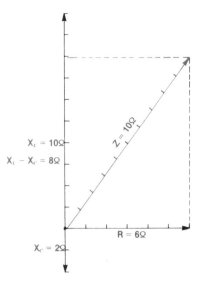

Figure 10-7. Parallelogram of RCL circuit in Figure 10-6.

10.3 PHASE ANGLE

In AC circuits, reactance and resistance change the phase differences between current and voltage. The phase angle is 90 degrees with reactance alone. But, we are no longer working with R alone, so the phase angle is somewhat less than 90 degrees.

Refer back to the circuit in Figure 10-6. A resistor, a capacitor, and an inductor are connected in series. The phase angle for this circuit is graphed into a vector diagram in Figure 10-8. The phase angle is between the lines for Z and R. It is marked as angle *theta* (θ).

Obviously, the phase angle is less than 90 degrees. The circuit represented in the vector diagram has more inductive reactance than capacitive reactance. So, the circuit tends to be inductive. Therefore, the current lags the voltage.

of the parallelogram used to figure the value of Z. In this vector sum diagram, Z has a value of 10.

Mathematical Solution

The mathematical solution for impedance in this type of circuit is similar to the previous example. The equation presented above was for an equation with resistance and capacitance only. The addition of inductive reactance adds a step. This is because there are two reactances. So, it is necessary to add the reactances.

Inductive reactance is a positive quantity. Capacitive reactance is negative. So, the two reactance values are added together. If the resulting sum is negative, it is preceded by a minus sign ($-$).

The formula for impedance in circuits with resistance, capacitance, and inductance is:

$$Z = \sqrt{R^2 + (X_L - X_C)^2}$$

With this formula, the value of Z in Figure 10-6 is developed as follows:

$$Z = \sqrt{36 + 64}$$
$$Z = \sqrt{100}$$
$$Z = 10$$

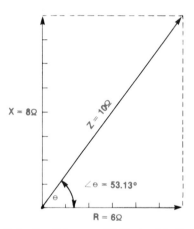

Figure 10-8. Parallelogram of RCL circuit in Figure 10-6 with phase angle represented.

10.4 SERIES CIRCUITS WITH RESISTANCE, CAPACITANCE, AND INDUCTANCE

Refer to Figure 10-9. This diagrams a circuit with a resistance of 1000 Ω connected in series with an inductor of 10 H and a capacitor of 10 μF. The current source is 120 V at 60 Hz.

Mathematical formulas are available to determine several circuit factors. You can calculate the current in the circuit. You can compute the voltage drop across each element. And you can find the value of θ, or the phase angle.

First, use the formula for **inductive reactance**. (Remember that $\pi = 3.14$. Therefore, $2\pi = 6.28$.)

$X_L = 2\pi fL$

$X_L = 6.28 \times 60 \times 10$

$X_L = 3768 \ \Omega$

The next step is to determine capacitive reactance. (In the formula used below, C is in farads. To convert to microfarads, you must multiply by 10^{-6}.)

$X_C = \dfrac{1}{2\pi fC}$

$X_C = \dfrac{1}{6.28 \times 60 \times (10 \times 10^{-6})}$

$X_C = \dfrac{1}{3768 \times 10^{-6}}$

$X_C = 265.4 \ \Omega$

To calculate total impedance of this circuit, use the figures for inductive reactance and capacitive reactance.

$Z = \sqrt{R^2 + (X_L - X_C)^2}$

$Z = \sqrt{1000^2 + 3502.6^2}$

$Z = \sqrt{13,268,206.79}$

$Z = 3642.55 \ \Omega$

The value of Z can be substituted in an Ohm's law formula for R. This makes it possible to determine current (I) for the circuit.

$I_T = \dfrac{E_A}{Z}$

$I_T = \dfrac{120}{3642.55}$

$I_T = 0.0329 A$

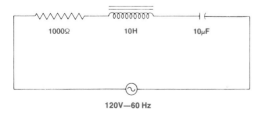

120V—60 Hz

Figure 10-9. Schematic of RCL circuit with values of components given.

REVIEW QUESTIONS FOR SECTION 10.4

1. What is the formula for determining the current in an RCL circuit?
2. What is the formula for determining capacitive reactance in an RCL circuit?
3. What is the formula for determining total impedance in an RCL circuit?

10.5 MAXIMUM VOLTAGE OF EACH PART OF A CIRCUIT

The current is the same through all components of a series circuit. Because of this, it is possible to compute the voltage across each of the circuit components. To do this, Z is substituted for R in Ohm's law. The formula becomes:

$E = I \times Z$

The current, as determined above, is used in the formula. The value of Z is used for the impedance value in the circuit. When the voltage drop across the resistor is determined, the formula used is:

$E = I \times R$

In the calculations below, Ohm's law is applied to the components of the circuit in Figure 10-9:

$I_T = I_R = I_L = I_C$

$E_L = I_L \times X_L$

$E_L = 0.0329 \times 3768$

$E_L = 123.97 \ V$

$E_C = I_C \times X_C$

$E_C = 0.0329 \times 265.4$

$E_C = 8.73 \ V$

$E_R = I_R \times R$

$E_R = 0.0329 \times 1000$

$E_R = 32.9 \ V$

Note that the total of the voltages does not equal the supply voltage (120 V). In fact, the voltage across inductor E_L is 123.97 V. This effect is common in AC circuits with reactive elements.

The voltages calculated represent the maximums across individual circuit elements. However, the circuit elements do not reach their maximum values at the same moment. Instead, voltage **leads** current in an inductor and **lags** current in a capacitor. So, maximum voltages across circuit elements should not be added and compared to source voltage. They are added vectorially for comparison with the source voltage.

10.6 SERIES CIRCUIT RESONANCE

Resonance is a specific circuit condition. In a resonant circuit, capacitive reactance and inductive reactance are equal ($X_C = X_L$).

The circuit in Figure 10-10 includes an inductor and a capacitor. The inductive reactance and the capacitive reactance of this circuit are equal. Under these conditions, the two reactances tend to cancel each other. Thus, the resulting impedance is equal to the resistance of the circuit.

A resonant condition can exist at only one frequency. This is true for any combination of inductance and capacitance. This becomes apparent in considering the reactance of the capacitor and the inductor as the frequency of applied voltage is changed.

In a capacitor, the reactance is increased as the frequency is decreased. Thus, for a high frequency of applied voltage, the reactance is lower than it would be for a low frequency. Figure 10-11 shows the curve of the value of X_C as the frequency is changed for a specific value of capacitance.

In an inductor, the reactance is increased as the frequency is increased. Thus, reactance increases as the frequency of the applied voltage increases. This relationship is shown as the straight line marked X_L in Figure 10-11.

Note that the values of X_L and X_C are the same at only one point. This is where the lines cross. This point of equal reactance will occur at a different location for each set of capacitors and inductors.

Inductance of the circuit in Figure 10-10 is 7.04 H. *Internal resistance* is 100 Ω. This is the resistance of the wire that forms the windings of the coil. The coil is connected in series with a 1-μF capacitor. The applied AC is rated at 120 V and 60 Hz.

The task is to solve this circuit for inductive and capacitive reactances, as well as impedance. The procedure used is the same as that applied earlier. The first step is to find inductive reactance (X_L).

$$X_L = 2\pi fL$$
$$X_L = 6.28 \times 60 \times 7.04$$
$$X_L = 2653 \ \Omega$$

The next step is to compute capacitive reactance.

$$X_C = \frac{1}{2\pi fC}$$

$$X_C = \frac{1}{6.28 \times 60 \times (1 \times 10^{-6})}$$

$$X_C = 2654 \ \Omega$$

The next task is to find the value of the impedance. To do this, the capacitive reactance is subtracted from the inductive reactance. The value of the resistance of the inductor is used for R. This is the only resistance shown in Figure 10-10. The calculation proceeds as follows:

$$Z = \sqrt{(X_L - X_C)^2 + R^2}$$
$$Z = \sqrt{1^2 + 100^2}$$
$$Z = \sqrt{10,001}$$
$$Z = 100.005 \ \Omega$$
$$Z = 100 \ \Omega$$

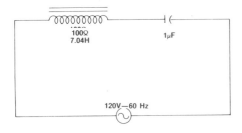

Figure 10-10. Schematic of series LC circuit.

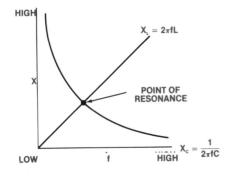

Figure 10-11. Curve showing point of resonance.

Activity: SERIES RESONANCE

OBJECTIVES

1. *To study the physical characteristics of a series resonant circuit.*
2. *To understand the electrical characteristics of a series resonant circuit as they apply to current, voltage, and impedance.*

MATERIALS NEEDED

1 Voltmeter, AC 0–500 volts

1 Lamp, 10-Watt, 120 V, and socket

2 Coils, 900 turns each

1 3/4″ Laminated core and inductor

2 Capacitors, 1.0 μF at 400 WVDC, nonelectrolytic

6 Wires for connections

PROCEDURE

1. Connect the circuit shown in Figure 1. CAUTION: Be careful, since the voltages can exceed the supply voltages in many instances. Do not work on the circuit while the power is on. Also, remember that this circuit contains capacitors, which can become charged and dangerous.

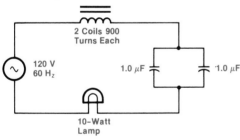

2 Coils 900 Turns Each

120 V 60 H$_z$

1.0 μF   1.0 μF

10-Watt Lamp

Figure 1

2. Be sure the laminated core is in place on the inductor frame. Set the voltage control of the power supply to the minimum value. *Turn on the power supply.* Slowly adjust the output to 120 volts AC. What happens to the lamp? (NOTE: Some transformerlike hum and/or chatter is normal for this activity since the bar across the top of the inductor frame is not permanently attached.)

3. The reaction of the lamp is caused by the combination of the inductance and the capacitance. (The two capacitors in parallel will be considered a single 2.0-μF capacitor.) To prove this, *turn off the power supply,* place a piece of wire across the leads of the inductor and *turn on the power supply again.* What happens to the lamp?

4. *Turn off the power supply.* Remove the lead placed in Step 3, and *turn on the power supply.*

5. Use the voltmeter and set it to read 0–500 volts AC. Measure the following voltages: voltage across the capacitor; across the inductor; across the lamp; of the supply.

6. The inductor, capacitor, and lamp are all connected in series. In series *resistance* circuits, the voltages add to equal the supply voltage. Do the voltages listed above add to equal the supply voltage? Why or why not?

7. This circuit is not quite at resonance, for it has been designed to prevent a buildup of voltage that might exceed the voltage ratings of the capacitors being used. Slowly slide the laminated core off the inductor, thus reducing the inductance and bringing the circuit toward resonance. What happens to the lamp when the core is removed from the top of the inductor frame? Why?

8. Place the laminated core on the inductor frame at the position in which the lamp was brightest. Use the voltmeter set to read 0–500 volts AC to measure the following voltages: across the capacitor; across the inductor; across the lamp. Does the voltage drop across the capacitor equal the voltage drop across the inductor? Why or why not?

9. With the laminated core completely removed from the inductor, again measure the following voltages with the voltmeter set to read 0–500 volts AC: across the capacitor; across the inductor; across the lamp.

10. How do the voltages measured in Step 9 compare with those measured in Step 8?

11. As the circuit is adjusted toward series resonance by removing the core, what happens to the current in the circuit as demonstrated by the changes in voltages?

12. *Turn off the power supply.* Discharge the capacitors and disconnect the circuits you have been using. Return all equipment and materials to their proper storage places.

SUMMARY

1. What happens to the impedance, the current, and the voltage across each element in a series resonant circuit?

2. What components are necessary to form a series resonant circuit?

10.7 RESONANT CURVE

The value of impedance at the frequency applied to the circuit in Figure 10-10 is shown in Figure 10-12. The small "x" marks indicate frequency readings in a range from 55 to 65 Hz. A curve drawn through the frequency marking points is known as the *resonant curve.*

The *impedance* for this circuit is at its *minimum* point *at the resonant frequency* of 60 Hz. This is a characteristic of the series resonant circuit. In such a circuit, the inductor and the capacitor must be carefully chosen. This choice must be correct, so that the components can withstand the current and the voltage present in the circuit at the resonant level. Impedance is at minimum level at the resonant point of 60 Hz. At this time, the current through the circuit is at its maximum, 1.2 A. Current is calculated through use of Ohm's law:

$$I_T = \frac{E_A}{Z}$$

$$I_T = \frac{120 \text{ V}}{100 \text{ }\Omega}$$

$$I_T = 1.2 \text{ A}$$

This means the inductor must be designed to pass a current of 1.2 A. It must be wound with a wire that will handle the amount of current. Accordingly, the windings should be made of a wire no smaller than #17.

A similar situation holds true for the capacitor. However, the main consideration for the capacitor is the voltage rating. The voltage rating for the capacitor is determined when it is manufactured. Capacitor ratings should be matched to circuits in which they are used. Keep in mind the peak voltage when selecting WVDC rating.

REVIEW QUESTIONS FOR SECTION 10.7

1. What is the resonant curve?
2. To what should capacitor ratings be matched?

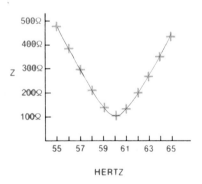

Figure 10-12. Resonant curve of series resonant circuit.

10.8 PARALLEL RL CIRCUITS

A resistor and inductor in parallel represent many circuits in homes, businesses, and industry. Such a circuit may include an incandescent light bulb (resistance) and a ballast of a fluorescent lamp (inductance).

The *power factor* and the *impedance* are of concern in a parallel RL circuit. Since in this circuit, a resistor and an inductor are in parallel, they both have the same voltage. Remember: voltage is the same across any devices connected in parallel. This means the variable element is current. Also remember that current is added in a resistive parallel circuit. Here, with the resistor and inductor, there is also an addition of the current. Since there is a phase angle to be considered, however, we must add vectorially to obtain the total current in the circuit.

Figure 10-13 shows a resistor with 120 volts and 10 amperes. That means the resistance (using Ohm's law) is 12 ohms. The circuit also has an inductor with 20 amperes flowing through it and 120 volts across it. That produces an X_L of 6 ohms.

Vectorially, the current is found by drawing the representations shown in Figure 10-14. Using the Pythagorean

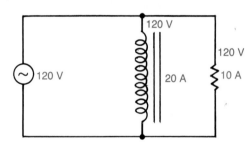

Figure 10-13. Parallel RL circuit.

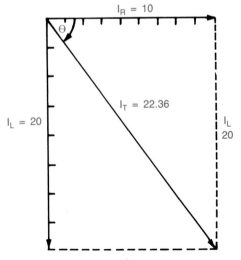

Figure 10-14. Total current is shown by hypotenuse.

Theorem again, you can obtain the formula that can be used mathematically to find the total current. Note also that the phase angle (θ) is also present and can be measured with a protractor if the units have been equally stepped off in the parallelogram.

Total current is found mathematically by:

$$I_T = \sqrt{I_R^2 + I_L^2}$$
$$= \sqrt{10^2 + 20^2}$$
$$= \sqrt{500^2}$$
$$= 22.36 \text{ amperes}$$

Total current can now be used to find the impedance, since $Z = E_A/I_T = 120/22.36 = 5.3667$ ohms.

Total current and current through the resistor are used to find the phase angle (θ).

$$< \theta = I_R/I_T = 10/22.36 = 0.4472$$

This can be converted to an angle by use of a trig table or calculator. (In the latter case, use either 2nd function $\boxed{\cos}$ or $\boxed{\text{inv}}$ $\boxed{\cos}$ to get the result.) The answer is that the current through the inductor is *lagging* the current through the resistor by 63.43°.

Power Factor

Power factor is the ratio of true power to apparent power. True power is actual power dissipated, such as when there is a resistive AC circuit or a DC circuit with resistance. True power is measured in watts. The formula used for finding true power is $TP = E \times I$. Apparent power is the power that appears to be used when the voltage and current are multiplied in a circuit that contains a resistor, capacitor, or inductor, or any combination of the three. Apparent power is equal to volts times amperes: $AP = VA$. The unit of measurement of apparent power is the volt-ampere.

The power factor in the example described above is also represented by the 0.4472 calculated. This means that the power being consumed as true power is 44.72 percent of what *appears* to be used when the total current (22.36 amperes) is multiplied by the applied voltage (120 volts). Note that $22.36 \times 120 = 2683.2$.

In order to find the actual power consumed, you must multiply the apparent power times the power factor: $TP = AP \times PF$.

If the power factor is kept near 1.00, or 100%, a generator has to work less. The power company in such a case can pass the savings along to the consumer. The small differences in power factor created by fluorescent lamps and electric motors in the home is hardly worth the expense of purchasing the equipment needed to bring the phase angle back to 1.00, or unity. In industry, where large motors draw large currents, it is necessary to try to correct the phase angle to be able to keep the expense of running the equipment down. So far, we have found only methods of finding the phase angle. Chapter 11 deals with bringing the value of this angle back to unity.

10.9 PARALLEL RC CIRCUITS

A parallel combination of resistor and capacitor presents some other interesting relationships, Figure 10-15. Since both are in parallel, it means the voltage across the resistor and the capacitor is the same as the applied voltage. The current will vary with X_C and R. To find the total current, you must change the formula used for the parallel RL circuits to reflect the capacitor instead of the inductor. However, note, in Figure 10-16, that the total current is above the I_R line, indicating a leading current, the opposite of the inductor circuit. The formula, then, is:

$$I_T = \sqrt{I_R^2 + I_C^2}$$

Impedance in an RC parallel circuit is the same as in a parallel RL circuit: $Z = E_A/I_T$.

Phase angle and power factor are found the same way as with the parallel RL circuits.

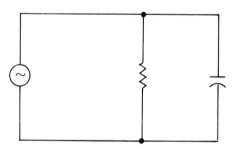

Figure 10-15. Parallel RC circuit.

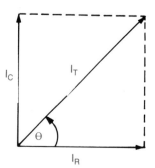

Figure 10-16. Parallel RC vectors.

10.10 PARALLEL RCL CIRCUITS

When a parallel combination of resistance, capacitance, and inductance are connected, a slightly different approach must be used. However, what was learned previously does enable us to calculate total current.

In this instance, the current in the inductor and the current in the capacitor are in opposite directions, or the current through the capacitor is leading and the current through the inductor is lagging, Figure 10-17. The current through the resistor is in phase with the voltage. It is therefore used as the point of comparison and for development of the appropriate vectors.

As you can see from the diagram in Figure 10-18, the inductive current is less than the capacitive current. That means the inductive current is negative (lagging) and the capacitive current is leading (positive). By adding these two algebraically, you obtain $(I_C - I_L)$ for the resulting leg of the triangle when the parallelogram is completed. This produces a formula that is basically the same as the one previously developed through the Pythagorean Theorem.

$$I_T = \sqrt{I_R{}^2 + (I_C - I_L)^2}$$

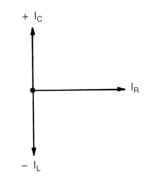

Figure 10-17. Vectors for a parallel RCL circuit.

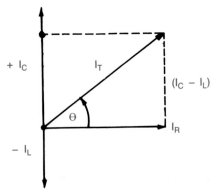

Figure 10-18. Vectors for finding I_T of a parallel RCL circuit.

Impedance is the same as for the other parallel combinations of inductance and resistance and, capacitance and resistance: $Z = E_A/I_T$.

Phase angle is also found the same way as in RC and RL parallel combinations: $\angle\ \theta = I_R/I_T$.

Power factor is found when phase angle θ is obtained. You can use it to obtain true power by multiplying it by AP.

10.11 DETERMINING RESONANT FREQUENCY

At resonance, the reactances X_L and X_C tend to be equal. They cancel each other. This fact is used to develop a formula for finding **resonant frequency**:

$$X_L = X_C$$

$$2\pi \times f \times L = \frac{1}{2\pi \times f \times C}$$

$$2\pi \times f \times L \times 2\pi \times f \times C = 1$$

$$\frac{2\pi^2 \times f^2 \times L \times C}{f^2} = \frac{1}{f^2}$$

$$2\pi^2 \times LC = \frac{1}{f^2}$$

Take the square root of each side of the equation. Invert both sides of the equation.

$$f_r = \frac{1}{2\pi\sqrt{LC}}$$

10.12 PARALLEL RESONANCE

For an example of a circuit with an inductor and a capacitor connected in parallel, see Figure 10-19. Suppose the conductor is broken. Assume this is done at point "X" in the diagram. If this happened, the capacitor and inductor would be in series. A series inductive–capacitive circuit would result.

The method for finding the resonant frequency of this parallel circuit is the same as for a **series resonant** circuit. The same formula is applied. The example below uses values from Figure 10-19, with some values substituted.

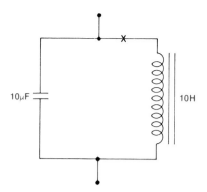

Figure 10-19. Schematic of parallel LC tank circuit.

Specifically, the values of the inductor (10 H) and the capacitor (10 μF) have been substituted.

$$f_r = \frac{1}{2\pi\sqrt{LC}}$$

$$f_r = \frac{1}{2\pi\sqrt{10 \times 10 \times 10^{-6}}}$$

$$f_r = \frac{1}{6.28 \times 0.01}$$

$$f_r = 15.92 \text{ Hz}$$

At the resonant frequency of a parallel resonant circuit, an action takes place that is known as the **flywheel effect**. The flywheel effect is a continuing **oscillation** in a resonant circuit between pulses of electrical energy. These circuits are sometimes referred to as **tank circuits**. They are used in tuned circuits in high-frequency transmitters and in receivers.

The operation of a tank circuit is shown in Figure 10-20. In Figure 10-20A, a current flows from the external circuit through the inductor and also onto the capacitor. The capacitor is charged. The inductor has a magnetic field that extends outward.

The external source is effectively removed, as no more current is demanded of it when the capacitor is charged. The capacitor now tries to discharge through the inductor, Figure 10-20B.

At the same time, the inductor will keep the current flowing through its windings, since its initial reactance to current flow has been overcome. The current that was supplied by the source overcame the opposition of the coil.

The collapsing magnetic field of the inductor supplies an emf for charging the capacitor, Figure 10-20C. This collapsing field sends current flowing in the same direction in the circuit. This charges the capacitor once again. However, the charge is of the opposite polarity. The inductive effect weakens. The capacitor takes another charge. It is ready to discharge through the circuit. Since current will flow from the negative to the positive, this is directly opposite to that occurring before. As it begins to flow, it meets opposition from the inductor. The opposition is there until the magnetic lines of force are extended outward once again. The capacitor is discharged.

The magnetic lines that built up around the inductor

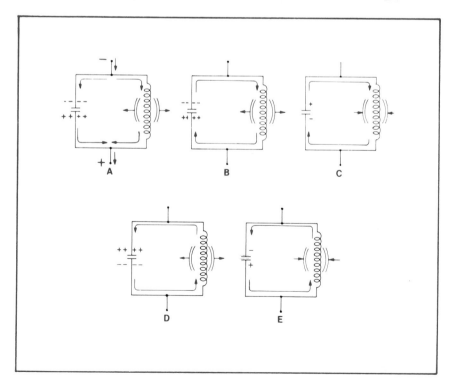

Figure 10-20. Diagram showing flywheel effect in tank circuit.

tend to keep the current flowing, as can be seen in Figure 10-20E. This, in turn, recharges the capacitor with the same polarity it had at the start.

The current then goes back and forth. This is called **oscillation**. This is the *flywheel effect* of a **parallel resonant** circuit. The only current drawn from the source is enough to make up for the losses due to the resistance of the coil, the wires, and the connections.

A graph of a current that produces the flywheel effect is shown in Figure 10-21. This is a pure sine wave. The number of oscillations resulting from the introduction of AC current is known as the *frequency* of a parallel resonant circuit.

Resistance within the circuit tends to limit the time that a parallel resonant circuit will oscillate. Consider the function of resistance. Suppose a circuit is designed for a resonant frequency of 20 Hz. Its current changes direction 40 times per second. The capacitor charges and discharges 40 times each second. This produces 20 complete hertz. But resistance within the circuit tends to act as a damping factor, limiting the flow of current that causes the oscillations. Eventually, resistance stops the oscillations.

A parallel resonant circuit containing resistance discharges in oscillations. This is called **damped oscillation**. A damped oscillation is similar to the wave shown in Figure 10-22. The wave for each hertz grows smaller. Finally, oscillation stops.

At the resonant frequency of a tank circuit, only a minimum of current is needed from the external source. Most of the circuit current is supplied by the circulating current of the tank circuit. Current from the external source is needed chiefly to overcome the current losses due to resistance.

Impedance is highest at the resonant frequency. Current required from the source is very small. The impedance curve for a parallel resonant circuit is shown in Figure 10-23.

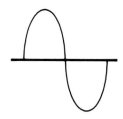

Figure 10-21. Graph of current that produces flywheel effect is a pure sine wave.

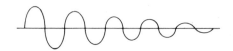

Figure 10-22. Waveform for damped oscillation.

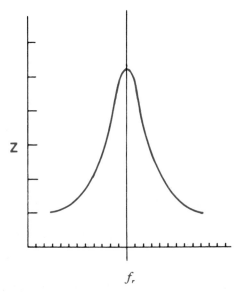

Figure 10-23. Impedance curve for a parallel resonant circuit.

REVIEW QUESTIONS FOR SECTION 10.12

1. How is the resonant frequency found in a parallel circuit?
2. What is the flywheel effect?
3. What is a tank circuit?
4. How does a tank circuit function?
5. What is damped oscillation?
6. At what point is impedance highest in a tank circuit?

10.13 COMPARISON OF SERIES AND PARALLEL RESONANCE

Compare the effects of series and parallel circuits at resonance. Note that in series resonance, total circuit impedance is at a minimum. In parallel resonance, the total circuit impedance is at a maximum.

In the series resonant circuit, the phase angle at resonance is zero. This is because of the cancellation of the reactances. Also, the series circuit appears to be a resistance of very low value.

In the parallel resonant circuit, the two reactances tend to cancel. The phase angle is zero. The circuit appears as resistive. This resistance has a high value.

The series resonant circuit offers a high impedance to frequencies other than the resonant frequency. The parallel resonant circuit offers a low impedance to all frequencies other than its resonant frequency. This factor accounts for most of the uses of the parallel resonant circuit.

Parallel resonant circuits may be used for filters. Filters separate one frequency from another. This is shown in Figure 10-24. The diagram is for a simple radio receiver.

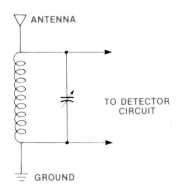

Figure 10-24. Schematic of simple radio-receiver input circuit.

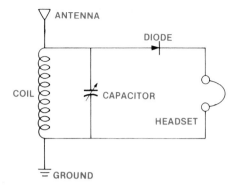

Figure 10-25. Schematic of simple crystal-set radio receiver.

This is a parallel resonant circuit. The antenna receives radio waves of sufficient amplitude and conducts them to the parallel resonant circuit. Frequencies other than the resonant frequency find a low impedance to ground. So, they are conducted to ground immediately. The resonant frequency meets with a high impedance. Only this frequency passes to the detector circuit.

The use of the parallel resonant circuit is valuable in electronics, particularly for radio and television equipment, Figure 10-25.

At series resonance the circuit:

- Has low impedance
- Resembles a resistance of *low* value
- Has high voltages developed across each element.

At parallel resonance the circuit:

- Has high impedance
- Resembles a resistance of *high* value
- Has high currents through each element.

SUMMARY

Resistance, capacitance, and inductance affect both alternating and direct current. Any or all of these elements may be present in any circuit. Each causes changes in voltage and current. Knowledge of the relationship of current and voltage within circuits is necessary for a complete understanding of electrical phenomena.

Total opposition to current flow in any circuit is called *impedance*. Impedance is abbreviated Z. Impedance is measured in ohms, but it is not measurable with an ohmmeter. Impedance is the *vector sum* of resistance and reactance.

Impedance is found in a series circuit with R, C, and L by calculation of the square root of the squares of the sums of the reactances: $Z = \sqrt{R^2 + (X_L - X_C)^2}$. The result is in ohms. The phase angle is between 0 and 90 degrees. Phase angle can be found by dividing R by Z. This also produces the power factor, or the percent of power actually consumed as compared to what appears to be consumed. The cosine of the angle *theta* is equal to the power factor.

Ohm's law for AC circuits can be used to obtain missing values, just as in DC circuits, with certain modifications of the values used in the formulas. X_C is used for R and X_L is used for R in some cases.

Resonance exists when the capacitive reactance is equal to the inductive reactance of a circuit and the resistance is the only power-consuming device left in the circuit. In other words, when $X_L = X_C$, resonance is achieved. This means the series circuit is equivalent to a short circuit if only an inductor and capacitor are in the circuit, or if the resistor is the only consuming device when there is a series RCL combination in the circuit. Therefore, it can be said that *impedance* is at its *minimum* point when *resonance* is reached in a circuit. A series resonant circuit has minimum impedance, but a parallel resonant circuit has maximum impedance at the signal frequency selected.

_____ USING YOUR KNOWLEDGE _____

1. Draw a parallelogram showing the relationship of $X_L = 10$ and $R = 10$. What is the impedance?
2. Draw a schematic of a series LC circuit.
3. Draw a schematic of a parallel LC circuit.
4. Draw a schematic for a crystal-set radio receiver.

KEY TERMS

inductance

capacitance

resistance

impedance

vector

Pythagorean Theorem

phase angle

inductive reactance

lag

lead

resonance

resonant frequency

series resonant

parallel resonant

oscillation

damped oscillation

flywheel effect

tank circuit

PROBLEMS

1. What is the impedance of an inductor of 10 H in series with 500 ohms on 50 Hz?

2. What is the impedance offered by an inductor and resistor combination that has an inductor of 10 H and a resistor of 5K ohms? Frequency is 100 Hz.

3. If the inductive reactance of a circuit is 1507.964474 ohms and the frequency is 60 Hz, what is the inductance?

4. What is the impedance of a series RL circuit with a resistor of 100 ohms and an inductance of 5 H? Frequency of the power source is 60 Hz.

5. What is the impedance of a parallel RL circuit with a resistor of 500 ohms and an inductance of 10 H on a power line with 60 Hz?

6. What is the impedance of a series RC circuit with a resistor of 100 ohms and a capacitance of 100 μF on a 60-Hz line?

7. What is the impedance of a parallel RC circuit with a resistor of 1000 ohms and a capacitance of 10 μF on a 60-Hz line?

8. What is the impedance of a parallel RCL circuit with a resistor of 1000 ohms, a capacitor of 100 μF, and an inductor of 10 H on a 60-Hz line?

9. What is the impedance of a series RCL circuit with a resistor of 2.2K ohms, 5 H, and 10 μF on 50 Hz?

10. What is the impedance of a series RL circuit if the applied voltage is 120 volts and the current through the inductor is 4 amperes?

11. What is the phase angle in problem 8?

12. What is the phase angle in problem 9?

ELECTROMECHANICAL DEVICES

INTRODUCTION

This chapter is about devices that are designed to control electricity. These devices are switches and relays.

Switches complete and break circuits. There are many types. In this chapter, you learn about the functions of switches. And you learn how to choose the right switch for a specific job.

Relays are electromechanical devices for routing electricity. They have contacts that are either opened or closed by circuit conditions. The opening or closing of relay contacts affects other electrical devices. This chapter deals with the operation of relays. The knowledge you build will also help you understand the semiconductors that are replacing relays in many jobs.

Electricity for Control

One of the major uses for electricity is for control. Electricity makes it possible to control many functions. These include processes, time, pressures, flow, and sequence of operations.

Electricity can be controlled by switches. **Switches** permit current to flow or prevent it from flowing. Electricity can also be controlled by varying the amount of current in a circuit. Another dimension of control is the direction of current flow.

11.1 USING SWITCHES FOR CONTROL

A circuit exists when there is a complete path for electrons. The flow is from the source, through the circuit,

and back to the source. When electrons flow through a circuit, it is said to be *on*. When electrons cannot flow, the circuit is said to be *off*. If the circuit is interrupted, the electron path is broken. A switch turns a circuit off and on when it breaks or completes the path along a conductor. A switch is, basically, a mechanical device. Its function is to make or break connections within a circuit.

Figure 11-1 illustrates a simple circuit that delivers current to a lamp. In this diagram, the switch is shown as open. So, the circuit is broken; the current does not flow. And the lamp will not glow.

Types of Switches

Switches are made in many sizes and types. Each is designed for a specific use. Some common types of switches include:

- The knife switch
- The toggle switch
- The momentary-contact, push-button switch
- The microswitch
- The slide switch
- The rotary switch
- The mercury switch.

Knife Switch. One of the simplest switches used in electricity is the **knife switch**, Figure 11-2. The closing of the switch resembles the closing of the blade of a partially opened pocket knife.

The knife switch contains a bar of metal. The bar is hinged at one end so that it may be moved up or down. When closed, the blade contacts a clip of similar metal. This completes the circuit.

Knife switches demonstrate the basic concepts of electrical control. They are often used in experimental work

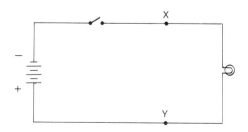

Figure 11-1. Diagram showing how switch breaks or completes a circuit.

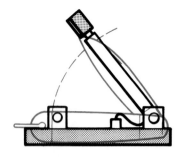

Figure 11-2. Single-pole, single-throw (SPST) knife switch operates like a pocket knife.

with low-voltage circuits. These units should not be used with high voltages. This is because the open contacts present a danger of accidental shock.

Toggle Switch. A switch used frequently in electronic circuits is the **toggle switch**, Figure 11-3. The operating handle causes a snap action to take place within the switch. This snap action moves a set of contacts from one side of the switch to the other. As the contacts move, they make or break the electrical connection.

Momentary-Contact, Push-Button Switch. Another frequently used type of switch operates from a push button. This type is also a **momentary-contact switch**. A switch of this type is illustrated in Figure 11-4.

Push-button switches are relatively simple. When the button is depressed, the switch is activated. Pressing the button can either break or complete the circuit. This depends on the construction of the switch (NO, Normally Open; NC, Normally Closed).

Usually, push-button switches are used in low-voltage, low-current applications. The ringing of doorbells, activation of small relays, and control of solenoid devices are examples. Other, larger push-button switches can be used for higher voltages.

Microswitch. For some jobs, it is desirable to activate a switch with a minimum of pressure. Special switches have been designed for this purpose. They are known as **microswitches**. These come in many shapes and sizes. Typically, a microswitch is enclosed in a plastic housing. A lever arm is attached. This arm permits the closing of the switch with very little pressure.

A microswitch is illustrated in Figure 11-5. This type of switch is used in a variety of devices. One example is the automatic garage door and the switch used in a car trunk. Another is the door in the store or office building that opens when you step onto a mat. Switches are set to be activated if the doors open too far. The pressing of a microswitch turns off the drive mechanism for the automatic opener. In other applications, microswitches automatically turn on lights or ring bells.

Slide Switch. The **slide switch** is a small unit, as shown in Figure 11-6. It is used widely in electronic devices. Advantages include low cost and practical, easy use.

The slide switch works on much the same principle as the knife switch. A bar of metal is moved to make the connection between two points. The slide switch has a bar of metal free to slide on a track. At the end of the track is a connector. The circuit is completed when the bar touches this connector. Typical uses for slide switches are in audio equipment, TV sets, and radios.

Figure 11-3. Toggle switch.

Figure 11-4. Push-button switch.

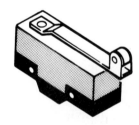

Figure 11-5. Microswitch, roller arm.

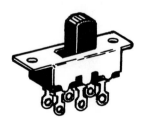

Figure 11-6. Slide switch.

Rotary Switch. **Rotary switches** may be large or small, Figure 11-7. A set of contacts is mounted on a small wafer. A wafer is a coin-shaped form made from an insulating material. Switch contacts are mounted on the wafer. A movable contact is mounted on a shaft. Rotating the shaft brings the movable contact and one of the wafer-mounted contacts together.

These switches are frequently constructed with a number of contacts on the switch wafer. This makes it possible to select one of several contacts for a circuit. Some rotary switches have several wafers. This makes it possible to complete several switching actions at the same time.

Rotary switches are used on TV sets, as selector switches for audio equipment, and as channel selectors for CB radios.

Mercury Switch. These are popular for household lighting circuits. A major advantage is that the mercury switch operates quietly.

In a **mercury switch**, a small amount of mercury is enclosed in a glass capsule. In one end of the capsule are two pieces of wire. These serve as connection points. When the switch is closed, the capsule is tilted. This causes the mercury to flow into contact with both wires. Mercury is a good conductor. So, the contact completes the circuit, Figure 11-8.

Types of Switch Contacts

Switches can break or complete single circuits. Or they can be constructed to make or break several contacts at the same time. The function of a switch depends upon its construction. Some types of contacts and actions for switches include:

- Single-pole, single-throw
- Single-pole, double-throw
- Double-pole, single-throw
- Double-pole, double-throw

Single-Pole, Single-Throw Switch. These switches have a limited, specific function. They break or complete single circuits. A bar-type contact is moved to break or complete the circuit. The contact is called a pole. The movement of the switch in making a contact is called a *throw*. Since there is only one pole, the switch is known as single-pole, single-throw. The abbreviation is **SPST**. The symbol for an SPST switch is shown in Figure 11-9.

SPST switches also come in push-button models. In these units, there are two positions. In one, the switch is *normally on*. In the other, it is *normally off*. On a normally-off switch, pressing the button completes the circuit. So, this is known as a *make-contact* switch. Another name for the same device is a *type-A contact*.

The other type of push-button SPST switch is the normally-on. Pressing the button breaks the circuit. So, this is known as a *break-contact* switch. Another name is the *type-B contact*.

Sometimes switches are marked *NO* for **normally open**, or *NC* for **normally closed**. Operation of NO and NC switches is diagrammed in Figure 11-10.

Figure 11-8. House wiring switch using a mercury contact capsule.

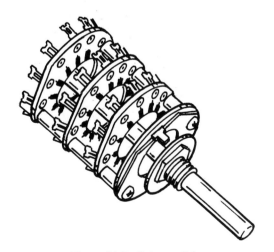

Figure 11-7. Rotary switch.

Figure 11-9. Symbol for SPST switch.

Single-Pole, Double-Throw Switch. This type of switch, abbreviated **SPDT**, is used to transfer the circuit from one position to another, Figure 11-11.

Many types of SPDT switches are available. The momentary contact switch is called type C. The switch case may be stamped with a code to identify each of the three leads. One set of letters used as a code is *A C B*. The A is the make-contact. The C is always the common terminal of the switch. The B is the break-contact. Some switches will be found with the abbreviation *NO* (normally open) in place of A. *NC* (normally closed) may be found in place of B.

Double-Pole, Single-Throw Switch. The abbreviation for this type of switch is **DPST**. The symbol and an illustration are shown in Figure 11-12. The main feature of this switch is that two poles are coupled together. Thus, two current paths can be opened or closed at the same time.

Double-Pole, Double-Throw Switch. The abbreviation for this type of switch is **DPDT**. The symbol for this switch, as well as an illustration of its operation, is shown in Figure 11-13. Because of the double-throw feature, these switches are used mainly for current-transfer functions. This contact arrangement is called the 2-C.

Safety Tips

1. Make sure your switches are rated at the right voltage and current for their circuits.
2. Don't use knife switches for voltages over 25 V.
3. Don't work on circuits that have knife switches for contacts. Make sure the power is removed before touching any part of the circuit.

Newer Types of Switches

Inasmuch as the printed circuit, semiconductors, and chips have caused a revolution in the electronics industry, it has become necessary to design switches that can do the job needed, yet fit the board and be lightweight and small.

Miniature Switches. There are miniature slide, rocker, and rotary switches made for mounting on printed circuit boards. Figure 11-14 shows how a slide switch's lower half is inserted into the PC board by hand or automatically. Wave solder or process soldering attaches the base as it does other components. The flux is cleaned from the board and when the latter is dry, a clean, uncontaminated upper switch half is snapped into place, by hand, onto the base. No tools are required for the operation.

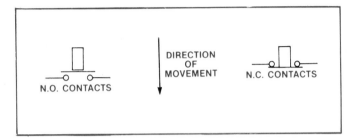

Figure 11-10. Symbols for NO and NC switches.

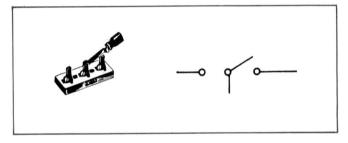

Figure 11-11. SPDT knife switch and symbol.

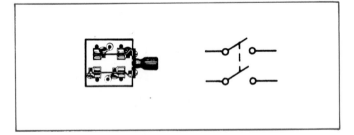

Figure 11-12. DPST knife switch and symbol.

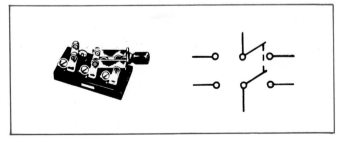

Figure 11-13. DPDT knife switch and symbol.

Figure 11-14. Miniature switch, which snaps into position after soldering to the board.

Subminiature Switches. There are subminiature switches designed for use with computers, calculators, and digital equipment. One of these types is the spacesaver rotary switch, Figure 11-15. These switches are about one-third or two-thirds of an inch high. They have a 0.36″ diameter. They can include up to eight contact positions brought out the bottom to fit into PC board holes for easy mounting. Glass-to-metal seal header prevents solder from entering onto contact pins. Such switches can handle 500 mA at 125 V AC.

Subminiature push-button switches are also available with a pilot light mounted in the top of the switch that may have a square or round bezel. The pilot light indicates the switch is activated.

Super Subminiature Switches. These very small slide switches are available in SPST, DPDT, SPDT, and SP3T, suited for printed circuit mounting, Figure 11-16. They can handle 0.4 VA maximum at 28 V AC/DC. This small switch is made in toggle, slide, and rocker models, as well as in momentary push-button and paddle types, Figure 11-17.

DIP Switches. The dual-inline switch resembles a chip or integrated circuit in appearance, with the exception of the top of the package. The switches must be turned on or off with the aid of a pencil or pen. They are low-profile subminiatures ideal for automatic insertion into a printed circuit board and for automatic immersion cleaning. They can handle 30 mA at 30 V DC, making them ideal for use with digital circuitry, Figure 11-18.

Switch Ratings

Switches are usually rated according to two factors. These are the maximum current capacity and type of circuit in which they should be used. An example: *10-A* **noninductive load**. This means that the switch can safely handle a circuit with 10 A. For the meaning of the rest of this specification, think back to the chapter on induction. When an inductive circuit is broken, current attempts to continue to flow. This flow takes place within the inductor. The continued flow will cause arcing in switch contacts. So, inductive circuits require switches with larger capacities than do many other types of circuits.

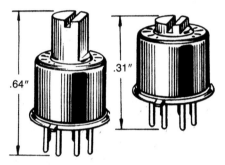

Figure 11-15. Subminiature rotary switches.

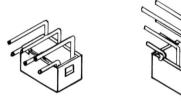

Figure 11-17. Micro-subminiature switches weigh only ¼ gram. They can handle 0.4 VA @ 28 V maximum.

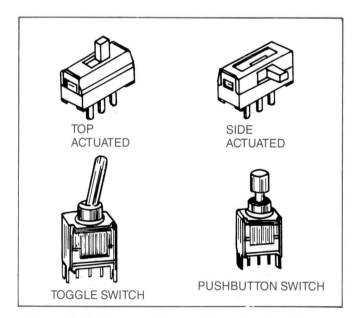

TOP ACTUATED

SIDE ACTUATED

TOGGLE SWITCH

PUSHBUTTON SWITCH

Figure 11-16. Super subminiature slide, toggle, and push-button switches. Note dimensions.

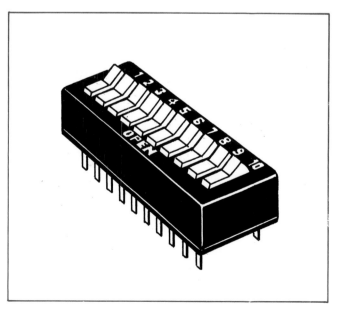

Figure 11-18. A DIP switch.

Activity: SWITCHES

OBJECTIVES

1. *To learn to operate switches of various types.*
2. *To wire switches into control circuits.*
3. *To study the operation and characteristics of switches.*

EQUIPMENT NEEDED

1 Power Supply, 6 volts DC	1 Switch, push-button, normally-open
1 Lamp, 6 volts, with socket	1 Switch, push-button, normally-closed
1 Switch, SPDT, on–off	3 Wires for connections

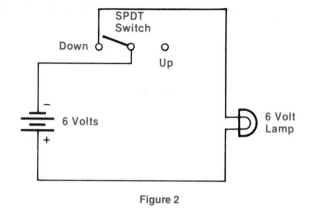

Figure 2

PROCEDURE

1. Connect the circuit shown in Figure 1.

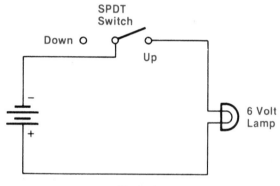

Figure 1

2. Set the voltage control of the power supply for minimum output. Turn on the power supply and slowly adjust the output to 6 volts DC. With the switch in the down position, does the lamp light?
3. Move the switch to the up position. Does the lamp light? Is the circuit now complete?
4. Compare the circuit of Step 1 with the circuit of Figure 1 in the Activity on page 40. In what ways are they similar? For what purposes was the circuit used in the first experiment? Is the purpose the same in this experiment?
5. Turn off the power supply. Disconnect the circuit you have been using. Connect the circuit shown in Figure 2.
6. Turn on the power supply. Adjust the output to 6 volts DC. With the switch in the down position, does the

lamp light? With the switch in the up position, does the lamp light? Why would the action of the circuit be the opposite of the action in Step 1?

7. Turn off the power supply. Disconnect the circuit you have been using and connect the circuit shown in Figure 3.

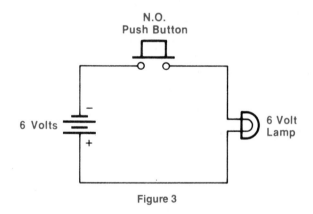

Figure 3

8. Turn on the power supply. Adjust the output to 6 volts DC. Does the lamp light? Press the push button. Does the lamp light? Is this the action of a switch that is normally open or normally closed?
9. Turn off the power supply. Disconnect the circuit that you have been using. Connect the circuit shown in Figure 4.
10. Turn on the power supply. Adjust the output to 6 volts DC. Does the lamp light? Press the push button.

Does the lamp light? Is this the action of a switch that is normally closed or one that is normally open?

11. Turn off the power supply. Disconnect the circuit you have been using and return all equipment and materials to their proper storage places.

SUMMARY

1. What is the difference between the operation of the push-button and the toggle switch?
2. Explain the difference between a normally-open (NO) and a normally-closed (NC) switch.
3. What is the basic purpose of a switch in a circuit?

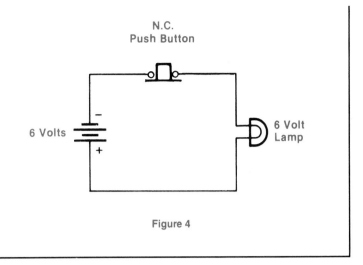

Figure 4

11.2 ELECTROMECHANICAL RELAYS

A switch controlled by the action of an electromagnet is a **relay**. With the electromagnet as a control, it becomes possible to operate the switch remotely. Remote switch control is one important application of relays. For example, a great amount of current may have to be controlled with a low-rated switch. Remotely controlled relays are standard in such situations.

How the Relay Works

Figure 11-19 diagrams the operation of a relay. Point 1 indicates that the unit has an electromagnetic coil. When the coil is activated, it attracts the armature, point 2. This breaks or completes the contact at point 3.

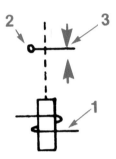

Figure 11-19. Diagram of relay switching action.

Types of Relay Contacts

Relays perform a number of different types of switching. The functions performed depend upon the types of contacts used. In relays, contacts are sometimes called *relay springs*. Two or more springs are used in each switching action.

The functions of relays are usually described in the same terms as those of switches. *Type A* is for make-contact. *Type B* is for break-contact. *Type C* is a single-pole, double-throw contact.

In addition, there is a *Type-D* relay. These units make the circuit from one source first, then break a circuit from another source. This is done with a single set of relay springs. Other types of relays are available. However, the discussion below is limited to A, B, and C contacts.

Rating of Relay Coils

Rating of relays is according to the current or voltage required to energize the coils. Relay coils are made in a variety of voltage ratings. Typical values range from 6 V AC and DC to 120 V AC and DC. Some coils are rated at 240 V AC.

In some instances, a different, indirect method for DC ratings is used. This applies when the resistance of a coil is indicated. If coil resistance is less than 2500 Ω, the unit is considered to be a normal-power relay. If resistance is more than 2500 Ω, the unit is considered to be a sensitive relay. Sensitive relays are used in vacuum-tube or transistorized circuits.

For relays that are rated by their current, two values of current are specified. These are for *pull-in* and for *drop-out*. Pull-in current is the amperage needed to cause the armature to move. At lower levels of current, the magnetic field is weakened. The armature is not held to the pole piece. This is referred to as the drop-out current. The drop-out current is always less than the pull-in current.

In a DC relay, operation is very quick and precise. A solid, soft-iron core is used for the relay magnet.

In an AC application, a great deal of relay chatter can be expected. Cores of AC relay magnets are made much the same as transformer cores. They are laminated. A laminated core has an advantage in AC operation. It reduces heat due to the effects of eddy currents. However, for economy, some small AC relays are manufactured with a solid core. In such relays, the armature is made to be a little more sluggish than the typical DC armature.

Generally, an AC relay will work with either AC or DC. However, a DC relay operates with direct current only.

Common Types of Relays

There are three common types of electromechanical relays:

- Common power relays
- Solenoid relays
- Stepping switches.

Common Power Relay. This type of relay is illustrated in Figure 11-20. Common power relays are made in many forms. All are similar in structure to the type shown.

Common power relays can be manufactured in two separate units. Some companies supply coils for a range of voltage. Separate switches are then available for different types of contacts. It is possible to buy the required coil and switch contacts as separate items. Relays are then assembled for specific applications. This greatly reduces the number of relay units that a supplier must stock.

Common power relays are used primarily in industry. Applications include burglar alarms and remote control of high-voltage circuits.

Solenoid Relay. This type of relay gets its name from the fact that it causes a physical movement. It does not have the usual moving armature. Instead, activation causes a set of contacts to be pulled into a closed position. Solenoid relays can handle large currents in industrial circuits. They can be designed to handle loads of hundreds of amperes.

Solenoid relays are frequently used as *actuators*. In these situations, they are the final link following a number of other relay functions. The solenoid relays then control the power to a motor or some other device. Typical applications include automotive starting and control over water flow in washing machines.

Stepping Switch. The **stepping switch** is similar to a rotary switch, Figure 11-21. A set of contacts is mounted on a wafer. A moving contact is made to shift from one contact to the next. This switching action takes place each time the relay coil is energized. The relay contact rotates by steps. This is the origin of the name *stepping switch*.

Stepping switches are used in specific types of circuits. These circuits are designed to initiate a fixed sequence of actions. Dial-telephone equipment is one example. In electromechanical telephone equipment, stepping switches select the lines being called. The pulses from a telephone dial activate the switches. The number of dial signals generated equals the number called.

Stepping switches are also used in some pinball machines. In this application, the switch serves as a kind of ele-

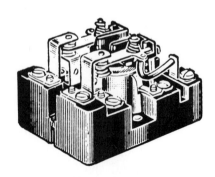

Figure 11-20. Power relay.

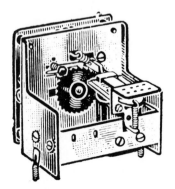

Figure 11-21. Stepping-switch relay.

mentary computer. Rotation is caused as balls strike contact points during play. Thus, the stepping switch adds the number of contacts made. The switch contacts then activate a display on the machine.

Relay Control Systems

Relay coils are controlled in several ways. Commonly used methods include:

- Direct control
- Shunt control
- Lock-up control.

Direct Control. Under this method, a switch is inserted between the source of power and the relay coil. Action of a direct-controlled relay is illustrated in Figure 11-22. When switch *A* is closed, current will flow through the relay coil. This closes contact *X*. When switch *A* is opened, the relay will be de-energized. Contact *X* is then opened.

This design is known as **direct control** of relay action. In actual applications, the circuits may be more complex. Several switches may have to be closed in a specific sequence before the relay is energized. Thus, it is possible for many switches to be involved in a direct-control system.

Direct-control relays are used for such jobs as remote operation of high-power equipment. They are also used for automatic controls in some machines. For example, many office copiers have direct-control relay systems. These stop the machines if paper fails to feed. In industry, a press operator must use two hands on the controls before the press will operate.

Shunt Control. This system for relay control is illustrated in Figure 11-23. In this circuit, the relay is always energized. This means switch *A* is normally open. The current path is easily traced from the battery through the relay coil. After that, the current goes through the shunt resistance *R* and back to the battery. Switch *A* is the shunt of the relay. When this switch is closed, it short-circuits the relay coil. Current flows from the battery, through the

resistance, then back to the battery. In this path, the current bypasses the relay coil. Under a **shunt control** system, closing the control switch de-energizes the relay.

One use of shunt-control relays is for automatic-paper-feed systems of computers. When a computer printer is out of paper, the shunt switch closes. The relay is activated and the machine is turned off. Shunt-control relays are also used in the start–stop controls of office dictating machines. This type of control makes instant starting and stopping action possible.

Lock-Up Control. Under this system, the relay itself enters into the control path. A circuit diagram of a lock-up control system is shown in Figure 11-24. In this circuit, make-contacts *Y* are connected through push-button switch *B*. Switch *B* is normally closed across the control switch, *A*. Without contact *Y* and switch *B*, this is a direct-control circuit like the one in Figure 11-22.

The action of this circuit is simple: The relay is energized when direct-control switch *A* (push-button) is closed. The relay closes contacts *X* and *Y*. Contact *Y* is part of the **lock-up control** circuit. It establishes a hold circuit through push-button switch *B*. Even when switch *A* is released, this relay will continue to be energized. It is released when push-button *B* is depressed. Pressing switch *B* opens the contacts. This breaks the circuit through lock-up contact *Y*.

Once the lock-up control is applied, switch *A* may be pressed as often as desired. It will not affect the operation of the relay. However, once push-button *B* is pressed, the relay is immediately de-energized. To re-energize the relay, it is necessary once again to depress push-button *A*.

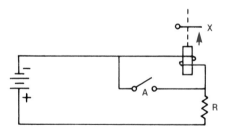

Figure 11-23. Circuit diagram showing operation of a shunt-control relay.

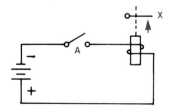

Figure 11-22. Circuit diagram showing operation of a direct-controlled relay.

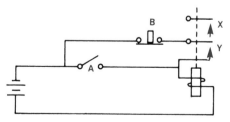

Figure 11-24. Circuit diagram showing operation of a lock-up control relay.

This approach is used in motor control. For example, look at Figure 11-25. This is the same as the circuit in Figure 11-24 except that the motor is connected to the relay, contact X. The source of the current is indicated as a 120-V line. When the motor start button (A) is depressed, the relay is energized. This closes contacts X and Y. Contact Y is the lock-up contact. X is the motor-control contact. The motor operates until push-button B is pressed, breaking the circuit. This is done through the lock-up relay contacts. The action de-energizes the relay, turning off the motor. The fuse symbol (C) in the diagram indicates the heating coil. This is the overload protection for the motor. If the motor overheats, this fuse element opens, de-energizing the relay. It is impossible to restart the motor until the overload or heating coils make contact again.

REVIEW QUESTIONS FOR SECTION 11.2

1. How does an electromechanical relay operate?
2. What are some of the main values of relays in electrical systems?
3. What is a relay spring?
4. What are some uses of the common power relay?
5. Describe the operation of the solenoid relay.
6. Describe the operation of the stepping switch.
7. What is the method of direct control for relay operation?
8. Describe shunt control for relay operation.
9. What is lock-up control for relay operation?

11.3 INDUSTRIAL CONTROL RELAYS

Industry uses relays for remote control of machines that are designed for rapid production of materials. The wiring diagrams are slightly different in industrial applications than in electronics. The diagrams, which are drawn to show how relays and contactors are wired into the circuit, are referred to as elementary diagrams, Figure 11-26. Note the two wires (L1 and L2) that bring power to the unit labeled. In the case of three-phase power, there is also an L3. However, here we deal only with single-phase power, so L1 and L2 are used exclusively. The abbreviations in Table 11-1 will help you understand what you are looking at.

M	Motor contactor coil*	NC	Normally Closed
F	Forward	NO	Normally Open
S	Stop	OL	Overload
PL	Pilot Light	Auto	Automatic
CR	Control Relay	Hand	by Hand, or manual

* This contactor coil energizes and handles heavy currents by means of contacts also labeled M. The switches or contacts controlled by the main contactor are labeled M as well. (See Figure 11-26.) In some cases, there is more than one contactor, so *M1* and *M2* labels are used where appropriate.

Table 11-1. Identification of Symbols Used in Industrial Control Circuits.

Unless otherwise specified, the circuits provide *undervoltage protection*, or *three-wire control*. In the event of a power failure, these circuits are designed to protect against automatic restarting when the power returns. This type of protection should be used where accidents or damage might result from unexpected starts.

In circuits using *undervoltage release*, or *two-wire control*, the motor starts automatically after a power failure. Typically, these circuits involve automatic pilot devices such as thermostats or float switches.

Figures 11-27 through 11-30 illustrate the various types of hook-ups available for different types of industrial processes. Each is labeled as a single station with its basic circuit and then more-complex switching arrangements (encountered in sequential operations requiring a number of switching actions to complete the task) are shown.

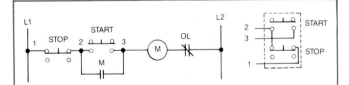

Operation — Pushing the start button energizes coil M; hold-in contacts M close, and maintain the circuit after the start button is released. Pushing the stop button breaks the circuit, de-energizing coil M; contacts M return to their normally open position.

Overload Protection — Operation of the overload relay contacts breaks the circuit and thus opens contacts M. To restart the motor, the overloads must be reset and the start button must again be depressed.

Undervoltage Protection — If a power failure de-energizes the circuit, hold-in contacts M open. This protects against the motor starting automatically after the power returns.

Figure 11-26. Ladder diagram with start-stop buttons and contactor coil M.

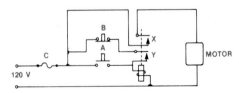

Figure 11-25. Circuit diagram showing operation of a motor-control relay.

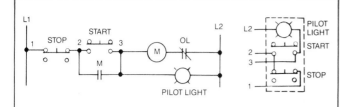

Whenever the motor is running, the pilot light is lit. Except for this modification, the circuit and its operation are the same as in the basic single station.

Figure 11-27. Ladder diagram with start-stop buttons, contactor coil and pilot light.

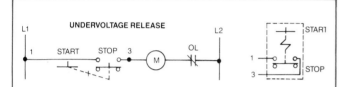

The start button mechanically maintains the contacts that take the place of hold-in contacts. Depressing the start button maintains the circuit; depressing the stop button breaks the circuit by opening the start contacts.

If the contactor is de-energized by a power failure or overload operation, the start contacts are unaffected. The motor restarts automatically.

Figure 11-28. Ladder diagram showing undervoltage release operation.

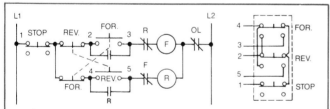

Operation — Depressing the forward button begins the following sequence: 1. Coil F is energized. 2. NO contacts F close to hold in the forward contactor; NC interlock contacts F open to prevent the reverse contactor from being energized.

Changing the Direction of Rotation — Use of the NC contacts in the forward and reverse push-button units makes it unnecessary to depress the stop button before changing the direction of rotation. Depressing the reverse button while running forward: (1) de-energizes the forward control circuit, and (2) energizes and holds in the reverse contactor in a manner similar to the forward operation outlined above. This results in "plug-reversing," that is, the motor acts as a brake until rotation stops, then the motor immediately starts turning in the opposite direction.

Figure 11-29. Ladder diagram showing push buttons used to reverse direction of rotation of motor.

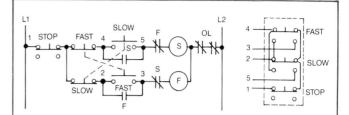

Operation — Depressing the slow button begins the following sequence: 1. Coil S is energized. 2. NO contacts S close to hold in the slow contactor; NC interlock contacts S open to prevent against the fast contactor being energized.

Changing Speeds — Use of the NC contacts in the slow and fast push-button units makes it unnecessary to depress the stop button before changing speeds. Depressing the slow button while running fast: (1) de-energizes the fast control circuit, and (2) energizes and holds in the slow contactor as outlined above.

Figure 11-30. Ladder diagram showing push buttons used to change motor speeds.

REVIEW QUESTIONS FOR SECTION 11.3

1. What is an elementary diagram?
2. What do L1 and L2 designate?
3. What does NO mean?
4. What does NC mean?
5. What does *two-wire control* mean?

SUMMARY

This chapter is about devices that are designed to control electricity. Some of these devices are electromechanical in nature and are called relays. Switches complete and break circuits. There are many types of these.

Relays are electromechanical devices used to route electricity to various places for different purposes. It is well to understand that semiconductors are replacing relays in many applications. However, relay is still reliable and is useful in special applications.

Switches are described through the use of terms such as *knife, toggle, momentary-contact, push-button, micro-switch, slide switch, rotary switch,* and *mercury switch.* Newer types include miniature switches, subminiature switches, and super subminiature switches. They are used primarily in computer and calculator circuits. These switches may be turned on and off with the aid of a pencil or pen. They are low-profile and are ideal for automatic

Activity: THE RELAY

OBJECTIVES

1. *To study the physical structure and operational characteristics of a simple relay.*
2. *To study the basic control methods utilized in relay circuits.*

EQUIPMENT NEEDED

1 Relay, 6 volts DC, 150-ohm coil, double-pole, double-throw
1 Power supply, 0–6 volts DC
1 Lamp, 6 volts, with socket
1 Resistor, 39 ohms, 5%, 2-watt
1 Switch, push-button, normally-open
1 Switch, push-button, normally-closed

PROCEDURE

1. Connect the circuit shown in Figure 1.

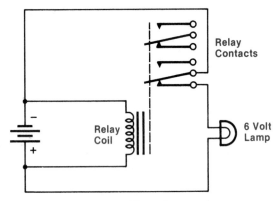

Figure 1

2. *Turn on the power supply.* Adjust the voltage slowly. Note and record the place where the voltage causes the relay to energize and the lamp to light. This is the energizing voltage of the relay. It is one of the important characteristics of a relay.
3. Slowly decrease the output of the power supply. At what voltage does the relay de-energize, causing the lamp to go out? This is the de-energizing voltage of the relay.
4. *Turn off the power supply.* Disconnect the circuit you have been using. Connect the circuit shown in Figure 2. Turn on the power supply. Press the push button. Does the lamp light?
5. Use a pencil to trace the circuit for the lamp. Does the lamp circuit include the push-button switch? Is the lamp controlled only by the relay contacts?

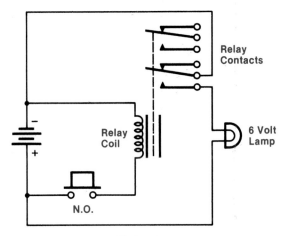

Figure 2

6. *Turn the power supply off.* Disconnect the circuit you have been using. Connect the circuit shown in Figure 3.
7. This is the shunt control for a relay. *Turn on the power supply.* Set the output voltage of the power supply for this circuit to 6 volts DC. Does the relay energize? What controls the lamp? Press the push button. What happens to the relay and the lamp? Use a pencil to trace the control circuit for the relay on the schematic diagram. What causes the action that takes place when the push button is depressed? How do direct control and shunt control of a relay differ?
8. *Turn off the power supply.* Disconnect the circuit you have been using. Connect the circuit shown in Figure 4.
9. This is the lock-up control for a relay. *Turn on the power supply.* Set it for an output of 6 volts DC. Does the lamp light? Press the NO push button. Does the lamp light? Release the NO push button. What happens to the light? Why does this happen? Press the NC push button. What happens to the lamp? Why? Trace the control path for the relay on the schematic. How does this circuit operate?
10. *Turn off the power supply.* Disconnect the circuit you have been using. Return all equipment and materials to their proper storage places.

SUMMARY

1. What are the differences between the direct, shunt, and lock-up control methods for a relay?
2. Why is it important to know the energizing and de-energizing voltages for a relay?

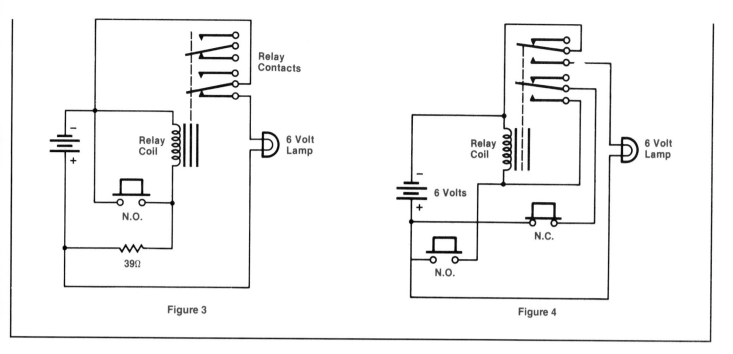

Figure 3

Figure 4

insertion into a printed-circuit board and for automatic immersion cleaning.

Switches are usually rated according to maximum current capacity and type of circuit in which they should be used.

A switch controlled by the action of an electromagnet is a relay. Use of the electromagnet as a control makes it possible to operate the switch remotely. Remote switch control is one important application of relays. Relay contacts and coils have various ratings according to the job they are expected to perform. Relays are controlled by direct control, lock-up control, and shunt control.

Relays are used in industry for remote control of machines and for sequencing of operations. Elementary diagrams are used for simplification of the wiring of relays and their switching arrangements. *M* stands for the motor-contactor-coil solenoid in these diagrams. *CR* stands for the control relay. There are other symbols utilized, as well as the terms *three-wire control* and *two-wire control*.

Three-wire control has an added safety feature: a machine will not start operating again if the power has failed and returned. In the case of two-wire control, the motor starts automatically after power failure.

USING YOUR KNOWLEDGE

1. Draw the symbol for a single-pole, single-throw switch.
2. Draw the symbol for a double-pole, double-throw switch.
3. Draw the symbol for a slide switch (SPST).
4. Draw a circuit schematic showing shunt control of a relay.
5. Hook up dry cells of the proper voltage to control a relay. Place a push-button switch in the circuit so it controls the relay.

KEY TERMS

switch	toggle switch
relay	momentary-contact switch
normally-open	mercury switch
normally-closed	knife switch
SPST	noninductive load
SPDT	microswitch
DPST	stepping switch
DPDT	lock-up control
slide switch	shunt control
rotary switch	direct control

Chapter 12

ELECTRICITY AND ELECTRONICS FOR HEAT AND LIGHT

OBJECTIVES

After studying this chapter, you will know

- *how a carbon arc produces light and heat.*
- *how resistance creates heat for a number of electrical applications.*
- *how radiant heat is produced by a quartz heater.*
- *the principles at work when a microwave oven heats food.*
- *the need for and principles behind the operation of safety fuses.*
- *the functioning of incandescent, fluorescent, mercury-vapor, and neon glow lamps.*
- *how a laser works and the uses for laser light.*

INTRODUCTION

Electricity is used to produce both heat and light. People depend on electricity for these purposes. So, it is important for you to learn about the electrical applications in this chapter. These include the use of electricity for quartz heaters, resistance heaters, welding, and microwave ovens.

Historical Background

One of the first uses of electricity was to produce heat. The arc furnace was used to melt metals as early as 1850, Figure 12-1. This paved the way for modern methods of steel making. The Chicago World's Fair of 1893 displayed the new and growing use of electricity. It was here that the electric iron was introduced. Also at this fair, the Weston enclosed arc was introduced, Figure 12-2.

12.1 HEAT BY ELECTRIC ARC

An **electric arc** is produced by current flowing through two carbon rods. The rods are brought into contact with each other. An arc is struck. Once the electric arc is established, the rods are moved apart slightly.

The arc vaporizes small quantities of carbon. To vaporize a substance is to turn a solid or liquid into a gas. The carbon vapor provides a path for current to flow. The current maintains the arc, which produces light and heat, Figure 12-3.

There are many uses for carbon arcs. One is brazing. Brazing is the joining of two pieces of metal through use of heat. The heat melts a high-temperature material, such as brass. The melted metal forms a bond to join, or braze, the two pieces of metal.

Another application is **welding**. In welding, two pieces of metal are also joined through use of heat. Welders use rods of materials that are similar to those being joined.

A carbon arc also produces intense light. Carbon arcs are light sources for projectors used in theaters. They are also used as main lights on movie sets. Production of

Figure 12-2. Weston enclosed arc lamp of 1893. (Westinghouse)

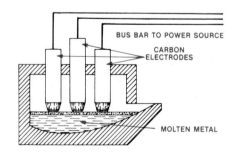

Figure 12-1. Drawing of early electric-arc furnace.

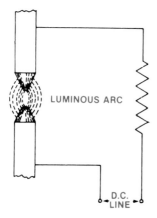

Figure 12-3. Diagram of carbon arc.

Some materials offer more opposition to the flow of electrons than others. This opposition to current is called resistance. Materials that resist current are called resistors. It takes more pressure to force electrons through a resistor than through a good conductor. This added pressure generates heat.

Many appliances and tools use resistance heat. Examples of household devices that use resistance and heat include electric ranges and irons. A heat gun, or blow dryer, uses resistance as a source of instant heat. A heat gun can produce heat at 1000 degrees F. A similar device is used in hair dryers. However, the temperature produced is much lower.

color motion pictures requires bright lights. This is because color film is less sensitive to light than is black-and-white film.

Welding

Heat plays a part in the manufacture of many welded metal products. For example, resistance welding is widely used in making automobiles, railroad rails, and many appliances.

Electrical welders are widely used. Some function with AC, Figure 12-4. Others utilize DC, Figure 12-5. With either type of welder, current is passed through a rod of resistor material. The rod strikes an arc against the metal to be joined, producing heat.

The electrical arc generates high temperatures. The heat is great enough to melt the rod and the metal pieces being joined. The melting metals flow together. When they cool, they have been joined.

In addition to giving off heat, arc welding also produces intense light. This light is rich in ultraviolet rays. It is so bright that it can damage the human eye and produce a "sunburn" on exposed skin. So, special goggles must be worn by welders. These goggles protect the eyes by filtering the light.

REVIEW QUESTIONS FOR SECTION 12.1

1. Describe a carbon arc.
2. How does a carbon arc produce heat?
3. How does a carbon arc produce light?

12.2 HEAT BY RESISTANCE

Movement of electrons through a conductor creates heat. Electrons are moved from their orbits by pressure. The source may be any number of methods that produce potential differences. The potential may come from a battery or a generator. These sources produce pressure that causes electrons to flow.

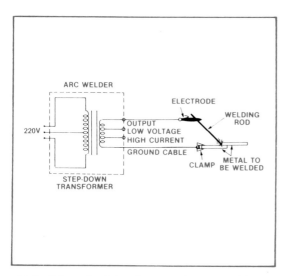

Figure 12-4. Schematic of step-down transformer and AC arc welder.

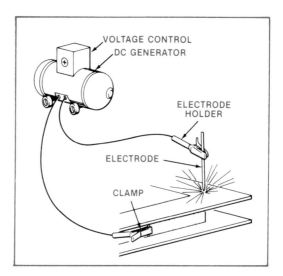

Figure 12-5. Functional drawing of DC welder.

Activity: HEAT BY RESISTANCE

_____ OBJECTIVES _____

1. *To investigate the production of heat caused by electrons flowing through resistance.*
2. *To observe the results of heat buildup in certain applications of electrical currents for heating.*

EQUIPMENT NEEDED

1 Power supply, AC, capable of 3 volts @ 8 amperes and 0–120 volts
1 Nichrome wire, 15″, #24

1 Lamp, 25-watt, 120 volts, and socket
1 Asbestos pad, 3″ × 3″
1 Voltmeter, AC, 0–150 volts

PROCEDURE

1. Connect the circuit shown in Figure 1.
   NOTE: *Coil about 5″ of #24 nichrome wire around a pencil. Withdraw the pencil and clip ends of wire to test clips. Be sure that the wire ends are firmly fastened to the test clips. The asbestos pad should be placed underneath the wire. Be careful not to touch the wire while it is hot.*

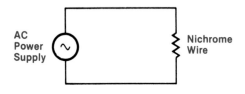

Figure 1

2. *Turn on the power supply.* Adjust the output of the power supply to 2 volts AC. What happens to the resistance wire? Why?

3. Increase the output of the power supply to 3 volts AC. How is the temperature of the resistance wire affected? Why?

4. Decrease the output of the power supply to 1 volt AC. How is the temperature of the wire affected? Why does the amount of voltage applied to the resistance wire affect its temperature?

5. *Turn off the power supply.* Allow the resistance wire to cool completely before attempting to remove it from the test clips. Obtain a 10″ length of nichrome wire and wind it around a pencil to obtain a coil. Remove the pencil. Clip each end of the coil in a test clip so that the coil is suspended between the test clips. Place the pad of asbestos cloth under the nichrome coil to protect the surface of the table on which you are working.

6. Set the voltage control of the power supply to minimum output. *Turn on the power supply.* Slowly increase the output of the power supply until the coil of wire is a cherry-red color. What happens to the coil as it becomes hot? *Turn off the power supply.* What happens to the coil as it cools?

7. Disconnect the circuit you have been working with and set it aside to cool down before dismantling it.

8. Connect the circuit shown in Figure 2.

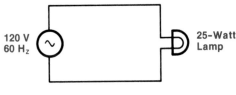

120 V
60 Hz

25-Watt Lamp

Figure 2

9. *Turn on the power supply* and adjust it to 120 volts AC. Allow the lamp to glow for a minute. Turn off the power supply. Carefully touch the top of the incandescent lamp. Is it warm to the touch? Why?

10. Disconnect all circuits you have been using and return all equipment and materials to their proper storage places.

SUMMARY

1. How does resistance in a circuit provide heat?
2. Why is nichrome wire used for heating elements in appliances?
3. What type of material is used for the filament of an incandescent lamp?
4. Why must ventilation be provided for incandescent lamps in lighting fixtures?

Another method for joining two pieces of metal through use of electric heat is spot welding. This technique is diagrammed in Figure 12-6. Two pieces of metal to be joined are placed between two electrodes. Large amounts of current pass through the metal. The resistance offered by the metal causes intense heat. The heat melts the metal. This causes the two pieces of metal to become attached to each other permanently. The heat results from resistance. This method is known as resistance welding.

REVIEW QUESTIONS FOR SECTION 12.2

1. How does electrical resistance generate heat?
2. Describe how brazing is done.
3. Describe how an electric arc can cause welding.
4. Describe how spot welding is done.

12.3 HEAT-PRODUCING DEVICES FOR THE HOME

Many devices that produce heat from resistance are used in the home. A few examples include water heaters, clothes dryers, and toasters. Other uses of resistance heat include the electric iron, space heaters, and radiant heating systems. Many special-purpose heaters are used in coffee pots, deep-fat fryers, and other convenience items.

The Quartz Heater

A somewhat unique device that uses electricity to produce heat is the **quartz heater**. Quartz is a crystal material. It is formed around a heating coil of resistance wire. The surrounding quartz makes it possible to increase the heat of the resistance wire. In a quartz heater, the heating element can reach temperatures of 2000 degrees Fahrenheit. [This is 1093 degrees Celsius.] Ordinarily, heating elements are limited to 1500 degrees F [816 degrees C].

Quartz absorbs heat from the resistance wire. The heat is intensified and radiated. The result is much the same as using a magnifying glass on light. Focusing a magnifying glass intensifies, or increases, the heat of the sun.

A quartz heater can produce up to 5120 BTUs of heat per hour. (*BTU* stands for *British Thermal Unit*. A BTU is the heat necessary to raise the temperature of 1 pound of water 1 degree Fahrenheit.) A shiny reflector is often placed behind the quartz rod. This can reflect heat for a distance of up to 30 feet, Figure 12-7.

The heat is delivered to the person or object requiring the warmth. The air in between is not heated by the quartz unit. This effect is known as *radiant heat*. Radiant heat works much the same as the heat of the sun. It moves through air and focuses on solid objects or liquids. The heat is absorbed by the solids or liquids. The surrounding air is not heated by the quartz unit.

Conduction

Another method for transferring heat is known as *conduction*. In conduction heating, the object receiving the heat comes into contact with the source of the heat. The pan on an electric range is heated by conduction. The contact is directly with a heating element of the range.

Convection

Another method of heat transfer is *convection*. With the convection method, heat is transferred through a liquid such as water or a gas such as air. Currents of water or

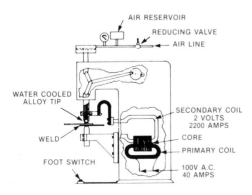

Figure 12-6. Diagram of spot welder.

Figure 12-7. Vertical quartz heater. (Arvin)

liquid transfer the heat. An example can be seen in warm-air furnaces. Air is blown over a source of heat and circulated through ducts. The entire house is warmed by the heated air.

Convection currents can be seen readily in a pan of water being heated on a stove. Convection causes a motion in the material being heated.

Microwave Ovens

Many electronic devices utilizing semiconductors provide useful heat sources for the home. One of the most popular is the **microwave** oven. Microwave ovens heat food through the use of radio-frequency waves. The radio frequency causes the molecules of water in the food to move rapidly. (Molecules are groups of connected atoms that form matter.) As the molecules move, they create friction, which causes heat.

Microwave ovens operate at 2450 megahertz (MHz). They use a device known as a magnetron to generate these high frequencies. These frequencies are in the same range normally used by radar. The oven cavity is part of the resonant circuit of the electron tube. To function properly, only materials that permit the passage of microwaves should be placed in the oven cavity. If metal is placed in the oven, it will reflect the microwave energy. The power supply and magnetron tube can be damaged as a result.

Before buying a microwave oven, be sure it meets federal safety standards. In the past, there have been problems with wave leakage from ovens. Radiation at these wavelengths can be harmful to humans. Control measures are easily applied. Just make sure the unit you use has been certified.

The dials for setting the time and operating functions of a microwave oven are electronic. They employ micro-processor technology.

Effects of Heat

Heat has always been a factor in the design and use of electronic devices. So, design of these units should provide a means for dissipating (carrying off) the heat they generate. Dissipation of heat is critical to the functioning of many electronic devices. This dissipation is often handled through fins or heat sinks. The idea of these devices is to conduct the heat through a series of strips, or fins. This increases the contact surface between metal and air. The air carries the heat away from the electronic devices.

Figure 12-8. Symbol for a fuse.

REVIEW QUESTIONS FOR SECTION 12.3

1. Describe the operation of a quartz heater.
2. What is radiant heat?
3. How is radiant heat distributed?
4. What is conduction heating?
5. Give an example of conduction heating.
6. What is convection heating?
7. Give an example of convection heating.
8. How does a microwave oven heat food?
9. Why is it important to keep metal out of micro-wave ovens?
10. Describe the purpose and operation of heat sinks or fins.

12.4 USING HEAT FOR SAFETY DEVICES

The heat produced in electric circuits could become dangerous. So, special safety devices are built into circuits to protect against overheating. These devices are **fuses**. The symbol for a fuse is shown in Figure 12-8.

A fuse is designed to open a circuit when too much current is present. A *link* within the fuse is designed to conduct only a specific amount of current. When the current level goes beyond this limit, the fuse link becomes overheated. The material in the link melts. The melted portion drops away, opening the circuit. Figure 12-9A shows a fuse link. This link is bending under heat. In Figure 12-9B, the link melts and the circuit is broken.

Links such as the one shown in Figure 12-9 are usually part of *plug* fuses. A plug fuse is the type used in homes. Many older homes still have this type of fuse. These screw into sockets in fuse boxes. Plug fuses resemble the drawing in Figure 12-10A.

The remaining drawings in Figure 12-10 show the operation of a special type of plug fuse. This is the *dual-element* fuse. Fuses of this type are designed to accept a brief circuit overload and still keep functioning. Dual-element fuses are recommended for use on circuits

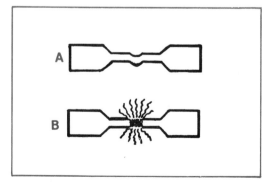

Figure 12-9. Reaction to overload of fuse link.

with electric motors. Current often surges when the motor starts. This can temporarily overload a fuse. A normal fuse burns out and breaks the circuit. A dual-element fuse is designed to withstand the surge and keep operating.

A fuse element is the same as a link. The drawing in Figure 12-10B illustrates the construction of a dual-element fuse. The first link is the curved wire. This is connected from the side of the fuse to a terminal in the center. Connected to the first link is a spring. This spring forms the second element. Figure 12-10C shows what happens when the link burns out. A surge of current overheats the link. It burns away from the terminal at the bottom of the fuse. However, current still keeps flowing across the secondary circuit. This secondary circuit is formed by the link and the spring, Figure 12-10D. Should the overload continue, the second element will burn out. So, the circuit is still protected. But the fuse can survive a brief overload and continue to provide electric service.

Still another type of plug fuse is designed to protect against circuit overloading. Normal plug fuses are made for a variety of loads. It is possible, therefore, to replace a burned-out fuse with one that has greater capacity. Under these conditions, the circuit could become overloaded.

This practice is prevented in homes set up to use *tamperproof* fuses. These are plug fuses with a special safety feature. Each size of fuse fits a different size of socket. For example, fuses rated at 15 A, 25 A, and 30 A are all different in size. They all fit into different sockets. So, it is impossible to replace a blown-out fuse with one that has a bigger rating. That's why these are called tamperproof fuses.

Many industrial and electronic devices use *cartridge* fuses. Some units have replaceable links. When a link burns out, the ends of the cartridge are unscrewed and a new link is inserted. A cartridge fuse is shown in Figure 12-11. Figure 12-12A shows a time-delay fuse. It can withstand current surges. Figure 12-12B shows a quick-acting link. This is designed to burn out immediately on overload. The quick burnout protects the equipment in which the fuse is used.

Fuses like those shown in Figure 12-12 are sometimes placed in clips within circuits. This is common in automotive electric circuits. A fuse holder for cartridge fuses is shown in Figure 12-12C.

Circuit Breakers

Recently built homes, office buildings, and factories almost always have circuit breakers in place of fuses. Many appliances and other electrical or electronic devices also have circuit breakers. A circuit breaker is a mechanical device that performs the same function as a fuse.

When they blow, or break the circuit, circuit breakers can be reset. This is done simply. Resetting a circuit breaker is like throwing a switch. However, before the circuit breaker is reset, all of the safety measures recommended for replacement of fuses should be followed.

Circuit breakers and their functions are discussed further in the next chapter.

___ **REVIEW QUESTIONS FOR SECTION 12.4** ___

1. How does a fuse protect against circuit overloading?
2. What is a plug fuse?
3. What is a fuse link?
4. What is a dual-element fuse?
5. What is a tamperproof fuse?
6. What is a cartridge fuse?
7. What is a circuit breaker?

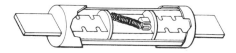

Figure 12-11. Replacement link fuse.

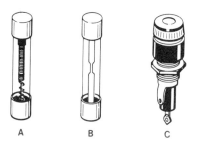

Figure 12-12. Cartridge fuses and fuse holder.

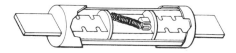

Figure 12-10. Operation of a dual-element plug fuse.

Activity: FUSES

_____ OBJECTIVES _____

1. *To become familiar with the type of material used in the manufacture of fuses.*
2. *To observe the operation of a fuse in an electric circuit.*

EQUIPMENT NEEDED

1 Power supply, 0–6 volts DC	1 Switch, on–off, toggle
4 Fuse wires, 5″ pieces, 1-amp size	7 Wires for connections
1 Lamp, 6 volt, and socket	1 Insulated pad, 3″ × 3″
	1 Candle, 1/2″ diameter × 5″ long

PROCEDURE

1. Connect the circuit as shown in Figure 1.
   NOTE: *The fuse wire must be secured firmly in the test clips and a small insulating pad placed beneath the two test clips.*

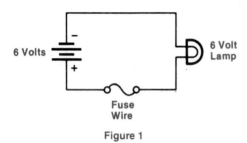

Figure 1

2. *Turn on the power supply.* Adjust the output of the power supply to 6 volts DC. What happens to the lamp in the circuit? Is the fuse wire a part of the circuit? What happens to the fuse wire? What would happen if the fuse wire were broken in two?
3. *Turn off the power supply.* Break the piece of fuse wire with your fingers. You will discover that the fuse wire is quite soft and easily broken. Turn on the power supply with the same voltage setting. Does the lamp light now? Why?
4. *Turn off the power supply.* Install a fresh piece of fuse wire in the test clips. *Turn on the power supply.* Does the lamp light now?
   NOTE: *Do not change the setting of the power-supply voltage control in this experiment. All parts of this experiment will be performed with 6 volts DC.*

5. Use the candle or a match to heat the fuse wire until it melts. What happens to the light? Does the fuse wire melt easily?
6. *Turn off the power supply.* Disconnect the circuit you have been using. Connect the circuit shown in Figure 2. Place a fresh piece of fuse wire in the test clips.

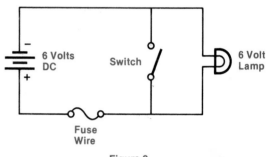

Figure 2

7. Make certain that the switch in the circuit is in the off position. *Turn on the power supply.* Adjust the output of the power supply to 6 volts DC. Does the lamp light?
8. You will notice that the switch is connected in the circuit so that when it is in the on position, it will short-circuit the lamp and place the fuse wire across the supply voltage. Move the switch in the circuit to the on position. Does the lamp continue to operate? Does the fuse wire melt in two and open the circuit? Why?
9. *Turn off the power supply.* Move the switch to the off position. Replace the fuse wire with a fresh piece.
10. *Turn on the power supply.* Does the circuit operate normally? Once again, move the switch to the on position. From this observation, can you safely say the fuse wire is reliable enough to be used to guard against short circuits or overloads in circuits such as you are using?
11. *Turn off the power supply.* Disconnect all circuits that you have been using. Return all equipment and materials to their proper storage places.

SUMMARY

1. What is a fuse?
2. Why does the fuse "blow"?
3. What type of material is used to make fuses?
4. How does the size of the fuse wire limit the current in a circuit?

12.5 ILLUMINATION AND ELECTRICITY

Electric light is a normal part of modern life. When the light bulb was introduced, it changed the way people live.

Thomas Edison introduced the carbon-**filament** lamp in 1878. One of the first light bulbs is illustrated in Figure 12-13.

The Chicago World's Fair of 1893 introduced the use of electric lighting on a large scale. The just-developed alternating-current generator was put to work at the same time. Some 20,000 electric lamps lighted the fairgrounds.

In 1912, tungsten filaments were substituted for carbon. The vacuum-type light bulb was modified to a gas-filled bulb in 1913. The use of electric lighting expanded rapidly after that. Use of gas greatly lengthened the life of the filament.

A fluorescent lamp was introduced to the public in 1939 at the New York World's Fair.

REVIEW QUESTIONS FOR SECTION 12.5

1. When did Edison introduce his first lamp?
2. What event marked the first large-scale use of AC?

12.6 INCANDESCENT LAMPS

Lamps using filaments that give off light are **incandescent**. When properly connected, the filament glows white hot. This glowing is called *incandescence*.

Incandescence is produced by a flow of electrons through the filament. The tungsten filament opposes this electron flow. As the current is forced through the resistance, the metal heats, then glows. The glowing filament gives off light.

Edison's first **incandescent** lamp lasted only 40 hours. Today's lamps burn much longer. For example, a 60-W lamp burns for an average of 1000 hours. Lamps rated at 75 W, 100 W, and 200 W burn for an average of 750 hours. In addition, there are special long-life bulbs rated at 5000 or 10,000 hours. A "lifetime" bulb is also offered.

Hot filaments would burn up if oxygen were present in the bulbs. For this reason, air is drawn out of the glass bulb, or envelope. The bulbs are filled with an inert gas that will not support combustion. The base is attached during manufacture. The glass is sealed at the base. This keeps the gas in and air out. The parts of a modern light bulb are shown in Figure 12-14.

When a bulb is lighted, the temperature of the filament reaches 4750 degrees F [2621 degrees C]. This temperature causes the glass to become hot. If moisture contacts a hot lamp, the glass may break. Hot bulbs can also start fires if lamps are too close to combustible materials. Safety should always be considered in the placement of lamps.

Types of Incandescent Lamps

There are nine types of incandescent lamps:
- General service
- Three-light
- White indirect
- Reflector
- Projector
- Tubular
- Daylight
- Reflector-type
- Severe service.

Each type of lamp has its own uses.

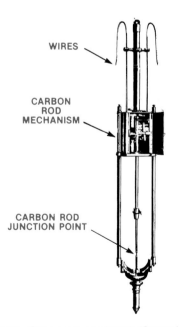

Figure 12-13. Open arc lamp of 1878. (General Electric)

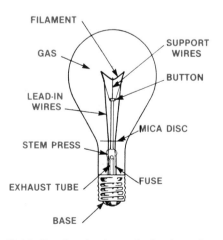

Figure 12-14. Drawing showing parts of an incandescent lamp.

General-Service Lamps. These are usually used in the home. They are available in sizes ranging from 7 W through 1500 W. The 300-W bulb is usually the largest size found in the home.

General service lamps may have frosted, white, or clear glass envelopes. Frosted glass is etched. Acid is applied in etching. This spreads, or *diffuses*, the light passing through the glass.

Even greater diffusion is provided by white lamps. The envelopes of these lamps are coated on the inside, rather than being etched.

The purposes of etching and coating are the same: Glare is controlled. Diffusion of light also softens shadows. This produces better visibility and a more pleasing appearance in working or living areas.

Three-Light Lamps. These lamps have two filaments, Figure 12-15. The two filaments are wired so that each can be used separately. Or, both can be lighted together. The light switch, operating a special socket switch, makes the selection of which filaments are connected. Three levels of wattage are thus available in a single lamp. Typical sizes are 50-100-150 W and 100-200-300 W.

White Indirect Lamps. These are designed for use in table and floor lights that do not have diffusing bowls. They are available in a number of watt sizes. The envelope or bulb has a heavy white coating on the sides. This permits only 20 percent of the light to be cast down. Some 80 percent of the light is directed upward for general illumination of the room.

Reflector Lamps. These are shaped like the white indirect lamps. Instead of a white coating, the sides of these bulbs are coated with an opaque silver finish. All of the light is directed out of the flat, wide end of the bulb. The end of the bulb may be patterned to focus the light. The envelope end may also be frosted or colored. The most common use of these bulbs is as floodlights or spotlights. Common sizes for reflector bulbs include 75 W, 150 W, and 300 W. Reflector lamps are designed for inside use.

Projector Lamps. These are the outdoor equivalent of reflector lamps. Projector bulbs are made of hard glass. These lamps are available in 75-W and 150-W sizes. They are used in floodlights or spotlights.

Tubular Lamps. Envelopes for these lamps are formed in long, tubelike shapes. They are made in two different base shapes. All sizes are available with clear, inside-

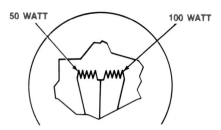

Figure 12-15. Drawing showing filaments of three-light lamp

frosted, or colored envelopes. A coil filament extends the full length of the tube. Small tubular lamps are known as showcase lights and are available in 25-W and 40-W sizes.

Daylight Lamps. These are standard lamps with blue-green glass envelopes. The coloring absorbs some of the red and yellow rays of light from the filament. The resulting room light is whiter than for a normal lamp.

Reflector-Type Lamps. These have a silver-colored or aluminum reflector built into the lamp. The reflector may be coated either on the inside or the outside of the glass envelope. Reflector coatings are available in more than 150 different lamps.

Severe-Service Lamps. These have filaments that will take rough handling. Auto trouble lamps use them.

Advantages of Incandescent Lamps

- Incandescent lamps are inexpensive to install.
- Incandescent lamps give out a relatively bright light from a small device.
- Light bulbs are relatively inexpensive.
- Colors appear relatively natural under incandescent light.

Disadvantages of Incandescent Lamps

- Incandescent lamps have comparatively high power consumption for the light they emit.
- Incandescent lamps tend to have a high level of glare. This is particularly true in comparison with fluorescent lights. They can lead to eyestrain.
- Incandescent lamps give off a great deal of heat. In hot weather, they can make a room uncomfortable.
- Incandescent lamps can implode if they are wet. They also implode if they are dropped.
- If a lamp is knocked over, the hot filament can be a fire hazard.
- The life of the average lamp is short in comparison with that of a fluorescent lamp.

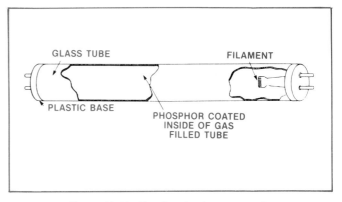

Figure 12-16. Drawing showing construction of typical fluorescent lamp.

12.7 FLUORESCENT LAMPS

The same process that takes place when light is given off by lightning is used in **fluorescent lamps**. These lamps are said to be an **electric-discharge light source**. The process used has been described as man-made lightning produced on a small scale. Of course, the discharge is also controlled. The parts and functions of a fluorescent lamp are shown in Figure 12-16.

How the Fluorescent Lamp Works

The filaments in the fluorescent lamp are heated by a small current. Once they become hot, they emit electrons. These electrons do not produce much light. Rather, they collide with the mercury-vapor atoms inside the tube. The mercury vapor then produces a gaseous discharge with considerable ultraviolet light. The ultraviolet radiation strikes the phosphor coating inside the tube. The phosphor coating then takes on the fluorescent glow.

When the mercury vapor becomes active, electrical resistance within the tube is reduced. This will create a high current drain on the power supply—unless it is limited.

The inductor is an excellent device for limiting current. An inductor is inserted in series with the fluorescent tube. This helps to stabilize the amount of current drawn to produce light. The size of this inductor, or *ballast*, varies with the wattage rating of the lamp.

Voltage is important in producing an arc within a fluorescent lamp. There must be sufficient voltage to cause electron emission. Heaters are used to avoid extremely high voltages. The filament (cathode) is heated to a temperature of 950 degrees Celsius. The heat causes the filament to give off electrons. In turn, the gaseous discharge is started among electrons of the mercury vapor. This is the hot cathode lamp. See the diagrams of the hot cathode lamps in Figure 12-17.

Once the arc (discharge) has been started, the filaments are removed from the circuit. The difference between the

voltages applied to each end of the tube maintains the operation of the lamp. To prevent overheating of the tube filaments, a *starter* automatically removes the filament from the circuit once the discharge has begun. At this point, the filament is no longer needed. This is illustrated in Figure 12-18.

Some fluorescent lamps can be started without preheating. These are known as cold-cathode lamps. Start-

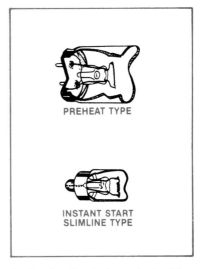

Figure 12-17. Drawing showing hot cathode of fluorescent lamp.

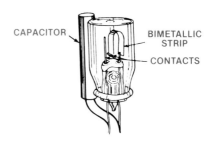

Figure 12-18. Drawing showing internal elements of fluorescent starter.

175

ing is accomplished by applying enough voltage to move electrons. However, lowering of the voltage usually results in a lost arc and a failure to produce light. Figure 12-19 diagrams a cold-cathode tube.

Uses of Fluorescent Lamps

Fluorescent lamps put out more light per watt of power than do incandescent lamps. Fluorescent tubes are about 2.5 times more efficient.

Further, fluorescent lamps have a useful life of between 12,000 and 20,000 hours. This is 15 to 30 times as long as for average incandescent lamps. The main factor in fluorescent lamp life is the coating over the filament. This coating is heated to produce the necessary electron emission needed to form the discharge. The coating material is gradually used up during the starting period. So fluorescent lamps last longer if their starts are kept to a minimum. The ratings on hours of life generally figure one start for each three hours of use.

You may see fluorescent tubes with blackened ends. If this happens early in the life of the tube, there may be a problem. This condition indicates that the active material on the cathode is being sputtered off too rapidly. This may be caused by:

- Frequent starting of the lamp
- Improper operation of starters
- A short-circuited ballast
- Installation of the wrong ballast
- High or low voltage
- Incorrect wiring of the unit.

A blinking lamp may cause extra heating. This can damage both the starter and the ballast.

Replace all lamps when you see that they do not operate properly. In replacing lamps, be sure to turn off the circuit first.

To use fluorescent lamps on DC circuits, special ballasts and circuits are needed.

Fluorescent lamps are designed for indoor use. They operate best in temperatures of 65 to 90 degrees F [18 to 32 degrees C].

Types of Fluorescent Lamps

There are four basic types of fluorescent lamps:

- General-line (preheat) lamps
- Rapid-start lamps
- Slimline lamps
- Circline lamps.

These lamps are available for both home and commercial use. A number of bases are used on fluorescent lamps. Their lampholders vary in size and shape, Figure 12-20.

Some lampholders have a starter socket mounted at one end, Figure 12-21.

Advantages of Fluorescent Lamps

- Operating costs are low.
- Operating temperatures are low.
- Glare is lower than with incandescent lighting.
- A diffused light is produced.
- Many colors are available.
- Tubular shapes give uniform light distribution.
- Lamps have long lives.

Figure 12-19. Drawing showing cold cathode of fluorescent lamp.

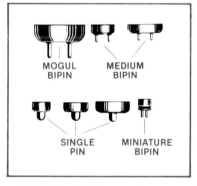

Figure 12-20. Standard bases of fluorescent lamps.

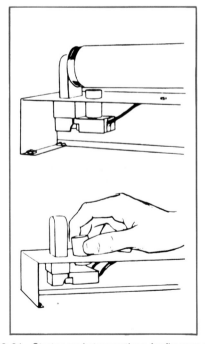

Figure 12-21. Starter-socket mountings for fluorescent lamps.

Disadvantages of Fluorescent Lamps

- Frequent starting shortens life of lamp.
- Preheat types are slow in starting.
- Operation on DC is inefficient.
- Operation is poor at very low or very high temperatures.
- Special equipment is required for starting.
- Installation is more costly than for incandescent lighting.

REVIEW QUESTIONS FOR SECTION 12.7

1. Describe the operation of a fluorescent lamp.
2. What is the function of the inductor in a fluorescent-lamp circuit?
3. What is the function of the filaments in fluorescent lamps?
4. How is light emitted from a fluorescent lamp?
5. What is a hot cathode?

12.8 MERCURY-VAPOR LAMPS

Mercury-vapor lamps are used commercially. The most common examples are for lighting highways, streets, and bridges. Mercury lamps also have a number of uses in the commercial and industrial fields. These include lighting for foundries and high-bay factories. In addition, mercury lighting is also used for floodlighting building exteriors, parking lots, recreational areas, and railroad yards. This type of lighting can also be found in auditoriums and airports, and at building construction sites.

One type of mercury-vapor lamp has a screw-type base. This is illustrated in Figure 12-22.

Most of the rays emitted by mercury-vapor lamps are ultraviolet. Several special lamps have been developed to use this ultraviolet radiation. They include sunlamps, black-light lamps, and germicidal lamps.

Advantages of Mercury Lighting

- Lamps have several thousand hours of life.
- Lamps are efficient. They produce about twice the amount of light as do incandescent lamps of equal wattage.

Disadvantages of Mercury Lighting

- There is a high initial cost.
- Special ballasts (transformers) are required [except when a self-ballast-type bulb is used].
- When turned on, the lamps are slow in reaching full output.
- If interrupted, lamps cannot be restarted until they cool sufficiently to allow the cycle to begin again.

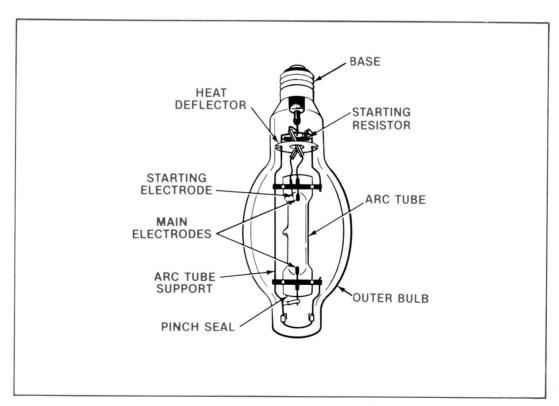

Figure 12-22. Drawing showing parts of a 400-watt mercury-vapor lamp.

- The lamps give off a bluish-green light. The light is almost completely lacking in red values. If natural color is desired, mercury-vapor and incandescent lights are usually mixed.
- If the outer bulb is broken, a high level of ultraviolet light is emitted. This can be a hazard in some locations.

REVIEW QUESTIONS FOR SECTION 12.8

1. Describe the operation of a mercury-vapor lamp.
2. What is the color of the light given off by a mercury-vapor lamp?
3. Why are mercury-vapor lamps used primarily for street and industrial lighting?

12.9 OTHER TYPES OF LAMPS

Neon Glow Lamps

Neon signs are used to identify and advertise businesses. They have been used commercially for many years. In addition, small lamps using neon or argon gases are used for many purposes. They serve as pilot lamps, current indicators, or signal lamps. Several sizes and shapes of neon glow lamps are illustrated in Figure 12-23.

Neon lamps are low in current consumption. For example, the largest available lamp uses only 0.03 A. These lamps are rugged, reliable, and long-lived. They have a rated life of between 3000 and 25,000 hours.

Operation of neon glow lamps is based on a difference of potential between two electrodes. One electrode of a neon lamp may appear to glow. This identifies the negative terminal of a DC power source. If both electrodes appear to glow, this indicates the presence of AC.

For normal line-voltage operation, most neon lamps require a resistor to be connected in series. These lamps usually ionize, or glow, at about 55 V.

Lasers

A laser is a device that has turned light into an advanced tool of science and industry. The word **laser** is an acronym. This means that the word is built from the initial letters of a series of words. The words are *Light Amplification by Stimulated Emission of Radiation*.

All lasers use light, but the type of light used is important. All light is in the form of electromagnetic radiation. Ordinary light, such as that produced by a light bulb, consists of electromagnetic radiations of different frequencies. Visible light is electromagnetic radiation at 430 to 730×10^{12} Hz—far above the frequency of microwaves.

Visible light is referred to as *incoherent*. Incoherent light is a random, out-of-phase emission of energy. It diffuses rapidly and cannot be precisely or accurately transmitted without a significant loss of energy. In short, it scatters and loses its ability to be seen after a short distance, Figure 12-24A.

Light generated by the laser is coherent light. It consists of radiation waves that are in phase, Figure 12-24B. Because of this characteristic, coherent light does not diffuse appreciably and can be transmitted over long distances without a signficant loss of energy.

Many types of different materials have been successfully stimulated to exhibit laser action. Figure 12-25 shows how the ruby rod (solid-state laser) is used to produce a coherent light for use as laser beams. The brilliant red, green, blue, and yellow beams seen at laser-light shows are produced by gaseous materials.

The devices used in communications are members of the light-emitting-diodes family (LED). LED lasers are created by current injection into a diode laser. When it is below a critical value (the threshold), the diode behaves just like an LED and emits a relatively broad spectrum of wavelengths in a wide radiation pattern. As the current increases to the threshold, however, the light narrows into a distinct beam and is confined to a very narrow spectrum. The device then functions as a laser.

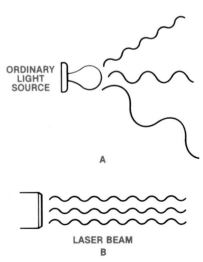

ORDINARY
LIGHT
SOURCE

A

LASER BEAM
B

Figure 12-24. Drawings show waves of incandescent lamp and laser beam.

Figure 12-23. Neon glow bulbs come in a variety of sizes and shapes. (General Electric)

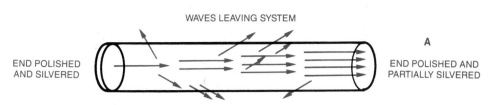

WAVES LEAVING SYSTEM

A

END POLISHED
AND SILVERED

END POLISHED AND
PARTIALLY SILVERED

The ends of a ruby rod are flattened (so they are parallel), and silvered to form mirrors. The mirror at one end is made to reflect only part of the light so that when there is a buildup in energy between the mirrors, the beam can escape.

WAVE GROWTH BY STIMULATED EMISSION

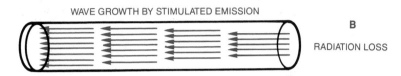

B

RADIATION LOSS

Soon after the chromium atoms in the ruby crystal are pumped by a flash lamp to a higher energy level, they drop to another level, and stimulated emission takes place. Waves moving at angles to the crystal's axis leave the system, but those traveling along the axis grow by stimulated emission of photons.

C

LASER BEAM

The parallel waves are reflected back and forth between the mirrors and the wave system grows in intensity. A pale red glow indicates a certain amount of light being lost at the mirror, but, beyond a critical point, the waves intensify enough to overcome this loss and an intense red beam flashes out of the crystal's partially silvered end.

Figure 12-25. Turning incoherent light into a laser beam.

Special types of semiconductor diodes are made to withstand continuous operation. The stripe geometry of the DH laser (Figure 12-26) illustrates the point. Gallium arsenide (GaAs) and aluminum gallium arsenide (AlGaAs) are used. This confines the generation and emission of light to the PN junction region sandwiched between the two or more semiconductor layers. This combination of events will allow for a high-efficiency operation at low threshold currents. These devices generate several milliwatts of laser light and are used for fiber-optic communications, video-disk readouts, and computer printing systems. This type of laser is called a DH (from *Double Heterojunction*). *Heterojunction* is another term for *combining two dissimilar semiconductor materials*. Semiconductors are covered in Chapter 16.

There are many uses for lasers. Some have been used extensively, and others are still experimental. This is a good field for those who like to experiment and produce new products. The following are just a few of the uses for lasers:

- Industrial welding
- Surveying equipment
- Medical applications, such as eye surgery
- Military applications such as the "smart bomb," which follows a laser beam to the target
- The production of holograms for three-dimensional photography and *Automan*® on television.
- Pickup and playback devices for video and audio compact discs (see Figure 12-27).

METAL
CONTACT

SiO
INSULATION

(p)AlGaAs

(p)GaAs

(n)AlGaAs

PN
JUNCTION

Figure 12-26. Stripe geometry of the DH laser.

REVIEW QUESTIONS FOR SECTION 12.9

1. Describe the operation of a neon lamp.
2. What is a laser?
3. What are the special characteristics of laser light?

Figure 12-27. Compact-disc player using 3-beam laser with remote control. (Courtesy of Tandy Corporation)

12.10 LIGHT CONTROL

Consumer demands for more household light have led to the introduction of larger bulbs. In addition, many lamps and fixtures have been developed to improve control over home lighting. With modern methods, it is possible to reduce glare and brightness while improving the distribution of light. Techniques for controlling light in the home include:

- Reflection
- Transmission
- Diffusion
- Refraction
- Absorption
- Polarization.

All of these methods rely on physical properties of materials used in lighting.

Reflection uses shiny surfaces. Metals and plastics are good examples of reflective materials used in home lighting.

Light is transmitted by clear glass. The clear envelope of a bulb provides a good example of this property.

Diffusion is the spreading of light by restriction of its passage. Etched glass or white-coated bulb envelopes provide examples of diffusion.

Refraction is the bending or guiding of light to focus illumination. Examples are the spotlight front surfaces built into some reflector lamps. Spotlights used in theaters refract and focus light through the use of lenses.

Absorption is the filtering of transmitted light to alter the color of the light. All colors except the desired color are absorbed by the filter. Colored or tinted glass envelopes on light bulbs provide examples of absorption.

Polarization is done with special glass that reduces and directs the transmission of light. The effect is to control the direction or vibration of light rays passing through glass. Polarized sunglasses are a common example of the use of this method.

These lighting-control methods are illustrated in Figure 12-28. Most light-control equipment uses a combination of two or more of these methods.

REVIEW QUESTIONS FOR SECTION 12.10

1. How does reflective lighting work?
2. What is diffused lighting?
3. What is the meaning of *absorption*?
4. Why is diffusion used in home lighting?
5. What is polarization of light and what does it do?

SUMMARY

An electric arc is produced by current flowing through two carbon rods. The rods are brought into contact with each other. An arc is struck. Once the electric arc is established, the rods are moved apart slightly. The arc vaporizes small quantities of carbon. The carbon vapor provides a path for current to flow. The current maintains the arc, which produces light and heat.

Materials that resist current are called resistors. It takes

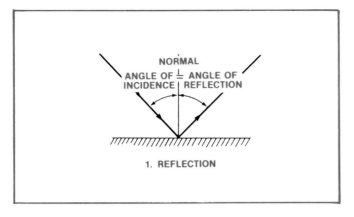

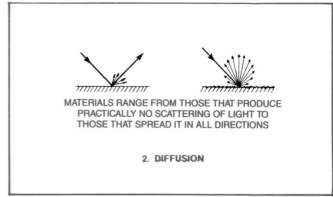

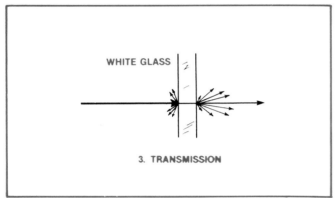

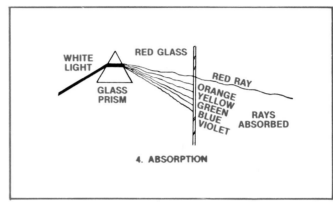

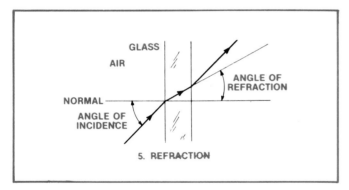

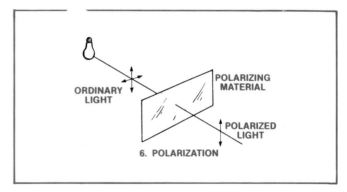

Figure 12-28. Illustrations of six ways to control light.

more pressure to force electrons through a resistor than through a good conductor. This added pressure generates heat. Welding is a good application of heat utilized to melt metals and fuse them together. A spot welder is capable of binding two pieces of metal together at a small point, using large currents and low voltage.

The quartz heater produces heat by utilizing a crystal material. Microwave ovens operate at 2450 megahertz. They heat material by agitating the water molecules in the material. Microprocessor technology is used in the control or timing of the microwave oven.

Fuses are safety devices. If too much current is drawn through a fuse, it melts and opens the circuit. Large industrial-type fuses have replaceable links.

Incandescent lamps use filaments that give off light when heated. These filaments glow white-hot. This glowing is called incandescence. There are eight types of incandescent lamps and each type has its own uses. Use of incandescent lamps has advantages and disadvantages.

Fluorescent lamps are said to be an electric-discharge light source. The process used has been described as man-made lightning produced on a small scale. Filaments in a fluorescent lamp are heated by a small current and become hot. They emit electrons. These electrons do not produce much light; rather, they collide with the mercury vapor atoms inside the tube. The mercury vapor then produces a gaseous discharge with considerable ultraviolet light. The ultraviolet radiation strikes the phosphor coating inside the tube. The phosphor coating then takes on a fluorescent glow. Fluorescent lamps have their advantages and disadvantages. These tubes come in at least seven colors and also include ultraviolet black-light tubes.

Mercury-vapor lamps are used commercially. They also emit ultraviolet rays. These lamps have several thousand hours of life. They are efficient and produce about twice the amount of light as an incandescent lamp of equal wattage. There are also some disadvantages of this type of lamp.

Neon glow lamps are used to identify and advertise businesses. They are also low-current devices used for various applications, such as in warning lights and in some television circuits.

A laser is a device that has turned light into an advanced tool of science, industry, and medicine. The word *laser* is an acronym. This means that the word is built from the initial letters of a series of words. The words are *L*ight *A*mplification by *S*timulated *E*mission of *R*adiation. A laser emits what is known as coherent light. ·

Light can be controlled by reflection, transmission, diffusion, refraction, absorption, and polarization.

_____ USING YOUR KNOWLEDGE _____

1. Draw the symbol for a fuse.
2. List the important steps that should be followed before changing a fuse or resetting a circuit breaker.
3. Draw a circuit showing the operation of a carbon arc.
4. Draw an AC circuit showing the operation of an incandescent lamp. Be sure to place a fuse and a switch in the circuit.

_____ KEY TERMS _____

electric arc
welding
quartz heater
microwave
fuse
illumination
filament
electric-discharge light source

fluorescence
fluorescent lamp
incandescent lamp
laser
diffusion
refraction
reflection

Safety Tips

1. If at all possible, turn off the power before changing a fuse.
2. Never touch the metal portion of the socket while you are screwing in a plug fuse.
3. Never change a fuse when your hands are wet or you are standing on a wet floor.
4. Do not try to repair a blown fuse unless it is the renewable type.
5. Before replacing a fuse, trace the circuit. Unplug any devices or lamps that could be causing overloads.
6. Never replace a fuse with one of a larger capacity that exceeds the rating for the circuit.
7. Never restore a circuit to service by placing a coin in the fuse socket.
8. Make sure that all fuses are inserted tightly.
9. If a motor is being used on a circuit, use of a dual-element fuse is recommended. This will protect against fuse blowout resulting from current surges. Motors cause current surges when they start.

Chapter 13

GENERATORS

OBJECTIVES

After studying this chapter, you will know

- *how a generator works.*
- *how DC generators are connected.*
- *the basic differences between single-phase and three-phase power.*
- *how a sine wave is generated.*
- *the difference between a salient-pole and a turbo-type generator.*
- *how DC is obtained from an AC generator.*
- *how three-phase power is produced.*
- *the differences between delta and wye connections.*

INTRODUCTION

Electrical energy can be produced a number of ways. However, the most efficient and the most economical way to do so is by means of a generator. The alternating-current generator is referred to as an alternator. The DC generator has been replaced on the automobile, and the AC generated by the alternator is converted to DC before it is allowed to leave the generator housing. It is easily seen that both types of generators are needed, inasmuch as AC and DC are both utilized in electrical motors and electronics equipment.

Most commercial power generated in the United States and in other countries is produced as **three-phase**. It is distributed as three-phase and in some cases as **single-phase**. The industrial world utilizes the three-phase power to operate equipment because such power use does not require start mechanisms on electric motors.

Delta and **wye** are terms to be dealt with when working with three-phase power. The terms will be used in reference to three-phase generators, transformers and motors. Each type of connection, whether delta or wye, has advantages and disadvantages. Some of these are explored, as is the conversion from one to another.

13.1 DC GENERATORS

A battery is one method for supplying direct current. However, a more economical source is the mechanically driven DC generator.

In a DC generator, mechanical force is used to rotate a wire loop in a magnetic field. This generates electricity. The magnetic field is generated by a current-carrying wire looped around a core.

These principles are demonstrated in Figure 13-1. The drawings show a magnetic field generated by a current-carrying coil. The electrical source is a battery. The north and south poles of the magnet are in the positions indicated. As a conductor moves upward between the magnetic poles, current flows. The direction of flow is indicated in Figure 13-1A. As the conductor moves downward, the direction of flow reverses, Figure 13-1B.

The voltage generated depends on a series of factors. One factor is the intensity of the magnetic field. Another is the number of turns of the wire. A third factor is the speed with which the wire passes through the magnetic field.

A simple generator is shown in Figure 13-2. A loop of wire is wrapped around an iron core, called an **armature**. Copper segments are attached to the ends of the loop of wire. These segments are insulated from the core and from each other. These copper segments form what is known as the *commutator. Brushes* are metal or carbon devices positioned so that they contact the commutator. These brushes then carry the generated electricity to the load.

To produce electricity, the armature must be mounted between the *field coils*. This means the magnetic force is

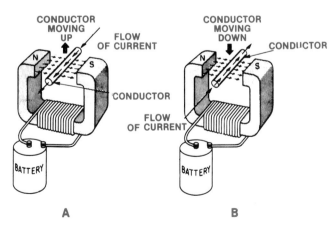

Figure 13-1. Principles of DC generation.

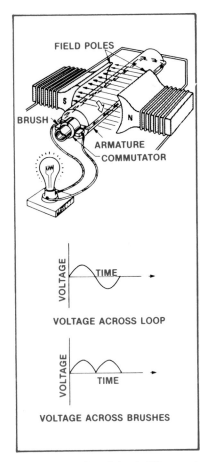

Figure 13-2. Drawing of simple DC generator.

Construction

A practical generator uses electromagnets, instead of permanent magnets. This produces a much stronger magnetic field without increasing the physical size of the magnets.

Figure 13-3 shows the pole piece bolted to the yoke. The field winding is wrapped around the poles. Note the north–south poles and the lines of flux.

The same basic generator, but with four poles, is shown in Figure 13-4. Note the north–south pole orientation here, and the way the field windings are connected. Figure 13-5 shows a more complicated arrangement with an eight-section, ring-type armature.

There are more commutator segments on a practical generator design, Figure 13-6. Note the coils and slots in

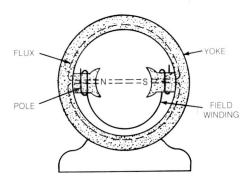

Figure 13-3. Two-pole assembly.

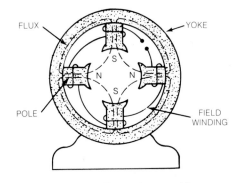

Figure 13-4. Four-pole assembly.

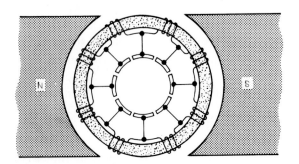

Figure 13-5. An eight-section, ring-type armature.

cut by the rotating armature. The field coils form an electromagnet. These coils, in turn, are wrapped around soft iron cores called *field poles*.

Note the position of the armature in the diagram in Figure 13-2. The armature loop is cutting directly across the magnetic field. In this position, it is generating maximum output. When the armature rotates a quarter turn, the loop is parallel to the magnetic field. In this rotated position, there is no output from the generator. As the armature moves through a full, 360-degree rotation, generation levels change. The maximum and minimum points are reached twice during each rotation. Maximums are reached when the armature passes both the north and south poles.

The voltage across the loop is shown in Figure 13-2 as alternating current. Voltage across the brushes is shown as *pulsating direct current*, or *PDC*. The commutator acts as a reversing switch. The switching action is applied as the armature rotates in different fields. The resulting output is a series of maximums and minimums. But the current flow is only in one direction.

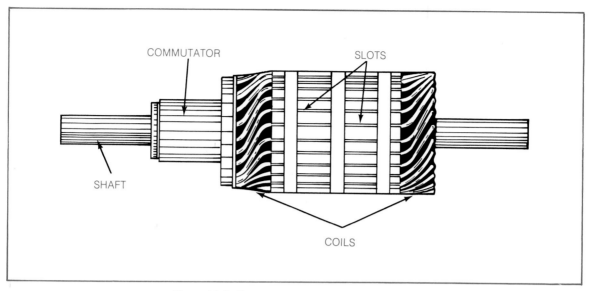

Figure 13-6. Drum-type armature, partly assembled.

this drum-type armature. Mica is used to insulate the copper segments of the commutator from one another. High-grade carbon brushes ride on the surface of the commutator. The brushes form the electrical contact between the armature coil and the external circuit. The brushes are held in place by brush holders that are insulated from the frame. Brushes are permitted to slide up and down in their holders so that they can follow the irregularities in the surface of the commutator. Pressure of the brushes may be varied and their position on the commutator is adjustable.

REVIEW QUESTIONS FOR SECTION 13.1

1. What is the role of a magnetic field in DC generation?
2. What is the function of the armature in a DC generator?
3. Why do generators use electromagnets instead of permanent magnets?
4. How are brushes held close to and against the surface of a commutator?

13.2 TYPES OF DC GENERATORS

DC generators are classified according to the method used to supply the exciting current to the field windings. When the field current is obtained from a separate source, the generator is said to be separately excited. The self-excited generator has the excitation current supplied by the generator itself. Self-excited generators are further divided into at least three groups: shunt, series, and compound.

Shunt Generator

The poles of a generator retain some magnetism, known as residual magnetism, when not in operation. Residual magnetism produces a weak magnetic field. That means that when the generator is started, a small voltage is induced in the armature and appears at the output terminals. Then, because the armature output voltage is connected across the field windings, a small current flows in the windings, Figures 13-7 and 13-8. This field cur-

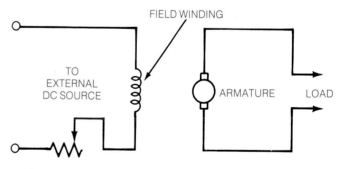

Figure 13-7. Shunt generator, separately excited.

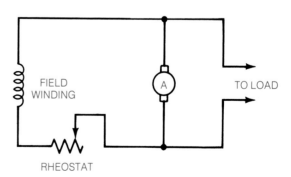

Figure 13-8. Shunt generator, self-excited.

rent, in turn, strengthens the magnetic field (the output voltage increases accordingly), and a larger current flows in the windings. This action is cumulative. The output voltage continues to rise to a point where the field is saturated; then no further increase in output voltage occurs.

If the initial direction of the armature rotation is wrong, the cumulative action does not occur, since the small induced voltage opposes the residual field and there is no build-up of output voltage.

When the field windings of a shunt generator are excited from an external source, the machine is said to be separately excited, Figure 13-7. This means the terminal voltage of the generator can be controlled by insertion of a rheostat in series with the field windings. As the resistance is increased or decreased, the voltage is increased or decreased by the amount of excitation furnished to the field winding.

Shunt generators are used primarily in such applications as battery charging, which requires a constant voltage under varying current conditions. Separately excited shunt generators are often used in certain speed-control systems.

Series Generator

The field winding of a series generator is connected in series with the armature output voltage, Figure 13-9. The field coils of the series generator are made up of a few turns of heavy wire. All current flowing through the field coil also flows through the armature. Series generators have very poor voltage regulation under changing load conditions. The greater the current through the field coils to the load, the greater will be the induced emf, and the greater will be the output voltage of the generator. Therefore, when load is increased, voltage will increase; when load is decreased, voltage will decrease. Because the

series generator permits such poor regulation, however, only relatively few are in actual use.

Compound Generator

A compound generator has both a series and a shunt field. Both windings are mounted on the same pole structure. The series field may be connected to aid or oppose the shunt field, Figures 13-10 and 13-11.

There are several types of compound generators:

Cumulative compounded: Has the series field aiding the shunt field, Figure 13-10.
Flat-compounded: Voltage remains contant for all loads.
Over-compounded: Voltage rises with increased load.
Under-compounded: Voltage drops with increased load.
Differentially compounded: Has the series field opposing the shunt field, Figure 13-11.

Compound generators are usually designed to be over-compounded. This permits the degree of compounding to be varied by connection of a variable shunt across the series field. This shunt is called a *diverter*. Compound generators are used where voltage regulation is of great importance.

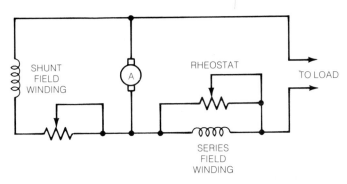

Figure 13-10. Cumulative compound connections.

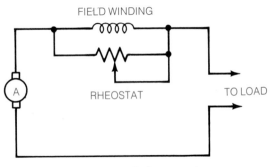

Figure 13-9. Series generator.

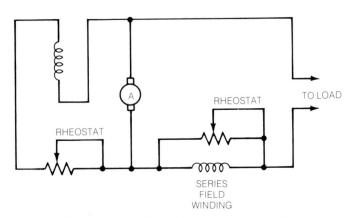

Figure 13-11. Differentially compound connections.

Differential generators have somewhat the same characteristics as series generators, in that they are essentially constant-current machines. However, they generate rated voltage at no load, with the voltage dropping as the load increases. This constant-current characteristic makes them ideally suited as power sources for electric arc welders, which are the principal application for them.

Short-Shunt, Long-Shunt Generators

A long shunt is produced by connecting the shunt field of a generator across both the armature and series field. If the shunt field is connected across the armature alone, it is called a short-shunt connection. Short-shunt and long-shunt generator connections produce similar results.

Excitation and Voltage Regulation (DC Generator)

Excitation refers to the process used to cause a magnetic field to be generated. This magnetic field is used to produce electricity; whether AC or DC depends upon how the generator is connected. There are two main types of excitation: separately excited (or independently excited) and self-excited.

Self-Excited Generators. Self-excited generators supply their own excitation. They rely upon residual magnetism to produce the magnetic field needed to generate an electric current when the armature first starts to rotate. After the armature starts to rotate, part of the output is fed back to excite the armature further, according to the generator load. This weak magnetic field from the residual magnetism in the core of the armature is capable of producing enough current to excite the armature further once the normal speed of the generator has been reached. The fed-back current is then of sufficient level to provide any output demanded, plus the excitation.

Self-excited generators are classified according to the type of field connection, of which there are three general types: series-wound, shunt-wound and compound-wound. Compound generators can also be broken down into other classifications, such as differential-compound and cumulative-compound, each with its own characteristics. It is possible further to classify shunt generators as long-shunt and short-shunt.

Separately-Excited Generators. Separately-excited generators have a DC generator attached to the shaft of the larger generator. This smaller generator produces the excitation needed to cause the generator to operate at its design level.

Larger, commercial-type AC generators usually have separate sources of excitation.

13.3 THE AC GENERATOR

An AC generator is called an alternator. As an example, see Figure 13-12. The power plant in this photo uses coal as its energy source. Coal is burned to heat water. Steam is produced, then super-heated to a very high temperature. Steam under pressure drives turbines, rotating cylinders. The turbines are connected to large AC generators. Figure 13-13 illustrates the size of AC generators.

Figure 13-12. Coal-burning electrical power plant. (Niagara Mohawk Power Corporation)

Figure 13-13. Power-plant turbine generator. (Niagara Mohawk Power Corporation)

Nuclear energy is also being used to produce electricity. Heat from controlled nuclear reactors is used to produce steam to drive turbines. The main difference in a nuclear power plant lies in the source of heat. Otherwise, power is generated in much the same way as in a coal-fired plant, Figure 13-14.

Other sources of heat used to drive power plants include oil and natural gas. However, attempts are being made to reduce use of these fuels. Recent energy shortages have made these fuels costly and scarce.

Falling water is also used to drive turbines and produce electric power. The water comes from reservoirs built up behind dams. This method is known as hydroelectric generation, Figure 13-15.

All alternating-current generators operate in the same basic way. The principles of an AC generator are illustrated in Figure 13-16.

Principles of Operation

Examine the four positions of the single-turn loop and slip rings as they rotate clockwise in a uniform magnetic field produced by the poles of a magnet, Figure 13-17.

Figure 13-14. Nuclear power plant. (Tennessee Valley Authority)

Figure 13-15. Hydroelectric power generation, Fontana Dam, Tennessee. (Tennessee Valley Authority)

When the loop rotates through the position 1 of Figure 13-17, the black coil side is moving toward the north pole and the white side is moving toward the south pole. Because the coil sides are moving parallel to the direction of the magnetic field, no flux lines are cut and the emf induced in the loop is zero. This condition may be verified by connection of a suitable meter across the output terminals shown by the arrows in Figure 13-17. (The center-scale position of needle indicates zero volts.) Note that the coil ends are connected to the slip rings. The ends rotate simultaneously with the loop. The stationary carbon brushes make contact with the slip rings. The former are used to conduct the generated voltage to the external load, which in this case is the meter.

After the loop has rotated 90° from its initial position and is passing through position 2, the black coil side is moving downward and the white side is moving upward. Both sides are cutting a maximum number of flux lines. The induced emf, as indicated by the meter, is at a positive maximum.

As the loop passes through an angle of 180°, in position 3, the coil sides are again cutting no flux lines. The generated emf is zero. Verify this point by referring to the sine wave shown in Figure 13-18.

As the loop passes through an angle of 270°, as in position 4, the coil sides are cutting a maximum number of flux lines. The generated emf is at a negative maximum.

The next 90° turn of the loop completes a 360° revolution. The generated emf falls to zero once again.

The complete rotation may therefore be summed up as follows: As the loop makes one complete revolution of 360°, the generated emf passes from zero, to a positive maximum, back to zero, to a negative maximum, and back again to zero. If a constant speed of rotation is assumed, the emf generated is a sine wave, Figure 13-18.

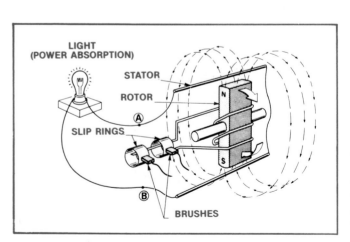

Figure 13-16. Simplified diagram of alternator.

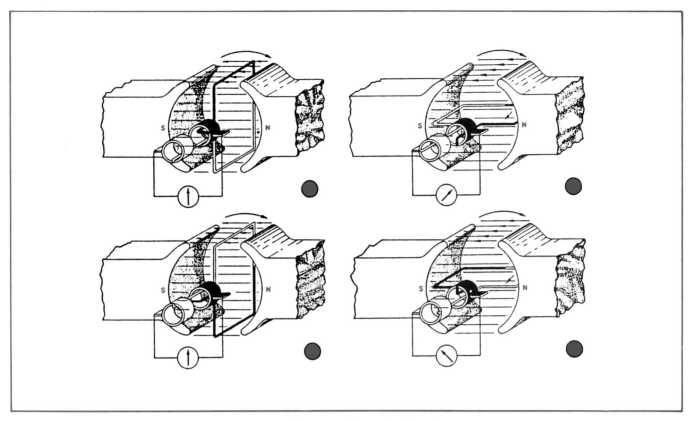

Figure 13-17. Generation of an emf with an alternator.

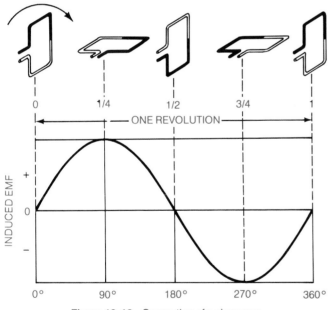

Figure 13-18. Generation of a sine wave.

The output shown from the generator is single-phase (ϕ). (The Greek letter **phi** (ϕ) is used to designate phase.) It is possible to generate three phases at the same time with the proper arrangement of the coils on the stator or on the rotor of the alternator. See Figure 13-24 for the three-phase (3ϕ) alternator output.

Types of Alternators

The salient-pole rotor, Figure 13-19, and the turbo-type rotor, Figure 13-20, are two types of alternators. A slow-speed diesel engine or a water turbine that runs at about 720 rpm can be used to drive the salient-pole alternator.

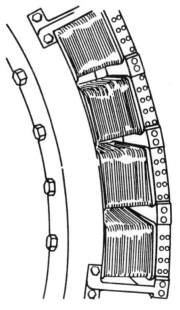

Figure 13-19. Salient-pole rotor.

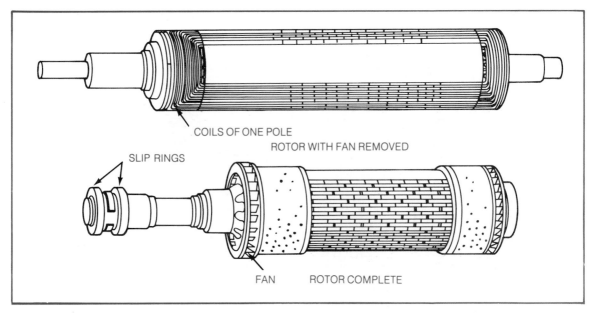

COILS OF ONE POLE

ROTOR WITH FAN REMOVED

SLIP RINGS

FAN   ROTOR COMPLETE

Figure 13-20. Turbo-type rotors.

Field poles are formed by fastening a number of steel laminations to a spoked frame, or spider. The heavy-pole pieces produce a flywheel effect on the slow-speed rotor. This helps to keep the angular speed constant and reduce variations in voltage and frequency of the generator output.

In the high-speed alternators (up to 3600 rpm), the smooth-surface, turbo-type rotor is used because it experiences less air-friction (heating) loss and because the windings can be placed so that they are better able to withstand the centrifugal forces developed at high speeds. The **stator** for the salient-pole alternator is shown in Figure 13-21. The stator for high-speed turbo generators is shown in Figure 13-22. Note that the stator laminations are ribbed to provide sufficient ventilation because the high temperature developed in the windings cannot be dissipated in the small air gap between the rotor and the stator. In some larger installations, alternators are totally enclosed and cooled by hydrogen gas under pressure, which has greater heat dissipating properties than air does. Stator coils in high-speed alternators must be well braced to prevent their being pulled out of place when the alternator is operating at heavy load.

Exciters

AC generators require a separate DC source for their fields. This DC field current must be obtained from an

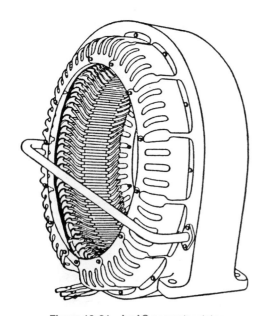

Figure 13-21. An AC generator stator.

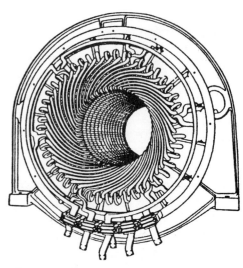

Figure 13-22. A turbo-type AC generator stator.

external source called an exciter. The exciter used to supply this current is often a flat, compound-wound DC generator designed to furnish from 125 to 250 volts. The exciter armature may be mounted directly on the rotor of the AC generator, or it may be belt driven.

Frequency

The frequency of the output alternating current is determined by the number of coils wound in the rotor or stator in most cases and by the speed of rotation of the rotor. How the windings are wound on the rotor or stator determines the phase of the output voltage. (See the discussion of three-phase generators later in this chapter.)

The frequency of the generator output is in hertz. Every complete rotation of the rotor produces a cycle, or hertz. The number of poles can also affect the frequency of the output voltage. The frequency is related to the number of poles and the speed of the generator. The formula for the frequency is expressed as follows:

$$f = \frac{P}{2} \times \frac{N}{60} = \frac{PN}{120}$$

where f equals frequency, P equals poles, and N equals speed of rotation in rpm.

Voltage Regulation

Inasmuch as the speed of most alternators (except for those used on automobiles) is constant, the voltage output is constant. The output voltage and the amount of current available depend on the exciter and its output. Therefore, the amount of excitation can have as much bearing on the voltage regulation as does the speed of the rotor.

13.4 AC-DC COMPARISON AND CONVERSION

A comparison of current output patterns for AC and DC generators is shown in Figure 13-23. This shows a comparison between outputs of an alternator and a DC generator in an automobile. The alternator has some important advantages. For one thing, it delivers more current at low speeds. In generating electricity for an automobile, this can be important. The greatest demand for power often comes at low speeds. At speeds of under 30 miles per hour, an alternator can keep up with demands for electric power. A DC generator cannot match this performance.

As another advantage, the alternator requires less maintenance. It delivers trouble-free operation longer than a DC generator does. It has slip rings instead of commutator segments. Commutator segments cause wear on the brushes.

An alternator, also called an alternating-current generator, provides AC as a source of electricity for an automobile. However, most of the electrical requirements in a car are for DC. Direct current is needed by the battery, by a number of DC motors, and by the electrical gages on the instrument panel. Alternating-current output is changed to DC before it leaves the alternator.

13.5 THREE-PHASE GENERATORS

All the commercially generated electrical power in the United States, Canada, and most other countries is generated as three-phase (3ϕ). Three-phase power is slightly different from the power you have been studying up to this point. Three-phase is power generated for use by industry and some commercial installations. It has many advantages

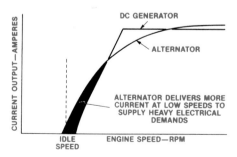

Figure 13-23. Graphs comparing outputs of AC and DC automotive generators.

over single-phase, which is normally used in residential wiring. Figure 13-24 shows how three-phase in Figure 13-25 differs from single-phase AC.

Electrical power generated for commercial and industrial use in the United States and Canada is produced as 60 Hz. In Europe, the power is generated as 50 Hz. Single-phase power used in homes is produced as three-phase but is obtained by properly connecting transformers in substations after the power has reached a local distribution center.

A three-phase generator is one in which the voltages have equal magnitudes and are displaced 120 *electrical degrees* from one another, Figure 13-25. The three windings are placed on the armature 120° apart, Figure 13-26. As the armature rotates, the outputs of the three windings are equal but out of phase by 120°, Figure 13-27.

Connecting Three-Phase Windings

Three-phase windings, both in generators and in transformers, are usually connected in either a delta or wye configuration. Each of these connections has definite

electrical characteristics, from which the designations *delta* and *wye* are themselves derived. Figure 13-28 shows the delta-connected schematic and symbol. Figure 13-29 shows the wye-connected schematic and symbol.

Properties of the Wye Connection

In the wye connection, the current in the line is in phase with the current in the winding. The voltage between any two lines is not equal to the voltage of a single phase, but is equal to the vector sum of the two windings between the lines. The current in line A of Figure 13-30 is the current flowing through the winding

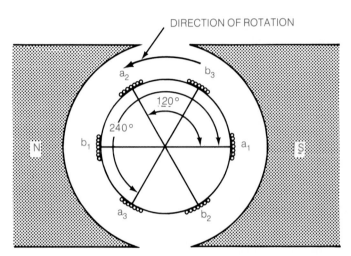

Figure 13-27. Basic representation of a three-phase alternator.

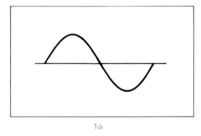

Figure 13-24. Three-phase and single-phase sine waves.

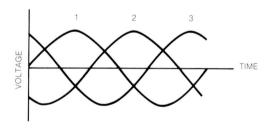

Figure 13-25. Three-phase voltage induced in stator windings.

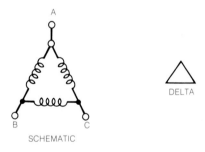

Figure 13-28. Three-phase delta schematic and its representation, the triangle.

Figure 13-26. Schematic representation of a three-phase Y stator.

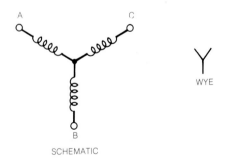

Figure 13-29. Schematic representation of a three-phase Y.

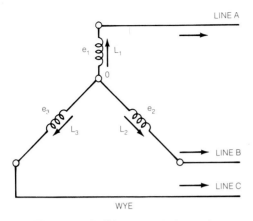

Figure 13-30. Wye-connected currents.

L_1; that in line B is the current through winding L_2; that in line C is the current through winding L_3. Therefore, the current in any line is in phase with the current in the winding that it feeds. Since the line voltage is the vector sum of the voltages across any two coils, the line voltage E and the voltage across the winding $E\phi$ are 30° out of phase. The line voltage may be found by multiplying the voltage of any winding $E\phi$ by **1.73**, or $\sqrt{3}$ (which is equal to **1.732050808**). This is the type of winding that produces 240 volts single-phase when properly connected. Note that the wye connection has a voltage advantage due to the fact that two windings are in series and in parallel with the remaining winding.

Properties of The Delta Connection

In a balanced circuit, when the generators are connected in delta, the voltage between any two lines is equal to that of a single phase. The line voltage and the voltage across any winding are in phase, but the line current is 30° or 150° out of phase with the current in any of the other windings, Figure 13-31. In the delta-connected generator, the line current from any one of the windings is found by multiplying the phase current by the square root of 3 (which can be rounded to 1.73). Keep in mind that the delta connection has a current advantage, whereas the wye connection has a voltage advantage.

The properties of delta connections may be summarized in the following manner. The three windings of the delta connection form a closed loop. The sum of the three equal voltages, which are 120° out of phase, is zero. This means that the circulating current in the closed loop formed by the windings is zero. The magnitude of any line current is equal to the square root of 3 (roughly 1.73) times the magnitude of any phase current.

Transfer Connections

The three-phase transformer is used when large power outputs are required. Either a single transformer or three separate transformers may be used, but the connection is generally delta or wye.

Practical three-phase transformers use the shell construction, Figure 13-32. This transformer is economical to construct and occupies less space than do three single transformers. However, if one phase burns out, the entire unit must be replaced.

Commercial three-phase voltage from power lines is usually 208 volts. The standard values of single-phase voltage can be supplied from the line, Figure 13-33. The windings represent the wye-connected transformer, which is generally an outside installation. Figure 13-34 illustrates the types of connections possible.

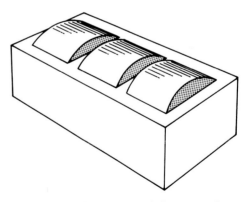

Figure 13-32. Three-phase, shell-type transformer.

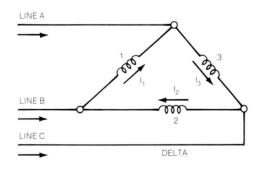

Figure 13-31. Delta-connected currents.

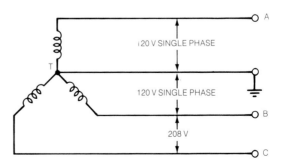

Figure 13-33. Commercial three-phase voltage from power line.

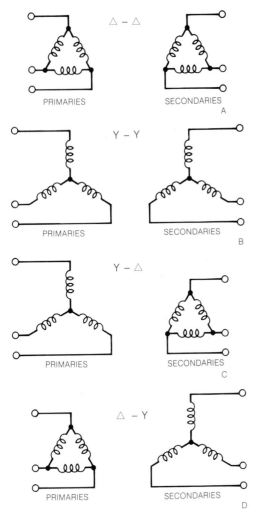

Figure 13-34. Methods of connecting three-phase transformers.

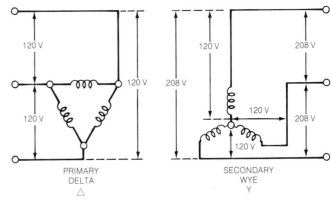

Figure 13-35. Voltages available from delta and wye transformers.

Figure 13-35 shows the voltages available from the two types of three-phase transformer connections.

The ratio between line voltage and secondary voltage in the delta–wye transformer connection is not the transformer turns ratio, but equals $\sqrt{3}$ times the turns ratio. The output voltage across any two windings is greater than that of a single winding. Therefore, the single winding need be insulated for only $E_Y\sqrt{3}$ volts. (E_Y is voltage across a single winding.)

REVIEW QUESTIONS FOR SECTION 13.5

1. What is the frequency of the AC power generated in the U.S. and Canada?
2. Compare three-phase with single-phase AC.
3. Where does the wye connection get its name?
4. Where does the delta connection get its name?
5. What is the voltage available commercially from three-phase lines?

13.6 THREE-PHASE MOTORS

In general, single-phase motors are more expensive to purchase and to maintain than are three-phase motors. The former are less efficient, and their starting currents are relatively high. All run at essentially constant speed. Nonetheless, most machines using electric motors around the home, on the farm, or in small commercial plants are equipped with single-phase motors.

Those who select a single-phase motor usually do so because three-phase power is not available. See Figure 13-36 for the simple methods used in the construction of a three-phase motor. Note this is a half-etched, squirrel-cage rotor. The bearings are not sealed ball bearings, but are of the sleeve type, with the oil caps placed so that oil may be added occasionally to keep the bearings lubricated.

A three-phase motor is shown in the cutaway view in Figure 13-37. Note the simple rotor and fan blades, the windings, and the sealed ball bearings. This is an almost totally maintenance-free motor. One of the advantages of three-phase motors is their ability to start without any start mechanism. Such start mechanisms in single-phase motors cause many maintenance problems.

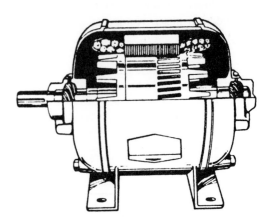

Figure 13-36. Three-phase motor with a half-etched, squirrel-cage rotor.

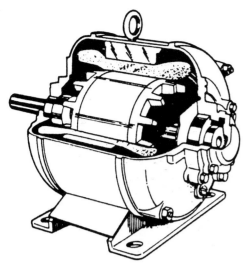

Figure 13-37. Three-phase motor with cast rotor.

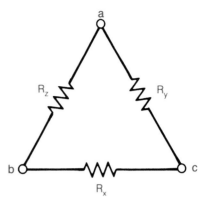

Figure 13-39. Delta resistor arrangement.

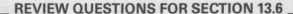

REVIEW QUESTIONS FOR SECTION 13.6

1. Why do we use single-phase power instead of three-phase in homes?
2. Why are single-phase motors more expensive to operate?
3. Which motor—three-phase or single-phase—is more maintenance free?

13.7 DELTA AND WYE RESISTOR CIRCUITS

Since most power is generated by commercial generators, it is necessary for persons working with electricity to be familiar with how the loads work in a three-phase system. We have already discussed three-phase delta and wye windings for transformers and generators. Now we shall see how to figure the load on each winding of a transformer.

A three-terminal resistor network can be connected as either a delta or a wye network, Figures 13-38 and 13-39.

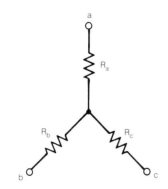

Figure 13-38. Wye resistor arrangement.

In the *wye network*, the resistances between the terminals are easily determined:

$$a \text{ to } b = R_a + R_b$$
$$a \text{ to } c = R_a + R_c$$
$$b \text{ to } c = R_b + R_c$$

"Y" Wye Network

In the *delta network*, the resistance between the terminals may be determined by combining the formulas for series and parallel resistance:

$$a \text{ to } b = \frac{R_z \cdot (R_x + R_y)}{R_z + R_x + R_y}$$
$$a \text{ to } c = \frac{R_y \cdot (R_x + R_z)}{R_y + R_x + R_y}$$
$$b \text{ to } c = \frac{R_x \cdot (R_z + R_y)}{R_x + R_z + R_y}$$

"△" Network

To simplify **resistor networks** for determining **equivalent resistances**, it is frequently necessary to convert a delta network to a wye, or a wye to a delta. The mathematics for such conversions is simple and is based upon the formulas for series and parallel resistance.

To convert from **delta to wye**, the formulas to be used are:

$$R_a = \frac{R_y \times R_z}{R_x + R_y + R_z}$$
$$R_b = \frac{R_x \times R_z}{R_x + R_y + R_z}$$
$$R_c = \frac{R_x \times R_y}{R_x + R_y + R_z}$$

To convert from **wye to delta**, the formulas to be used are:

$$R_x = \frac{(R_a \cdot R_b) + (R_b \cdot R_c) + (R_c \cdot R_a)}{R_a}$$

$$R_y = \frac{(R_a \cdot R_b) + (R_b \cdot R_c) + (R_c \cdot R_a)}{R_b}$$

$$R_z = \frac{(R_a \cdot R_b) + (R_b \cdot R_c) + (R_c \cdot R_a)}{R_c}$$

A couple of examples will serve to give a better understanding of how these formulas aid in obtaining conversions and equivalents.

EXAMPLE 1

Convert the resistors in the wye connection to their delta equivalents. (See Figure 13-40.)

$$R_x = \frac{(5000 \cdot 10,000) + (10,000 \cdot 15,000) + (15,000 \cdot 5000)}{5000}$$

$$R_x = \frac{50,000,000 + 150,000,000 + 75,000,000}{5000}$$

$$R_x = \frac{275,000,000}{5000}$$

$$R_x = 55,000 \ \Omega$$

Since the formulas for R_y and R_z use the same numerator as does the formula for R_x, the solutions for R_y and R_z are as follows:

$$R_y = \frac{275,000,000}{10,000} = 27,500 \ \Omega$$

$$R_z = \frac{275,000,000}{15,000} = 18,333 \ \Omega$$

EXAMPLE 2

Convert the resistors connected in the delta configuration to their equivalent in a wye. (See Figure 13-41.)

$$R_a = \frac{(10,000)(5000)}{15,000 + 10,000 + 5000}$$

$$R_a = \frac{50,000,000}{30,000}$$

$$R_a = 1667 \ \Omega$$

Since the formulas for R_b and R_c use the same denominator as does the formula for the solution for R_a, the solutions for R_b and R_c are as follows:

$$R_b = \frac{(15,000)(5000)}{30,000} = 2500 \ \Omega$$

$$R_c = \frac{(15,000)(10,000)}{30,000} = 5000 \ \Omega$$

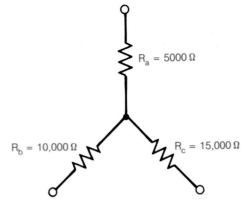

Figure 13-40. Wye-connected resistors.

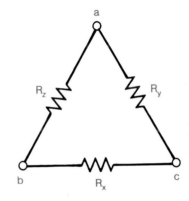

Figure 13-41. Delta-connected resistors.

____ **REVIEW QUESTIONS FOR SECTION 13.7** ____

1. What is a wye network?
2. What is a delta network?
3. What is the formula used to convert from a wye to a delta configuration?
4. What is the formula used to convert from a delta to a wye configuration?

Safety Tips

1. Keep in mind that the 3ϕ power for local use has a low voltage of 208, which is sufficient to kill. Stay clear. Do not touch any 3ϕ power circuits.
2. Do not climb a pole that holds or is connected to transformers or any other electrical equipment or wiring. Do not fly kites near transformers or high-voltage lines.
3. Stay clear of platforms with three-phase transformers mounted between two poles. These are sometimes located near schools.

Activity: DC AND AC GENERATORS

_____ OBJECTIVES _____

1. *To investigate the mechanical construction of generators.*
2. *To understand the principles of generator operation.*

EQUIPMENT NEEDED

1 Motor–generator unit
   (see Figure 1)   1 Meter, 1-0-1 mA, DC
2 Bar magnets, to fit the
   motor–generator unit

PROCEDURE

1. Set up the motor–generator unit as shown in Figure 1.

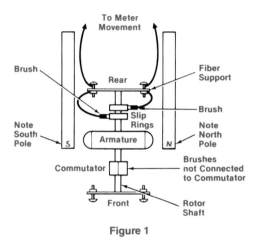

Figure 1

2. Connect the rear terminals of the armature to the meter (1-0-1 mA). Inspect the brass brushes and make sure they are contacting the slip ring assembly of the armature. Use your finger to turn the armature slowly but uniformly. Is there any indication of a generated emf on the meter? What is the action of the meter as the armature is rotated? Is this a characteristic of AC or DC? Why?

3. Spin the armature faster. What is the action of the meter? Does the faster rotation of the armature increase or decrease the frequency of the generated current?

4. Disconnect the meter from the output terminals of the motor–generator unit and connect the voltmeter to the same terminals. Set the voltmeter to read 0–1 volt AC. Again, spin the armature slowly. What is the voltage output? Spin the armature faster. What is the voltage output? **NOTE:** *If the meter is driven off scale, change the setting of the voltmeter to read 0–5*

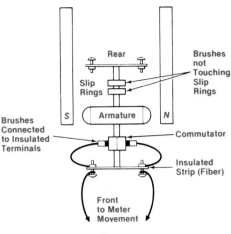

Figure 2

volts AC. Digital voltmeters don't work! Use analog type meter.

5. Disconnect the voltmeter from the motor–generator unit. Change the circuit to that shown in Figure 2.

6. You will notice that the connections to the motor–generator unit have been changed to the front terminals. Connect the terminals to the meter (1-0-1 mA). Inspect the front brass brushes and make sure they are contacting the commutator assembly of the armature. Turn the armature slowly but uniformly. Is there any indication of an emf being generated? What is the action of the meter as the armature is turned? Is this characteristic of AC or DC?

7. Spin the armature faster. What is the action of the meter?

8. How does the speed of rotation affect the amount of emf generated?

9. Inspect the commutator. Describe how it functions and how it causes this simple generator to produce direct current.

10. Disconnect all circuits you have been using and return all equipment and materials to their proper storage places.

SUMMARY

1. What is the major difference between an AC and a DC generator?

2. How do the strength of the magnetic field, speed of rotation, and the number of turns on the armature determine the output of a generator?

3. In the DC generator, does the direction of rotation have anything to do with the polarity of the output voltage?

SUMMARY

In a DC generator, mechanical force is used to rotate a wire loop in a magnetic field. This generates electricity. The magnetic field is generated by a current-carrying wire looped around a core.

A generator uses a field coil and an armature if DC, and an armature (rotor) and a stator if AC. AC generators are called alternators. They have slip rings instead of commutators. Brushes make contact with the slip rings in AC generators and with the commutator segments in DC generators.

Series, shunt, and compound are three types of DC generators. Each type has its own characteristics that are useful for various types of loads. Shunt generators are used primarily for battery-charging types of work. Separately excited generators are used in certain speed-control systems.

There are several types of compound generators: cumulative, flat-compounded, over-compounded, under-compounded, and differentially compounded.

Differentially compounded generators have somewhat the same characteristics as do series generators, in that they are essentially constant-current machines.

Most electrical power generated in the United States is produced as three-phase. Single-phase power is used in homes and schools. Three-phase is of greater use in larger installations, where more efficiency and economy is essential.

The turbo-type AC generator is used where high speeds are encountered. The salient-pole type of alternator is used where the power source is slow moving, usually under 720 rpm.

Frequency of the output of an alternator is determined by the number of poles and speed of rotation.

In the wye connection, the current in the line is in phase with the current in the winding.

In delta-connection generators, the connection has a current advantage. The line current is equal to the phase current times the square root of 3, or roughly 1.73.

Commercial three-phase voltage from power lines is usually 208 volts; the standard values of single-phase voltage can be supplied from the line.

A three-terminal resistor network can be connected as either a delta or a wye network. In the networks, the resistances between the terminals can be converted from delta to wye or from wye to delta.

USING YOUR KNOWLEDGE

1. Draw a delta-to-wye transformer configuration.
2. Draw a three-phase sine-wave representation.
3. If the current from one winding of a delta-connected generator is 100 amperes, what is the total current available from the generator?
4. Draw a schematic of wye and delta resistor network configurations.

KEY TERMS

three-phase power	delta
single-phase power	$\sqrt{3}$, or roughly 1.73
ϕ (Greek letter *phi* for phase)	equivalent resistance
armature	resistor network
stator	delta to wye
wye	wye to delta

Chapter 14

HOUSE WIRING

After studying this chapter, you will know

■ *how electricity is delivered from a generator to your home.*

■ *the purposes and importance of the* National Electrical Code.®

■ *the need for UL labels on equipment used in house wiring.*

■ *how to identify the service entrance and distribution panel.*

■ *the purposes and methods of connecting light fixtures, convenience outlets, and types of switches.*

■ *how a branch circuit in a house is wired.*

■ *how 120 V is obtained from 240-V supply.*

■ *the rules for household electrical safety.*

INTRODUCTION

House wiring is important to today's needs. It gives us light and heat. It provides us with home entertainment and cooked food. It is part of our high standard of living.

This chapter introduces you to the electrical system in your home. It also shows you how house wiring is done. Introduced are the tools, equipment, and wire used in household electricity.

Particularly important is the content of this chapter that stresses safety in the use of electricity.

14.1 DELIVERING ELECTRICITY

Because of the speed at which it travels—approximately 186,000 miles per second—electricity can be used the instant it is generated. This is true even if a generator is hundreds of miles away from your home.

Figure 14-1, page 200, diagrams the delivery of electricity from a generator to homes and offices. Notice that there are several changes in voltage levels between the generator and the average home. Electricity leaves the generating plant at 13,200 V. It goes through a transformer at a substation. The voltage is stepped up to 132,000 V. Electricity moves over long-distance transmission lines at 132,000 V. Notice that this step-up is used only for long-distance transmission.

As electricity approaches the point where it is to be used, it goes through a substation. This steps down the

voltage. Power leaves the substation at voltages of between 2300 and 4000 V. As it continues on its route to users, the electricity goes through additional transformers. These reduce the voltage level to 240 V. Electricity is delivered to your home, and to most offices, at 240 V.

Figure 14-1, page 200

REVIEW QUESTIONS FOR SECTION 14.1

1. What is the voltage of AC produced at a generating station?
2. What voltage is commonly used for long-distance transmission of electricity?
3. What is the voltage at which electricity is delivered to the home?

14.2 THE HOUSEHOLD ELECTRICAL SYSTEM

Electricity in each household is delivered through a system. Within each home, the electrical system is designed and installed according to a plan. The development of a household wiring plan begins with the needs of the house's occupants. The plan recognizes that electricity is essential. But electricity is also a convenience. The electrical plan should be designed to meet both needs and conveniences. In doing this, the planned electrical system should protect the occupants and the home. It should meet safety standards that apply to household wiring. It should also be designed with future needs in mind.

The average home has more than 175 devices that use electricity. These are designed for work saving, time saving, convenience, and entertainment. All use energy. In 1930, there were only 19 electricity-using units in the average home. That's one reason why the wiring systems in older homes often need special attention.

REVIEW QUESTIONS FOR SECTION 14.2

1. What is an electrical plan?
2. Why may it be necessary to develop a new electrical plan for an old house?

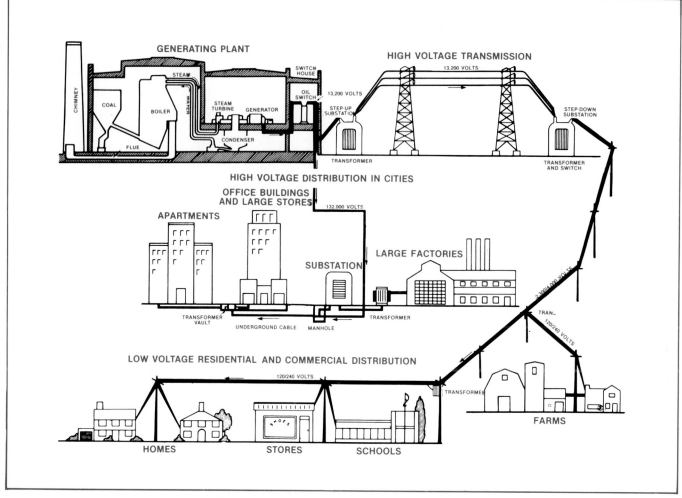

Figure 14-1. Diagram showing distribution of electricity.

14.3 USING THE SYSTEM

Each wiring system needs a **service entrance**. This is the point at which electricity enters the home. The wires from the power company's distribution network are connected to the service entrance.

The wires from the service entrance are hooked up to a **distribution panel**. The distribution panel sets up the circuits for use within the home. Lights and appliances are connected to **branch circuits**. These branch circuits start at the distribution panel.

A good, usable electrical system has branch circuits that are suitable for the house and its occupants. All equipment and lights should operate reliably. There should be no excessive voltage drops when lights or appliances are turned on.

A well-planned electrical system should have spare circuits. These will permit expansion of service as needed. Electrical use keeps growing, even in the face of energy problems. So, the electrical system in each house should be able to take care of the future needs of its occupants.

Extension cords should not be necessary. Fuses should not be blown or circuit breakers tripped in a properly wired house. These problems can be eliminated with planning. Electrical planning is done before a new home is built. Existing houses require some careful planning when electrical service is expanded.

One of the elements of electrical planning is the meeting of safety and service standards. Special safeguards are required under laws in most cities or counties. There are also protections available to consumers. In planning for and using an electrical system, these laws should be followed. And available aids and safeguards should be used.

National Electrical Code®

Most laws or regulations covering the wiring of houses are based on the **National Electrical Code®**. This is a publication on wiring standards. It is issued by the National Board of Fire Underwriters. This code has grown with the development of housing and industry in the United States.

The first rules designed to assure safety in electric wiring were published in 1881. These were issued by the New York Board of Fire Underwriters. Underwriters are companies that issue (or write) insurance policies. Fire underwriters insure against fire losses by their policyholders.

The initial effort in New York City was expanded soon afterward. The efforts grew into the *National Electrical Code.* The *Code* is revised every three years. Each revision reflects the latest developments in electrical technology and safety.

Many cities publish their own electrical codes. Most of these include all of the rules of the *National Electrical Code.* Some add extra safety regulations for their own areas.

Underwriters' Laboratories, Inc.

An important contribution to the safety of electrical products is made by **Underwriters' Laboratories, Inc.** (UL®). As with the *Code*, this is an effort started by the insurance industry. UL® was founded by the National

Board of Fire Underwriters. It is a nonprofit organization. UL® tests electrical products to determine that they meet *Code* and safety standards.

UL® approval is important both to manufacturers of electrical equipment and to consumers. For manufacturers, there is assurance that parts or components purchased meet safety standards. For consumers, there is assurance that products will be safe. Without UL® approval, product safety is open to question. So, you should always look for the UL® seal on electrical products you buy or use, Figure 14-2.

In Canada, testing and approval of electrical products are handled by the Canadian Standards Association (CSA), Figure 14-3.

Electrical Plans

Original wiring or rewiring of houses begins with planning. As a tool for knowing what electrical service will be available, an electrical plan is drawn, Figure 14-4. The

Figure 14-2. Underwriter's Laboratories, Inc. symbol.

Figure 14-3. Canadian Standards Association symbol.

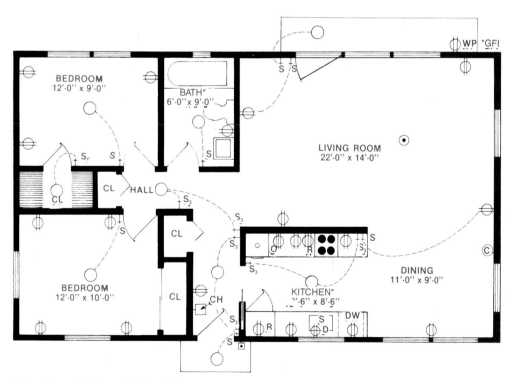

Figure 14-4. Household electrical plan. (*Bathrooms, kitchens, and outside outlets require ground-fault circuit interrupters.)

plan shows the location of each electrical service point within the house. Occupants can then match the plan with their expected needs.

In a modern home, there are certain standards of electrical service expected. Convenience outlets should be available where they are needed. Extension cords should not be necessary in a properly wired house. Lighting should be adequate for reading or other activities. Most important, light switches should be located so that they can be turned on and off easily. Lighting should be planned so that people can move around a house safely.

To help develop and read electrical plans, standard symbols are used. All architects and electricians know and use the same symbols. These form the basis for developing and understanding electrical plans, Figure 14-5.

An electrical plan details location of all service points. With a plan, it should be possible to locate the service entrance for electricity into the home. It should then be possible to follow electrical service through the distribution panel and into branch circuits.

REVIEW QUESTIONS FOR SECTION 14.3

1. What is an electrical service entrance?
2. What is a household electrical distribution panel?
3. What is a branch circuit?
4. What is the purpose of the *National Electrical Code®*?
5. What is the function of Underwriters' Laboratories, Inc.?

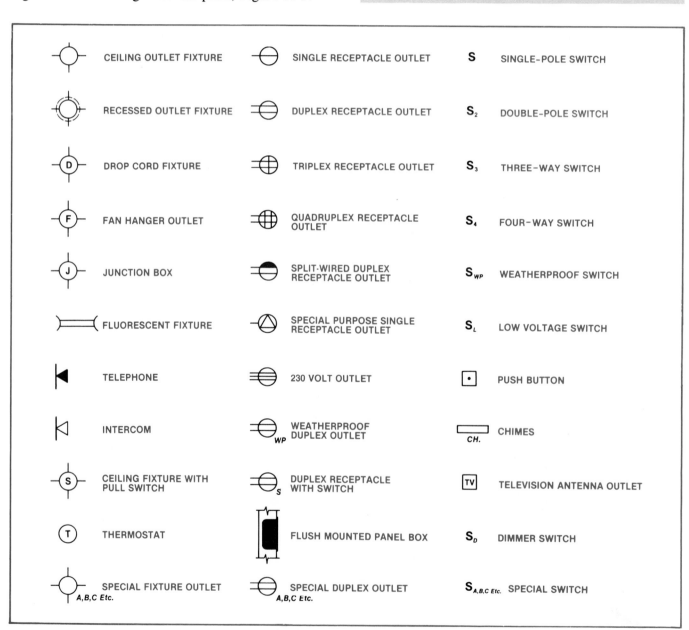

Figure 14-5. Symbols used in electrical planning.

14.4 HOUSEHOLD DISTRIBUTION

The power company's connection of wires to a home is called a **service drop**, Figure 14-6. Service drops include the wires that carry current to the meter in the home. These can come from a nearby power-line pole. Or the power may be delivered from cables buried in the ground. Some companies just deliver to the weather head. The customer is responsible for all but the meter hub.

In any case, there will be three wires conducting electricity into the home. They deliver 240-V AC current. From the electric meter, these three wires lead into a distribution panel. This distribution panel is the central point for the home's electrical service. Figures 14-7A and 14-7B (pages 204 and 205) are drawings of distribution panels.

Branch Circuits

An average home may have two types of electrical requirements. One is 120 V. The other is 240 V. The 240-V circuits are used for heavy-duty equipment or appliances. Included are electric heaters, electric kitchen ranges, and water heaters. Current at 240 V is also needed for clothes dryers, and some large air conditioners.

Lighting and small appliances run on 120 V. Most of the branch circuits in the average home are rated at 120 V. These include light fixtures and convenience outlets. **Convenience outlets** are the wall sockets into which lamps and appliances are plugged.

Both 240-V and 120-V branch circuits are set up at the distribution panel. Three wires go into the service panel. Two of these are "hot" lines. The other is neutral. The three wires together provide 240-V service. Thus, for 240-V branch circuits, three wires leave the distribution panel. For 120-V branch circuits, two wires leave the distribution panel. One of these is a "hot" line. The other is neutral.

An additional wire is added to power circuits, as necessary. This makes for a total of three wires in a 120-V branch circuit, four to provide 240-V service. This additional wire has green insulation. If the wires are in protected cable, the extra wire may be bare. Its function is to ground each convenience outlet, lighting fixture, or unit of electrical equipment.

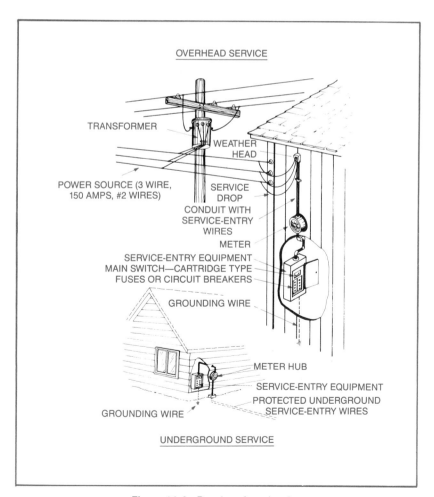

Figure 14-6. Drawing of service drop.

An equipment ground is a lead that conducts unwanted current to the earth (ground). A grounding connection may go directly to the ground. Or it may go to a metal object that is grounded. Thus, the green wire does not play an active role in delivering electricity. However, it provides a high degree of safety assurance. All active household power circuits must include a grounding wire. This is now required by the *National Electrical Code*.®

Circuit Breakers and Fuses

Figure 14-8 is a photograph of a circuit-breaker box. A box of this type would be included in a household distribution panel. This is the type of unit found in many modern homes.

The function of a distribution panel is to establish circuits that will deliver service throughout the home. The distribution panel also provides safety and protection.

Should circuits become overloaded or shorted, considerable heat can be generated. Without the protection of **circuit breakers**, the heat could cause fires. Circuit breakers are inside the distribution panel.

Within the box in Figure 14-8, there are two rows of circuit breakers. Each of these circuit breakers has an amperage rating. When the load on a circuit exceeds the rating of the breaker, the circuit is tripped. Thus, current stops flowing to the overloaded appliances or into the short-circuited connection.

Figure 14-9 is a drawing of an individual circuit breaker. There are a number of breakers of this type in a single panel. Notice that the breaker in Figure 14-9 contains a switch with the number 15. This indicates that the breaker has a 15-A rating.

If a circuit becomes overloaded, the circuit breaker is activated. Its switch is thrown automatically to the "off"

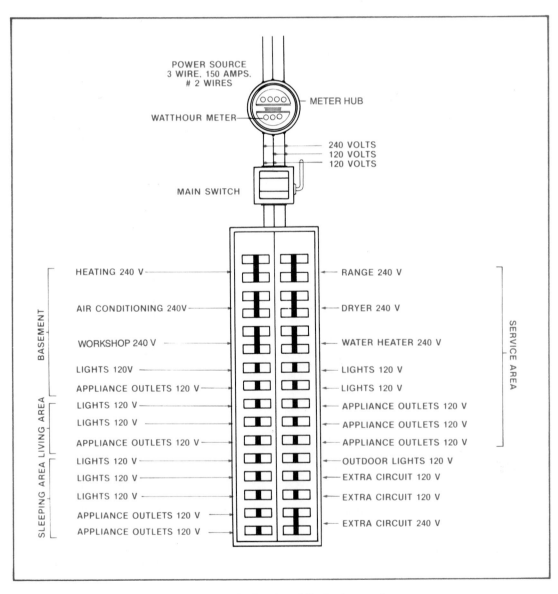

Figure 14-7A. Drawing of distribution panel.

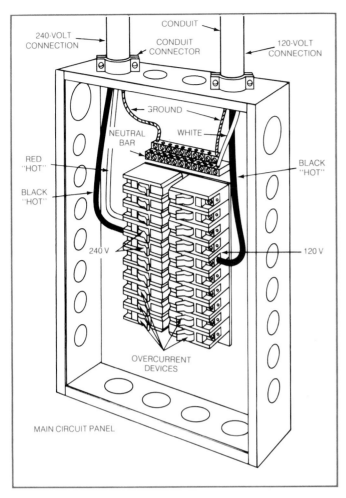

Figure 14-7B. Distribution box with cover removed. Note location of ground wire.

Figure 14-8. Photograph of circuit-breaker box.

position. Any time this happens, check the load on the circuit. Unplug any appliances or devices that might be causing the overload. Then reset the switch. After the switch has been reset, immediately check the lamps or devices receiving current. Look for any problems that may have caused a current overload.

Most modern buildings have circuit breakers of the type illustrated in Figure 14-9. However, some older buildings may have fuses for their branch circuits. A fused distribution panel is shown in Figure 14-10. In place of switched

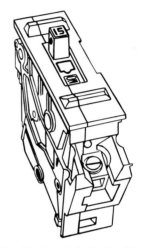

Figure 14-9. Drawing of single circuit breaker.

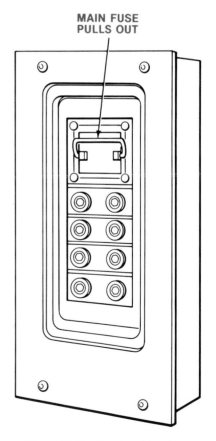

Figure 14-10. Drawing of fuse box.

devices that close circuits, fused panels have sockets. **Fuses** screw into the sockets. If a circuit becomes overloaded, the fuse "blows." That is, the fusing element burns out, causing the circuit to break open.

The principle is the same as for circuit breakers: If the circuit becomes dangerous, it is opened automatically. Many household fuse boxes are set up for plug fuses, Figure 14-11. To modernize service panels with fused circuits, you may find it best to buy circuit-breaking adapters. One of these is illustrated in Figure 14-12. This is a circuit breaker that screws into a fuse socket. To reset the breaker, simply push the button in the middle.

Conduits and Cables

As indicated above, all circuits originate at the distribution panel. From there, wires are run throughout the house. To distribute electricity safely, these wires must be protected.

Several methods of protection are used. One level of protection is the insulation on the wires themselves. In addition, there are wires that are packaged in special protective coatings. For some forms of electrical distribution, it is necessary to provide special casings, or tubing. These tubing materials are called *conduit*. Conduit is no longer in common use for wiring homes. However, you may find conduit in older homes. Conduit is still used in commercial buildings and factories.

Figure 14-13 illustrates installation of one common form of conduit. This is rigid **conduit**. It is a pipe made of a material such as aluminum or steel. Conduit of this type is usually installed in walls during construction, Figure 14-13. To fit rigid conduit into spaces within walls,

it is frequently necessary to bend or shape the material. This is done with a special tool, Figure 11-14. The tool is called a **conduit bender**.

Another form of conduit is called *flexible conduit*. This conduit can be bent by hand into any shape needed to complete electrical installations, Figure 14-15. Flexible,

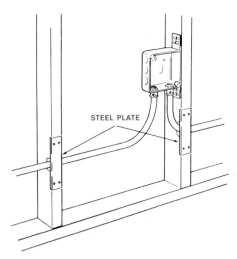

Figure 14-13. Rigid-conduit installation.

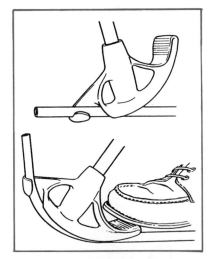

Figure 14-14. Bending rigid conduit.

Figure 14-11. Household plug fuse.

Figure 14-12. Circuit-breaker adapter to replace plug fuse.

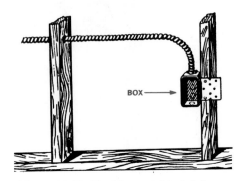

Figure 14-15. Flexible-conduit installation.

or *flex*, conduit is made of overlapped metal wrapped in a continuous form, Figure 14-16. This form is known as a *helix*, or spiral. Such conduit is said to be made of helically wrapped metal.

Both rigid and flexible conduit have the same basic function. They provide hollow spaces through which electricians can pull wires. This is a major method for distributing electricity.

Another distribution method for electricity is through specially wrapped cables. One of these distribution wires is illustrated in Figure 14-17. This is known as **armored cable**. Notice that armored cable looks something like flexible conduit. Wires within the cable are helically wrapped with a plastic covering. Armored cable is available with two, three, and four wires. Figure 14-18 shows fittings and techniques used to connect armored cable to electrical service boxes.

Another form of protected wire is *Romex*. Figure 14-19 shows a piece of Romex cable. Three wires are encased in heavy plastic. Where zoning laws permit, Romex is also used for electrical distribution. Romex cables are available with two or three wires and a bare ground wire.

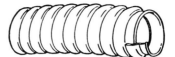

Figure 14-16. Flexible conduit.

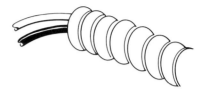

Figure 14-17. Armored cable.

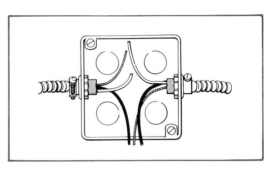

Figure 14-18. Armored-cable installation.

REVIEW QUESTIONS FOR SECTION 14.4

1. What is an electrical service drop?
2. How many wires are in a 240-V branch circuit?
3. How many wires are in a 120-V branch circuit?
4. What is meant by grounding electric circuits?
5. How are circuits grounded?
6. What is electrical conduit?
7. What is armored cable?
8. What is Romex cable?

14.5 LIGHTING

Chapter 12 discusses the use of electricity to produce light. Now you are ready to learn about the wiring and operation of lighting fixtures.

Figure 14-20 shows the installation of a simple wall-mounted light fixture. An electrical box is mounted to the wall. Electric wire is delivered from a cable. A wall switch is mounted into the box from the other side. The fixture itself is spliced into the circuit through use of solderless connectors. These are plastic caps that can be screwed over the bare wire leads that are twisted together. There is a copper spiral inside each cap that helps make a tight, safe connection. Figure 14-21 illustrates the use of a solderless connector. These devices are sometimes called *wire nuts*.

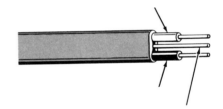

Figure 14-19. Romex cable.

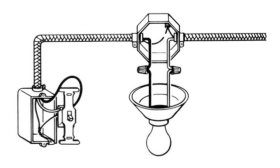

Figure 14-20. Diagram of wall-mounted light fixture.

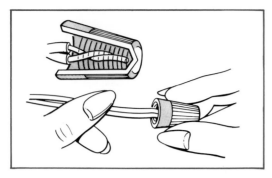

Figure 14-21. Using a solderless connector.

A schematic diagram of the lighting circuit is shown in Figure 14-22. This is the simplest form of lighting circuit. It uses a single-pole, single-throw switch. A typical switch used for this type of circuit is illustrated in Figure 14-23. The wires are attached to the screws at the side of the switch. Moving the switch up or down completes or breaks the circuit. This turns the lights on or off.

Figure 14-24 illustrates an important safety tip about connecting wires to switches. Always connect the wire so that it turns with the direction of the tightening screw. If the connection is made the other way, the tightening of the screw creates a problem. It actually pushes the wire away from the connection. Some switches have push-in holders and a wire-strip gage.

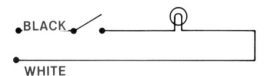

Figure 14-22. Schematic of lighting circuit using single-pole, single-throw switch.

Figure 14-23. Single-pole, single-throw switch.

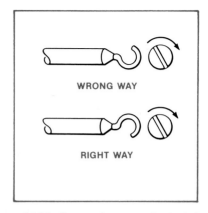

Figure 14-24. Proper wire-connection technique.

Three-Way Switching

The description above covers a circuit involving a single light operated by a single switch. However, there are many situations in which multiple wall switches are needed for individual lights.

For example, if you have an exterior light, you may want two controls. One would be outside. The other could be in the main hall inside the house. Another example is the light over a stairway. You want to turn it on before you start up the stairs. Then, when you get to the top, you want to turn the light off.

This requires two wall switches controlling a single light circuit. The **three-way switch** is an example of this type of control. The wiring for separated switches to control a single lamp is illustrated in Figure 14-25. The conductor between the two switches has three wires. The switches must touch a matching red or black wire to complete the circuit. The red wire between two switches is called a *traveler*. It carries current only when it is switched into the circuit. Except for the connection between the two switch boxes, the remainder of the circuit is served by two wires.

Note the identifications for the wires in Figure 14-25. The two-wire portions of the circuit are identified as 14/2 WG Romex. The three-wire conductor is identified as 14/3 WG Romex. The number *14* represents a number-14 wire. Wires are numbered according to thickness. They can be measured with standard wire gages. The number *2* means that there are two wires in the conductor. The number *3* indicates three wires in the conductor. The letters *WG* mean "with ground."

Figure 14-25 refers to wire-nut splices. Wire nuts, or solderless connectors, are used only when splices will be contained in electrical boxes. This circuit shows that the splices fit within boxes.

Figure 14-26 shows the arrangement of wires and three-way switches to control a single lamp. Feed is through the

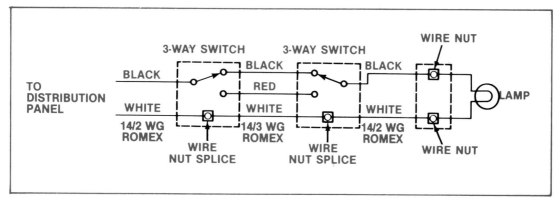

Figure 14-25. Schematic of three-way switching circuit.

center-lamp box. The terminals marked A and B are the light-colored points to which red and white wires must be connected. Terminal C is the dark-colored (brass screw) point to which the black wire must be connected. Figure 14-27 shows how two three-way switches are tied together with a three-wire Romex cable. The lamp and switch box are fed with a two-wire cable. The terminals marked A and B are the light-colored screws to which red and white wires must be connected. Terminal C is the dark-colored (brass screw) point to which the black wire must be connected.

Four-Way Switching

Sometimes, more than two locations must be available for switching a single lamp or light fixture. In these situations, a **four-way switch** is needed. This type of switch is used with two three-way switches, Figure 14-28. Note that the four-way switch is between the three-way units. It can be located at any point between the three-way

switches. The three-way switches are always at the ends of the circuit.

A special crossover action enables the four-way switch to close the circuit. This makes it possible for the four-way switch to operate at all times. It functions regardless of the positions of the three-way switches.

Figure 14-29 shows two three-way switches and one four-way switch in a circuit arrangement that makes it possible to control a lamp from three locations. Terminals marked A and B are the light-colored screws to which red and white wires must be connected. Terminal C is the dark-colored screw to which the black wire must be connected. Terminals AA show where the two ends of the red wire are connected between the four-way and the three-way switches on the right. Terminals BB show where the two ends of the white wire are connected to the four-way and the three-way switches.

If four locations are to be served, two three-way switches and two four-way switches are required. If five locations are to be served, only two three-way, but three four-way switches, are required. As you can see, control

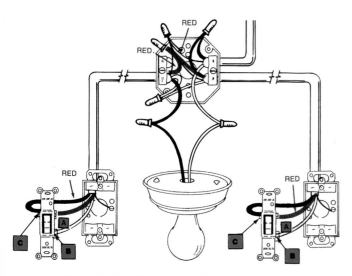

Figure 14-26. Three-way switch control of a lamp.

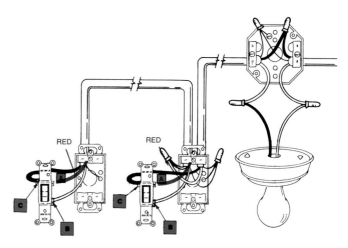

Figure 14-27. Alternative three-way switch control of a lamp.

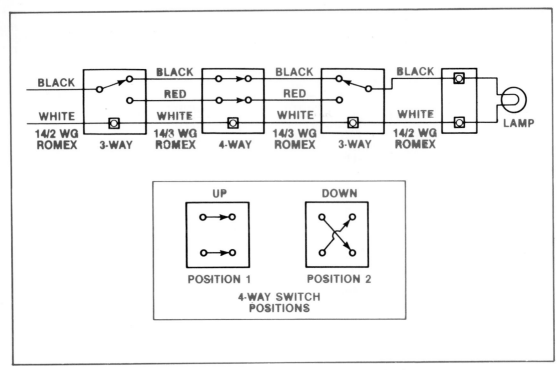

Figure 14-28. Schematic of four-way switching circuit.

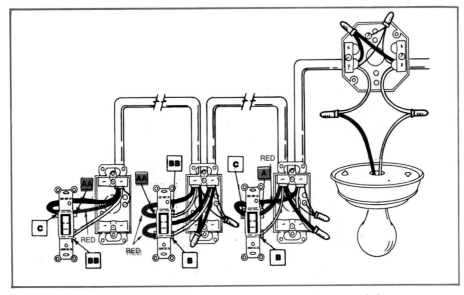

Figure 14-29. Two three-way switches and one four-way switch used to control a lamp from three locations.

from an additional location requires simply the addition of a four-way switch to the two already required three-way switches.

Light-Switch Locations

Light switches should be located conveniently. You should be able to find them comfortably, even in the dark. For convenience and safety, there is a standard for the placement of most household switches. They are located with their centerline at least 48 inches above the floor to make dry-wall installation easier.

REVIEW QUESTIONS FOR SECTION 14.5

1. Where can solderless connectors be used?
2. Where are three-way switching circuits used?
3. How does a three-way switching circuit work?
4. Where are four-way switching circuits used?
5. How does a four-way switching circuit work?
6. What rules should be followed for the placement of light switches?

Activity: *THE THREE-WAY SWITCH*

OBJECTIVES

1. *To understand the three-way switch and how it can be used to control a device from two different locations.*
2. *To study switching and control functions from a number of positions.*

EQUIPMENT NEEDED

1 Plywood square, 1/2", 36" × 36"
4 Wire nuts
2 Utility boxes
1 AC plug
1 5-ft 14/2 Romex cable
1 5-ft 14/3 Romex cable
5 Connectors
2 Switch covers
1 Octagonal box
1 Porcelain lamp socket
3 Staples
6 Sheet-metal screws, 1/2" #8

PROCEDURE

1. Wire the board as shown in Figure 1. This wiring should be mounted on a 36" × 36" piece of plywood, so you can work with it on the bench top in your lab.
2. Make sure both three-way switches are in the up position. Plug in the AC plug to a source of 120 volts AC. Does the lamp light?
3. Now flip the switch marked A to the down position. Does the lamp light now? Why?
4. Move to switch B and flip it to the down position. Does the lamp light now? Why?
5. Move to switch A. Flip it to the up position. Does the lamp light? Why?
6. Move to switch B. Flip it to the up position. Does the lamp light? Why?
7. How would you describe the action here?
8. Can you see any practical application for such a circuit? Where?
9. Why don't three-way switches have *on* and *off* marked on their handles?
10. Disconnect the circuit you have been using and return all equipment and materials to their proper storage places.

SUMMARY

1. What is a three-way switch?
2. Why does it take two three-way switches to turn a lamp off and on from two locations? Why couldn't two regular on–off types be used?
3. What are the names given to the black and red wires that connect the two switch boxes? Why do you think they have this name?
4. Why is the white wire spliced all the way through the circuit?
5. Why is the black wire switched?

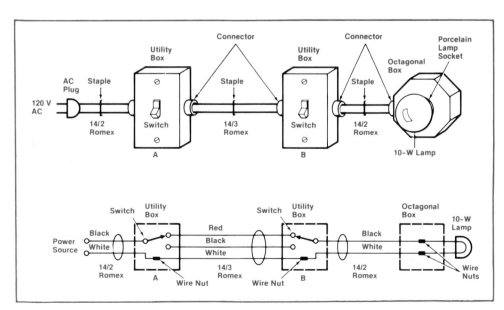

Figure 1

14.6 CONVENIENCE OUTLETS

The placement of outlets in a home should assure convenience and safety. Outlets should be placed so that the cords plugged into them do not block foot traffic. Guidelines for placement of outlets are designed to meet these goals. In general, any wall space that is more than two feet wide should have an outlet. Distance between outlets is usually 6 to 8 feet.

Usually, convenience outlets are placed 12 inches above the floor level. This is low enough so that cords can be dropped smoothly to the floor. People are less likely to trip over cords plugged into low outlets.

There are, however, some special outlet requirements. For example, in the kitchen, outlets are usually placed above working counters. In shops where they are used for machines, outlets are frequently placed at the same height as switches.

Outlet Installation

An outlet installation is illustrated in Figure 14-30. Note that the convenience outlet requires three wires. In this case, the bare wire is attached to the bottom screw of the outlet receptacle. This is the grounding terminal. In case of a short, current is automatically carried to ground from this connection. The black and white wires are attached to the other screws. Black goes to the brass screw and white goes to the silver screw or back pressure-grip hole.

REVIEW QUESTIONS FOR SECTION 14.6

1. What rules should be followed for the placement of convenience outlets?
2. In what situation would a convenience outlet be placed at the same height as a light switch?

14.7 ADDING ELECTRICAL CIRCUITS

Electrical service in a home should be planned in advance. However, there will still be times when extra lights or outlets are needed. It can be satisfying to do this work yourself. In addition, you can save a great deal of money if you know how to do your own electrical wiring.

CAUTION: In most areas, you cannot alter wiring in a home without getting a permit first. It is also necessary to have all work inspected before the walls are closed and the service is connected.

A typical type of project involving extension of electrical service is illustrated in Figure 14-31. The idea is to add a new convenience outlet. The additional outlet is needed on the side of a door opposite from an existing outlet. In preparation for this job, the door frame has been removed. Two sections of baseboard have also been removed. These are on either side of the door. Removing the baseboards and door frame provides access for placing the electric cable.

The wire is strung from the existing outlet to the new location. Not shown in this drawing is the attachment for the electrical box for the new outlet. This would have to be located at a stud. A stud is a vertical support in the wall. The electrical box is nailed to the stud to hold it in place.

In preparation for the wiring, leads must be trimmed on both ends of the new cable. To do this, you have to strip away the wrapping on the cable. Connectors must be attached to the ends. In some cases, clamps are already part of the box. Connectors or clamps are used to hold the wire in the electrical boxes.

When this work is completed, you have exposed ends of two or three insulated wires. It is then necessary to trim away the insulation. This gives you bare wire to attach to the outlet receptacle.

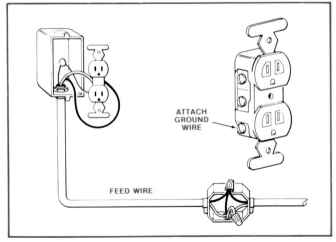

Figure 14-30. Diagram of convenience-outlet installation.

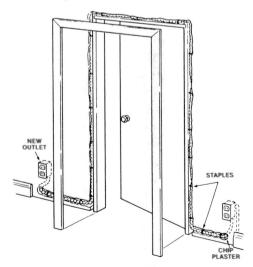

Figure 14-31. Electric-service expansion project.

Activity: CONVENIENCE OUTLETS

OBJECTIVES

1. *To study the convenience outlets in a house wiring situation.*
2. *To wire a convenience outlet when it is needed at the end of an existing circuit.*

EQUIPMENT NEEDED

1 Switch, toggle, on–off	2 Utility boxes
1 5 ft 14/2 Romex cable*	1 AC plug
1 5 ft 14/3 Romex cable*	5 Connectors
1 Octagonal box	3 Wire nuts
1 Porcelain lamp socket	3 Staples
1 Duplex receptacle cover	1 AC cord
1 Duplex receptacle	1 10-watt lamp, 120 volts
1 Switch cover	

***NOTE:** 12/2, 12/3, respectively, are used in house circuits with 20-amp requirements.

PROCEDURE

1. Wire up the board with the circuit shown in Figure 1. Follow proper safety precautions before you plug in the circuit. Have your teacher inspect your work before you place the covers over the switch and outlet.

2. Turn the switch to the *off* position. Plug in the AC plug. Does the lamp light?
3. Insert a test lamp into the receptacle where the convenience outlet was added. Does the test lamp work?
4. Flip the switch to the *on* position. Does the lamp now glow?
5. What happened to the test light when the switch was turned to the *on* position?
6. Flip the on–off switch to the *off* position. Does the lamp go out? Why?
7. Does the convenience outlet's test light still glow with the on–off switch in the *off* position? Why?
8. Remove the plug from the wall outlet and disconnect all parts of the circuit. Return all equipment and materials to their proper storage places.

SUMMARY

1. What is the purpose of a circuit like the one shown in Figure 1?
2. Why doesn't the on–off switch affect the outlet?
3. How does the switch work to make and break contact with the line to turn the lamp on and off?
4. What is a practical application of a circuit of this type?

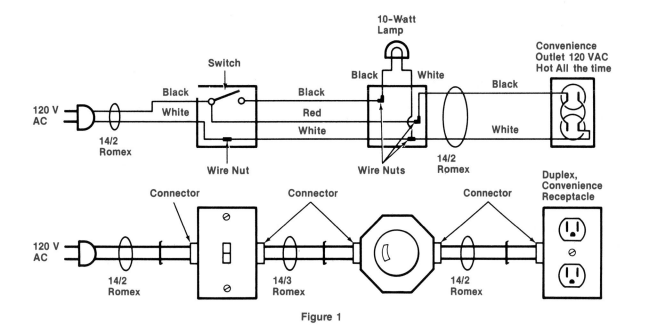

Figure 1

Use a lineman's pliers to trim the ends of wires, Figure 14-32. First, use the part of the pliers marked by the letter *A*. Place the end of the wire in this part of the pliers. Use this to crush the insulation. You will see the insulation flatten. Then use the cutter at point *B* on the pliers to cut the insulation and trim it away. The jaws of the pliers are at point *C*. These can be used to hold wire or to shape the ends of the wire to fit the connections.

CAUTION: It is important to crush the insulation before removing it. Do not simply cut through the insulation and pull it off. If you do this, you can cut into the wire. This causes a nick in the solid wire. Then, when you attempt to connect the wire to the receptacle, it could break off.

A properly trimmed wire lead is illustrated in Figure 14-33.

Note also that the electrical cable must be attached to the wall. The most common way of doing this is with staples like the one illustrated in Figure 14-34. The staples are placed across the cable and hammered until they hold.

IMPORTANT: Before covering up the cables for an extension project, make sure you call the inspector. It makes sense to have the inspection done before the door frame and baseboards are replaced. This can save the trouble of having to tear the walls out again. One of the points to be inspected will be "box fill." The electrical code for most areas limits the amount of conductors, splices, and other items that go into an electrical box. Make sure you follow regulations. See the latest issue of the *National Electrical Code®* handbook.

Adding New Outlets to Existing Wire

Being able to add to existing branch circuits makes it possible to obtain electrical power where needed. Figure 14-35 shows how to add a wall switch for control of a ceiling light at the end of the run. If a new convenience outlet is needed, it can be added from an existing junction box as shown in Figure 14-36. However, if you need to add

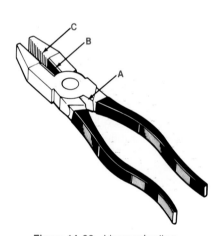

Figure 14-32. Lineman's pliers.

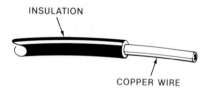

Figure 14-33. Trimmed wire lead.

Figure 14-34. Cable staple.

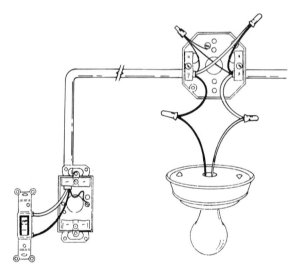

Figure 14-35. Adding a switch to control a ceiling light.

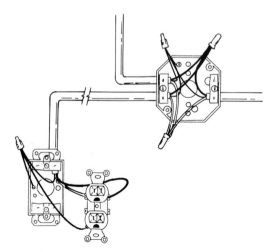

Figure 14-36. Adding a convenience outlet.

a wall switch to control a ceiling light in the middle of a run, you need to arrange the wiring as shown in Figure 14-37.

Adding a new convenience outlet beyond old outlets can be done as shown in Figure 14-38. Figure 14-39 shows how to add a switch and convenience outlet in one box, beyond the existing ceiling light. If the need arises to install two ceiling lights on the same line, one controlled by a switch, this can be accomplished by an arrangement similar to that shown in Figure 14-40.

Figure 14-41 shows the method for adding a switch and convenience outlet beyond an existing ceiling light. To install one ceiling outlet and two new switches, refer to Figure 14-42 on page 216.

To make sure the ground wire is making good contact with the metal box, a screw with a green head, or a slip-on clip may be used. With plastic boxes, solder or wire-nut the ground wires as the *Code* dictates.

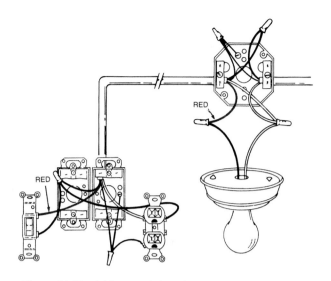

Figure 14-39. Adding a switch and convenience outlet in one box.

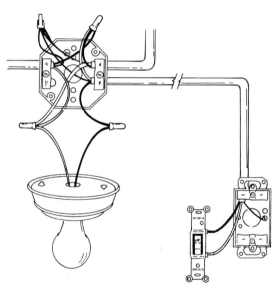

Figure 14-37. Adding a wall switch for a ceiling light.

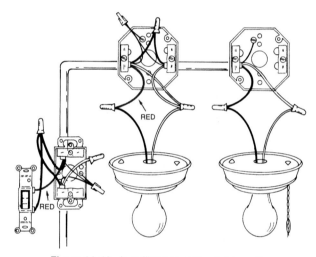

Figure 14-40. Installing two ceiling lights on the same line controlled by one switch.

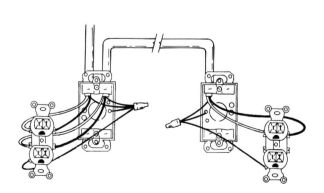

Figure 14-38. Adding a convenience outlet.

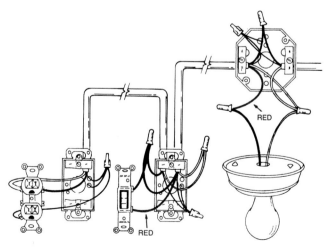

Figure 14-41. Adding a switch and convenience outlet.

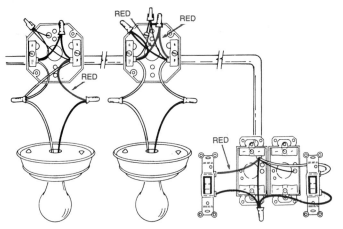

Figure 14-42. Installing a ceiling outlet and two new switch outlets.

REVIEW QUESTIONS FOR SECTION 14.7

1. How is an electrical outlet box held in place?
2. Describe the process for stripping leads for wires in an armored cable.
3. What is box fill and why is it important?
4. Why is it important not to nick solid wire when trimming leads?

14.8 HOUSEHOLD ELECTRICAL SAFETY

Electricity is one of the most useful tools of our civilization. But it can also be dangerous. So, whenever you use electricity or work on electrical wiring, think safety! Here are a few simple, valuable rules:

- Always disconnect a circuit before you work on it. The easiest way is to throw the circuit breaker. If your distribution panel has fuses, unscrew the fuse for the circuit. Of course, test first to make sure that you have the right circuit.
- Pay attention to the color of the insulation on wires you handle. Normally, you should not connect wires with differently colored insulation to each other. (In some switching situations, it may be necessary to connect differently colored wires. If so, make sure there is a good reason.)
- Prior to 1960, the white wire was "hot." Since then, the black wire is considered hot. Some older homes may have been wired with white as hot. Check your distribution panel to be sure.
- Don't overload your circuits. If you blow a fuse, replace it with one of the same capacity. Never short-circuit a fuse by placing a coin in the socket.
- Appliances with heavy loads should have their own circuits. Examples are refrigerators, heaters, kitchen ranges, and air conditioners.

Spotting Signs of Trouble

Use of electricity in the home has multiplied six times in 25 years. Partly because of this increased use, it is estimated that nine out of every 10 homes are improperly wired. About one in every eight home fires results from bad wiring or defective appliances.

So, it is important that you be able to recognize signs of defective, unsafe wiring. Here are some things to look for:

- Do the lights dim when an appliance goes on?
- Do motors slow down?
- Do fuses blow or circuit breakers trip frequently?
- Do toasters or electric irons fail to come up to their full heat?
- Does the TV picture shrink when something turns on?
- Are there too few outlets in the house for all your appliances?
- Do you need more circuits to support the devices you use?
- Are switches and outlets installed and operating properly?

Kinds of Circuits Needed. Different types of circuits are installed in the average home.

In bedrooms and living rooms, there should be general-purpose circuits. These usually carry 120 V AC and may be rated at 15 A. They are used to plug in lamps, radios, and small appliances. If you use one of these for a TV set, avoid plugging any other devices into the same circuit.

For areas where you use heavy appliances, you should have circuits of greater capacity. These can include 120 V AC circuits that carry 20 A or 30 A of current. Use these for appliances such as air conditioners, toasters, broilers, deep-fat fryers, dishwashers, and clothes washers. Be sure that there is only one major appliance of this type on each circuit.

Some special circuits with 240-V capacity may be needed. These are required for appliances such as electric ranges or dryers. Some air conditioners may also require 240-V circuits. Never use more than one device of this type on each circuit.

All circuits should have three-wire outlets. This means that all appliances should have three-prong plugs like the one in Figure 14-43.

Wiring should be installed and checked only by a qualified electrician. All wiring materials should have an Underwriters' Laboratories label.

Circuit Breakers and Fuses. Both fuses and circuit breakers react to overloads, or situations in which a circuit is drawing more current than it was designed to handle. Occasionally, a fuse or circuit breaker will react to a temporary surge in current. If trouble persists, however, you should find its cause.

Circuit breakers are reset in much the same way as light switches. Resetting a circuit breaker is no problem unless it pops off again frequently. Then you should remove some of the load from the circuit before trying again.

Changing a fuse requires some special care. Remember the safety suggestions on fuse changing that you read in Chapter 12.

Cords. The insulation should always be in good condition on electrical cords. Cracked insulation or frayed ends on electric cords can cause electrical fires. Short circuits or shocks can result from the use of faulty cords. Sometimes faulty cords will cause fires.

Be sure you have the right type and size of cord for every job. Heavy-duty cord should be used on tools. Moisture-resistant wire should be used for outdoor work. Check all cords for a UL® label.

Don't place electrical cords where they can trip people. Do not put cords under rugs. Cords should not pass through doorways or be placed near rocking chairs. Keep electric cords away from heat and water.

Never pull the cord to disconnect its plug. Pull on the plug to remove the cord from the wall outlet. Try to prevent twisting or kinking of electrical cords. Inspect cords for wear. Check plug connections carefully. Repair or replace the plugs or connections when there are signs of wear.

Appliances. Each large appliance should have its own circuit. This includes the range, the washer, the dryer, the dishwasher, and the freezer. Each air conditioner should also have a separate circuit.

For smaller appliances, such as vacuum cleaners and toasters, make sure the UL® label is visible. Operate each of these units according to its owner's manual.

Never turn on an appliance when you are standing in water. A wet floor is just as bad. Don't put any electrical parts into water for cleaning.

Don't touch plumbing or any metal object and an electrical appliance at the same time. Keep electric motors clean from lint, dust, and dirt. Teach small children not to touch appliances with wet hands. Use heat-resistant cords. Turn off appliances that spark or stall.

Always disconnect small appliances before cleaning. Connect appliances *directly* to wall outlets. Do not use extension cords for appliances.

Keep combustible materials away from lamps or heating devices. Clothing, curtains, and papers are very easily ignited. Be sure the electric iron is unplugged when you leave it. Never leave Christmas tree lights on when you leave home.

Power Tools. Many people use power tools at home. They use them for fun and for repairs around the house. This equipment should be adequately wired for the job. There should be enough circuits of the right size for their use. All circuits should be grounded. Hand tools with three-prong plugs should be used or the tool should be double insulated, Figure 14-43.

Use safety release switches to prevent accidental starting of power tools. Use and store in a dry place. Dampness increases shock hazard. Keep tools away from flammable vapors. This is because most hand tools have brushes that spark. Keep your shop clean. Keep the shavings, paper, and rags properly stored in metal containers.

Keep tools cleaned, oiled, and in good repair. Protect your cords from heat, chemicals, and oil. Coil them loosely and store in a dry place. Repair cord breaks. Shorten the cord by removing the portions with insulation breaks. Then splice. Or, better yet, get a new cord.

Inspect your tools often. Look for wear, grounding connections, and obvious defects.

Outdoor Outlets. Make sure you use weatherproofed fixtures outdoors. Wiring should be specially designed for outdoor use. Figure 14-44 illustrates an outdoor electric outlet. Outdoor wiring should have its own circuit.

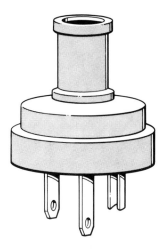

Figure 14-43. Three-prong plug.

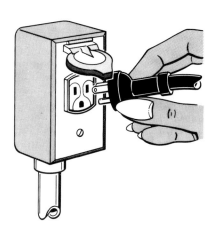

Figure 14-44. Weatherproof outdoor outlet.

Some safety tips for outdoor use of electricity:

- Keep cords out of water. Hang them over wooden pegs only. Never hang cords on metal.
- Disconnect equipment or turn off the circuit when changing bulbs. Many times, a switch is installed indoors to control outdoor outlets.
- Never use electric tools outside in the rain. Also, don't use electric tools on wet surfaces.
- Don't connect lights while you are adjusting them on outside Christmas trees.
- Wear shoes while using an electric lawnmower or hedge clipper.
- Use an electric power mower only in dry grass.
- Consider lightning rods for protection against lightning storms.
- Antennas for TV should be grounded. They should have lightning arrestors.

Emergencies. In case of electric fire in devices such as motors, pull the plug if you can. Use a recommended fire extinguisher. You can throw common baking soda onto the fire to put it out. Never use water on live wires.

Broken wires outside should never be touched. Call the police and the power company. Warn others to keep away. Always assume a loose wire is live.

Electric Shock. Keep in mind that it is the current that kills, not voltage alone. Current can:

- contract chest muscles. This interferes with breathing. Asphyxiation can result if the effect continues.
- paralyze nerve centers. This can cause breathing to stop.
- upset the normal heart rhythm. This can interrupt blood circulation.
- cause hemorrhages or destruction of tissues, nerves, and muscles. This effect comes only from the heat of heavy current.

The table in Figure 14-45 shows some of the effects of electric current on humans.

Emergency Actions. If an accident victim is in contact with live, low-voltage electricity, shut off the power. Pull the plug or supply cord. Turn off the switch, if possible.

If these actions can't be taken, free the victim by using a dry rope, dry board, or dry stick.

Make sure your own hands are dry and that you are standing on dry surfaces.

If the victim is in contact with a fallen live wire outdoors, the only safe procedure is to call the power company to turn off the power.

If the victim is asphyxiated (not breathing), use mouth-to-mouth or chest-pressure/arm-lift artificial respiration, but only after power is shut off.

Condition	Current	Results
Safe	1 mA or less	Causes no sensation. Isn't felt.
Safe	1 mA to 8 mA	Sensation of shock, not painful. Individual can let go at will. Muscular control is not lost.
Unsafe	8 to 15 mA	Painful shock. Individual can let go at will. Muscular control is not lost.
Unsafe	15 to 20 mA	Painful shock. Muscular control of adjacent muscles lost. Cannot let go.
Unsafe	20 to 50 mA	Painful. Severe muscular contractions. Breathing is difficult.
Unsafe	50 to 100 mA	Ventricular fibrillation—heart condition that may result in death.
Fatal	Over 100 mA	Certain death.

Figure 14-45. Electrical exposures and their effects.

If the victim is in shock, get medical aid. Keep the victim warm. Place the head lower than the feet. If medical aid is delayed for an hour or more, give fluids such as water, or a mixture of salt, soda, and water. However, do not give liquids to a victim who is unconscious or nauseated, or has a deep abdominal wound.

If the patient has burns, cut away loose clothing. Cover the burned area with sterile dressing. Keep out the air. Treat for shock and get medical aid at once.

(Information for this section has been made available by the National Safety Council, Chicago, Illinois.)

REVIEW QUESTIONS FOR SECTION 14.8

1. List at least five signs that could indicate an overloaded household circuit.
2. Name at least four appliances that should have their own electrical circuits.
3. What special care should be taken with outdoor wiring?
4. What steps should you take in case of an electrical fire at home?
5. What steps should you take to provide first aid for a victim of electric shock?

SUMMARY

Electricity can be used the instant it is generated. It travels at 186,000 miles per second, which means that it can be used instantly many miles from its generated site.

Houses are wired according to a plan. Most electricians utilize the *National Electrical Code®* as a guide for how to wire a house safely. Underwriters Laboratories and the Canadian Standards Association test equipment that is used in homes and industry and make sure it will operate safely in the environment it is intended to serve.

Symbols are used in planning an electrical system for a house or any other building.

Distribution of electricity in a house occurs from the distribution panel or circuit breaker box. Some older homes still use screw-in type fuses for circuit control.

Solderless connectors, called wire nuts, are used to make connections. The on–off switch, the three-way switch and the four-way switch are used to control the on–off functions in a house. Dimmers are used to control the intensity of the light within a room.

Convenience outlets are installed according to suggestions made by the *National Electrical Code*® The number of outlets on a circuit and the type of spacing around the room is also suggested by the *Code*.

Romex is the wire most frequently used in home wiring. It is a plastic-coated cable consisting of two insulated wires and a third wire that is uninsulated and used for a ground. Some Romex wire has three insulated wires and a ground wire.

Electric shock is always a possibility. Keep in mind that it is the current that kills, not the voltage alone. The emergency actions to take should be understood by all those who work with electricity.

USING YOUR KNOWLEDGE

1. Draw a circuit using two three-way switches. Show the two switches. The lamp should be turned on.
2. Draw a circuit with two three-way switches and one four-way switch. The lamp should be turned off.

KEY TERMS

service drop
service entrance
distribution panel
branch circuit
extension cord
fuse
circuit breaker
convenience outlet

National Electrical Code®
Underwriters' Laboratories, Inc.
armored cable
conduit
conduit bender
three-way switch
four-way switch

Chapter 15

ELECTRIC MOTORS

After studying this chapter, you will know

- *how a DC motor works.*
- *how cemf is produced in a motor.*
- *the characteristics and types of DC motors.*
- *the characteristics and types of AC motors.*
- *how to select the correct motor for a given job.*
- *how a start winding is taken out of a circuit.*
- *why a start-winding switch on a single-phase AC motor is necessary.*
- *how to identify the shaded-pole motor by its appearance.*

INTRODUCTION

Electric motors are the workhorses of industry. There are AC and DC motors. Each type of motor has its characteristics and has the ability to perform a particular job. Each motor is matched to the job by identification of what is needed to be done and how this type of motor can operate in its atmosphere and under the required load. Knowing how a motor works and how it is selected for a particular job is important.

15.1 DC MOTORS

Direct-current motors are basically the same as DC generators. However, their uses are different. A motor converts electrical energy to mechanical energy. A generator converts mechanical energy to electrical energy.

In DC motors, current is fed through an **armature** of coiled wire. This produces a rotary motion. This turning motion is mechanical energy. It is applied to turn gears, belts, or pulleys.

In DC generators, the cycle starts with mechanical energy. In a car, the energy comes from the engine. The energy rotates an **armature** of coiled wire in a magnetic field. Electrical energy results.

How a DC Motor Works

DC motors apply a principle of magnetism. The principle is that unlike poles attract and like poles repel.

To create a turning effect, a single-loop armature is mounted in a magnetic field. The armature is placed between the poles of a permanent magnet, Figure 15-1.

The field produced by a permanent magnet is applied over the shortest possible route. That is, the attraction is directly between the north and south poles. Current passing through the armature sets up a magnetic field around each conductor loop, Figure 15-2. The magnetic fields of the loop distort the main field. This causes flux pressure on one side of each conductor. On the other side, a flux vacuum is produced. This magnetic force acts upon the conductors. It tends to draw the conductors into the vacuum to straighten out the distorted magnetic field.

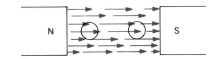

Figure 15-1. Single-loop armatures in a magnetic field are the basis for DC motors.

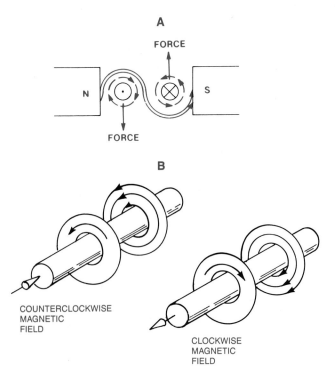

Figure 15-2.
A. In DC motors, conductors distort a magnetic field to create a flux action.
B. Note direction of current flow and arrowheads indicating the resulting magnetic field.

Imagine the magnetic field as a series of rubber bands. The loop is an object pressing upon them. The effect upon the loop is easy to see. Force is exerted, Figure 15-2. The drawing demonstrates that pressure is being exerted upon the top of the left loop. There is also pressure upon the bottom of the right loop.

Figures 15-3 and 15-4 demonstrate how the magnetic forces of the two loops cancel each other. If the armature loop is free to move, this pressure forces the left loop downward. The right loop moves upward under pressure. This causes rotation of the armature.

The armatures of all electric motors are freely suspended between the motor-pole pieces. Current passes through the armature coils. The flux of the field is distorted. **Torque**, or turning effect, causes the armature to rotate.

The direction of rotation of a DC motor is determined by the *right-hand* **motor rule**. This rule is illustrated in Figure 15-5. To determine rotation direction, use your right hand. Extend the thumb, forefinger, and middle finger at right angles to each other. Point the forefinger in the direction of the magnetic flux. Point the middle finger in the direction of the current flow. Then, the thumb is pointing in the direction of the motion of the conductor.

Counter-emf in a Motor (cemf)

The simple DC motor does not differ essentially from the generator. The conducting loop of the armature rotates in a magnetic field. The magnetic field induces a current in the rotating armature. The action of the armature is described by Lenz's law: The induced emf is opposed to the motion inducing it. This is **back emf**, or **counter-emf**. Keep in mind that *emf* stands for *electromotive force*, or *voltage*. Its strength depends upon the speed at which the armature rotates in the magnetic field. The difference between the counter-emf and the applied voltage determines the current in the motor. The line current necessary to operate the motor also depends upon counter-emf.

Counter-emf is an important force in operation of a DC motor. Consider what happens with a DC motor operates at full speed under no-load conditions. At this point, counter-emf is nearly equal to the applied voltage. This means that only a small current is required from the line.

A slowly turning armature generates a smaller counter-emf. The larger the existing voltage difference, the larger is the line current required. Then, as the motor gains speed, the increasing counter-emf cuts the need for line current.

A DC motor must be started with a resistance connected in series with the armature, Figure 15-6. The resistance is then removed gradually as the motor comes up to speed. At top speed, the resistance is removed

Figure 15-3. Cancellation by magnetic forces in two loops.

Figure 15-4. Reinforcement of magnetic field stimulates armature motion.

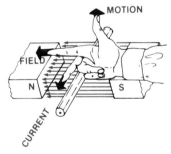

Figure 15-5. Drawing illustrates right-hand rule for motors.

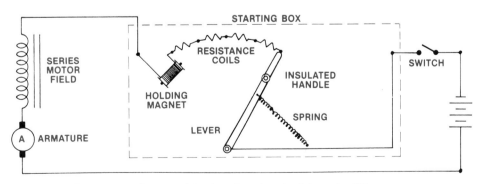

Figure 15-6. Functional drawing of starting circuit for series DC motor.

completely. DC starters are available for motors of any size or type.

Types of DC Motors

There are three basic types of DC motors:

- Series
- Shunt
- Compound.

Series Motor. A motor of this type has its field connected in series with the armature and the load, Figure 15-7 and 15-8. The series motor runs very slowly under heavy loads. Under light loads, it runs very fast. If the load is completely removed in a large series motor, the speed increases until the motor flies apart. **Series motors** *must never be run under no-load conditions.* They are seldom used with belt drives. This is because the motor could be damaged if the belt broke. Series motors are used in automotive starters. Other applications include constantly loaded equipment such as hoists or cranes.

Shunt Motor. In this type of motor, the field is connected in parallel, or shunt. That is, the field is across the line, as is the armature, Figure 15-9. Torque varies with the armature current. If the load on the motor increases, the motor slows down. If the load decreases, the motor speeds up. The variation of speed from no-load to normal load is about 10 percent of the no-load speed. **Shunt motors** are considered to be *constant-speed motors.* They are used to provide constant speed under a varying load. Therefore, their best use is under light or no-load conditions. Shunt motors are often used in machine tools, conveyors, and printing presses.

Compound Motor. This motor uses a combination of two sets of field coils, Figure 15-10. One set is connected across the armature. The other is connected in series with the armature. There are two separate types of **compound motors**, *cumulative* and *differential.*

The most commonly used type is the cumulative. On these units, the two sets of fields are arranged so that they aid each other. In this type of compound motor, an increase in load decreases speed greatly. But added load also increases the torque developed by the motor. The starting torque is also great. The cumulative motor is considered to have fairly constant speed. It also has good pulling power on heavy loads. Because of these features, cumulative compound motors are used on machine tools and continuously operating equipment.

In the differentially compound motor, the series set of **field coils** opposes the shunt field. The total field is weakened when the load increases. Within tolerances of safety, the speed increases with increased load. Starting torque is very low.

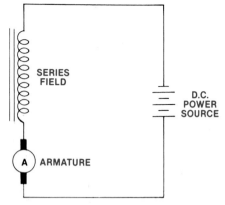

Figure 15-7. Circuit diagram for series DC motor.

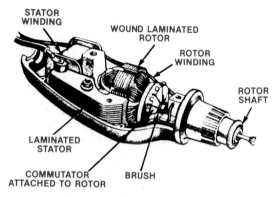

Figure 15-8. A series DC motor may also run on AC.

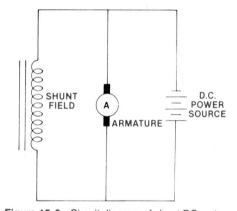

Figure 15-9. Circuit diagram of shunt DC motor.

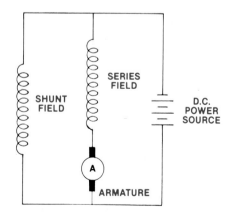

Figure 15-10. Circuit diagram of a compound DC motor.

Activity: THE DC MOTOR

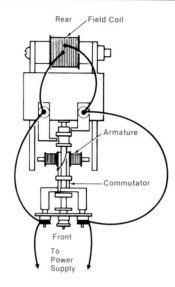

Figure 2

OBJECTIVES

1. *To study the basic types of DC motors—both series and shunt.*
2. *To understand those factors that determine the speed and direction of rotation of a motor.*

EQUIPMENT NEEDED

1 Power supply, 0–6 volts DC	1 Motor unit
	1 Field coil assembly

PROCEDURE

1. Set up the motor-generator unit, as shown in Figure 1.

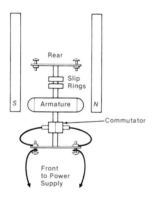

Figure 1

2. Connect the motor to the power supply as shown. Set the voltage control to a minimum. *Turn on the power supply.* Slowly adjust the output of the power supply to 4 volts DC. Start the armature turning with your finger. Does the armature now turn by itself? Look at the armature from the front. Is it turning clockwise or counterclockwise?
3. Increase the output of the power supply to 6 volts DC. Does the motor turn faster? Is it still turning in the same direction?
4. *Turn off the power supply.* Reverse the leads to the armature and *turn on the power supply.* Start the armature turning with your finger. When you look at the armature from the front, what is the direction of rotation?
5. *Turn off the power supply.* Remove the permanent magnets and set up the field coil in the position shown in Figure 2.

6. This is a shunt-connected motor. *Turn on the power supply*, with its output still adjusted to give 6 volts DC. Does the armature turn faster, slower, or at about the same speed as it did in Step 3? What is the direction of rotation?
7. *Turn off the power supply.* Reverse the connections to the field coil. *Turn on the power supply.* What is the direction of rotation? Has the speed of the armature been changed?
8. *Turn off the power supply.* Reverse the connections to the armature. *Turn off the power supply.* What is the direction of rotation?
9. *Turn off the power supply.* Remove the connections to the motor-generator unit and connect the unit shown in Figure 3.

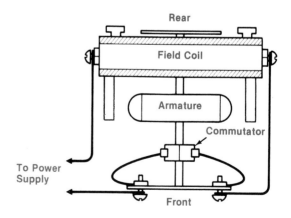

Figure 3

10. This is a series-connected motor. *Turn on the power supply.* The output of the power supply should be 6 volts DC. Start the armature turning with your finger. Does the armature turn faster, slower, or at the same speed as in Step 6? Should there be a change in the speed of rotation? Why?

11. *Turn off the power supply.* Reverse the connection to the field coil. *Turn on the power supply.* What change do you observe in the speed and direction of rotation?

12. *Turn off the power supply.* Disconnect all circuits you have been using and remove the field coil from the motor–generator unit. Return all equipment and materials to their proper storage places.

SUMMARY

1. What factors affect the direction of rotation of a DC motor?

2. What factors affect the speed of rotation of a DC motor?

___ REVIEW QUESTIONS FOR SECTION 15.1 ___

1. Compare DC motors and DC generators.
2. Describe the operation of a DC motor.
3. State the motor rule covering rotation direction for a DC motor.
4. Describe the effect of back emf, or counter-emf.
5. What are the requirements for starting a DC motor?
6. Name and describe three types of DC motors.

Safety Tip

Series motors can explode if they are run without a load. So, keep clear of an operating series motor. Never use a belt to connect a DC series motor to a load. If the belt breaks, the motor, running without a load, could blow up.

Figure 15-11. Universal motor in typical use.

15.2 AC MOTORS

There are three types of AC motors:

• Universal
• Induction
• Synchronous.

Universal Motor.   This is a small DC series motor that may operate on an AC source, Figure 15-11. There is a difference in function under AC operation. The field coil current and the armature current reverse direction at the same time. Because of this, torque maintains the same direction throughout the cycle. Special field windings and laminated pole pieces are used to reduce heat losses. These special fittings are used when motors are designed for use with both DC and AC. A variable speed control may be provided with small motors. This type of unit is shown in Figure 15-12. These motors are used to operate hand tools and small appliances. Examples include food mixers, sewing machines, and vacuum cleaners.

Figure 15-12. Small universal (DC–AC) motors often come with variable speed controls. (Radatron)

Induction Motor. This is the most widely used AC motor. Induction motors are relatively simple to build. They are rugged. And, they operate at constant speed if they are not overloaded. An **induction motor** has two basic parts. There is a set of field coils. And there is a rotor, Figure 15-13. The rotor is made of a laminated iron core. The core is slotted lengthwise at an angle. Copper bars are laid inside the slots. The ends of the copper bars are shorted by a copper ring. This ring forms a cylindrical cage. It is referred to as a *squirrel-cage rotor*, Figure 15-14.

Three-phase power may be used along with three pairs of poles. If so, the magnetic field of the stator will rotate electrically. The currents induced in the rotor are carried by the copper bars. These currents follow the rotating field. For torque to develop, the **rotor** must slip, or lag behind the field. A three-phase rotor is shown in Figure 15-15.

The rotor appears as a short-circuited secondary of a transformer. Lenz's law applies to the induced current. If the load is increased, a greater slip is present. As the load increases, a greater induced current is present in the rotor. As a result, a greater torque is produced. This helps to maintain the speed with the increase in load.

The induction motor depends upon a rotating magnetic field for its operation. A single-phase induction motor is not self-starting. To start a **single-phase motor**, some method is needed to start a rotating magnetic field.

Single-phase power merely reverses itself and doesn't rotate as does the three phase. A single phase motor can be started by hand. A starter may be built in the form of a centrifugal switch. A centrifugal switch is one that drops out automatically as the motor comes up to speed, Figure 15-16. For single-phase motors, a split-phase winding is used, Figure 15-17.

Figure 15-13. Induction motor, stator slots, insulation, and coils. (Spaulding)

Figure 15-15. Three-phase rotor with fan for use in induction motor.

Figure 15-14. AC induction rotor.

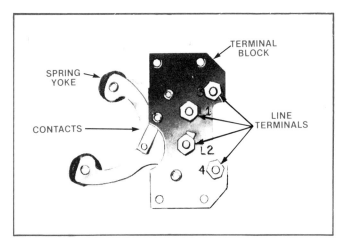

Figure 15-16. Centrifugal switch used to start single-phase induction motors.

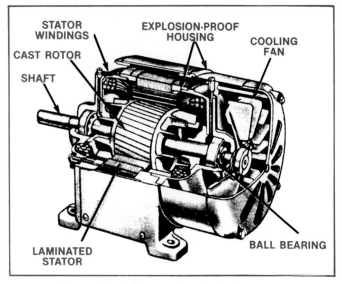

STATOR
WINDINGS

EXPLOSION-PROOF
HOUSING

COOLING
FAN

CAST ROTOR

SHAFT

LAMINATED
STATOR

BALL BEARING

Figure 15-17. Cutaway view of split-phase motor.

Other methods may also be used to start a **split-phase motor**. These include a capacitor, a shading pole arrangement, or a repulsion winding. The most commonly used starters are the split-phase winding and the capacitor. The **capacitor-start** motor will start under load, Figure 15-18. The split-phase motor, however, will not start under load. The shaded-pole motor is used in electric clocks as well as in other constant-speed applications, such as phonograph turntables and tape recorders. A shaded-pole motor is illustrated in Figure 15-19. Although there are better motors for these jobs, they cost more.

Synchronous Motor. This is a constant-speed AC motor, Figure 15-20. These motors are synchronized (in time) with the AC generator frequency. This type of motor is not self-starting.

In small **synchronous motors**, the rotor is usually a

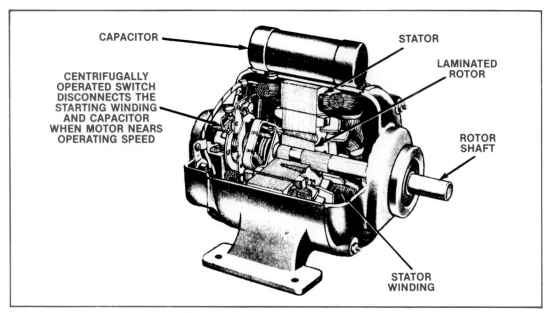

CAPACITOR

STATOR

CENTRIFUGALLY
OPERATED SWITCH
DISCONNECTS THE
STARTING WINDING
AND CAPACITOR
WHEN MOTOR NEARS
OPERATING SPEED

LAMINATED
ROTOR

ROTOR
SHAFT

STATOR
WINDING

Figure 15-18. Capacitor-start motor.

Figure 15-19. Shaded-pole motor.

Figure 15-20. Synchronous motor keeps time with AC frequency. (Superior)

permanent magnet. Wound rotors are used in larger motors. They must be excited with DC to produce a north–south polarity. Such a motor must be brought up to speed by some type of external force. Once started, the motor will synchronize with the AC frequency. It keeps in step with any changes in frequency.

Improved synchronous motors are now used for accurate control of speed. Other uses involve varying the power factor and driving DC generators at a constant speed. These motors are also used in computer-tape transports and some high-quality audio equipment.

<hr>

REVIEW QUESTIONS FOR SECTION 15.2

1. What are the key features of universal motors?
2. What are the differences between AC and DC operation for universal motors?

SUMMARY

DC motors are basically the same as DC generators. However, their uses are different. A motor converts electrical energy to mechanical energy. A generator converts mechanical energy to electrical energy.

DC motors apply a principle of magnetism. The principle is that unlike poles attract and like poles repel.

The armatures of all electric motors are freely suspended between the motor-pole pieces. Current passes through the armature coils. The flux of the field is distorted. Torque, or turning effect, causes the armature to rotate.

Counter-emf (cemf) is an important force in the operation of a DC motor. A slowly turning armature generates a smaller counter-emf. The larger the existing voltage difference, the larger is the line current required to operate the motor.

Three types of DC motors are series, shunt, and compound motors. Each type has its own identifiable characteristics.

AC motors come in three basic types: universal, induction, and synchronous. The most widely used is the induction motor because it is relatively simple to build and is rugged. It can operate at constant speed if not overheated. The induction motor has two basic parts: a set of field coils and a rotor. Three-phase motors do not need a start winding or start-winding switch for operation on three-phase current. A centrifugal switch is used to take the start winding out of the circuit once a single-phase motor comes up to speed.

The synchronous motor is a constant-speed type. Its speed is determined by the frequency of the power source.

USING YOUR KNOWLEDGE

1. Draw the schematic for the series DC motor.
2. Draw the schematic for the shunt DC motor.
3. Draw the schematic for the compound DC motor.
4. Read the field coil resistance of an AC motor.
5. Inspect the centrifugal switch on an AC single-phase motor for pitting at the points.
6. Inspect the commutator of a DC motor for arcing when it is running. When it is standing still, check the commutator for pitting.

KEY TERMS

armature	three-phase power
field coil	split-phase motor
commutator	capacitor-start motor
rotor	torque
compound motor	back emf (counter-emf)
series motor	motor rule
synchronous motor	shunt motor
induction motor	differential
single-phase motor	universal motor

Project: DC MOTOR

A motor with a wound rotor can be operated on AC or DC, but runs better on DC than on AC. This motor is designed to operate from 6 volts to 24 volts. A step-down transformer may be used to power it. Alternatively, a battery charger may be used to provide the DC needed for proper operation.

This motor can be made to run in series or shunt. See the schematic drawings for proper hook-ups, Figure 1. The shunt motor can be made to rotate faster if the field is weakened by placement of a resistor in series with the field coil. Counter-emf is reduced, causing an increase in current in the armature, which offsets the loss of strength in the field coil.

If a series resistor is inserted in the series motor

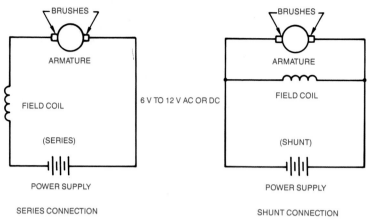

SCHEMATIC FOR SERIES AND SHUNT HOOK-UPS

Figure 1

hookup, the motor will rotate more slowly, since the resistor becomes a voltage-dropping device and produces a lower voltage across the field coil and armature.

To carry out this project, you will need the following materials.

1 Base, 1/2" × 3-1/4" × 4", wood
1 Support bracket (A), 1/8" × 1/2" × 3-1/8", band iron
1 Support bracket (B), 1/8" × 1/2" × 3-1/8", band iron
1 Field coil core, 1/8" × 3/4" × 7-1/2", band iron
1 Armature core, 1/8" × 3/4" × 2-1/2", band iron
1 Armature shaft, 1/4" dia. × 3", drill rod
1 Commutator, 1/2" I.D. × 1", brass or copper (copper plumbing pipe can be used)
2 Brushes, 1/4" × 3", brass or copper (#12 copper wire can be used)
1 Dowel, 1/2" dia. × 1", wood
10 Wood screws, 1/2" #6 R.H., steel
1 Machine screw (A), 3/4" 1/4-20 hexagonal, steel
1 Machine screw (B), 3/4" 1/4-20 hexagonal, steel
1 Nut (A), 1/4-20 hexagonal, steel
1 Nut (B), 1/4-20 hexagonal, steel
2 Fahnstock clips, 3/4" × 5/16", #6 mounting hole, nickel-plated brass
1 Field coil, 2 layers #22 copper wire (Formvar), approx. 26 feet
1 Armature coil, 2 layers #22 copper wire (Formvar), approx. 20 feet

Construction Hints

Field.   Bend the field coil core so that the two horizontal legs are 2 inches long. Wrap the core with electrician's tape. Wind the layers of #22 enameled wire very closely and neatly; bind the finished coil with electrician's tape

to prevent it from unraveling. Allow 4-inch leads from the coil. Start and end the coil at the bottom of the core. Allow 26 feet of wire, Figure 2.

Armature.   Round the ends of the core with a file so that a 2-inch circle is created when it rotates. Cover the core with insulating tape. The core should be drilled to receive the shaft. Start winding at the shaft, wind to the end, and continue winding in the same direction back to the shaft, allowing 2-inch leads. Do not change winding direction. This will produce the required two layers of wire on the armature. The armature should clear the field core by approximately 1/16 inch.

Commutator.   The commutator slits should be aligned so that both brushes touch both segments when the armature core is located vertically. A slight turn on the shaft is necessary to start the motor. The completed armature is ready for mounting in the machine screws A and B when the shaft has been ground to a point on both ends and the machine screws have been drilled to receive the pointed shaft ends. Machine oil can be used to reduce friction at these points. Do not allow the oil to run onto the commutator, because oil is an insulator.

Operation Hints

1. It may be necessary to enlarge the space gap in the commutator segments.
2. It may be necessary to turn the commutator to the right to put the armature field ahead of the field coil's magnetic field.
3. Tension on the brushes may need adjusting if they apply too much pressure on the commutator.
4. Do not allow the coils to overheat.

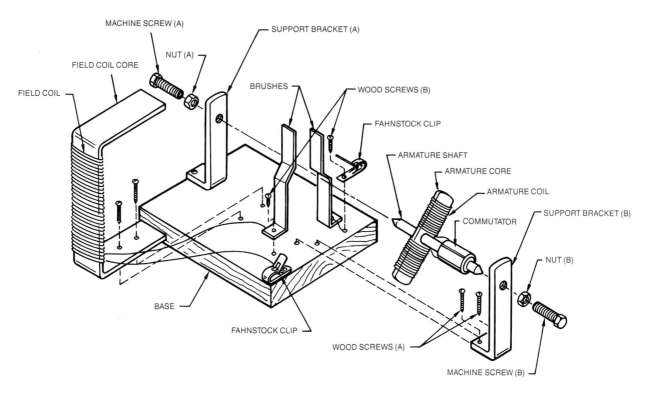

EXPLODED VIEW OF A DC MOTOR

Figure 2

Chapter 16

THERMIONIC DEVICES AND SEMICONDUCTORS

After studying this chapter, you will

- *understand the principle of the DeForest triode.*
- *identify the cathode, grid, and plate of a vacuum tube.*
- *identify various types of tubes according to the number of grids.*
- *understand the functioning of a vacuum-tube diode, triode, and other types of tubes.*
- *explain how a triode amplifies.*
- *explain the identification characteristics of tubes.*
- *understand the main advantages of transistors over vacuum tubes.*
- *understand that there are two types of materials used for the manufacture of semiconductor devices.*
- *understand the structure and functioning of semiconductor diodes, including forward- and reverse-bias alignments.*
- *understand how zener, tunnel, and other diodes operate.*
- *explain what the elements of a transistor are.*
- *identify the various types of semiconductor devices.*
- *identify various types of IC packaging.*
- *explain how transistors are manufactured and utilized in circuits.*

INTRODUCTION

Electron tubes are devices in which electrons are generated within glass envelopes. *Vacuum tubes* are units from which most of the air has been withdrawn. *Thermionic tubes* are devices in which electrons are produced by the heating of an electrode.

Many vacuum tubes have been replaced by transistors and other solid-state devices. However, the vacuum tube is still in wide use. Your TV picture tube is a vacuum tube. Some radio and TV broadcast equipment still uses vacuum tubes. Computer monitors are vacuum tubes.

Vacuum tubes will be around for many years. So, it is still important that you understand the principles of their operation.

The term **semiconductor**, as you know, describes a material that resists the flow of electrons. A semiconductor material lies between conductors and insulators in terms of electron movement.

The same word, *semiconductor*, also identifies a type of electronic device. Included are transistors and diodes. These units have played key roles in the development and growth of the field of electronics.

In this chapter, you learn about semiconductor devices, how they function, and how they are connected into circuits.

16.1 THE EDISON EFFECT

The discovery and early use of the vacuum tube were based on the work of Thomas Edison. The experiment that led to the vacuum tube took place in 1883. Edison discovered that a carbon filament tended to burn out near the positive end. When it did, a black deposit was produced inside the glass envelope. Also, Edison noticed that there was a shadow on the positive leg of the lamp filament.

In trying to remedy this situation, Edison used a metal plate. The plate was placed inside the glass, near the filament. The plate was connected through a **galvanometer** to the filament battery, Figure 16-1.

A **deflection** of the galvanometer was observed. This occurred only when the positive terminal was connected to the metal plate. When the metal plate was connected to the negative terminal, there was no deflection. Edison recorded these events in his notebook. He carried this work no further. However, this reaction came to be known later as the **Edison effect**. Figure 16-1 shows the bulb set up by Edison.

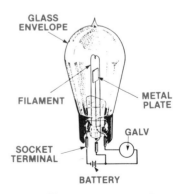

Figure 16-1. Diagram showing the Edison effect.

16.2 THE DeFOREST TRIODE

In 1906, Dr. Lee DeForest produced a vacuum tube that would amplify weak signals. He placed a third element, a *grid*, between the negative filament (cathode) and the plate. This grid was positively charged. The cathode was negatively charged. The cathode emitted electrons when it was heated. DeForest's tube is illustrated in Figure 16-2.

DeForest varied current to the grid. He found that current in the plate circuit also varied. The variation in the plate was found to be greater than that in the grid. This meant that weak signals could be amplified by DeForest's **triode**.

With this tube, the signals were first amplified (increased), then detected, using a diode tube. In this situation, detection involved separation of audio and radio frequencies.

DeForest's triode opened the way for the development of more-complex vacuum tubes.

16.3 THERMIONIC EMISSION

A basic principle of the vacuum tube is the generation of free electrons. The electrons are emitted from the cathode by heat. For this reason, the cathode is also called an *emitter*. The process of emitting electrons is known as *thermionic emission*. A more complete definition is: **Thermionic emission** *is the escaping of electrons from the surface of a hot substance.*

Types of Emitters

There are two basic types of emitters:

* Directly heated cathodes
* Indirectly heated cathodes.

Directly Heated Cathodes. These cathodes have an electron-emitting material on the filament, Figure 16-3. They are referred to as filament cathodes. They are also called filaments. High-power tubes use this type of **cathode**.

Indirectly Heated Cathodes. These have an insulated filament inside a cathode sleeve, Figure 16-4. They are

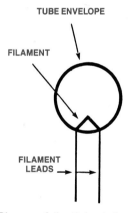

Figure 16-3. Diagram of directly heated emitter with the filament and cathode in the same element.

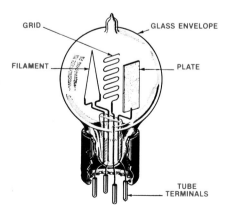

Figure 16-2. Diagram of DeForest's triode.

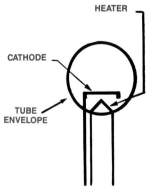

Figure 16-4. Diagram of indirectly heated cathode with separate cathode and filament.

referred to as heater cathodes or simply as cathodes. Practically all signal-receiving equipment uses indirectly heated cathodes. In this book, assume all identified cathodes are indirectly heated unless another description is provided. All directly heated cathodes will be identified specifically.

REVIEW QUESTIONS FOR SECTION 16.3

1. What is thermionic emission?
2. What is an emitter?
3. Describe a tube that has a directly heated cathode.
4. Describe a tube that has an indirectly heated cathode.

16.4 THE DIODE

Vacuum-tube **diodes** have two electronic elements: the cathode and the plate. Electrons flow from the negative cathode to the positive plate. Figure 16-5 diagrams a vacuum-tube diode. Included is the external circuit through which electrons are returned to the cathode through the power supply.

The space around the cathode becomes negatively charged as electrons are emitted. This is known as a **space charge**. The space charge opposes further electron emission by the cathode. The number of electrons emitted is dependent upon the temperature and size of the cathode.

When the cathode is at a low temperature, electrons near the plate are attracted to it. This flow produces a small current. As the plate voltage is increased, more electrons are attracted. The space charge is thus reduced and the plate current is increased.

REVIEW QUESTIONS FOR SECTION 16.4

1. How does a diode operate?
2. What is the space charge in a diode?

16.5 THE TRIODE

The symbol for a triode is shown in Figure 16-6. It is based on DeForest's discovery that a grid between the cathode and plate could control plate current. A change in grid voltage causes a greater change in plate current.

Within a triode, the grid is a third electrode. It is called the **control grid**. This is because it controls the flow of electrons between the cathode and plate. Control is the key. The grid does not obstruct the flow of electrons.

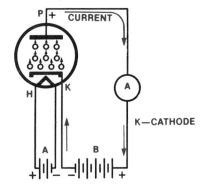

Figure 16-5. Diagram of diode circuit.

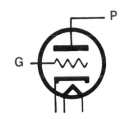

Figure 16-6. Symbol for triode.

Within a triode, a small AC voltage is applied to the grid. The result is a large variation in AC current. This is called **amplification**. AC plate current varies at the same rate (frequency) as the input AC signal applied to the grid.

The triode has a plate load resistor. It is placed in series with the plate of the tube and the DC power source for plate voltage. The resistor has the varying plate current (AC) flowing through it. Therefore, it produces a voltage drop at the same frequency as that applied to the grid.

The voltage drop across the load resistor is much greater than the AC signal voltage applied to the grid. The output voltage is greater than the input voltage. (The output voltage is taken from across the load resistor.) This increase is *amplification*.

Operating Characteristics of the Triode

Varying the grid **bias** voltage produces varying effects. Suppose a negative grid bias is less than the cutoff voltage value. Some electrons pass through the grid and reach the plate. Thus, there is some current. As the negative grid bias decreases, plate current increases. At zero grid bias, the cathode and grid have the same potential in reference to each other. The plate current then depends upon plate voltage.

No grid bias exists when the grid is made positive with respect to the cathode, Figure 16-7. The grid cannot repel electrons if it has a positive charge. Instead, the positive charge accelerates electrons toward the plate. Plate current increases.

When the grid is positive, it, too, can attract electrons.

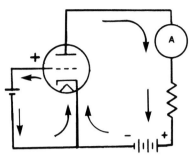

Figure 16-7. Diagram of triode circuit with grid drawing current.

This means there is current flow through the grid circuit. If this occurs, it represents a waste of power (in most vacuum-tube circuits). Therefore, the grid is usually kept at some negative potential in respect to the cathode.

___ REVIEW QUESTIONS FOR SECTION 16.5 ___

1. How does a triode operate?
2. What is the function of the grid in a triode?

16.6 OTHER TYPES OF VACUUM TUBES

So far, the diode and triode have been described. These were forerunners among vacuum tubes. Through the years, other, more-complex tubes have been developed. Some of these are described below.

Tetrode. The root of the name for this tube comes from the word *tetra*. This is Greek for *four*. The suffix, *-ode*, comes from *electrode*. So, the word means *four electrodes*.

The fourth electrode is a *screen grid*. The screen grid reduces capacitance between the other electrodes. Capacitance is reduced between the control grid and the cathode. A reduction also takes place between the control grid and the plate.

Pentode. The root comes from the Greek word *penta*, meaning *five*. The stem comes from *electrode*. Thus, the pentode is a five-element tube.

Safety Tips

1. Remember that vacuum tubes use high voltages. You can be shocked, even killed, by them. Use the standard safety precaution of keeping one hand in your pocket while working on equipment containing vacuum tubes.
2. When measuring high voltages, place the meter probe on the point to be measured before you turn on the power. Remove your hand from the probe first. Then turn on the power.

A pentode has three grids. These are the control grid, screen grid, and suppressor grid. The *suppressor grid* is added to eliminate secondary emission. Secondary emission results when electrons bounce off the plate. The bouncing results because the electrons strike the plate at high speeds, having been accelerated by the screen grid.

Beam Power Tube. The **beam power tube** also has five elements. In these units, the suppressor grid is replaced by a *beam-forming plate*. This beam-forming plate is usually connected internally to the cathode. It concentrates, or beams, the electrons to the plate. This occurs because the connection with the cathode causes a negative potential.

These tubes are used when output is needed in terms of power rather than voltage.

___ REVIEW QUESTIONS FOR SECTION 16.6 ___

1. How many elements does a tetrode have?
2. Describe a pentode.
3. Describe how a beam power tube forms a beam.

16.7 TUBE BASES

Since the first vacuum tubes were made, the internal complexity of the tubes has increased. A system of locating the many connections was needed. This was done by establishing a series of standard bases for connecting tubes into circuits, Figure 16-8. The connector-pin layouts illustrated are common for many receiving devices. These are the seven-pin miniature, the nine-pin miniature, and the eight-pin base.

Note that, from the bottom, the number of the pins is read in a clockwise direction from the keyway. The keyway is the guiding point used in locating the tubes in their connectors.

Figure 16-9, page 234, contains a series of tube diagrams. These show vacuum-tube symbols. The drawings also identify electrodes with pin numbers.

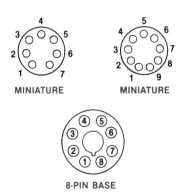

Figure 16-8. Diagram of standard bases for vacuum tubes.

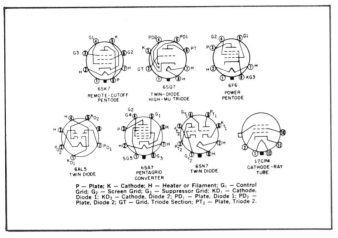

Figure 16-9. Vacuum-tube symbols showing identification of electrodes with pin numbers.

___ **REVIEW QUESTIONS FOR SECTION 16.7** ___

1. How do you read the connector-pin layout of a vacuum tube?
2. What is the keyway of a vacuum tube?

16.8 VACUUM-TUBE IDENTIFICATION

Although there is not a strict standard, the numbers assigned to vacuum tubes usually have some meaning. As an example, a common tube is numbered 6SK7. The *6* indicates a 6.3-V filament. The *S* represents the word *sans* (Latin for *without*). This means the tube is without a grid cap. In other words, the grid connection is now made to one of the tube base pins. The *K* means that it is an RF amplifier tube. The *7* means that seven of the eight pins are used for connections to the elements. These elements are the cathode, filaments, plate, grid, and the two other grids.

If the tube is designated 12SK7, it has a 12.6-V filament. All other characteristics are the same as for the 6SK7. With the increase of voltage from 6.3 to 12.6, current decreases from 0.3 to 0.15 A. Other filament designations include the following.

- 1R5 means the filament has 1.2 V.
- 35Z5 means the filament is rated at 35 V.

The letter designations *Z, W, X, Y,* and *U* have a special meaning. These letters indicate a tube is a rectifier type. The letters *L, V, C,* and *B* mean the tube is an audio power amplifier.

There are many exceptions to these designations. The most accurate information on any tube can be found in a tube manual. RCA, General Electric, and Sylvania all publish tube manuals.

16.9 CATHODE-RAY TUBES (CRT)

The oscilloscope is a good troubleshooting tool for electronic technicians. The oscilloscope utilizes the cathode-ray tube to display various patterns created by signals within an electrical or electronic circuit. Figure 16-10 shows how the electron beam is used to strike a fluorescent screen to produce a pattern according to the signal-voltage variations.

The electron gun is constructed so that the heated cathode gives off electrons by thermionic emission. The cathode is surrounded by a negatively charged cylinder with a hole in one end. The electrons are repelled, since they too have a negative charge. They stream through the hole toward a less negatively charged accelerator electrode. This speeds the beam of electrons toward the screen, so that when the screen is struck by the electrons the phosphorous material on the screen glows as it absorbs

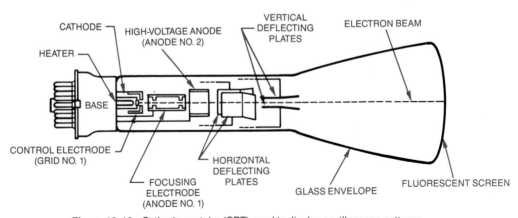

Figure 16-10. Cathode-ray tube (CRT) used to display oscilloscope patterns.

the energy of the speeding electrons. Horizontal and vertical deflecting plates cause a pattern to be traced on the screen that duplicates the changes in the voltages applied to the plates.

Cathode-ray tubes use these same principles to produce television pictures and to reproduce the graphics displayed on *computer monitors*. Until charged coupled devices (CCD) are improved a great deal, the vacuum-tube type of monitor and picture tube will be utilized to their limits.

REVIEW QUESTIONS
FOR SECTIONS 16.8 AND 16.9

1. What is the meaning of the letter *S* in a vacuum-tube identification?
2. What is the meaning of the letter *K* in a vacuum-tube identification?
3. What type of vacuum tube includes the letter *Z, X, Y,* or *U* in its identification?
4. What type of vacuum tube includes the letter *L, V, C,* or *B* in its identification?
5. What does *CRT* stand for?

16.10 HISTORY OF TRANSISTORS

Michael Faraday made an important contribution to research in semiconductors in 1833. He experimented with silver sulfide. Faraday found that silver sulfide has a resistance that varies **inversely** with temperature. That is, as temperature increases, the resistance of silver sulfide decreases. Such a property is unusual. Most conductors increase resistance with an increase in temperature.

More than a century later, Faraday's work paved the way for discovery of the **crystal amplifier**, or *transistor*. In June, 1948, three scientists at Bell Laboratories discovered the possibility of amplification by a crystal. The three scientists were John Bardeen, William Shockley, and W. H. Brattain. They later won a Nobel Prize for this work.

A transistor uses solid semiconductor materials. Current flow through these materials produces amplification and switching. Materials such as **germanium** and **silicon** are used for semiconductors.

Transistor Advantages

Transistors have been used largely to replace vacuum tubes. As compared with vacuum tubes, they have many advantages. One is their size. Transistors are small. This

has been an advantage in reducing the size of radios, computers, and other electronic devices.

Transistors have useful lives running into many years. This is because no heat is necessary to cause free electrons to move. The transistor is ready for use immediately upon application of an emf. No warmup period is necessary.

Required operating voltages are small. Transistors can be operated with power ratings of from 1 milliwatt to more than 50 W at normal operating temperature. This temperature is 77 degrees F [25 degrees C]. In addition, the transistor requires no filament or heater. This reduces power requirements and operating temperatures.

REVIEW QUESTIONS FOR SECTION 16.10

1. How is the resistance of silver sulfide affected by heating?
2. How does the reaction of silver sulfide to heat compare with that of other conductors?
3. What is the effect of current flow through a transistor?
4. Why are transistors used to replace vacuum tubes?
5. How does the useful life of a transistor compare with that of a vacuum tube?
6. How does heat dissipation of a transistor compare with that of a vacuum tube?
7. How does the warmup time for a transistor compare with that of a vacuum tube?
8. How do sizes of transistors compare with those of vacuum tubes?

16.11 SEMICONDUCTORS

A number of semiconductor devices are shown in Figure 16-11.

Semiconductor technology is usually referred to as **solid state**. This simply means that the materials used are solid or continuous. This is in contrast with vacuum tubes that consist of a series of assembled parts. Of course, the substances from which semiconductors are made are not really solid. They have atomic structures consisting largely of empty space. These spaces are essential for the movement of electrons.

There are many semiconductor materials. The two used most commonly are germanium and silicon. In their pure forms, both germanium and silicon have properties close to those of insulators. By nature, both of these materials

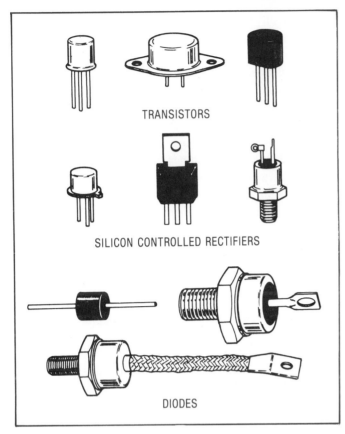

Figure 16-11. Semiconductors are available in a number of sizes and shapes.

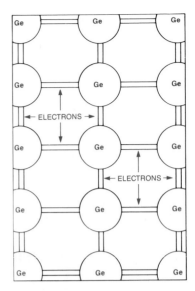

Figure 16-12. Diagram showing crystal-lattice structure of germanium.

are hard, brittle crystals. They have a type of structure known as a *lattice*, Figure 16-12.

The outer (valence) electrons of individual atoms of these materials have special traits. They are bonded in pairs to adjacent atoms. This type of structure has no free electrons. Therefore, semiconductor materials are normally poor conductors. To free outer electrons, it is necessary to apply higher temperatures or strong electron pressure.

For use in electronic devices, it is necessary to reduce the bond on valence electrons. This helps to make free electrons available. To achieve this effect, materials are blended. Small amounts of other elements with different atomic structures are added to silicon or germanium. These added materials are called *impurities*. The impurities are selected specifically for their ability to provide free electrons.

The table in Figure 16-13 lists a number of elements used in the manufacture of semiconductors. For each, the table gives its abbreviation and atomic number. Remember that the atomic number is equal to the number of protons in the nucleus of the atom.

Figure 16-14 diagrams the helium, hydrogen, and oxygen atoms. The helium atom is *inert*. This means that

helium does not react with other elements. This is because the two electrons in a helium atom are *stable*. They are stable because the ring into which they fit is complete. The first ring around the nucleus of any atom contains only two electrons. When a ring is complete, it is hard to free its electrons.

By contrast, note the structure of the hydrogen atom in Figure 16-14. There is only one electron in a ring that can hold two. So, the ring, or *shell*, of a hydrogen atom is incomplete. This means that the single electron in a hydrogen atom will be less stable than the two electrons in a helium atom. It is more available as a free electron.

Now look at the diagram of the oxygen atom in Figure 16-14. There are two electron rings. The first shell is complete. It has its limit of two electrons. The second shell has six electrons. The total number of electrons is eight. They have a negative charge. The negative charge is balanced by the positive charge of the eight protons in the nucleus.

The second ring in atomic structures can hold up to eight electrons. The oxygen atom has only six electrons

Element	Symbol	Atomic Number
Germanium	Ge	32
Silicon	Si	14
Antimony	Sb	51
Arsenic	As	33
Indium	In	49
Gallium	Ga	31
Boron	B	5

Figure 16-13. Materials used in transistor manufacture.

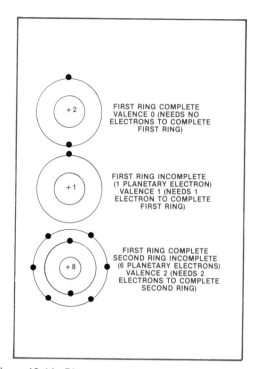

Figure 16-14. Diagrams showing structures of helium, hydrogen, and oxygen atoms.

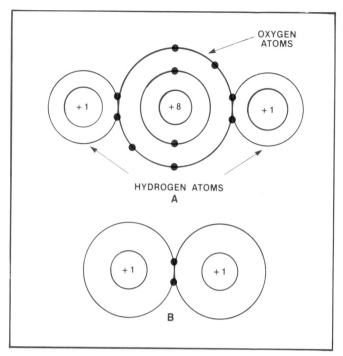

Figure 16-15. Diagrams showing the sharing of valence electrons. **A.** One oxygen and two hydrogen atoms share valence electrons to form a molecule of water. **B.** Two hydrogen atoms share valence electrons.

in its second ring. Therefore, the second ring of the oxygen atom is incomplete. To be complete, the ring would need two more atoms. Because this ring is incomplete, the outer electrons in an oxygen atom are less stable than those in the helium atom.

When atomic structures have space in their outer rings, nature seeks to fill these spaces. One method is for two or more atoms to share electrons in their outer rings. This is demonstrated in Figure 16-15. Figure 16-15A shows an oxygen atom that has completed its outer ring. This has been done by sharing of the outer ring with the electrons of two hydrogen atoms. When atoms are combined in this way, they lose their individual identities. The atoms linked through the sharing of electrons become **molecules**. The joined atoms have a *molecular structure*. This molecular structure is unique for a given atomic combination. For example, Figure 16-15A shows a molecule that consists of two atoms of hydrogen and one of oxygen. This is the most common molecule on earth. It is known as water.

The sharing of outer electrons by atoms is called a *covalent bond*. The name comes from the fact that the atoms share valence electrons. Figure 16-15B shows a covalent bond. Two atoms of hydrogen are joined to share valence electrons.

Conductors

Conductors are made up of atoms whose outer-orbit electrons are **loosely held**. At room temperature, there is

enough heat to free large numbers of electrons. The free electrons drift from one atom to another. Applying a voltage to the ends of a conductor directs the flow of electrons. Electron movement is from the negative (excess) side of the conductor. Movement is toward the positive (deficiency) side. The movement of electrons along a conductor is a flow of current.

Insulators

Materials made from atoms with tightly held valence (outer ring) electrons are **insulators**. The outer-orbit electrons are highly stable. They are not easily forced free. This means very little current flows. When electrical pressure or voltage becomes extremely high, some electrons are loosened. At this point, there is current flow. In some instances, this flow will take the form of electric arcing. It appears as though lightning is striking the insulator.

Semiconductors

Semiconductors fall between insulators and conductors. They conduct better than insulators. But they do not carry current as well as conductors. The two materials most frequently used as semiconductors are germanium and silicon.

Germanium has 32 protons in its nucleus. It has four shells containing 32 electrons. The three rings nearest the nucleus contain 28 electrons. This means there are four electrons in the outer ring. These four valence electrons determine the chemical properties of the atom.

Because there are only four valence electrons, they are loosely held. They can escape from the binding force of their atoms at room temperatures. On escaping, they become free electrons. When a valence electron escapes, it leaves a "**hole**," or open position. When a negative electron leaves, the electrical balance of an atom is upset. On losing the negative charge of an electron, the atom takes on a positive charge. It is then called a positive *ion*. An ion is a charged atomic particle. Any unit of matter that is smaller than a full, balanced atom is an atomic particle. Thus, an ion can be as small as a single proton with a positive charge. An ion can also be an atom that has lost an electron and taken on a positive charge.

Positive "Holes"

A positive ion attracts negative electrons. Thus, a free electron that has left an atom is attracted by positive forces. When an electron leaves an atom, a hole, or open position, is created. The electron then fills an open hole in another atom. The overall effect is the same as though the hole were moving. Holes created by movement of negative electrons have positive charges. So, these holes naturally move in the direction of a negatively charged terminal.

Crystals

Crystals of an element are groups of atoms. The neighboring atoms can share valence electrons, in which cases the atoms make up a molecule.

Figure 16-16 shows the molecular pattern of a germanium crystal. Each circle is identified with the marking "Ge." This is the abbreviation for germanium. Thus, each circle represents an atom. The shared valence electrons are represented by the lines that join the circles. This diagram is flat. Actual crystals are three-dimensional. That is, they have depth as well as the height and width shown. Within a natural or "pure" molecular crystal structure, atoms are electrically balanced. They have the same numbers of protons and electrons.

Impurities

Impurities are added to the germanium crystals to be used in commercial semiconductors. These impurities may add extra electrons to some atoms. They may also cause positive holes in other atoms. Examples of added impurities are arsenic and boron.

Arsenic. This element has five electrons in its outer ring. When an atom of arsenic is added to a germanium

crystal, an extra, or free, electron is present. A germanium crystal with a free electron is called an N-type crystal, Figure 16-17.

Boron. This is another impurity material that can be added to a germanium crystal. Boron has only three electrons in its outer shell. This leads to a shortage of one electron. In other words, a positive hole is created. The result is known as a P-type crystal, Figure 16-18.

These two types of crystals, N and P, are used to make both diodes and transistors.

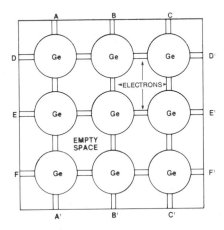

Figure 16-16. Diagram of germanium crystal structure.

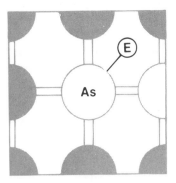

Figure 16-17. Diagram of N-type crystal.

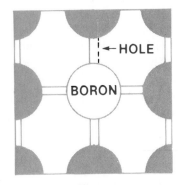

Figure 16-18. Diagram of P-type crystal.

1. Why are semiconductors called solid-state devices?
2. How does the complete or incomplete state of its valence orbit affect the conductivity of an atom?
3. What is the purpose of adding impurities to germanium or silicon used in semiconductors?
4. How do impurities change the conductivity of semiconductor materials?
5. What is meant in referring to an element as stable or unstable?
6. What is a molecule?
7. What is a molecular structure?
8. What is a positive hole in a semiconductor material?
9. What is a negative hole in a semiconductor material?
10. What are N and P crystals?

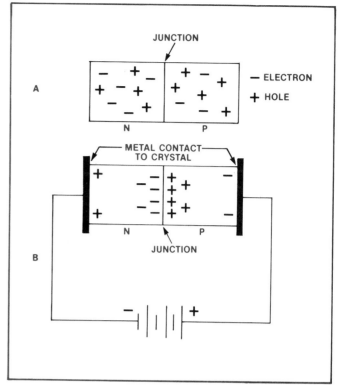

Figure 16-19. Diagrams of semiconductor diodes.
A. NP junction diode.
B. NP junction diode in forward-bias circuit.

16.12 SEMICONDUCTOR DIODES

Figure 16-19 shows what happens when two crystals are joined. In Figure 16-19A, an N and a P crystal have been joined. These crystals have opposite structures. The N crystal has two holes and six electrons. In the P crystal, there are two electrons and six holes. The electrons and holes line up opposite each other. This produces a rectifying device, a semiconductor diode.

In Figure 16-19B, a voltage is applied to the crystals. This causes the holes and electrons to line up according to the polarity of the power source. The negative terminal of the power source attracts the positively charged holes. The negative free electrons are attracted to the positive terminal.

Note the direction of electron movement. This represents the current flow in the circuit. The electrons move toward the positive terminal. Yet the electrons have remained within their own crystals. This alignment is said to represent a **forward bias** in a semiconductor diode. The path of the electrons represents the least opposition to current flow. It is known as the *low-impedance* direction.

In Figure 16-20, the same crystal structure is in place. But the power source is reversed. This is called the **reverse bias** direction. Once the voltage has been applied, the electrons move toward the positive terminal. And holes move toward the negative terminal.

When power is applied, there is an initial rush of electrons and holes. They align themselves toward the

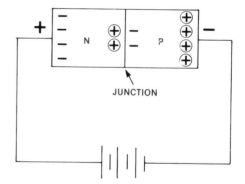

Figure 16-20. Diagram of NP diode in reverse-bias circuit.

opposite, attracting terminals. After this alignment, there are very few electrons and holes left to move across the junction. This causes a state of high impedance with a reverse bias. The reverse-bias direction represents a nonconduction crystal arrangement. A low impedance in one direction and a high impedance in the other produces a semiconductor diode. Note the symbol for a semiconductor diode in Figure 16-21.

Figure 16-22 shows the effects of applying AC across a PN junction. Current flows easily in a forward direction. However, there is high impedance when polarity is reversed.

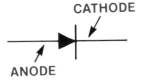

Figure 16-21. Symbol for diode.

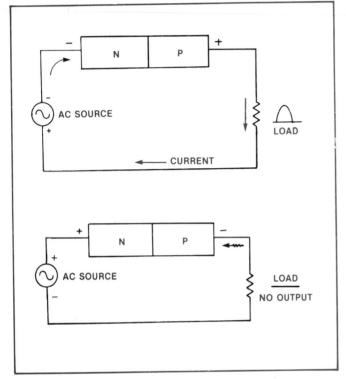

Figure 16-22. Diagrams showing forward **A.** and reverse **B.** AC flow across a PN junction.

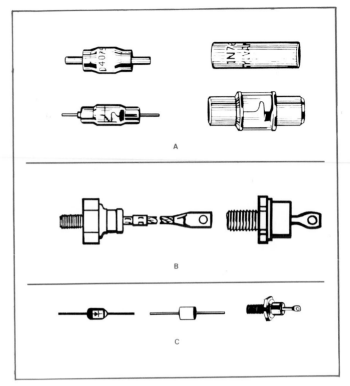

Figure 16-23. Special-purpose diodes for
A. radar and computer systems,
B. larger-current circuits, and
C. sensitive circuits.

A number of factors affect the voltage and current ratings of a PN **junction diode**. As you work in electronics, learn to use a transistor handbook. This will provide the information needed on forward and reverse bias. Ratings for peak and inverse AC currents are also given. This information is available for both germanium and silicon diodes.

Figure 16-23 shows a number of specialized diodes. This illustration demonstrates that diodes are designed for specific jobs. Both internal and external construction are determined by circuit requirements. For example, Figure 16-23A shows diodes used for radar and computer circuits. The diodes in Figure 16-23B are designed to carry large currents. So, their cases are heat conductors. As the diodes become warm, the generated heat is transferred to the air. Figure 16-23C shows zener diodes. These units protect sensitive meter movements and regulate voltage.

___ **REVIEW QUESTIONS FOR SECTION 16.12** ___

1. How is a semiconductor diode formed?
2. What is forward bias?
3. What is reverse bias?
4. What is a PN junction diode?

16.13 ZENER DIODES

Reversing voltage can destroy semiconductor diodes under some conditions. In other situations, the effect of a reverse-bias voltage can make diodes useful. Zener diodes are devices built around the effects of reverse voltage bias.

There are several important conditions and functions that you should understand in connection with zener diodes. These include:

• Avalanche breakdown
• Zener effect
• Voltage regulation
• Switching.

Avalanche Breakdown. This is a condition in which the boundary within a diode breaks down. As a result of the breakdown, electrons pour across the boundary.

The effect results from an increase in the reverse bias of the voltage. This increase is beyond the normal rating for a diode. As a result, the minority-carrier electrons acquire energy. This energy is sufficient so that they can cross the boundary region. These electrons collide with electrons in the N region of the germanium crystal. The collisions release valence electrons.

Figure 16-24 is a curve showing the effect of an increase in reverse bias. In this example, the breakdown occurs in the area of 500 to 600 V. At the point of breakdown, the current increases rapidly. This rapid increase is the **avalanche breakdown**.

Following avalanche breakdown, a diode may be ruined. This happens if the diode is not large enough physically to dissipate the heat. At this point, there is hole flow across the barrier into the P region, Figure 16-25 on page 242. When a diode reaches the avalanche point, holes are quickly neutralized. Neutralization results as electrons from the negative bias terminal quickly fill the holes. This means that the opposition represented by the barrier region has broken down. The diode then acts as

a path of low impedance. Large currents are allowed to flow in either direction.

Look at Figure 16-24. Note that voltage is applied in both forward and reverse bias. Also note the milliamperes of current in the upward direction. Once this avalanche breakdown occurs, a regular diode may be destroyed. However, there are diodes that are designed for a controlled avalanche point. They take advantage of this condition. The breakdown in the negative direction is used for voltage regulation. Other applications also make use of this property. These include switching and current control.

Zener Effect. A breakdown can occur as a result of reverse bias on a diode. The breakdown takes place when reverse bias exceeds a diode's rated value. At the point of breakdown, the reverse current increases rather rapidly.

Refer to Figure 16-25. Electron-hole pairs are created in the ion layers at the diode junction. This takes place at the time of breakdown of the barrier region. Electrons in the barrier region are torn from their covalent bonds. This results from the force of the reverse-bias voltage. Once there is sufficient voltage to cause a break in one

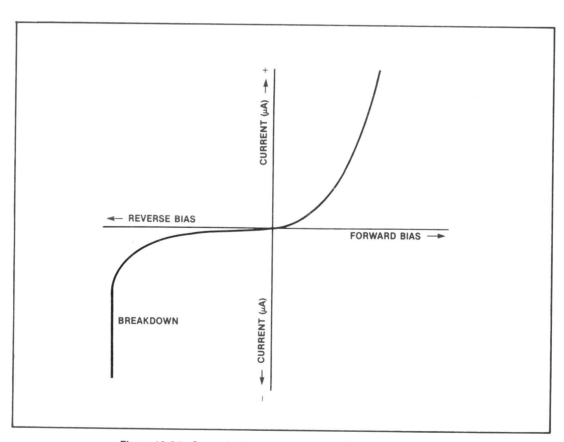

Figure 16-24. Curve showing average current-voltage of PN-junction germanium diode. (Curve has been exaggerated to show biasing.)

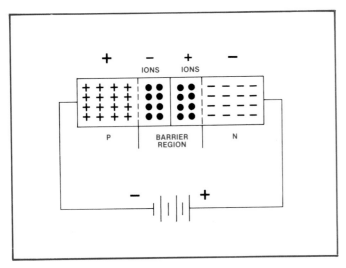

Figure 16-25. Schematic of reverse-biased PN junction.

covalent bond, others follow. As a result, holes move into the P region and electrons move into the N region. These changes create an increase in the reverse current through the junction.

Once the zener breakdown point has been reached, the current increases rapidly. This happens with little or no voltage increase.

A zener diode is not damaged if it is operated within its rating. The zener diode has been designed to utilize this breakdown point. It is rated accordingly. Chemical control during manufacture determines the rating of a zener diode. Zener diodes are available with ratings of 2 to 200 V. Wattage ratings are from 400 mW to 50 W. The wattage rating is important. This is because the physical size determines the amount of heat dissipated. If the heat is not dissipated rapidly, the junction impedance is lowered. If this happens, the current increases more rapidly than specified. Operating temperature is usually specified by the manufacturer.

Voltage Regulation. Zener diodes can be useful for voltage regulation. This is because the current through these devices varies over a wide range. At the same time, however, the voltage drop is insignificant in most circuit applications.

Look at Figure 16-26. This shows the simplest of circuits for voltage regulation with a zener diode. A resistor is needed in series with the diode. The resistor's value limits the current through the diode. This prevents the diode from reaching its breakdown point. Also prevented is a short-circuit condition, which would appear at the breakdown point.

Review the circuit in Figure 16-26. Consider what happens when a load is connected to the output terminals. The current demanded from the source is increased. How-

ever, the voltage drop across the diode remains the same. The current through the zener increases only after the input voltage is increased.

This behavior is due to breakdown characteristics. More current flows through the diode when input voltage is increased. Under this condition, a larger voltage drop appears across the resistor. But the voltage available at the output terminal remains constant.

The output load can demand more current without upsetting the diode. The only time the diode changes its current is when the voltage—and load—across the diode increase. The diode shunts the load. It conducts more heavily at a voltage above its rated value. By shunting excess current through the diode, the excess voltage is dropped across the resistor, R. Thus, the load in series with resistor R does not get the increased voltage.

Now assume a voltage decrease across the diode. This means that the diode does not conduct as much current. The circuit returns to its voltage distribution condition. This is the same condition that existed before the increase in voltage occurred.

Zener diodes may be connected in series to increase voltage ratings. For instance, two 200-V diodes connected in series can regulate a 400-V source.

Zener diodes are used in automobile transistor radios. They regulate the applied voltage. And they also prevent damage to transistors from voltage surges. Surges are

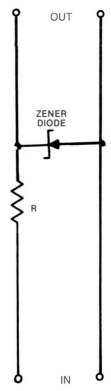

Figure 16-26. Schematic showing circuit with zener diode used for voltage regulation.

produced by the varying engine speed. They can also result from rapid changes in load across the alternator.

Switching. To function as a zener, a diode must be reverse biased. The zener point is referred to as E_z, or zener voltage.

An advantage of a zener diode lies in its current-handling ability. It can absorb a large increase in current without an increase in voltage drop across the diode.

This capability is useful in computers. Switching requires rapid changes at high speeds. To illustrate, some switching operations are in the 2.5-MHz range. This requires a device that can complete its switching function in 0.4 microsecond. Within this time, the diode must return to its original condition. A zener diode can complete this cycle in 0.1 microsecond. Thus, a zener diode is useful for switching at high frequencies.

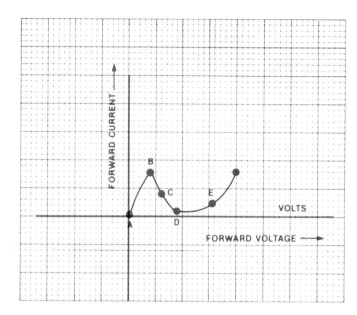

Figure 16-27. Characteristic curve for tunnel diode.

16.14 TUNNEL DIODES

The behavior patterns of tunnel diodes are somewhat different from those of junction diodes. **Tunnel diodes** may resemble junction diodes physically. But the effect of tunneling can make a big difference. Tunneling can be an amazing thing to witness.

Look at the diagram in Figure 16-27. Note the strong forward bias and the absence of reverse bias.

A tunnel diode is created by doping silicon with an unusually large amount of impurities. The doping leads to high concentrations of charge carriers. This leads to a reduction of the critical voltage for an avalanche effect. Actually, the avalanche breakdown point is reduced to below the zero-voltage level. The breakdown point moves into the region of small forward-bias voltages.

Electrons are said to tunnel through the barrier almost instantly. This is partly because the barrier is extremely thin. Its thickness is less than 0.000001 inch

[0.0000254 cm]. Electrons move through the barrier at rates approaching the speed of light.

Note the behavior of current in Figure 16-27. Current is reduced with increases in forward bias over a portion of the diode's operating range. This indicates a negative resistance characteristic. The same behavior is also observed in a tetrode vacuum tube.

Negative resistance is shown by the down curve in Figure 16-27. This takes in the line from points *B* to *D*. This behavior is not normal for a junction diode. It has been produced intentionally in the tunnel diode. There is a reason for introducing this behavior: Such a diode can be used as an amplifier, an oscillator, or as a very fast switching device. From point *E* upward, the tunnel diode behaves in the same way as a normal junction diode.

A number of doping compounds are used to make tunnel diodes. They include gallium arsenide, gallium antimonide, and indium antimonide.

Doping changes the reaction of tunnel diodes. It produces the operating patterns diagrammed in Figure 16-27, including the following.

- When forward voltage is increased, current flow rises to a maximum, or peak-current, value (point *B*).
- A greater increase in forward-bias voltage causes the current to decrease. The movement is from point *C* to point *D*. This is the negative-resistance feature of the tunnel diode.
- After the current reaches a minimum at point *D*, it again starts to increase. Behavior is the same as for any junction diode. When an increase in forward-bias voltage is applied, behavior is as shown from point *E* upward.

Activity: *THE ZENER DIODE*

OBJECTIVES

1. *To investigate the operational characteristics of a zener diode.*
2. *To design a circuit with a zener diode for voltage regulation.*

EQUIPMENT NEEDED

1 Diode, zener, 6.2 volts, 1-watt
1 Resistor, 1500 ohms, 10%, 2-watt
1 Meter, 0–10 mA DC
1 Power supply, 0–10 volts DC
6 Wires for connections
1 Voltmeter, 0–10 volts DC

PROCEDURE

1. Connect the circuit shown in Figure 1.

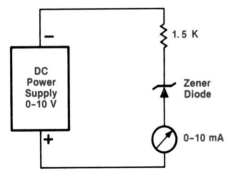

Figure 1

2. *Turn on the power supply*, with its output voltage set to zero. Gradually increase the voltage up to 10 volts.
3. Adjust the power supply voltage one volt at a time. Read the milliammeter. Record your reading in Figure 2. Do this for each volt adjustment from 0 to 10 volts.
4. *Turn off the power supply* and reverse the diode in the circuit.

5. *Turn on the power supply*, again set at zero. Slowly adjust the voltage output one volt at a time. Record the readings on the left of zero on the Figure 2 chart. In other words, use the negative values this time. Plot the values on the graph.

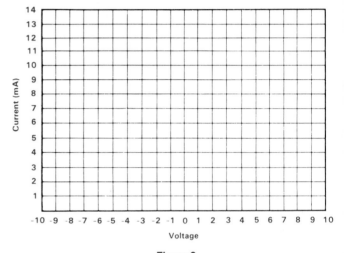

Figure 2

6. *Turn off the power supply*. Disconnect the circuit. Return all equipment and materials to their proper storage places.

SUMMARY

1. When the zener was at forward bias (connected as shown in Figure 1), did the current increase in a linear fashion with increases in voltage?
2. When the zener was reverse biased, did the current increase in a linear fashion with increases in voltage? If not, what occurred?
3. What can you say about the breakdown point of a zener?
4. For what purposes can the zener effect be used in electronic circuits?

- The voltage range over the negative resistance portion is called the *voltage swing.*
- The ratio of peak current (point *B*) to the valley current (point *D*) is called the **peak-to-valley ratio.**

Remember the negative resistance portion of the curve. These changes take place with small voltage variations, in millivolts. The peak-to-valley current ratio is usually around 10.

The tunnel diode can operate at much higher temperatures than other semiconductors. Germanium diodes stop working at about 200 degrees F [93.3 degrees C]. Silicon diodes stop at about 400 degrees F [204.4 degrees C]. But tunnel diodes can operate at temperatures of up to 650 degrees F [343.3 degrees C]. The resistance to radiation and temperature gives the tunnel diode great potential.

Tunnel diodes are used in amplifiers, oscillators, and computers. They function at extremely high radio frequencies. Frequency ranges can run to between 2 and 10 GHz. *GHz* is the abbreviation for *gigahertz*. This is equivalent to 1000 MHz.

Tunnel diodes can be incorporated into integrated circuits that present extremely small space requirements.

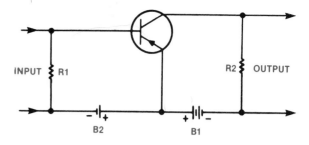

Figure 16-29. Schematic of common-emitter PNP circuit.

REVIEW QUESTIONS FOR SECTION 16.14 ___

1. What is the tunnel effect?
2. How is a tunnel diode produced?
3. What is the avalanche breakdown point for a tunnel diode?
4. What is the initial effect on a tunnel diode when forward voltage is increased?
5. What is the effect of a continuing voltage increase after a tunnel diode has reached its peak-current point?
6. How does a tunnel diode react to radiation?
7. What are the high-temperature performance characteristics of tunnel diodes?

16.15 TRANSISTORS

Transistors are made from N- and P-type crystals. Once joined, the two different types of crystals produce junctions. Transistors are identified according to emitter junction and collector junction.

A PNP transistor is formed by a thin N region between two P regions, Figure 16-28. The center N region is called the *base*. This base is usually 0.001 inch [0.00254 cm] thick. A *collector* junction and an *emitter* junction are also formed.

The transistor is used most frequently with a common emitter. A **common-emitter** *circuit* is shown in Figure 16-29. In this type of circuit, the current through the load flows between the emitter and collector. The input signal is applied between the emitter and base. In its

normal operation, the collector junction is reverse biased by the supply voltage, B_1. The emitter junction is forward biased by the applied voltage, B_2. Electrons flow across the forward-biased emitter into the base. They diffuse through the base region and flow across the collector junction. Then they flow through the external collector circuit.

Battery B_2 voltage is applied in the forward direction. This means the voltage is positive to the emitter P-type crystal. It also means that voltage is negative to the N-type crystal. Thus, the emitter-base junction has a low impedance.

The voltage of Battery B_1 is applied in the reverse direction. This means voltage is positive to the N-type crystal. It also means that voltage is negative to the P-type crystal. Thus, a collector–base junction has a high impedance.

REVIEW QUESTIONS FOR SECTION 16.15 ___

1. What types of crystals are used to produce transistors?
2. What is the construction of a PNP transistor?
3. What is the base of a transistor?
4. What type of material is used for the base in a PNP transistor?

16.16 TRANSISTOR IMPEDANCES

Remember that the impedance of the emitter junction is low. So, electrons flow from the emitter region to the base region. At the junction, the electrons combine with the holes in the N-type base crystal. If the base is thin enough, almost all the holes are attracted to the negative terminal of the collector. They then flow through the load to Battery B.

The collector current is stopped by application of a positive voltage to the base and a negative voltage to the emitter. In actual transistors, however, this cannot be done. This is because of several basic limitations. Some of the electrons in the base region flow across the emitter junction. Some combine with the holes in the base region. For

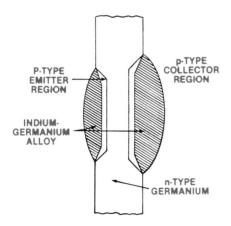

Figure 16-28. Schematic of PNP transistor.

this reason, it is necessary to supply a current to the base. This makes up for these losses.

The ratio of the collector current to the base current is known as the *current gain* of the transistor. Current gain, called *beta*, is found by dividing base current into collector current. At high frequencies, the fundamental limitation is time for carriers to diffuse across the base region. They move from the emitter to the collector. This is why the base-region width or thickness is so important. The thinner the base region, the less time is required for the carriers to diffuse across it. This causes the transistor to operate faster.

Figure 16-30 shows a **common-base** *circuit*. The signal is introduced into the emitter–base circuit. The output signal is extracted from the collector–base circuit.

An important advantage of the transistor is its ability to transfer impedances. This is where it obtains its name. *Transistor* is a combination of *trans*fer and re*sistor*.

The low-impedance circuit of the emitter allows current flow. This current flow then creates a current through the **collector circuit**. The emitter has low impedance and low current. The collector has high impedance and even slightly less current than the emitter. However, more power is the result in the collector. This is because $P = I^2R$ or, in this case, $P = I^2Z$. Impedance (Z) is high and the current is squared. Thus, the collector circuit has more power than the emitter circuit with its low impedance.

A common emitter circuit has about 1.3K ohms of input impedance. This is compared to an output impedance of 50K ohms. Thus, there is an increase in impedance from emitter to collector of about *39 times*. The junction transistor amplifies in this way. It acts as a *power amplifier*. A small change in emitter voltage causes a large change in the collector circuit. The impedance differences cause this reaction.

Popular Types of Transistors

NPN and PNP transistors are the two most popular types. The main difference between the two transistors is in polarity. Polarity can be recognized by pin locations on

transistors, Figure 16-31. (Pin designations for specific transistors are found in a transistor handbook.)

Symbols for representing transistors in circuits are shown in Figure 16-32.

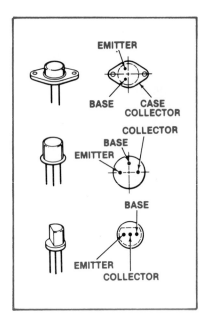

Figure 16-31. Diagrams showing transistor pin locations.

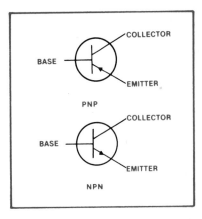

Figure 16-32. Symbols for transistors. (GE)

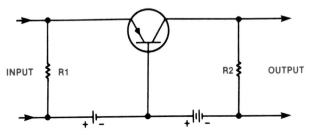

Figure 16-30. Schematic of common-base PNP circuit.

1. How is the current gain of a transistor determined?
2. Why is the thickness of the base important to a transistor's operation?
3. What is a common-base circuit?
4. How does a transistor achieve power amplification?
5. What are the differences between NPN and PNP transistors?

Activity: *THE TRANSISTOR*

OBJECTIVES

1. *To study the transistor circuit arrangements in reference to bias voltages.*
2. *To make a transistor curve, in order to determine its operating characteristics.*

EQUIPMENT NEEDED

1 9-volt battery
1 Potentiometer, 500K, linear taper
1 Meter, 0–100 microamperes DC
1 Meter, 0–100 milliamperes DC
1 PNP transistor, 2N1098, with proper heat sink or equivalent

1 PNP transistor, 2N1048 or equivalent
1 Power supply, 0–20 volts DC
9 Wires for connections
1 10K 1-W Resistor

PROCEDURE

1. Connect the circuit as shown in Figure 1. Be sure to observe the polarities of the meters and the transistor elements.

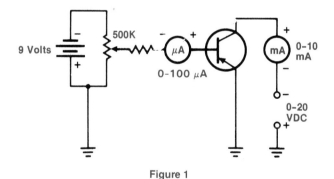

Figure 1

2. *Turn on the power supply*, set to zero. Slowly adjust the potentiometer to obtain 20 microamperes on the base meter. **NOTE:** *If in running this experiment the current through the transistor slowly increases when the voltage and base current have been set to their proper values, discontinue those readings with either increased base current or collector voltage. This slow increase in collector current indicates that the transistor is heating and could possibly go into a condition of thermal runaway.*

3. Adjust the power-supply voltage and check the change in collector current as the collector-to-emitter voltage is changed Record the changes in the chart in Figure 2.
4. Adjust the base current to read 30 microamperes. Adjust the power-supply voltage to cause the collector current to change accordingly. Record your observations in the chart in Figure 2.

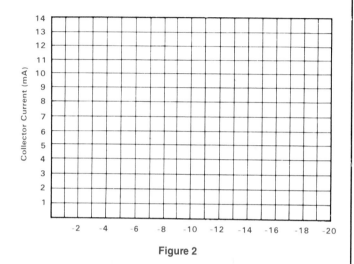

Figure 2

5. Continue to change the base current by adjusting the potentiometer to 40 microamperes, then 50 microamperes, then 60 microamperes, and then 70 microamperes. Each time you change the base current, go back and change the supply voltage and record the change in collector current at various collector-to-emitter voltages.
6. *Turn off the power supply.* Disconnect the circuit and return all equipment and materials to their proper storage places.
7. Plot the curves in Figure 2.

SUMMARY

1. Does the family of curves for the transistor appear to be similar to those published for the pentode tube? If there is a difference, explain. How could these curves be of value to you in your future experimentation and circuit designing?
2. How does the increase in voltage between the collector and emitter affect the current in the collector circuit?

16.17 BASIC TRANSISTOR CIRCUITS

Transistors are used in three basic types of circuits, Figures 16-34, 16-35, and 16-36. The table in Figure 16-33 lists these circuit types and indicates comparative characteristics. These characteristics apply for Class A amplifiers.

Decibel, dB

In electronics, power ratios are measured in decibels (dB). The gain or loss of power expressed in decibels is 10 times the logarithm of the power ratio. Numerical values that are used are not proportional to the actual power level. Rather, these figures represent a rough measurement of the sensation on the ear when the electrical power is converted into sound. A one-decibel change in sound is probably the smallest change the trained ear can detect. Mathematically, it is equal to $10 \times Log_{10}$.

Suppose a circuit has a power input of 5 mW and an output of 500 mW. Divide the input into the output to determine the ratio. Dividing 5 into 500 produces a quotient of 100. The ratio is 1:100. This ratio is equal to a power gain of 20 dB.

To find the ratio of a power loss, divide the output power into the input power. Suppose there is an input of 1 mW and an output of 0.005 mW. Dividing 0.005 into 1 produces a quotient of 200. The power-loss ratio is 200:1.

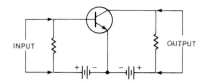

Figure 16-34. Schematic of common-emitter circuit.

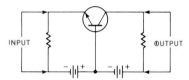

Figure 16-35. Schematic of common-base circuit.

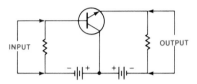

Figure 16-36. Schematic of common-collector circuit.

16.18 OTHER TYPES OF TRANSISTORS

In addition to the types of transistors identified above, you should know about three others. These are described below.

Alloy Transistors

These transistors are made by alloying metal into opposite sides of a thin piece of semiconductor. The process produces emitter and collector regions. This procedure is illustrated in Figure 16-37A.

To achieve uniformity of the transistor characteristics, the thickness of the metal pellet must be controlled. Also critical is the quality of the metal. Further, the area of

Circuit	Characteristics*	
Common Emitter	Moderate input impedance	1.3 K
	Moderate output impedance	50K
	High current gain (increase)	35
	High voltage gain (increase)	−270
	Highest power gain (increase)	40 dB
Common Base	Lowest input impedance	35 ohms
	Highest output impedance	1 meg
	Low current gain	−0.98
	High voltage gain	380
	Moderate power gain	26 dB
Common Collector	Highest input impedance	350K
	Lowest output impedance	500 ohms
	High current gain	−36
	Unity voltage	1.00
	Lowest power gain	15 dB

*These characteristics have been given for a 2N525 at audio frequencies with bias of 5 volts and 1 mA. Load resistance was 10K. Generator resistance was 1K. (Courtesy of GE.)

Figure 16-33. Comparative characteristics of transistor circuits.

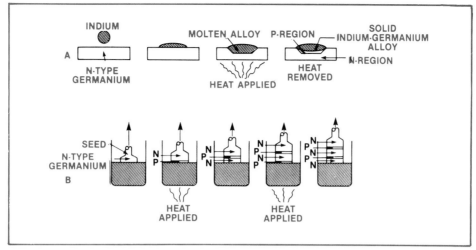

Figure 16-37. Diagrams of transistor junction formations.
A. Junction formation of microalloy transistor.
B. Junction formation by rate growing of crystal.

contact between metal and semiconductor and the alloying temperature must be carefully controlled. Each of these variables affects the electrical characteristics of the transistor.

Grown-Junction Transistors

Development of this type of transistor is illustrated in Figure 16-37B. The grown-junction transistor is different in an important way from the alloy transistor: The junctions are created during the growth of the crystal rather than by alloying after the crystal is grown.

Field-Effect Transistors

These transistors are small in size and mechanically rugged. Another advantage is low power consumption. These units have a high input impedance similar to that of a vacuum tube. *Field-effect transistor* is abbreviated **FET**.

FETs are *unipolar*. This means they operate as a result of one type of charge carrier. These are holes in the P-channel types. In the N-channel types, the charge carriers are electrons. Other types of transistors are bipolar. They require the presence of both hole and electron carriers.

The term **MOS** is most often used to describe the FET. MOS means *metal-oxide semiconductor*. This describes the method of construction used. The metal control *gate* is separated from the semiconductor *channel*. Separation is through use of an insulating oxide layer, Figure 16-38. An FET is not affected by the polarity of the bias on the control gate. Changes in temperature affect the FET. They are used in voltage amplifiers, RF amplifiers, and voltage-controlled attenuators.

There are two basic types of FET. The *depletion* type has charge carriers in the channel when no bias voltage is applied to the gate. However, a reverse gate voltage is used to reduce the channel conductivity. Forward bias on the gate causes more charge carriers to become introduced into the channel. This increases the channel conductivity.

In the *enhancement* type, the forward-biased gate produces active carriers. This permits conduction through the channel. At zero or reverse gate bias, there is no useful channel conductivity.

There are four distinct types of MOS transistors. These classifications are based on sources of conduction. The units can make use of either electrons (N channel) or

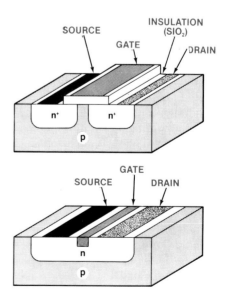

Figure 16-38. Drawing showing structures of PN junctions of field-effect transistors.

holes (P channel) for conduction. Classification is also based on whether the transistor is of the enhancement or depletion type. Symbols for the four types of MOS transistors are shown in Figure 16-39. The direction of the arrowhead in the symbol differentiates between N- and P-channel types. A solid channel line in the symbol (depletion) indicates "normally on." A dotted line (enhancement) indicates "normally off."

MOS field-effect transistors are used for low-power operation of broadcast-band receivers. They have been used in RF amplifiers, conversion stages, IF stages, and first audio stages. Figure 16-40 presents a number of MOS-type transistors.

FETs have been used in FM receivers. They are used for RF amplifiers, IF amplifiers, and limiters. They can also be operated as oscillators and phase splitters. The schematic diagrams in Figures 16-41 and 16-42 illustrate their versatility.

REVIEW QUESTIONS FOR SECTION 16.18

1. How is an alloy transistor made?
2. How is a grown-junction transistor made?
3. What is the meaning of the term *MOS*?

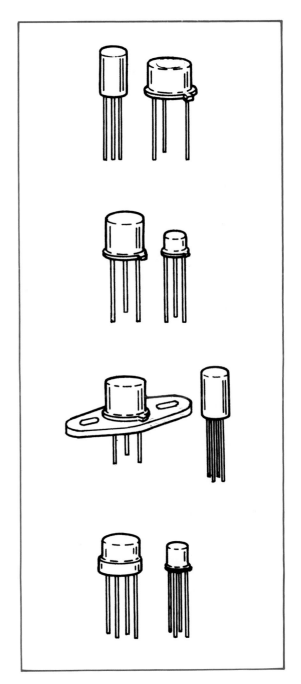

Figure 16-40. MOS transistors.

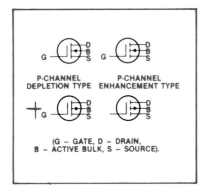

Figure 16-39. Symbols for transistors.

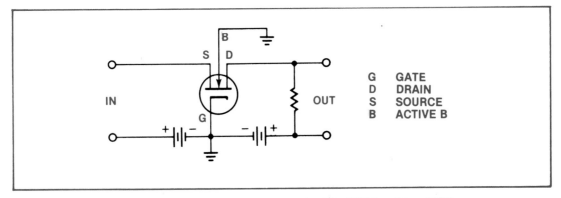

Figure 16-41. Schematic of common-gate circuit for MOS transistors. (RCA)

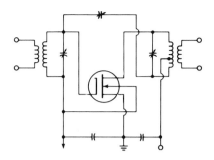

Figure 16-42. Schematic of RF amplifier stage using MOS field-effect transistor. (RCA)

16.19 TRANSISTOR ENCLOSURE

Transistors need protection from harmful impurities and mechanical damage. This protection is provided by placing them in caps or other types of enclosures. Heat ratings and stability of a transistor's electrical characteristics are determined by its case.

The transistor structure must be prepared for encapsulation. This is done by etching the surface metal to dissolve any impurities acquired during manufacture. Following the etching process, the transistor is handled in a controlled atmosphere. This prevents surface contamination or oxidation. Then the transistor is raised to a high temperature to remove all moisture. It is then refilled with a controlled atmosphere and capped. A getter may be injected before the cap is welded in place.

___ **REVIEW QUESTIONS FOR SECTION 16.19** ___

1. What are the functions of the enclosure of a transistor?
2. How is a transistor prepared for encapsulation?
3. What is the function of the heating process during the encapsulation of a transistor?

16.20 SILICON CONTROLLED RECTIFIERS (SCRs)

A specialized type of semiconductor used for control of electrical circuits is the *silicon controlled rectifier* (*SCR*). This is a four-layer device. The structure can be either NPNP or PNPN.

An SCR conducts current in a forward direction only. The symbol for an SCR is shown in Figure 16-43. Current always flows through an SCR from the cathode (C) to the anode (A). The illustration indicates that the SCR also has a gate (G).

The function of an SCR is shown in the circuit diagram in Figure 16-44. The most typical use of an SCR is for a controlled circuit. Examples include a light dimmer and a speed control for a motor. This type of circuit is illustrated in Figure 16-44. The resistor in this circuit, R_1, is a rheostat, or adjustable resistor. This is used to control the amount of voltage delivered to the gate of the SCR. The more voltage delivered, the greater is the flow. Thus, adjusting the rheostat can serve to control the circuit. If the circuit illuminates a lamp, lowering the voltage to the rheostat dims the bulb. If the load is a motor, its speed is slowed.

Figures 16-45 and 16-46 are diagrams showing typical SCRs.

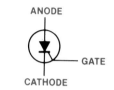

Figure 16-43. Symbol for SCR.

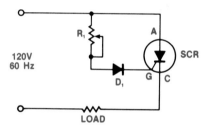

Figure 16-44. Schematic of SCR-controlled circuit.

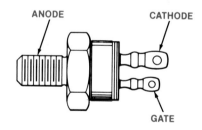

Figure 16-45. Drawing of typical SCR.

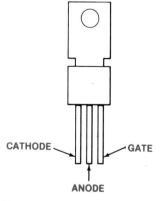

Figure 16-46. Drawing of typical SCR.

251

16.21 INTEGRATED CIRCUITS

A single, monolithic chip of semiconductor was developed in 1958. J. S. Kilby was responsible for its fabrication. Active and passive circuit elements were successively diffused and deposited on a single chip. Shortly thereafter, Robert Noyce made a complete circuit on a single chip. This led the way to the modern, inexpensive **integrated circuit (IC)**.

Resistors, capacitors, and transistors can be placed on a chip. Diodes can be made in many groups to do different things. Photolithography, a combination of photographic and printing techniques, has been used. Photolithography has made it possible to mass-produce sophisticated devices with high reliability.

Safety Tips

1. Be careful of the heat generated by a transistor. It is possible to get burned by touching the enclosure of a transistor or diode.
2. Semiconductor circuits should be properly fused. Without fuses, these circuits represent a fire danger.

Figure 16-47A shows how a resistor is placed on an IC chip. It is made in the form of a thin filament of conductive material. Contacts are made. It is isolated by a reverse-biased junction. It is possible to create a few ohms or up to 20,000 Ω of resistance on ICs.

Figure 16-47B shows how a low-loss capacitor is formed. This is done by depositing a thin insulating layer of silicon dioxide on a conducting region. It is then given a metalized layer to form the second plate. It is possible to reverse-bias a diode to form capacitances up to 50 pF. Capacitors require more space than resistors do. Diodes, transistors, and FETs need little space on the chip. Connections from one device to another on the chip are made by a layer of aluminum. The aluminum is deposited, masked, and etched to the proper pattern, Figures 16-47C and 16-47D.

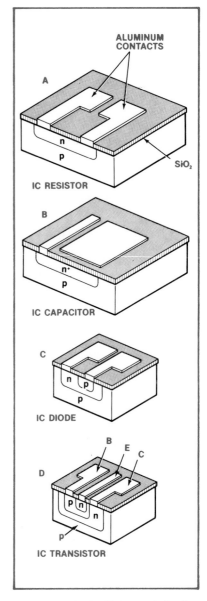

Figure 16-47 (A–D). Methods of integrated-circuit construction.

Figure 16-48. Integrated-circuit package.

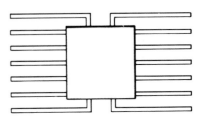

Figure 16-49. Flat-pack (14-pin) integrated circuit.

Packaging ICs

Integrated circuits are available in a number of standard packages. The *multipin circular* type is the same size as a regular transistor package. A chip can hold many hundreds of transistors in the space formerly needed for one transistor, Figure 16-48.

The *flat pack* is hermetically sealed. This means it is vacuum packed, Figure 16-49. The ceramic flat pack has either 10 or 14 pins.

The **dual-inline package** (**DIP**) is easy to use. This is because it fits standard sockets and printed-circuit holes. This is important, since standard sizes mean accuracy in manufacturing. ICs are usually inserted by machine into holes in printed-circuit boards. DIP units are illustrated in Figure 16-50.

Sometimes the cost of its package is greater than the cost of the IC itself. See Figure 16-51.

Making ICs

The IC uses the same technology as the planar transistor. A thin slice of P-type silicon provides a substrate, or base. It is usually 2 to 4 inches in diameter and 10 mils (0.01 inch) thick. [This is 50 to 100 mm in diameter and 0.25 mm thick.]

Active and passive components are built within the N-type epitaxial layer on top. An epitaxial layer is a crystal surface. The crystal is oriented in the same direction as the substrate, Figure 16-52 on next page. The surface is oxidized first. This is done with heat. Oxygen is used as the atmosphere during heating.

The cooled wafer is coated with photosensitive material. *Photosensitive* means that the material reacts to light. The material used is a photoresist. This means that the portions exposed to light will develop photographically. Then, after development, the coating material will resist the action of acid. The resist forms a mask in the proper pattern and size for the IC elements. In this way, the

material covered by the resist will not be etched away. The undesired coating of silicon dioxide (SiO_2) is then etched away. This leaves a pattern of windows. Diffusion takes place through these windows.

Many thousands of IC units can be made at the same time. This cuts down on the expense of manufacturing,

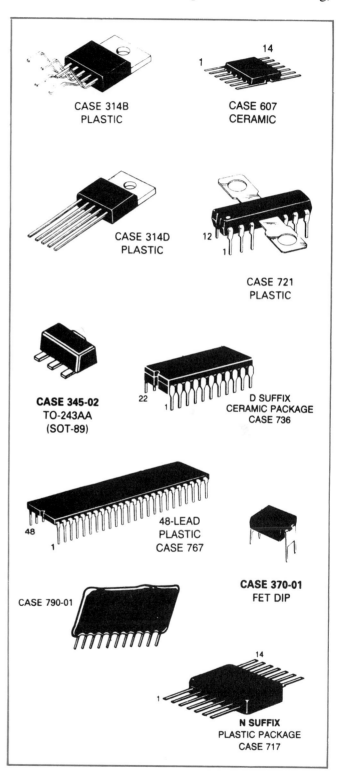

Figure 16-51. Packaging for integrated circuits takes many forms. (Motorola)

Figure 16-50. Integrated circuits in dual-inline packages (DIPs).

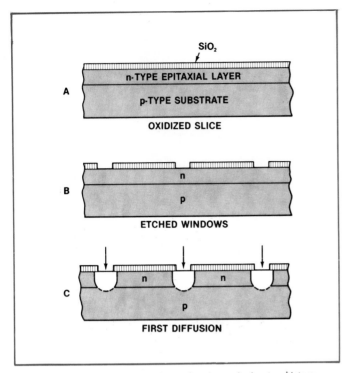

Figure 16-52. Drawings show steps in producing transistors on integrated circuits.

Hundreds of chips, each with up to 100 components, are made in single manufacturing operations. This helps to reduce costs of semiconductors.

IC Uses

ICs are relatively standardized. This is done partly to take advantage of mass-production techniques. Standard components and connections are needed if the finished electronic equipment is to be low in cost.

ICs are used in large numbers in digital computers. They are small and consume little power. They are very reliable. It is possible to obtain ICs with large memory capacities. Thousands of electronic elements may be placed on a single chip. This is done through methods known as **large-scale integration (LSI)**.

In addition to computers, hearing aids and space vehicles use ICs. Amplifiers are fabricated as complete units. Figure 16-54 is a simple amplifier circuit. There is no limit to the possibilities of this type of circuit. It has many applications in toys and calculators. Future potential is great.

Figure 16-53. Masks for making the desired circuits are made about 500 times larger than the finished ICs. They are reduced photographically. This provides a piece of film that is used for exposing the resist. The pattern in the resist represents the semiconductor materials on the IC. After the semiconductor images are printed and etched, the large wafer is cut into chips. The chips are then packaged.

REVIEW QUESTIONS FOR SECTION 16.21

1. What is an integrated circuit?
2. What methods are used to produce integrated circuits?
3. What types of components can be formed on integrated circuits?

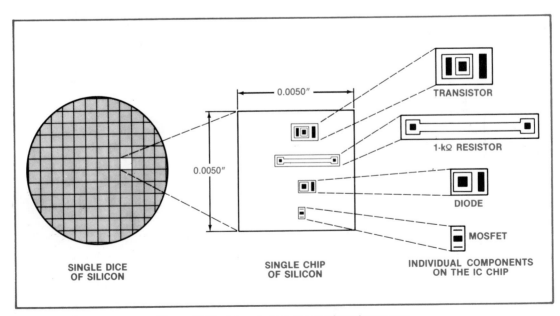

Figure 16-53. Integrated circuit manufacturing process.

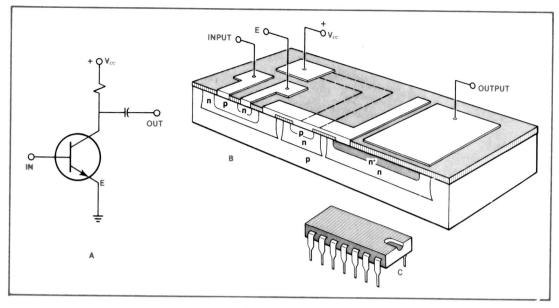

Figure 16-54.
A. Schematic of circuit with transistor, resistor, and capacitor.
B. IC with all three components on one chip.
C. Packaged IC.

16.22 THE LIGHT-EMITTING DIODE (LED)

Visible light is emitted by a specialized semiconductor known as the light-emitting diode (LED). These units have many uses. They are capable of operation at very low voltages, usually under 5 V. They draw from a few milliamperes to as high as 100 mA.

LEDs are made of gallium arsenide junctions, a semiconductor material. Creation of electron-hole pairs is a reversible process. Energy is released when an electron recombines with a hole. In gallium arsenide, an electron drops directly into a hole and a photon of energy is emitted. The gallium arsenide junctions provide the best conditions for the generation of radiation in the visible range.

LEDs are used as indicator lamps, Figure 16-55. In most instances, they must be used in series with a resistor. They are also used as logic indicators for computer circuits. Symbols for the LED are shown in Figure 16-56.

When it is reverse biased, the LED is nonconducting. It is capable of conducting current when it is forward biased. It emits light when conducting a forward-bias current. An LED usually operates on 1 to 3 V. Excessive current will destroy an LED.

LEDs are also used in seven-segment displays, Figure 16-57. These units are used to display numbers on many types of electronic equipment. Each segment consists of two LEDs in series. Current passing through a pair of such series diodes energizes certain segments. This causes the LED to give off light. Combinations are used to provide the numbers 0 through 9. Figure 16-58 shows a driver circuit that operates an LED display.

Among the convenient and commonly used devices for numeric displays are the seven-segment LEDs, Figure 16-58A. Such diodes, when passing current in the forward direction have their junction region illuminated. By arranging the junctions properly, upon excitation of appropriate combinations of diodes, any one of the integers 0 to 9 may be displayed. LEDs are available with

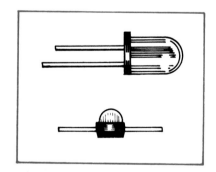

Figure 16-55. Light-emitting diodes.

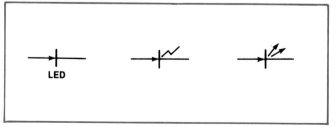

LED

Figure 16-56. Symbols for light-emitting diodes.

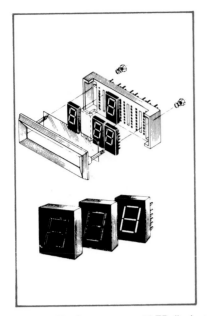

Figure 16-57. Seven-segment LED displays.

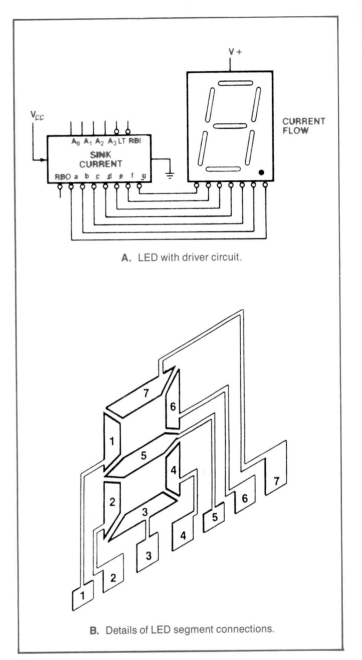

A. LED with driver circuit.

B. Details of LED segment connections.

Figure 16-58.

either common cathodes or common anodes. In the cathode case, particular segments are energized when a positive voltage is applied to them with the remainder down. In the anode case, unenergized portions are high, and those portions to be energized low—that is, at zero volts.

The numerical information to be displayed is in binary-coded decimal (BCD) form. To convert this information into appropriate combinations of seven-segment excitations, a decoder must be used. One that is commonly used has the designation *7447*, Figure 16-58B.

16.23 LIQUID CRYSTALS

Liquid-crystal displays (LCDs) are currently used in watches, signs, and some portable computers and calculators. Such display systems do not require as much energy as LEDs do. The image seen on LCDs is silver, whereas the LED is a red display. Images for LCDs generated by lasers have potential for multicolored video displays of information systems. The LCD offers many possibilities for easier information retrieval.

A liquid-crystal cell consists of a kneematic messophase liquid-crystal material between two glass plates that are glued or fused together, Figure 16-59. The thickness of the LC material is about 10 to 25 micrometers. The two glass plates have transparent electrodes deposited on their inside faces. The electrodes are made of a transparent and conducting material such as tin or indium oxide, Figure 16-60. This type of display needs ambient light to be seen. That is, it does not give off light of itself, but relies upon the room light to give it visibility.

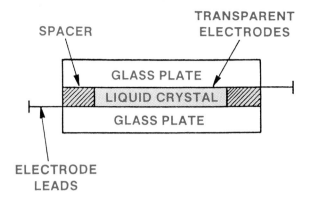

Figure 16-59. Cross-sectional view of the LCD.

__ REVIEW QUESTIONS FOR SECTION 16.23 __

1. What color image is seen on LCDs?
2. What color image is seen on LEDs?
3. What conducting material is used to make the electrodes on liquid-crystal cells?

16.24 SILICON PHOTODETECTORS

A variety of silicon photodetectors are available for a wide range of light-detecting applications. Their advantages over phototubes are high sensitivity, good temperature stability, and proven silicon reliability. Applications include card and tape readers, pattern and character recognition devices, shaft encoders, position sensors, and counters. Maximum sensitivity occurs at approximately 800 nm (nanometers).

Photodiodes are used where high speed is required (1.0 ns, or nanosecond, where *nano* is one-thousandth of a millionth, or 0.000000001). Phototransistors are used where moderate sensitivity and medium speed (2 ns) are required.

__ REVIEW QUESTIONS FOR SECTION 16.24 __

1. When does maximum sensitivity occur in silicon photodetectors?
2. When are photodiodes used?
3. When are phototransistors used?

SUMMARY

Many vacuum tubes have been replaced by transistors and other solid-state devices. Vacuum tubes are units from which most of the air has been withdrawn. Thermionic tubes are devices in which electrons are produced by heating an electrode. Computer monitors and picture tubes of television sets are still generally vacuum-tubes. Liquid-crystal display (LCD) is not available on a wide scale as of this date.

The word *semiconductor* is used in describing the diode and transistor. The crystal amplifier (transistor) dates back to 1948, but Faraday's work a hundred years earlier was the basis of experimentation that led to development of the transistor.

Semiconductor devices utilize impurities doped into pure silicon or germanium to cause them to conduct electrons or holes and thus rectify, switch, or amplify.

Semiconductor diodes allow current to flow in one direction only. P- and N-type materials are used to make a diode. Special-purpose diodes are made for use with radar and other devices. Zener diodes are made so that they break down at a given voltage. They can then be used as part of a voltage-regulation circuit. Avalanche breakdown is a condition in which the boundary within a diode breaks down. However, the zener does recover from breakdown.

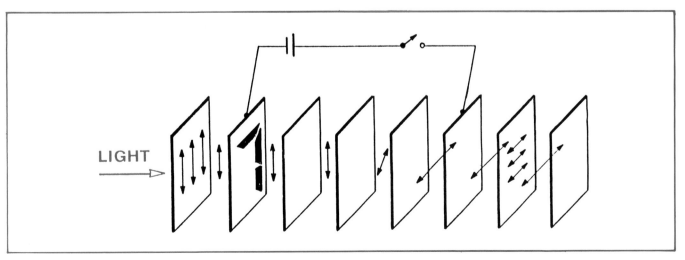

LIGHT

Figure 16-60. LC material.

Tunnel diodes exhibit behavior patterns somewhat different from those of junction diodes. A tunnel diode is created by doping silicon with an unusually large amount of impurities. The doping leads to high concentrations of charge carriers. This leads to a reduction of the critical voltage for an avalanche effect. Tunnel diodes can operate at higher temperatures than can other diodes. The former are used in oscillators, amplifiers, and computers.

Transistors have PNP or NPN designations. This designation indicates the types of materials used in construction. The transistor has emitter, base, and collector leads. The transistor used most often has a common-emitter configuration.

In electronics, the decibel, dB, measures power ratios. The gain or loss of power is expressed in decibels. A 1-decibel change in sound is probably the smallest change the trained ear can detect. Mathematically, it is equal to $10 \times \text{Log}_{10}$.

Transistors are also classified as: alloy, grown-junction, and field effect. The field-effect transistors (FET) are unipolar. This means they operate as a result of one type of charge carrier. These are holes in the P-channel types. In the N-channel types, the charge carriers are electrons. Other types of transistors are bipolar. They require the presence of both hole and electron carriers. FETs have been used in FM receivers.

MOS field-effect transistors are used for low-power operation of broadcast-band receivers. They are also used for RF amplifiers, conversion stages, and IF stages, and in first audio stages.

Transistors use a number of types of enclosures to protect the material within.

An SCR, or silicon controlled rectifier, is also referred to as a thyristor. It is especially made for controlling the action of the diode and can be found in many devices in which speed control is needed.

Integrated circuits or ICs have put capacitors, transistors, diodes, and capacitors on chips. They are a part of microelectronics. IC packaging is a specialized field, as is the manufacturing of the devices for special applications. The uses for ICs or chips are unlimited. Computers, calculators, and microwave ovens all utilize ICs for their operation and control circuits.

LEDs are light-emitting diodes that can be utilized in infrared circuits and as indicator lamps.

_____ USING YOUR KNOWLEDGE _____

1. Draw the schematic symbol for a triode tube.
2. Draw the schematic symbol for a diode tube.
3. Diagram a tube socket with seven pins. Label the pins. Assume your drawing is a bottom view of the tube socket.
4. Diagram a directly heated cathode.
5. Diagram an indirectly heated cathode.
6. Draw the outline of a cathode-ray tube.
7. Draw the symbol for a PNP transistor.
8. Draw the symbol for an NPN transistor.
9. Draw the symbol for a semiconductor diode.
10. Draw the symbol for a zener diode.
11. Draw the circuit for a common-emitter transistor amplifier.
12. Draw the circuit for a common-base transistor amplifier.
13. Draw the circuit for a common-collector stage.

_____ KEY TERMS _____

semiconductor	FET
inversely	MOS transistor
solid state	IC
germanium	dual-inline package (DIP)
silicon	large-scale integration (LSI)
crystal amplifier	Edison effect
molecule	deflection
insulators	galvanometer
loosely held	cathode
holes	triode
forward bias	diode
reverse bias	thermionic emission
junction diode	space charge
avalanche breakdown	control grid
zener effect	amplification
tunnel diode	bias
peak-to-valley ratio	saturation
common emitter	tetrode
common base	pentode
common collector circuit	beam-power tube

Chapter 17

POWER SUPPLIES

OBJECTIVES

After studying this chapter, you will know

- *that a dry cell generates electricity chemically, on demand.*
- *the uses for dry cells and how to assure that they provide dependable service and have a maximum useful life.*
- *what happens when cells are connected in series and in parallel.*
- *that primary and secondary cells are different in their ability to be charged.*
- *the purpose of a power supply.*
- *the process for converting current from AC to DC.*
- *how a zener diode is used in a voltage-regulation circuit.*
- *how a voltage-doubler circuit operates.*

INTRODUCTION

This chapter is about sources of electricity. You will learn about the dry cell and how it is used to change chemical energy into electrical energy. You will learn that batteries are sources of electrical energy. Batteries are, in effect, electric generators that operate on demand. They produce electricity through chemical reactions. Batteries produce an even, reliable source of DC.

This chapter covers commercial generators and devices that can change AC to DC for use in electronics equipment. Full-wave and half-wave rectification are covered as methods utilized to produce power from alternating-current sources to operate devices that require DC.

The term *power supply* is used rather freely to mean any number of pieces of equipment capable of producing a variety or a single source of electrical power. It may be a 120-volt AC source, a 120-volt DC source, or any combination of these two with stepped-up or stepped-down voltages and conversion of current types. A power supply supplies power to electrical or electronics equipment. It can be made by combining the right component parts, or it can be purchased.

Historic Background

Alessandro Volta was the first person to produce electricity through chemical action. He developed the first electric cell in 1798. The term *volt* honors his achievement.

Volta used acetic acid (vinegar) as an electrolyte. An electrolyte is a nonmetallic conductor. His electrodes were copper and zinc strips placed in the liquid. Today, we think of this type of cell as crude. But such units remained in use until 1868. In that year, Georges Leclanche developed the first practical cell.

Leclanche used carbon for the positive electrode. The liquid electrolyte was ammonium chloride (sal ammoniac). A **depolarizer** of powdered manganese dioxide was placed in a cup surrounding the carbon rod. This prevented the coating of the carbon rod with hydrogen-gas bubbles. The depolarizer also helped to prolong the life of the cell. In addition, the depolarizer increased the shelf life, or the amount of time the battery could be stored before use.

The first "dry" cell was produced in 1888 by Dr. Gassner. Zinc was used as the container for the entire cell. The zinc also formed the negative electrode. The electrolyte was mixed into a porous material. Then the top of the cell was sealed to make this the first "dry" cell. Commercial production of dry cells began in 1890.

17.1 DRY CELL CHARACTERISTICS

Almost everyone uses batteries at some time. Understanding how they function will enable you to use them properly. Through proper use, you can avoid problems and get the most energy for your money.

The symbol for a dry cell is shown in Figure 17-1. Remember that dry cells produce DC. Thus, circuit diagrams showing dry cells as sources represent DC circuits.

Figure 17-1. Symbol for a dry cell.

Voltage

Each dry cell has its own voltage. This voltage is not dependent upon the size of the cell. Rather, voltage depends upon the materials used, Figure 17-2.

Dry cells use sal ammoniac for an electrolyte. Carbon and manganese dioxide are used for the positive electrode. Zinc is used for the case and negative electrode.

The initial *open-circuit voltage* of a dry cell is rated at 1.56 V. The actual voltage of cells now on the market may vary. When new, they range from under 1.5 V up to 1.6 V. This variation has nothing to do with the quality of the cell.

Cells in Series

When cells are connected in series, their individual voltages are added. Thus, for example, the total voltage available from four 1.5-V cells would be 6 V. The symbol for cells in series is shown in Figure 17-3.

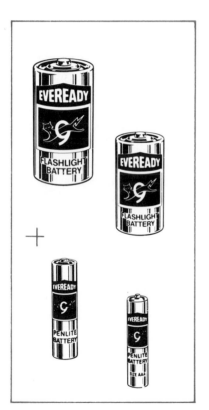

Figure 17-2. Dry cells come in many sizes. Here are some examples. (Union Carbide)

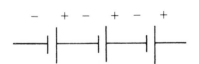

Figure 17-3. Symbol for dry cells in series.

Connecting two or more dry cells together forms a *battery*. This applies whether the connection is in series or parallel. It takes two or more connected cells to form a battery. Single cells are referred to simply as *cells*.

Thus, a battery is composed of a group of dry cells. The battery produces electrical energy by chemical reaction within its cells. This energy is delivered as a steady, reliable DC voltage. Current from a battery is generated chemically as required. The battery is a DC generator.

Internal Resistance

New and unused dry cells have very low internal resistance. This means that internal resistance can be omitted in most situations involving batteries. However, internal resistance of batteries must be considered in the design of some electric circuits.

Amperage

Current capacity is determined by the physical size of the cell. Amperage is generally higher in larger cells. But there may be variations among grades and makes. Often, a cell with lower amperage may have better total electrical capacity. However, the reverse may also be true.

In summary, cells are rated according to voltage. The amount of current delivered can vary widely.

Cells in Parallel

Connecting cells in parallel increases the current available. Total current of cells connected in parallel is equal to the sum of their currents. To find total current, add current of the individual cells. The symbol for cells connected in parallel is shown in Figure 17-4.

Shelf Life

As dry cells become older, they grow weaker. This happens whether or not they are used. Such gradual deterioration is unavoidable. It results from very slow chemical reactions and moisture changes. These changes occur inside the cell.

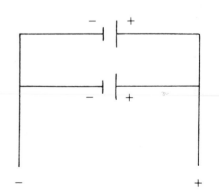

Figure 17-4. Symbol for dry cells in parallel.

These effects are referred to as shelf deterioration. They gradually reduce the output of the cell. For example, suppose a flashlight is left in the glove compartment of your car. In time, the batteries deteriorate. After a few months, the light output falls off. Your flashlight will last longer if you turn it on occasionally. A few minutes per week of use will extend battery life.

The term *shelf life* describes the life of a cell that is not used. Shelf life of mercury cells is much longer than for carbon–zinc dry cells. In some instances, mercury cells have shelf lives of up to six years. By comparison, carbon–zinc cells deteriorate in a matter of months.

Temperature

Dry cells are designed to operate best at normal room temperature. This is approximately 70 degrees F (21.1 degrees C).

At temperatures above 100 degrees F [37.7 degrees C], problems can occur. These result because the materials in the battery expand. Prolonged heat can destroy a dry-cell battery.

Low temperatures cause a slowing of the chemical process. The output of a battery drops as the battery becomes colder. At −10 degrees F [−23.3 degrees C], dry cells stop operating. However, exposure to low temperatures does not cause permanent damage. Cells that have been frozen return to their original condition when warmed.

The shelf life of cells is extended if they are stored at low temperatures. It is a good idea to store unused dry cells in a refrigerator. Just allow at least two days for them to return to room temperature before using them.

Dry-Cell Structure

The structure and makeup of a carbon–zinc dry cell are illustrated in Figure 17-5. Note the construction features. A zinc wall within the case is one of the electrodes. A carbon rod serves as the other electrode. Electrical demand triggers the chemical reaction. Thus, electricity is generated when needed. Note also that space is allowed within the case for expansion of the chemicals at elevated temperatures. This helps prevent leakage. The cell is carefully sealed to provide a self-contained unit.

Other Types of Dry Cells

In addition to carbon–zinc units, several other types of dry cells are manufactured. Each has separate advantages and disadvantages. Some common additional types of cells are described below.

Mercury Cell. The mercury cell, Figure 17-6, utilizes 80 to 90 percent of its active materials. This rate is more efficient than that of other primary cells. A **primary cell** is one that cannot be recharged. Mercury cells are comparatively small in size. These units operate under wide ranges of temperature, pressure, and humidity. Mercury

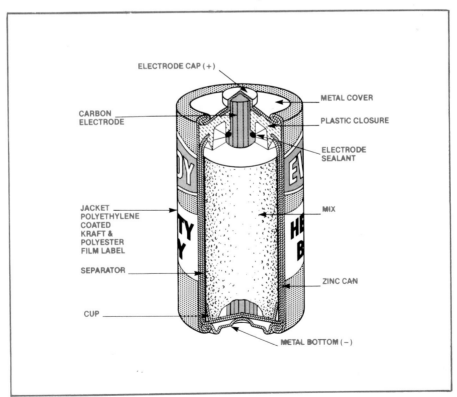

Figure 17-5. Cutaway drawing of carbon–zinc dry cell. (Union Carbide)

Figure 17-6. Mercury cells come in a variety of sizes. (Union Carbide)

cells have voltage and ampere-hour rates that are nearly constant. They also have low internal resistance and long shelf life. Uses for mercury cells include hearing aids, portable radios, electronic heart pacers, and cameras.

Alkaline Cell. Under certain conditions, alkaline cells, Figure 17-7, deliver more than 10 times the service of carbon–zinc cells. These units have very low internal resistance. They are sealed in steel cases. Uses include continuous-duty and heavy-duty jobs.

Nickel–Cadmium Cell. The nickel–cadmium cell, Figure 17-8, can be recharged. Because of this, it is called a **secondary cell**. The cells can be recharged many times. They have a relatively constant voltage during discharge. There are few maintanance problems. The cells are

rugged; they will stand more abuse than any other cell. Performance at low temperatures is good. Long storage has no effect.

Experimental Fuel Cell. The fuel cell, Figure 17-9, represents a different approach to producing electricity through chemical reaction. In the fuel cell, the chemical reaction produces electricity. But the chemicals are not stored within the cell. Instead, they are fed in as needed from outside sources.

Solar Cell. This contains photosensitive silicon cells. The semiconductor units generate voltage output when exposed to light. The symbol for a solar cell is shown in Figure 17-10.

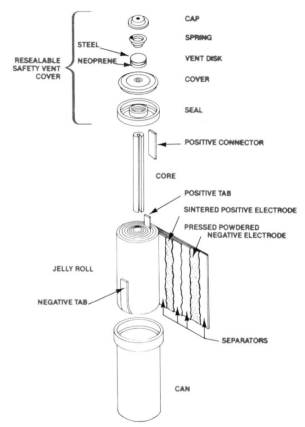

Figure 17-8. Exploded view of sealed nickel–cadmium rechargeable cell. (Union Carbide)

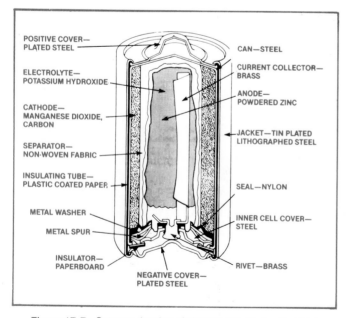

Figure 17-7. Cutaway drawing of alkaline cell. (Union Carbide)

Figure 17-9. Experimental fuel cell. (Union Carbide)

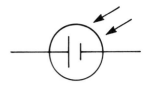

Figure 17-10. Symbol for solar cell.

17.2 LEAD–ACID CELLS

Probably the most widely used type of battery is made with lead–acid cells. This is the type of battery used in most automobiles. It furnishes electrical power to start the car. It also meets electrical needs before the car's generator or alternator begins to function. In turn, the battery is recharged by the output of the alternator, Figure 17-11.

A **lead–acid unit** is a secondary cell. This means it is rechargeable. In the lead–acid cell, the electrolyte is sulfuric acid. The positive plate is lead peroxide. The negative plate is lead.

When discharging, the acid becomes less concentrated. Both plates change chemically to lead sulfate. Recharging the cell changes the electrolyte back to a more concentrated acid. The plates also regain their chemical properties.

Hydrogen, a highly explosive gas, is given off during recharging. So, open flames and cigarettes should be kept away from lead–acid batteries.

Lead–acid cells can provide large amounts of power for short periods of time. They are more economical than other chemical sources of electrical power.

Like other secondary cells, lead–acid units do not store electricity. Rather, they store chemical energy. This

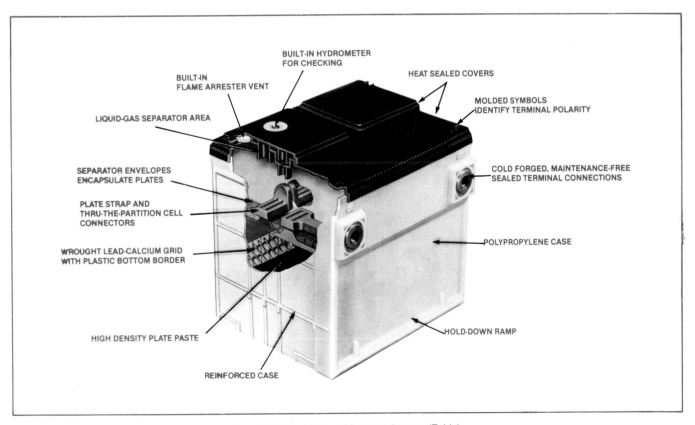

Figure 17-11. Lead–acid storage battery. (Exide)

Activity: PRIMARY CELLS

OBJECTIVES

1. *To understand the action of a primary cell.*
2. *To observe the combination of cells to form a battery.*

EQUIPMENT NEEDED

1 Voltmeter,
  0–10 volts DC
1 Copper strip,
  5″ × 3/4″
1 Zinc strip, 5″ × 3/4″
1 Carbon strip,
  5″ × 3/4″

1 Pint saltwater solution
1 Pint sal ammoniac
  solution
1 Cell unit
  (See Figure 1)
4 Dry cells, 1.5 volts
  (D-cells with holders)

PROCEDURE

1. Set up the cell unit shown in the drawing in Figure 1.

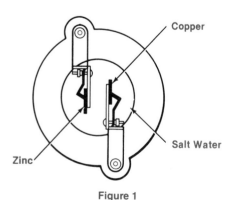

Figure 1

2. Fill the glass two-thirds full of saltwater solution. Lower the electrodes into the solution. Set the voltmeter to read 0–5 volts DC and connect the meter to the terminals of the cell. Which electrode is the positive pole? How do you tell which pole is positive? What is the voltage output of the cell? **NOTE:** *The concentration of the electrolyte and the impedance of the meter can affect all these readings.*

3. Remove the electrodes from the saltwater solution and pour the solution back into its jar. Rinse off the electrodes and glass to remove any remaining trace of the saltwater solution. Use the voltmeter, set to read 0–5 volts DC, to determine the polarity and voltage output of the cell once the glass has been filled two-thirds full of sal ammoniac. Which is the positive electrode of the cell? What is the output of the cell? What is the voltage reading?

4. Remove the electrodes from the solution. Rinse the electrodes to clean them. Remove the copper strip and replace it with the carbon strip, as shown in Figure 2.

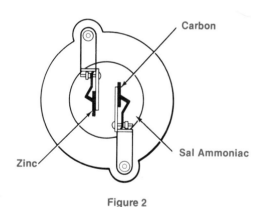

Figure 2

5. Place the electrodes in the sal ammoniac solution. Use the voltmeter as before to determine the polarity and voltage output of the cell. Which electrode is positive? What is the voltage output of the cell?

6. As a result of your findings, how can you justify the use of zinc, carbon, and sal ammoniac as the active elements of commercial dry cells?

7. Remove the electrodes from the solution of sal ammoniac and pour the solution back into its jar. Carefully rinse the electrodes and the glass in running water to remove any traces of sal ammoniac that

might remain. Dry both the glass and the electrodes thoroughly. Connect the circuit shown in Figure 3.

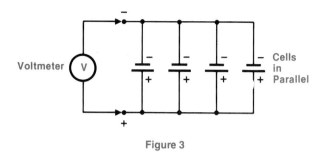

Figure 3

8. You have connected the cells in parallel. Use the voltmeter, set to read 0–5 volts DC, to determine the voltage of the circuit. What is the reading of the voltmeter? **NOTE:** *In parallel operation, only the current capacities of the individual cells are added.*
9. Disconnect the circuit you have been using and connect the circuit shown in Figure 4.
10. You have connected the cells in series. Use the voltmeter to determine the voltage of the circuit. Set the meter to read in the 0–10 volt range. **NOTE:** *In series operation, only the voltages of the individual cells are added.*

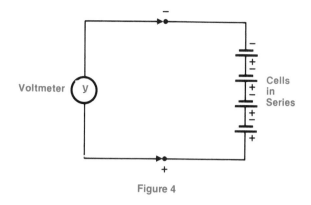

Figure 4

11. Disconnect all the circuits you have been using. Clean and return all equipment and materials to their proper storage places.

SUMMARY

1. How can a dry-cell battery be made from a group of cells to produce an output voltage of 9 volts?
2. Explain the name "dry cell."
3. Can two pieces of metal produce electricity without an electrolyte?
4. What role does the electrolyte play in a cell's operation?
5. How does a change in the electrodes affect the voltage output of a cell?

chemical energy is converted to electricity as needed. The chemical reaction takes place between the electrodes and the electrolyte. During this reaction, electron movement begins at the negative electrode. The path of the current is through the outside load to the positive electrode, Figure 17-12.

Normal chemical action would cause the battery to run down. For continuous use, the battery must be recharged. The car's electrical system supplies a current for this purpose. With recharging, the electrons within the battery return to their original positions. This is demonstrated in the series of drawings in Figure 17-12. External units are also available for battery recharging, Figure 17-13.

Lead–acid batteries produce between 2 V and 2.2 V per cell. Six cells in series produce a 12-V battery. The 12-V battery is standard in most cars.

The ampere-hour rating of a battery is determined by a number of factors. Included are the physical size and number of plates. The amount of electrolyte also affects the ampere-hour rating. An ampere-hour is the ability to furnish one ampere of current for one hour. Thus, 45 ampere-hours could mean delivery of 45 A for one hour. But it could also mean delivery of 1 A for 45 hours.

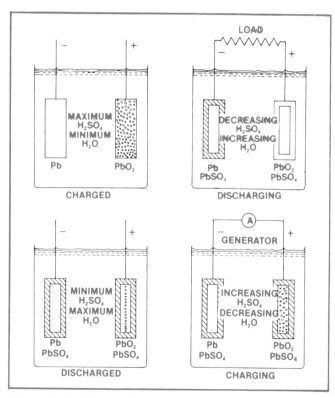

Figure 17-12. Charge–discharge cycle of lead–acid storage battery.

Figure 17-13. Battery charger. (Delco Remy)

17.3 CHANGING AC TO DC

The Power Supply

In most electrical equipment that uses DC for power, the power is furnished by an AC source. A transformer is usually employed in reducing the voltage, and a diode or combination of diodes is used to *rectify* the AC to produce DC, Figure 17-14. Inasmuch as it would be prohibitive to produce an electronic device with many batteries to supply the various voltage requirements of any given circuit, it is then necessary to utilize the various power-circuit arrangements to produce the desired voltages and currents. In some cases, a **pulsating DC** (PDC) is usable. In other instances, it is necessary to have as near pure DC as possible. This calls for filtering the PDC after it is obtained by rectification.

Power Transformers

A power transformer can step up or step down voltage. Secondary windings can furnish a number of different voltages. These voltages may be either higher or lower than the primary voltage. This type of transformer is used in electronics equipment in which a number of different voltages are needed. (See Chapter 7 for a review of transformers.)

Power-Supply Rectifiers

There are three types of basic power-supply circuits. They are the **full-wave**, **half-wave**, and **bridge rectifier**. Each has its advantages and disadvantages.

Half-Wave Rectification. A half-wave rectifier is a device that changes AC to pulsating direct current (PDC) by allowing current to flow through during one-half of the power-supply cycle, or hertz. This one-way current is controlled by a semiconductor device known as a diode. Figure 17-15 shows a number of different packages for diodes. These are general-purpose **rectifiers**.

Figure 17-16 shows how a diode is inserted in a circuit with a transformer. This is a simple half-wave rectifier circuit. The main reason for the transformer is to increase or decrease the 120 volts to a higher or lower level to be rectified. Note that the resistor is connected in series with the (+) end of the diode. The load resistor is there for a purpose.

The operation of the half-wave rectifier circuit is shown in Figure 17-16. The alternations of the input voltage e_1 are reproduced by the transformer with an increase in voltage e_2 in the secondary windings. The waveforms indicate a 180° difference in phase between e_1 and e_2. The

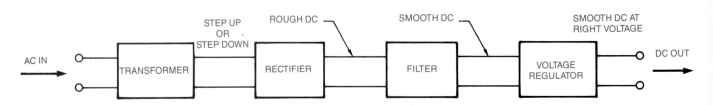

Figure 17-14. Electronic equipment power supply.

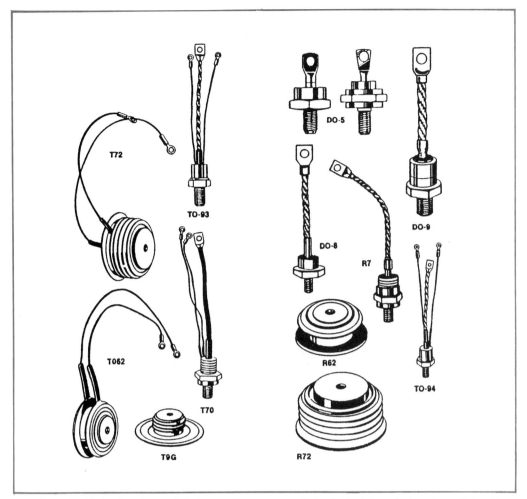

Figure 17-15. Various types of diodes used in power supplies, with their case designations. (Motorola)

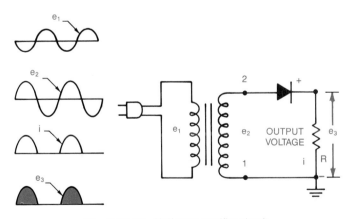

Figure 17-16. Half-wave-rectifier circuit.

difference is characteristic of induced voltages. The induced secondary voltage e_2 is impressed across the diode and its series load resistance. This voltage causes current i to flow through the diode and its series resistor on the positive half of the hertz. The resultant voltage e_3 across the load resistor has a pulsating waveform, as shown in the illustration. This pulsating waveform is referred to as a ripple voltage.

The half-wave rectifier uses the transformer during one-half of the hertz. Therefore, for a given size transformer, less power can be developed than with both halves of the hertz. If any considerable amount of power is to develop in the load, the half-wave rectifier's transformer must be relatively large compared with what it would have to be if both halves of the hertz were utilized. This small disadvantage limits the use of the half-wave rectifier to applications that require a very small current drain. The half-wave rectifier is widely used in small commercial receivers and was used in some early television receivers and oscilloscopes. It is also used in some battery-charging circuits, since the battery has a tendency to act as a **filter** and smooth out the pulses.

Full-Wave Rectification. A full-wave rectifier is a device that has two or more elements so arranged that the current flows in the same direction during each half-hertz of the AC power supply. Full-wave **rectification** may be accomplished by use of two diodes as shown in Figure 17-17. The end of the load resistor is connected to the center tap of the transformer. Note the waveform of the

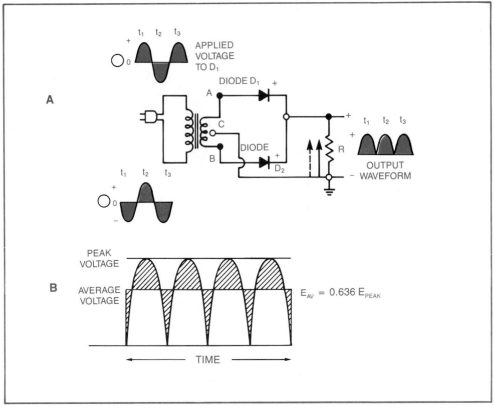

Figure 17-17. Full-wave rectification.

power output in rectified form. Notice the appearance of the peak voltage and average voltage.

The two halves of the secondary winding, AC and BC, may be a center-tapped winding, as shown, or they may be separated windings. In either case, the load circuit is returned to a point midway in potential between A and B so that the load current is divided equally between the two diodes.

The part of the secondary winding between A and C may be considered a voltage source that produces a voltage and of the shape shown in 1 of Figure 17-17. This voltage is impressed across D_1 in series with the load resistor R. During the half-hertz marked t_1, the electrons flow in the direction indicated by the arrow. This flow of electrons from the ground up through the load resistor R makes one end of the diode (cathode) positive with respect to ground. Thus, the load voltage is developed across R between the cathode and ground. During this same half-hertz, the voltage across BC is negative, as shown in Figure 17-17A, and D_2 is not conducting. A half-hertz later, during interval t_2, the polarity of the voltages on the diodes is reversed. D_2 now conducts, and D_1 is not conducting. The electron flow through D_2 is in the direction indicated by the dotted arrow. This current flows also from the ground up through R and makes the cathode positive with respect to ground. Thus, another half-hertz of load voltage is developed across R. Keep in mind that only one diode is conducting at any given instant.

This means the full-wave rectifier uses the output of the transformer in its entirety instead of half the time as was the case in half-wave rectification. That means the full-wave rectifier is more efficient than the half-wave. It also means that the full-wave has less ripple to filter. This type of rectifier is utilized in a wide range of electronic devices.

Bridge Rectification

A bridge rectification circuit is shown in Figure 17-18. This is a full-wave-rectifier circuit with four diodes. Note how the negatives and positives are tied together to produce a − and a + output from the arrangement of the four diodes. The − and + connection locations indicate that there is no polarity, and the AC can be input here. In reality, the diode will have a black body with a white ring around one end. The white ring indicates the + end. In some instances, the body of the diode may be white with a black ring around the one end that is used to denote the +, or cathode, connection. (**NOTE:** The *cathode*, in vacuum-tube terminology, is negative. However, engineers and physicists make use of conventional current-flow terminology and use the term *cathode* for the

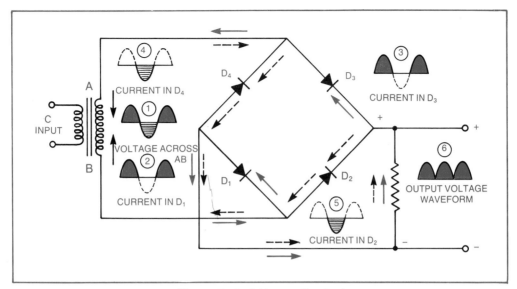

Figure 17-18. Bridge-rectifier circuit.

positive semiconductor end. Technicians still adhere to electron-flow and − to + conventions.)

Note how the current flow is indicated first in one direction by the solid arrow and then in the other by the dotted arrows. Note that the dotted and solid arrows both go upward through the load resistor, indicating that both halves of the hertz flow in the same direction, thus producing a pulsating direct-current output.

There are a few advantages to the bridge-rectifier over the regular semiconductor full-wave rectifier. First, the bridge arrangement allows for twice the voltage output from the same power transformer. Second, a bridge rectifier is so designed that it has only half the inverse voltage as does the full-wave rectifier with only two diodes. Peak inverse voltage (*piv*) is a negative voltage applied across the diode when there is no current flowing through the diode.

Another advantage is the packaging of the bridge rectifier circuit. It is available in molded units for ease in handling and replacement.

Filters

Power sources frequently need DC power that is free of the ripple that results from rectification. A **filter** is used to smooth out these variations produced when AC is changed to DC or pulsating DC (PDC).

The filter consists of a capacitor, inductor, and resistor, or a capacitor and an inductor, or a capacitor and a resistor. There may also be two capacitors used in a filter circuit. See Figure 17-19 for an example of a pi-type filter that smooths out rectified AC to produce a high level of DC voltage. If the current draw from the power source is rather high, it is necessary to filter out the current variations by using a filter, such as that shown in Figure 17-20,

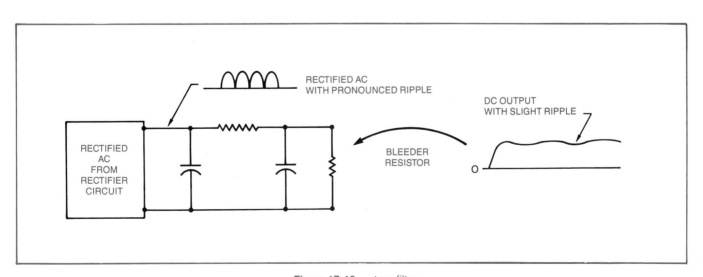

Figure 17-19. π-type filter.

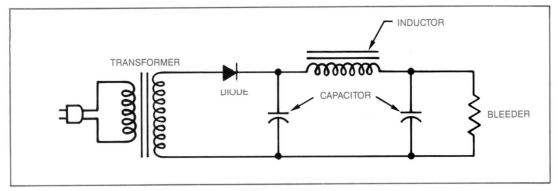

Figure 17-20. π-type filter with inductor.

with an inductor. A bleeder resistor is sometimes utilized to aid in the regulation of the voltage output.

Voltage Regulation

Inasmuch as power sources fluctuate in their voltage outputs, it is necessary to design circuits that will produce a steady or regulated voltage. This is done through a number of means, but the zener diode is one of the most commonly used devices for this purpose.

Zener Diode. The zener diode is a reverse-bias diode, as explained in Chapter 16. An increase in the number of impurities in the diode causes it to break down at different levels of reverse voltage. The zener diode is made in a wide range from 2.4 volts to 200 volts. Power ratings range from 0.25 W up to 50W.

Figure 17-21 shows an example of how the zener diode is used as a simple regulator. As current flows through R_1, the excess voltage is dropped across it and only the desired voltage appears at the regulated output terminals. Once the voltage has dropped to below the set point of the diode, the diode no longer conducts. This means it can keep the voltage down but has nothing to do with keeping the voltage from dropping below its set point. Thus, in most instances, the zener is used in a circuit with a higher-than-rated voltage so it can conduct most of the time and keep the voltage down to its rating.

Doublers

Peak rectified voltage can be increased without increasing the peak transformer voltage. The doubler circuit uses two diodes and two capacitors to increase the output voltage from the network.

The circuit operates by utilizing both halves of the rectified AC power. During the positive portion of the input AC, D_1 will conduct and D_2 will be reverse biased. This causes capacitor C_1 to charge to the peak voltage of the rectified pulse, Figure 17-22. During the negative portion of the input, D_2 conducts and D_1 is not conducting, Figure 17-23. The output peak voltage is then the sum of that applied and the voltage held by the capacitor C_1, and results in two times the peak voltage across C_2.

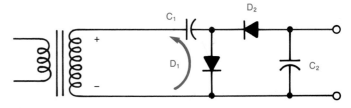

Figure 17-22. Voltage-doubler circuit, with D_1 conducting first and allowing C_1 to charge to the voltage of the transformer.

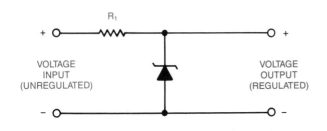

Figure 17-21. Zener diode E regulator circuit.

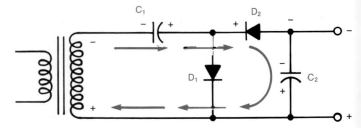

Figure 17-23. Doubler circuit, with D_2 conducting and charging C_2 to two times the voltage of the transformer.

By extension of the circuit, it can become a tripler or a quadrupler, as shown in Figure 17-24. The rating of the diodes used has to be at least twice the peak voltage.

Floating Ground

In most laboratory DC supply sources, the voltage can be taken between + and −, or + and ground, or − and ground, Figure 17-25. In most supplies, the output between + and − is said to be *floating*, since it is not connected to a common ground or potential level of the network. *Ground* is a term that simply refers to a zero earth potential level. The chassis, or outer shell, of most electrical equipment is grounded through the power cable to the terminal board. The third prong on the power cord is the ground connection.

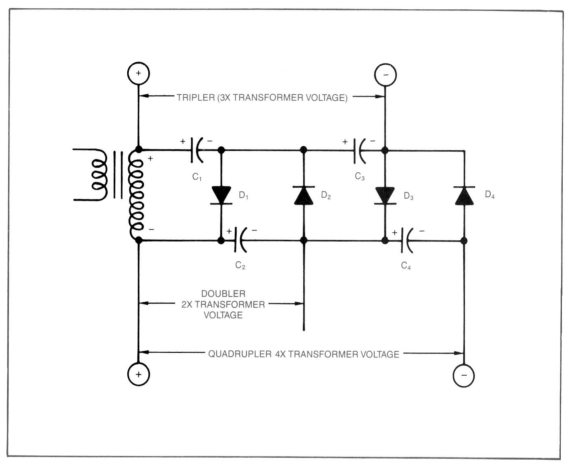

Figure 17-24. Voltage-tripler or -quadrupler circuit. Voltage depends on where it is taken off the circuit.

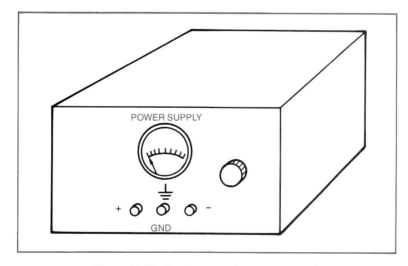

Figure 17-25. Floating ground on a power supply.

Changing DC to AC

There are many applications in which it is necessary to change DC to AC. Automobiles, boats, and aircraft have AC applications. Radar and other electronic equipment on aircraft use AC. However, the regular power supply on aircraft is 24 V DC.

The process for changing DC to AC is called *inverting*. Systems that change DC to AC are called *inverters*. Semiconductor circuits are used to invert DC to AC. No moving parts are required.

___ REVIEW QUESTIONS FOR SECTION 17.3 ___

1. What are the major advantages of AC over DC in automotive systems?
2. What are the components of a power supply?
3. What is rectification?
4. What is a rectifier and what does it do?
5. Why is inverting necessary?
6. How can voltage be regulated in a power supply?
7. What is the purpose of a filter in a power supply?
8. What is pulsating direct current and how is it created?

Safety Tips

1. Make sure the power is turned off when a power supply is not in use.
2. Check to make sure the fuse in the power supply is the correct size.
3. Do not discharge the capacitors in a power supply when the power supply is still plugged into its AC source.
4. Keep in mind that diodes can cause electrolytic capacitors to explode if they are not properly placed in the circuit. Check the polarity of the voltage source and the polarity of the capacitor before plugging in and turning on the power source.
5. Some primary cells will explode when you attempt to charge them. Be careful. Make sure the cell is one that can be charged. Also make sure the voltage of the charger is designed for that particular cell.

SUMMARY

Dry cells have been around for many years. When placed in series, their resultant voltage is equal to the sum of the voltages of the individual cells. When placed in parallel, they add their available currents. It takes two or more cells to make a battery.

There are a number of types of cells used to power a variety of electronics devices. The mercury cell, alkaline cell, nickel–cadmium cell, solar cell, and experimental cell all have different voltage outputs. The lead–acid cell is the workhorse of cells, since it is used to power submarines and automobile equipment.

Power supplies provide the source of power to electrical and electronic equipment. They usually consist of a transformer, rectifier, filter, and voltage regulator.

Power transformers are available in a variety of sizes and shapes. Transformers can be used to supply various voltages to electronic circuits. Their output is rectified by a half-wave, full-wave, or bridge-rectifier circuit to produce DC. Each type of rectification has its own advantages and disadvantages.

A filter is used to smooth out the variations produced when AC is changed to DC or pulsating DC.

A power-supply voltage can be doubled, tripled, or quadrupled by use of a number of diodes and capacitors to take advantage of the peak voltages of the power source.

The zener diode is used for its ability to break down and recover. It is utilized in many voltage-regulation circuits.

___ USING YOUR KNOWLEDGE ___

1. Figure out how many 1.5-V dry cells would be needed to produce a battery that delivers 300 V.
2. Read voltages for a number of batteries. Try to find and check a mercury cell, an alkaline cell, and a lead–acid cell. Record your findings.

___ KEY TERMS ___

primary cell	voltage regulation
secondary cell	voltage doubler
depolarizer	floating ground
lead–acid cell	bridge rectifier
nickel–cadmium cell	half-wave rectifier
alkaline cell	full-wave rectifier
pulsating direct current	rectification
rectifier	filter

Project: STROBE LIGHT

Basic Principle

The basic principle of the modern *stroboscope* is generally considered to date back to J.A.F. Plateau in 1836. However, Michael Faraday used it in some of his experiments even earlier.

Strobe Effect

The strobe furnishes a single flash of light lasting for thousandths or millionths of a second. The burst of intense light can be controlled and, if synchronized with rotating machinery, can make the machinery appear to stand still. Try it with an electric fan and watch the blades seem to stop when the flash rate is properly adjusted.

If the rate of motion is slightly more or less than the flash rate of the strobe, the motion will appear to go forward or reverse according to the flashing rate. Movies use a series of flashing pictures for the illusion of motion. They produce a "strobe effect" when the wheels of a stage coach appear to be turning in the direction opposite to that in which the coach is traveling. The eye is easily tricked.

Rapidly moving objects can be photographed in a "stopped-motion" fashion with a series of strobe-light flashes. This is why a strobe in a dimly lighted room makes movements of people dancing or moving about resemble the motion in old-fashioned "flickers" or movies. The rate of the flashes can slow down or speed up the resulting action, since the eye sees the brightly lighted objects in a series of halting movements.

The strobe light can be used to determine the speed of rotation of high-speed motors and as warning lights on the wing tips of airborne craft. In psychiatry and medicine, the strobe is used in the diagnosis of epilepsy.

The circuit used to produce the strobe flash is a simple one. A capacitor stores energy. A xenon tube then shorts the capacitor and produces a brilliant flash of light as the energy of the capacitor is dissipated in ionizing the gas of the tube. The xenon tube has a cathode, anode, and trigger electrodes. A low-energy, but high-voltage, trigger pulse (around 6000 volts in this circuit) causes the gas between the cathode and anode to ionize. Once ionized, the gas presents a very-low-resistance path for discharge of the capacitor (16 μF). The rapid discharge of the capacitor produces a high-energy light source. A reflec-

tor is usually employed to direct the light in the desired direction. CAUTION: Do not stare into the light source. Eye damage may result.

The Circuit

The two diodes, D_1 and D_2, the 300-ohm resistor, R_1, and the two electrolytic capacitors, C_1 and C_2, make up the voltage-doubler circuit. They increase the voltage and change it to DC. The 120 volts AC from the line cord is rectified or changed to DC by the diodes. The capacitors give a doubling effect to the voltage, and this results in a DC voltage of around 340 volts across the 16-μF electrolytic capacitor. This energy is used to flash the xenon bulb at the proper time. This timing is the important part and requires another capacitor, a variable resistor (for control of flashing rate), and an SCR and trigger coil. See the schematic, Figure 1.

Resistors R_2, R_5, R_4, and R_3 make up a voltage-divider network for triggering the flash tube. The potentiometer (R_5) is the variable resistor used to control the number of flashes.

C_2 possesses a charge of sufficient level (around 340 volts) to keep the gas in the xenon tube agitated or ionized at a low level. The trigger pulse is fired at the center terminal of the tube and this causes the gas to become very agitated. The ionized gas particles collide with one another, producing a bright flash of light. In ionizing, the gas causes a very-low-resistance path across the 16-μF capacitor. It therefore discharges through the tube, providing the energy necessary to cause the brilliant flash.

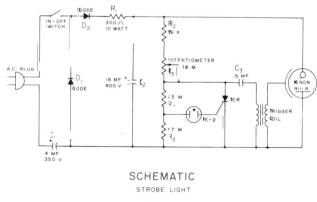

SCHEMATIC
STROBE LIGHT

Figure 1

The timing of the flash is controlled by the potentiometer resistance in conjunction with R_2 and C_3.

Once the xenon tube fires, C_2 is completely discharged. C_2 starts to recharge (drawing energy from the line cord) through the 300-ohm resistor. The 0.5-μF capacitor starts to charge too, limited by the 56K resistor in series with the value of resistance in the potentiometer. At the same time, the 3.3-meg and 4.7-meg resistors have current through them. As the voltage drop across R_3 reaches 90 volts, the neon tube fires. Once the neon tube fires, this causes the SCR to conduct from cathode to anode with very little internal resistance. This very low resistance of the SCR causes the 0.5-μF capacitor to discharge through the primary coil and the SCR. Once the trigger coil has energy pulsed through the primary, it steps up the voltage by means of a turns ratio and produces 6000 volts in the secondary coil.

This 6 kilovolts is used as the trigger for causing the xenon tube to fire. Firing of the tube completely discharges C_2 so the whole process begins again. If the flash rate is set for the lowest number of flashes, the neon bulb can be seen to glow just before the xenon tube flashes. The 0.5-μF capacitor discharges up to 200 volts through the SCR and primary of the trigger coil. Note that the trigger coil is really a transformer and not a coil in the true sense of *inductor* or *coil*.

To carry out this project, you will need the following materials.

MATERIALS NEEDED

1 Case with reflector
1 Back
1 Printed-circuit board
1 Resistor, 300 ohms, 10 W (R_1)
1 Capacitor, 4 μF, 350 VDC (C_1)
1 Capacitor, 16 μF, 400 VDC (C_2)
2 Diode, 1 amp, 1000 PIV (D_1, D_2)
1 Xenon bulb
1 Neon bulb, NE-2
1 Trigger coil
1 Resistor, 56K, 1/2 watt, 10% (R_2)
1 Resistor, 4.7 meg, 1/2 watt, 10% (R_3)
1 Resistor, 3.3 meg, 1/2 watt, 10% (R_4)
1 Potentiometer, 1.8 meg (R_5)

1 SCR, GE C103B
1 Capacitor, 0.5 μF 200 VDC, 20% (C_3) (0.47 μF may be substituted)
1 AC line cord (zip cord)
1 Slide switch, SPST
2 Screw, #6, 1/2" long, self-tapping
2 Screw, #6, 3/8" long, self-tapping
3 Wire, insulated, 6" long
1 Solder, rosin-core, 60/40, 12"
2 Screw, #4-40, machine, 1/2" long
2 Nut, #4-40, hex
1 Round, red, cardboard backing for xenon tube

ASSEMBLY INSTRUCTIONS

1. Make the printed circuit board using the patterns shown in Figure 2.
2. Insert components in the pc board, following the steps illustrated in Figures 3–17.
3. Place the reflector inside the case, Figure 18. Attach the completed pc board to the case as shown in Figure 19.
4. Check to make sure it has been properly inserted and attached by screws, Figure 20.
5. By this time, you should have checked each component against the parts list and have checked each for proper operation. Now, examine the back cover. Notice the locations provided for mounting the on–off switch, Figures 21 and 22.
6. Attach a 6" piece of insulated hookup wire to one terminal of the switch. Solder in place, Figure 22.
7. Remove the nut from the potentiometer. Place the potentiometer onto the back cover of the strobe light.
8. Slide the shaft of the potentiometer through the hole from the inside of the back cover. Place the nut removed from the "pot" on the other side of the board. Tighten to hold the pot in place, Figure 21.
9. Check Figure 22 for proper location of the two 6" pieces of insulated wire. Make sure you have a good mechanical connection and solder. Check the quality of the solder joints after they have cooled by exerting a slight tug on the loose ends of the wire.

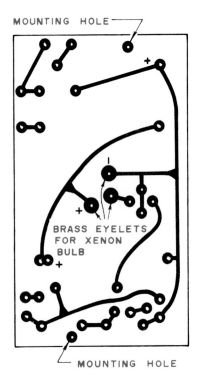

MOUNTING HOLE

BRASS EYELETS FOR XENON BULB

MOUNTING HOLE

Figure 2

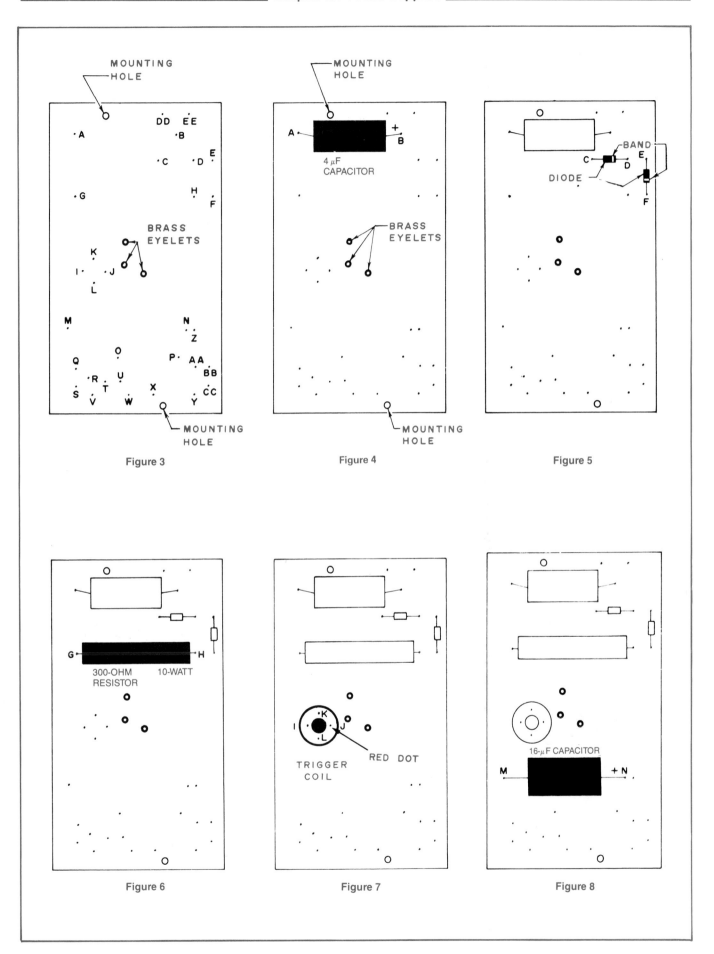

Figure 3

Figure 4

Figure 5

Figure 6

Figure 7

Figure 8

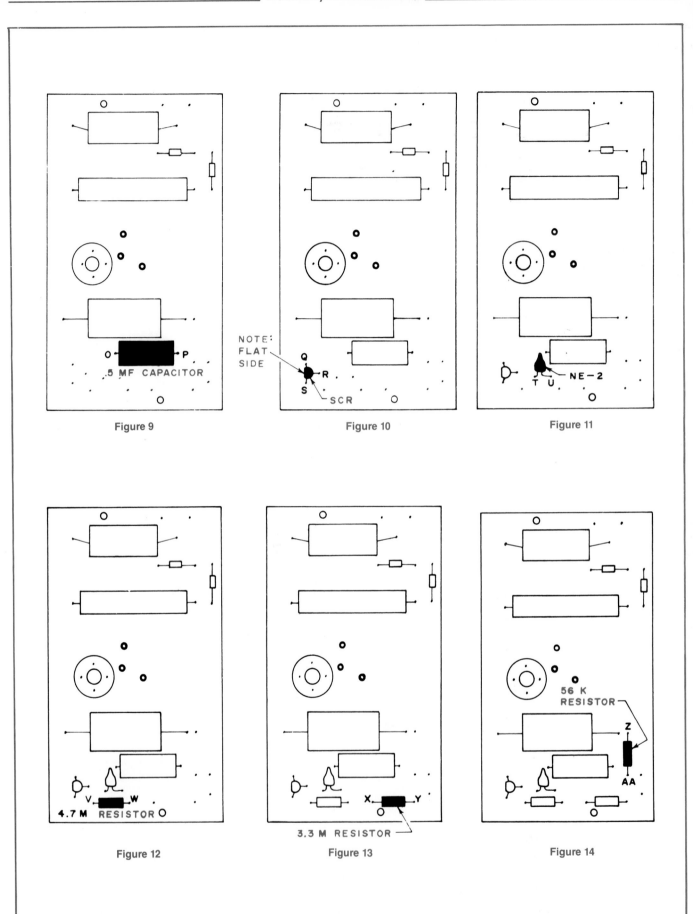

Figure 9

.5 MF CAPACITOR

Figure 10

NOTE: FLAT SIDE

Q R S SCR

Figure 11

NE-2

T U

Figure 12

4.7 M RESISTOR

V W

Figure 13

3.3 M RESISTOR

X Y

Figure 14

56 K RESISTOR

Z AA

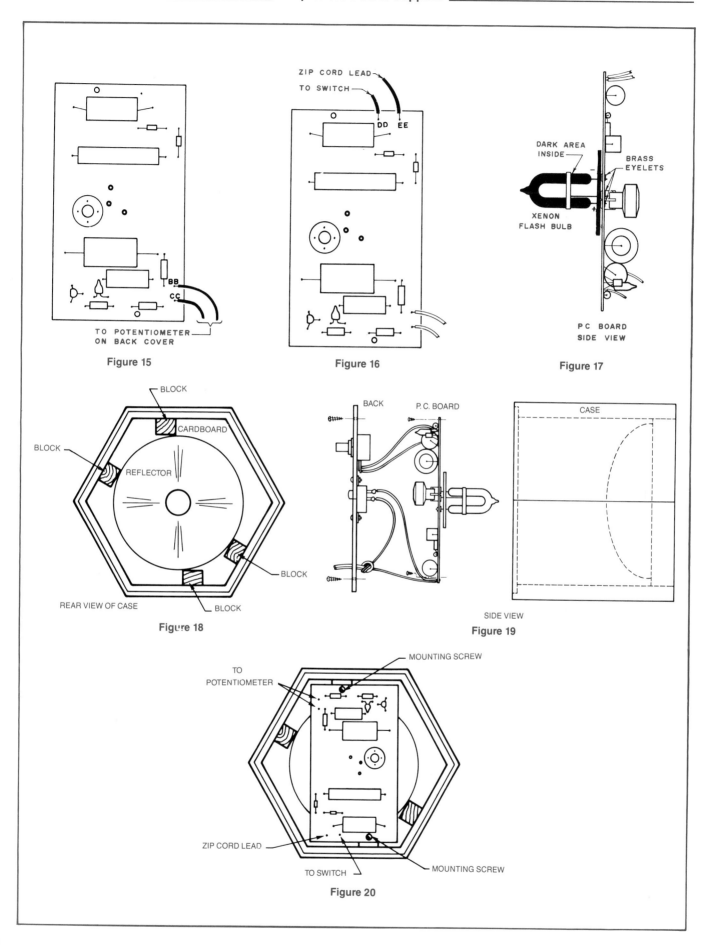

Figure 15

Figure 16

Figure 17

Figure 18

Figure 19

Figure 20

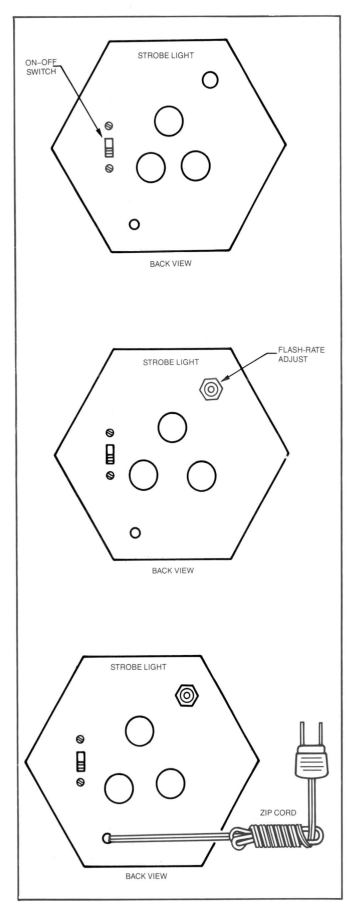

ON–OFF SWITCH

STROBE LIGHT

BACK VIEW

HEX NUT

SWITCH

6" WIRE

HEX NUT

INSIDE VIEW OF BACK COVER

FLASH-RATE ADJUST

STROBE LIGHT

BACK VIEW

POTENTIOMETER

WIRE

6" WIRE

INSIDE VIEW OF BACK COVER

STROBE LIGHT

ZIP CORD

BACK VIEW

SEPARATED ZIP CORD LEAD

KNOT

SEPARATED ZIP CORD LEAD

INSIDE VIEW OF BACK COVER

Figure 21

Figure 22

10. Note, in Figure 21, the location and arrangement of the AC line cord.

11. Thread the free end of the line cord through the back cover. Check Figure 22 for location of the stress-relief knot in the lamp cord. CAUTION: Remember that the strobe has a very high voltage and can cause shock. Stay clear of the exposed circuitry when the power cord is in the wall. Discharge the electrolytic capacitor when the power is removed to prevent shocking anyone working with the circuitry.

12. Test for operation before final assembly:
    (a) Place the back cover and pc board combination on a nonconductive surface (such as a wooden table top).
    (b) Insert the AC plug into a source of 120 volts AC.
    (c) Turn on the slide switch.
    (d) Adjust the potentiometer to vary the rate of flash.
    (e) If the xenon tube does not flash, see the "Trouble-shooting" section below.
    (f) If the xenon tube does flash, fill out the voltage chart so you'll have a record of the proper operating voltages in case of trouble later.

13. Proceed with the final assembly only if the unit flashed as designed.

14. Place the back cover onto the case. Insert the proper screws to hold the back cover on the case.

15. Test for proper operation:
    (a) Insert the AC cord into a source of 120 volts AC.
    (b) Turn on the switch.
    (c) Adjust the potentiometer for 1 to 12 flashes per second.
    (d) If the unit performs as specified, rate it *OK* and, if you like, use it at your next party!
    (e) If the unit does not perform as specified, see the "Troubleshooting" section below and follow the suggestions there.

Troubleshooting

If the strobe light malfunctions in any of the ways below, follow these procedures for locating the possible source of trouble. (This assumes you have checked the pc board for hairline cracks and bad solder joints.)

1. If NOTHING HAPPENS (no flashing tube), then:
   (a) Pull power plug from wall socket.
   (b) Remove back cover and pc board from the case.
   (c) Place on a nonconductive surface.
   (d) Discharge the capacitors.
   (e) Give the pc board and solder connections a visual inspection.
   (f) Look for solder-bridging across copper strips on the foil side of the pc board.

   (g) Check for burned or discolored components.
   (h) Recheck location of parts.
   (i) Double-check polarity of the capacitors and diodes; make sure the cathode end is in the proper hole for each component marked with a +.
   (j) Resolder connections that look questionable or resemble a cold-solder joint.

2. If NOTHING HAPPENS and you have checked all the previous steps, then:
   (a) Observe the neon lamp (NE-2). If it glows dimly:
       (1) Check the trigger coil with an ohmmeter after you have removed the power cord and discharged the capacitors.
       (2) Check C_3 for short. Remove one end from the pc board for checking.
       (3) Check C_3 for open. Remove one end from the pc board for checking.
   (b) If the neon tube does not glow:
       (1) Check the SCR.
       (2) Check R_3.
       (3) Check R_5.   VOLTAGE OFF. DISCHARGE CAPACITORS
       (4) Check R_2.
       (5) Check for proper solder connections of wires from R_5 to pc board.
       (6) Check C_2.
       (7) Check R_1.   VOLTAGE OFF. CHECK CAPACITORS LEFT CHARGED.
       (8) Check D_1.
       (9) Check D_2.

3. If NOTHING HAPPENS and the first two steps have been carried out, then:
   (a) Put in the AC plug.
   (b) Turn on the switch.
   (c) Turn the potentiometer fully clockwise (to high flash rate).
   (d) Pull the AC plug.
   (e) Short across C_2 (16-μF) capacitor. If there is then no pop or spark:
       (1) Check D_1 and D_2.
       (2) Check on–off switch.
       (3) Check AC power line.
       (4) Make sure there is AC at the wall plug.
       (5) Check R_1 for open.
       If there is a loud pop or spark:
       (6) D_1 and D_2 are OK.
       (7) R_1 is OK.
       (8) AC line cord is OK.
       (9) On–off switch is OK.
       (10) Then concentrate your search on the trigger circuit: SCR, trigger coil, C_3 (0.5 μF), xenon tube.

4. If STILL NOTHING HAPPENS, even after the previous steps have been followed, then:

(a) Pull the AC plug.

(b) Isolate the strobe light by plugging it into an isolation transformer. (An isolation transformer is not necessary if you use a VOM meter for measurements). The transformer is an absolute must if you use an oscilloscope or a VTVM.

(c) Using a meter or scope, check the voltages across the components shown in the voltage chart below.

Copy the chart and record your readings.

(d) Isolate the defective component through comparison of your readings with those on the Voltage Chart. Allow ±20% for meter error and human error.

(e) Replace the defective component and recheck for proper operation.

	VOLTAGE CHART*	
Location	Readings taken with a V.O.M.	Your Readings (with a V.O.M.)
Across R_1 Points G and H+	About 4 volts, but only when flash occurs	
Across R_2 Points Z+ and AA	2 volts and up to 4 volts when flash occurs	
Across R_3 Points V and W+	72 Volts — VOM across R_3 causes flash to stop	
Across R_4 Points X and Y+	2 volts to 4 volts — shows rapid build-up and drop	
Across R_5 Points Back Cover	125 volts to 230 volts rapid rise and drop	
Across C_1 Points A and B+	160 V drops to 140 V when flash occurs	
Across C_2 Points M and N+	325 V drops to 270 V when flash occurs	
Across Trigger Coil — Primary Points I+ and K	Can't be measured with V.O.M.	
*Across Trigger Coil — Secondary Points J+ and K	Can't be measured with V.O.M.	

*Potentiometer adjusted for 1 flash per second during voltage tests
+ denotes positive probe of meter to this point

NOTE: *If a V.T.V.M. or scope is used for voltage checks, make sure an isolation transformer is utilized for the strobe-light power supply.*

Chapter 18

OSCILLATORS

After studying this chapter, you will know

- *how to identify various types of oscillators.*
- *how feedback keeps oscillators operating.*
- *how to draw circuits for Hartley, Colpitts, and crystal-controlled oscillators.*
- *how oscillations are produced in various types of oscillators.*
- *the types of oscillators used in microwaves.*
- *how a magnetron and a klystron work.*
- *how a Gunn oscillator is used.*
- *the difference between a Wein Bridge and a Wheatstone bridge.*

INTRODUCTION

To oscillate means *to vibrate or to swing back and forth, like a pendulum*. Oscillation is the act of vibrating or fluctuating. An oscillator is a device that oscillates. In the field of electricity/electronics, the term *oscillator* is restricted to the application of oscillations produced by electrical means.

Oscillators can be mechanical or electronic. The alternator is a good example of a mechanical oscillator. It produces frequencies as high as several thousand hertz. However, it is limited in its higher range due to its inability to rotate at high speeds without self-destruction.

Frequencies higher than those safely produced by a mechanical oscillator are produced by electronic means. There are a number of types of electronic oscillators used in various types of electrical and electronic equipment.

18.1 PRINCIPLES OF OPERATION

It is well to go back to the concept of the resonant LC circuit to explain the generation of oscillations that can be controlled and utilized in radio communications. Any combination of inductor and capacitor will produce a frequency if it is properly energized and a source of feedback is provided. Producing oscillations by a parallel LC tank circuit is just one method of obtaining very-high-frequency oscillations.

A quick review of the operation of the parallel LC tank circuit serves to illustrate the basic principles used to produce electrical vibrations or oscillations, Figure 18-1.

In Figure 18-1, a charged capacitor is placed across a coil or inductor. The capacitor discharges through the inductor as shown in A. Once charged, the capacitor acts as a source of emf and forces current through the circuit. The flow of current through the inductor creates a magnetic field about the coil. This magnetic field opposes the rise in current through the circuit. As the capacitor discharges, the current reaches a maximum and then declines. Between the time the current is zero and the time the current is maximum, the magnetic field builds up. Energy is thus stored in the magnetic field.

Once the capacitor is discharged, the current has a tendency to fall to zero. This change in current flow is opposed by the inductance. The magnetic field now acts as a source of emf and continues to force current through the circuit in the same direction, as in Figure 18-1B. This current serves to charge the capacitor to a voltage whose polarity is opposite to the polarity it had when it was placed in the circuit. As the magnetic field collapses, the current slowly falls to zero.

The capacitor in Figure 18-1C now discharges, but does so in the opposite direction, reversing the original path of current flow. This current flow is again opposed by the inductor. The inductor builds up a magnetic field that is opposite in direction to the original magnetic field. The current reaches a maximum and then declines.

Once the capacitor has discharged, the current attempts to fall to zero. This change is opposed by the inductor, which uses the energy stored in its magnetic field to prevent the current from dropping to zero at the instant the capacitor is discharged. Therefore, the current is sustained, and it serves to charge the capacitor to a voltage whose polarity is the same as it was originally, Figure 18-1D.

The entire process is then repeated again and again as long as energy remains in the circuit. The current in the LC circuit has the form of a sine wave, Figure 18-1E. It has changed from zero to a maximum in one direction, through zero to a maximum in the opposite direction, and back to zero. Several hertz of output are shown in Figure 18-1F.

This output is possible only if the oscillator circuit undergoes no loss in energy. If superconductors are used in the coil, the result approximates a no-loss condition.

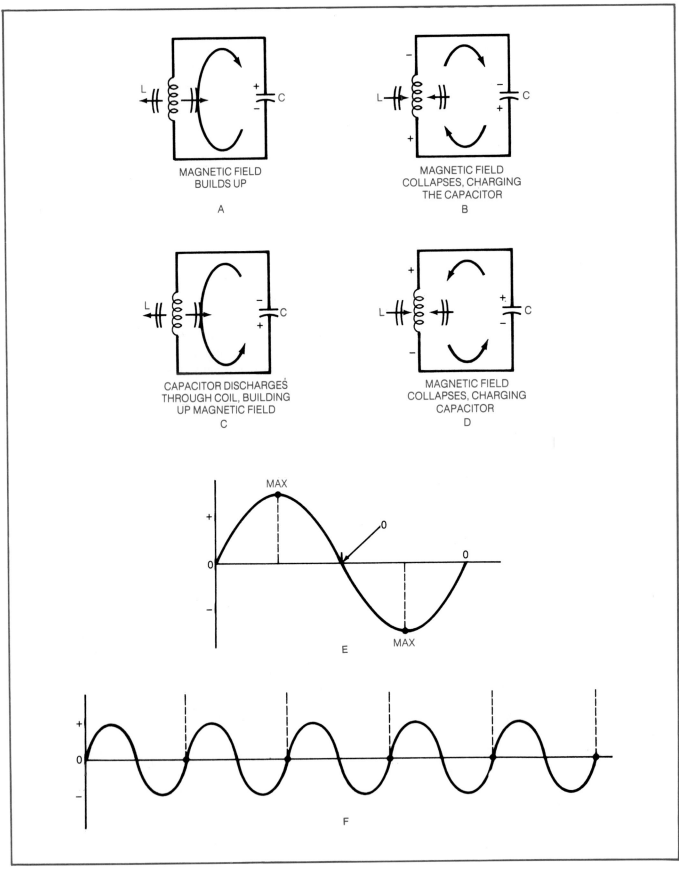

MAGNETIC FIELD
BUILDS UP

A

MAGNETIC FIELD
COLLAPSES, CHARGING
THE CAPACITOR

B

CAPACITOR DISCHARGES
THROUGH COIL, BUILDING
UP MAGNETIC FIELD

C

MAGNETIC FIELD
COLLAPSES, CHARGING
CAPACITOR

D

E

F

Figure 18-1. Tank-circuit operation.

However, this is not always possible, so the resistance of the coil wire has a tendency to expend the energy in the circuit. A damped oscillation results, Figure 18-2.

Oscillator Output

Oscillators must produce a usable output. The amount of output can be small, provided it is sufficient to drive the input circuit of the following stage. The stages that follow can be used to build up the oscillations, if necessary. This means that oscillations in the output circuit must be large enough to provide the proper amount of feedback and a useful output.

Feedback in Oscillators

Feedback is essential to the maintenance of oscillation in an **oscillator**. An oscillator cannot continue to oscillate unless it has feedback. It is also important to remember that the feedback has to arrive at the right time and in the proper direction. In order to ensure that feedback occurs at the right time, the resonant frequency of the LC tank circuit in the output stage is made approximately the same as the LC tank circuit of the input LC circuit.

If current in the input LC circuit is flowing in one direction and the feedback induces a current in the opposite direction, the oscillator current is damped more quickly than if no feedback were present. This type of feedback current is **degenerative** in nature, Figure 18-3A.

If the feedback induces a current in the same direction as the oscillatory current in the input LC tank circuit, the feedback is regenerative. Its effect is to sustain current oscillations. If the feedback is just sufficient to replace the losses, each oscillation is of the same size, Figure 18-3B. However, if the **regenerative feedback** is too small, the

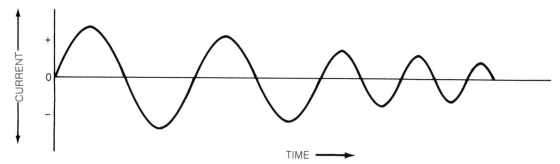

Figure 18-2. Damped oscillations.

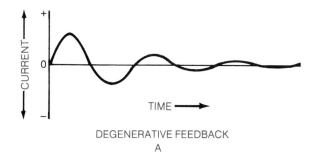

DEGENERATIVE FEEDBACK

A

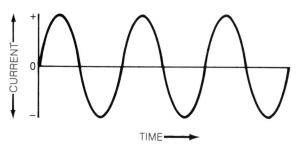

REGENERATIVE FEEDBACK

B

Figure 18-3. Feedback effects.

oscillations in the input tank circuit die out, although the oscillations are not damped as greatly as when there is no regenerative feedback. If the feedback is too large, the oscillations in the input tank circuit build up until the output LC tank current has swung alternately from zero to saturation. This results in a highly distorted oscillation. Regenerative feedback must therefore have the proper amplitude.

Transistors require a certain amount of energy to operate. This energy must be replaced by the energy taken from the collector circuit in an oscillator. Since the output also is taken from the collector circuit, the energy available in the output circuit must be greater than the energy needed in the input circuit. That means the total output must be greater than the input. It also means that the transistor must be capable of amplifying. Consequently, any transistor with an amplification factor greater than one can be used as an oscillator.

REVIEW QUESTIONS FOR SECTION 18.1

1. How can a resonant LC circuit be used in an oscillator?
2. How does an LC combination set up oscillations?
3. What is a damped oscillation?
4. What is feedback?
5. Why is feedback important in an oscillator circuit?

18.2 BASIC TYPES OF OSCILLATORS

There are many types of oscillators. We will be concerned with only three types here: the Armstrong, the Colpitts, and the Hartley.

The Armstrong Oscillator

The **Armstrong oscillator**, Figure 18-4, was used originally with regenerative-type receivers when little or no amplification of a signal was possible. The circuit utilized a tickler coil to provide the needed feedback to keep the circuit oscillating. It had a number of advantages over other types but its use diminished when the triode amplifier was introduced. However, it did make a comeback when Citizen Band radio was first introduced, since the ability to pick up a very weak signal was then necessary. The Armstrong oscillator was invented by Edwin Howard Armstrong (1890–1954). The oscillator was used to improve sensitivity of receivers in early days before the invention of heterodyning.

The Colpitts Oscillator

This type of oscillator can be easily recognized by its tapped capacitance, Figure 18-5. Feedback is accomplished by the capacitor at the bottom of the tank circuit. The amount of feedback depends on the ratio of capacitance between the feedback capacitor and the tank-circuit capacitor.

Hartley Oscillator

The **Hartley oscillator** uses a tapped coil instead of tapped capacitance. It has more frequency stability than the Armstrong oscillator, but less than the Colpitts oscillator. A coupling (C_2) capacitor is used for the path of feedback to keep the oscillator operating, Figure 18-8.

REVIEW QUESTIONS FOR SECTION 18.2

1. What type of oscillator is the Armstrong?
2. How can the Colpitts oscillator be easily recognized?
3. How can the Hartley oscillator be easily recognized?

18.3 TRANSISTOR OSCILLATORS

Oscillator circuits take many shapes and utilize vacuum tubes, transistors, and integrated circuits. Oscillators can be classified into many types, including tri-tet, dynatron, and transistor. The transistor can be grouped under the heading of negative-resistance oscillators. In ultra-high-frequency (UHF) work, such oscillators as the klystron and the magnetron are generally used.

The transistor can be used as the amplifier in the following types of oscillators: **tuned-base**, **Clapp**, Hartley, **crystal**, Colpitts, **Wein bridge**, and **multivibrator**.

Tuned-Base and Tuned-Grid Oscillators

The tuned-base and tuned-grid oscillators are similar. The tuned-grid type was utilized in vacuum-tube circuits. Figure 18-6 shows the tuned-base type of oscillator. Feedback is supplied by the transformer T_1. The common-emitter configuration is used. Resistor R_E is the emitter

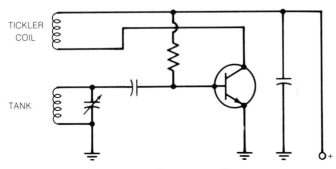

Figure 18-4. Armstrong oscillator.

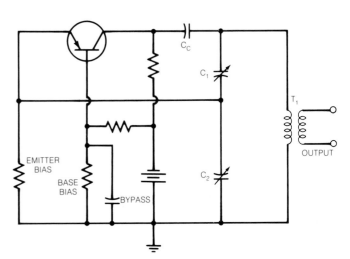

Figure 18-5. Colpitts oscillator.

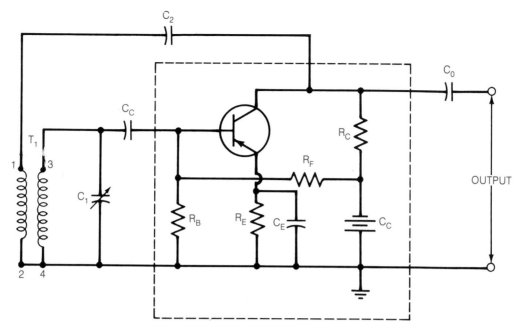

Figure 18-6. Tuned-base oscillator.

swamping resistor. Within the dotted lines is the amplifier stage. The other resistors within the box are used for biasing the transistor and making it operate properly. Feedback through the transformer makes the device an oscillator.

The LC tank circuit of the tuned-base oscillator is made up of the transformer secondary and the tuning capacitor C_1. Tuning the capacitor can generate a wide range of frequencies. C_C and R_B are RC-coupled to bring the output of the tank circuit to the base of the transistor amplifier. R_E is the emitter bias, with C_E serving as the feedback capacitor. C_2 and C_1 should be approximately equal to the ratio of the output impedance to the input impedance of the transistor. C_O serves as a coupling capacitor to the next stage.

Clapp Oscillator

The stability of the Colpitts oscillator can be improved by the addition of a capacitor, Figure 18-7. The **Clapp oscillator** has a variable capacitor C in series with the T_1 points 1 and 2. The variable capacitor tunes the output of the tank circuit over a wide range of frequencies. The shunting impedance of the LC series combination is at a minimum, thereby making the oscillating frequency comparatively independent of the transistor parameter variations. Compare this with the Colpitts oscillator in Figure 18-5.

Hartley Oscillator

The Hartley oscillator is similar to the Colpitts. The Hartley, however, has a split inductance instead of a split

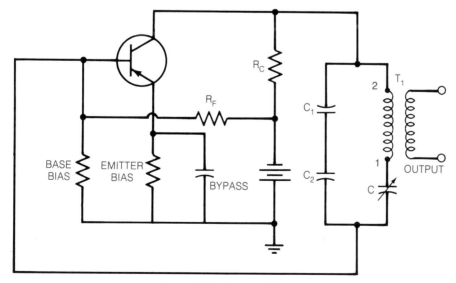

Figure 18-7. Clapp oscillator.

capacitance to obtain feedback. Both the series-fed and shunt-fed Hartley oscillators operate the same. They differ only in the way they obtain collector bias. A modified Hartley oscillator that provides greater power output is shown in Figure 18-8. This one is called the push–pull oscillator. Note that the tank circuit is made up of part of T_1 and C_1. Feedback is through the other part of T_1 located in the base circuit. After the feedback is accomplished through C_C, the operation of the circuit is the same as for a push–pull amplifier, Figure 18-9.

Crystal Oscillator

A quartz crystal is used to establish the operating frequency of the circuit in a **crystal oscillator**, Figure 18-10. The circuit is a crystal-controlled **tickler-coil** oscillator. This stage uses the series mode of operation of the crystal and functions similarly to the tuned-collector circuit. Better frequency stability is obtained by placing the crystal in series with the feedback path. However, the frequency is essentially fixed by the crystal. The crystal has to be changed to change the frequency. Each crystal is tuned to or ground to its own frequency. Regenerative,

or positive, feedback is through the mutual inductance of the transformer windings. The transformer action provides the necessary 180° of phase shift for the feedback signal. At frequencies above or below the series resonant frequency of the crystal, the impedance of the crystal increases and reduces the amount of feedback. This prevents oscillation at frequencies other than the series resonant frequency.

Crystal Colpitts Oscillator

The crystal-controlled Colpitts oscillator is a common-emitter configuration, with the feedback supplied from the collector to the base, Figure 18-11. The crystal shunts the tapped capacitors C_1 and C_2. Note how the center tap between the two capacitors is grounded. C_1 is used to feed its voltage back to the base of the transistor 180° out of phase and with the proper amplitude to keep the oscillator going. In this type of arrangement, the crystal determines the frequency of the oscillator. The resistors provide proper bias and stabilizing conditions for the circuit.

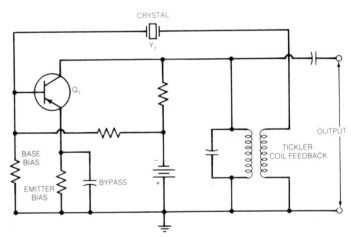

Figure 18-10. Crystal oscillator with tickler-cell feedback.

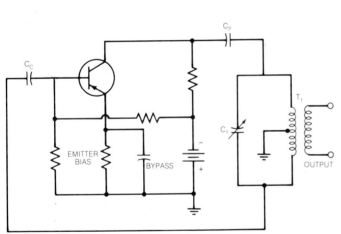

Figure 18-8. Hartley shunt-fed oscillator.

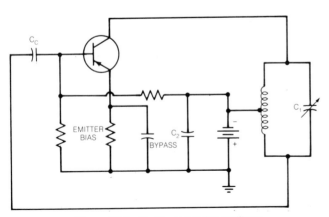

Figure 18-9. Hartley series-fed oscillator.

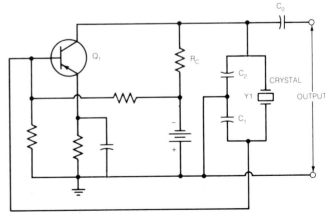

Figure 18-11. Crystal-controlled Colpitts oscillator.

Wein-Bridge Oscillator

Examine Figure 18-12. Note how the **Wein-bridge oscillator** uses a resistance–capacitance network for the development of a sinusoidal output (dotted lines). Both of the transistor circuits are identical. Figure 18-12 shows how the bias and feedback are accomplished, and also indicates how the bridge or *diamond-shaped* circuit looks when drawn in the shape of the *typical bridge* circuit.

The two transistors, Q_1 and Q_2, are connected in the common-emitter configuration. The second stage functions as an amplifier and phase-inverter and provides the feedback signal in proper phase. Regeneration is provided for oscillation. Degeneration is also provided to obtain frequency stability and a distortionless output.

The output of transistor Q_1 is coupled to the input of transistor Q_2. This is done through capacitor C_C and R_{F2}. Capacitor C_3 couples a portion of the output of the amplifier stage Q_2 to the bridge network to provide the necessary feedback (both negative and positive). The output to the load is coupled through capacitor C_0.

Multivibrators

The **free-running** (astable) **multivibrator** is essentially a nonsinusoidal two-stage oscillator. One stage conducts while the other is cut off. This continues until a point is reached at which the stages reverse their conditions. That is, the stage that had been conducting cuts off, and the stage that had been cut off conducts. This oscillating process is normally used to provide a square-wave output.

Note that the multivibrator shown in Figure 18-13 is a two-stage resistance–capacitance-coupled common-emitter amplifier with the output of the first stage coupled to the input of the second stage and the output of the second stage coupled to the input of the first stage. The signal in a collector circuit of a common-emitter amplifier is reversed in phase with respect to the input of that stage. That means that a portion of the output of each stage is fed to the other stage in phase with the signal on the base. This *regenerative* feedback with amplification is required for oscillation. Bias stabilization is the same with both transistors.

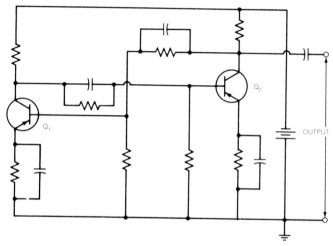

Figure 18-13. Multivibrator.

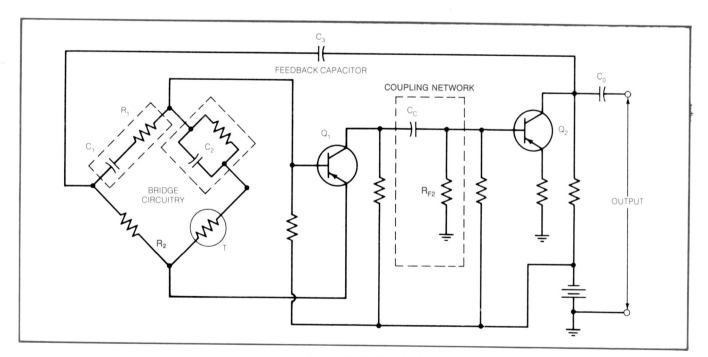

Figure 18-12. Wein bridge.

287

Activity: CODE PRACTICE

OBJECTIVES

1. *To learn the basic characters of the International Morse Code.*
2. *To observe the operation of a transistorized code-practice oscillator.*

EQUIPMENT NEEDED

1 Telegraph key
1 Headset, 2000 ohms
1 Power supply, 6 volts DC
1 Resistor, 10 K, 20%, 1/2-watt
1 Capacitor, 0.01 μF, any working voltage DC

1 Output transformer, 2000 ohms to 8 ohms
1 Speaker, 8 ohms
1 Transistor, 2N1048 or equivalent
6 Wires for connections

PROCEDURE

1. Connect the circuit shown in Figure 1. This is a transistorized code-practice oscillator. The major components for this circuit are the transistor, the telegraph key, the output transformer, the speaker or headphone, and the code network of resistor and two capacitors.

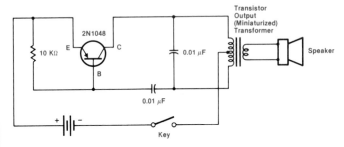

Figure 1

2. Use the 6-volt power-source set for DC output.
3. Press the telegraph key. Does a high-pitched sound come from the speaker? **NOTE:** *If there is no sound from the speaker, try changing the connections to the headphones. If the circuit still does not operate, ask your teacher to check your circuit.*
4. Now that you have your code-practice oscillator operating, you will learn to form each of the basic letters with the key. Go slowly! The dot should be a definite tone, not just a bit of static. The dash should

be three times the length of the dot. The space between dots and dashes within a letter should be equal to the length of a dot. Practice sending the following basic letters:

E .	T −	A . −
I . .	M − −	N − .
S . . .	O − − −	R . − .
H		

5. When you feel confident in sending these letters, have your lab partner send the letters to you, so that you can become accustomed to receiving these letters. Do not think of each letter as a series of dots and dashes, but rather as a rhythm pattern. You should train yourself to think of the letter each time your ear detects its rhythm pattern. If you try to catch each dot and dash and then translate it to the proper letter, you will become confused and will have difficulty in mastering the code.
6. When you feel you can send and receive the basic letters with ease, try sending words using these letters. The spacing between each letter should be the equivalent of three dots and the spacing between words should be the equivalent of five dots.

Try sending these simple words based on the basic letters:

tom	torn	ran	moat	mat	meat
tar	mash	rah	rain	rot	moan
man	mars	rat	earn	her	heat
sir	tone	hat	term	him	host
eat	toast	sat	tenth	tat	anise
nor	shine	nan	errata	sin	roses

7. When you finish with these and feel proficient, have your lab partner send them to you, so that you can practice receiving. You will find many words that you can add to the list. When you have mastered these basic letters, add more to your list until you can send and receive all letters of the alphabet.
8. Disconnect the circuits you have been using. Return all equipment and materials to their proper storage places.

SUMMARY

1. Describe the timing of code to form letters and words.
2. How should you think of the code when you are receiving it?

1. How are transistorized oscillators classified?
2. What types of oscillators use the transistor as an amplifier?
3. How does the Colpitts differ from the Clapp oscillator?
4. How do the series-fed and shunt-fed Hartley oscillators differ?
5. What is the most stable type of oscillator?
6. How does the Wein-bridge oscillator resemble a bridge rectifier circuit?
7. What is degeneration used for in an oscillator circuit?
8. What does *astable* mean?
9. What is regenerative feedback?

18.4 SOME NEWER OSCILLATOR TYPES

The advent of the integrated circuit added new ways to obtain and stabilize oscillations. The **RC phase-shift oscillator**, the **op-amp Wein-bridge oscillator**, and the VCO, or **voltage-controlled oscillator**, are examples of the newer types.

RC Phase-Shift Oscillator

One of the newer types of oscillators uses an op-amp (operational amplifier), Figure 18-14. This oscillator uses a resistor–capacitor network to determine the oscillator *frequency*. The op-amp provides the necessary gain for feedback. The RC phase-shift network provides the required conditions for circuit oscillation. Output frequency can be varied by changing either the resistance or the capacitance.

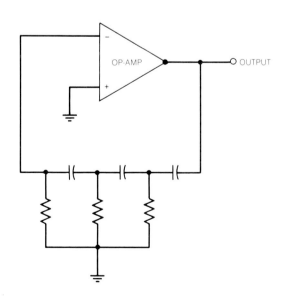

Figure 18-14. RC phase-shift oscillator.

Op-Amp Wein-Bridge Oscillator

Resistors and capacitors are used in an RC phase-shifting circuit in the Wein-bridge oscillator, Figure 18-15. The resistors determine the frequency at which the oscillator operates. (Remember: $T = R \times C$.) Inputs to the bridge are at points 1 and 2. The op-amp inputs are at points 3 and 4. The frequency of the output is sinusoidal. The output frequency can be calculated by use of the formula:

$$f_o = 1/2\pi\sqrt{R_1C_1R_2C_2}$$

or

$$f_o = \frac{1}{2\pi\sqrt{R_1C_1R_2C_2}}$$

Note the location of R_1C_1 and R_2C_2 in the formula. Feedback is from the output of the op-amp to point 1 of the bridge.

VCO, or Voltage-Controlled Oscillator

The voltage-controlled oscillator (VCO) uses a 566 IC chip. It can be used to generate both square-wave and triangular-wave signals. The frequency can be set or an external resistor and capacitor can be used to adjust the frequency. In this case, they are R_1 and C_1. The frequency variation is developed by an applied DC voltage, Figure 18-16. The output frequency is a linear function of the controlling voltage.

A Schmitt trigger circuit is used to switch the current source between charge and discharge in the capacitor. A triangular voltage is developed across the capacitor, with its charge and discharge. The square wave from the Schmitt trigger, along with the triangular wave form are produced as outputs through buffer amplifiers located within the chip or IC. Pin connections for the 566 chip correspond to the numbers on the schematic diagram.

Other types of oscillators are available for special applications. The IC allows many circuit variations and

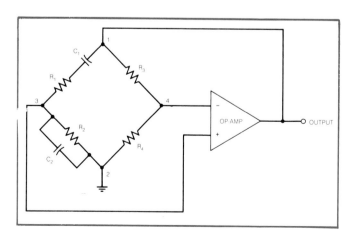

Figure 18-15. Op-amp Wein bridge.

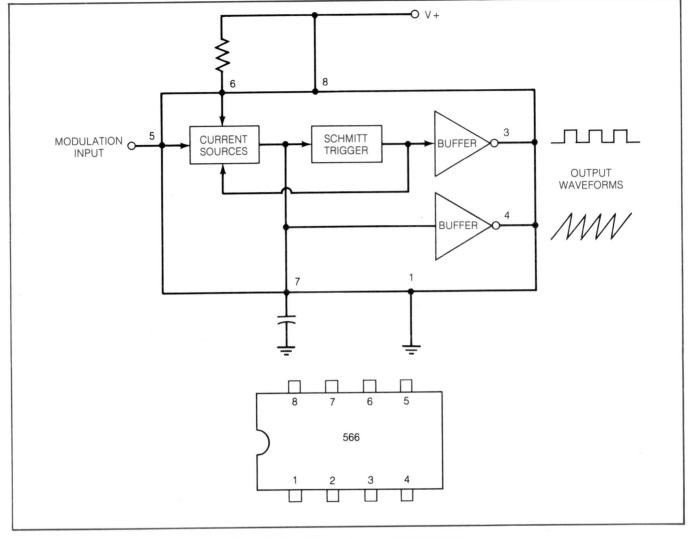

Figure 18-16. VCO, or voltage-controlled oscillator.

produces a wide variety of outputs. One good example of a source with a number of output frequencies, and with square-wave, triangular-wave, and other waveforms, is the synthesizer that produces organ music.

REVIEW QUESTIONS FOR SECTION 18.4

1. List some of the newer oscillator types.
2. What does *VCO* mean?
3. How is a Schmitt trigger circuit used?

18.5 MICROWAVES

Microwaves get their name from *micro* which means *small*. These waves are small, inasmuch as their wavelength is usually less than one-half an inch. That means pulses of the wave travel one-half inch apart. It also means that some interesting arrangements have to be made in

terms of capacitance and inductance. The generation of a wave of this superhigh frequency (SHF) has to be somewhat different from that from the crystal oscillator or coil-capacitor tank circuit. There are three microwave oscillators most commonly used. They are the **klystron**, the **magnetron**, and the **Gunn oscillator**.

Microwaves are divided into bands. The **S band** is well known for its first use as a police radar speed-trap device. The device was typically set on a tripod alongside a road, and a police car monitored its resulting speed readouts. The S band was radar at 2450 MHz. This is the same frequency used today in microwave ovens. There were some inherent problems with this frequency range, so police radar was moved to the **X band**, where it operates at 10,525 MHz or 10.525 GHz. This is the type of radar now used to check for speeding on highways. It is usually mounted on the rear window of the police car or on the dashboard. It too has some limitations when used in areas with a great deal of electrical interference. And, of course, since it uses the Doppler effect, its readouts can be affected by the speed of the wind where it is operating, and can be erroneous as a result. Microwaves do have some limitations that have to be taken into effect when used under different conditions.

The **Ku-band** radar [24.15 GHz] is used for hand-held radar speed guns. Its use presents some problems during rainy or foggy conditions. It appears that the wave is absorbed or dispersed by the water droplets in much the same way as light waves are. This is due mainly to the high frequency of the microwaves, which approaches that of visible light.

The Klystron

Superhigh frequency energy for a radar unit uses the klystron as the source. This device can be classified as a mechanical oscillator inasmuch as the shape and size of the cavities in the unit determine the frequency at which it oscillates, Figure 18-17.

The principle of bunching of electrons and **velocity modulation** was discovered in 1935 by two German scientists named Heil and Heil. The principle of the cavity resonator was discovered by William W. Hansen of Stanford University in 1938. The klystron was developed and constructed in 1938 by Hansen and also by Russell and Sigurd Varian, brothers. The klystron and the principle of bunching were put together, and the klystron produced a signal that was extremely high in frequency. Figure 18-17 is a cutaway view of the klystron.

Figure 18-18 shows how the klystron works. The schematic of a single-cavity, reflex klystron and the polarity of the voltages required for operation are shown in that figure. The various elements that make up the tube are the cathode, a focusing electrode at the cathode potential, a resonator that also serves as an anode, and a repeller or reflector that is at a negative potential with respect to the cathode.

The combination of the cathode, focusing electrode, and anode beams the electrons through the resonator gap and out toward the reflector. Since the reflector element is negative with respect to the anode, the electrons are turned back toward the anode, where they pass through the gap a second time. When the klystron is oscillating, an alternating voltage appears across the gap of the resonator. As electrons pass through the gap, they are either accelerated or decelerated as the voltage across it changes in magnitude with time.

This results in accelerated electrons leaving the gap at an increased velocity, and decelerated electrons leaving it at a reduced velocity. Because of this, electrons leaving the gap at different parts of the gap voltage cycle take different lengths of time to return to the gap; that is, they have different transit times. As a result, the electrons bunch together as they return through the gap. This variation in velocity of the electrons is called *velocity modulation.*

As a bunch of electrons passes through the gap, the electrons react, with the voltage appearing across the gap. If the bunch passes through the gap at a time in the gap voltage cycle such that the electrons are slowed down, then energy will be delivered to the resonator, and oscillations

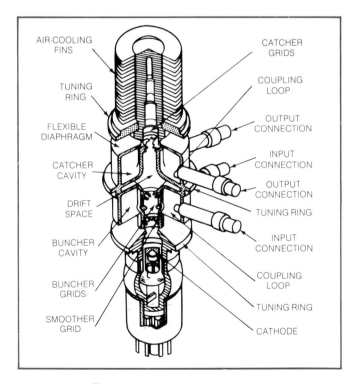

Figure 18-17. Klystron, cutaway view.

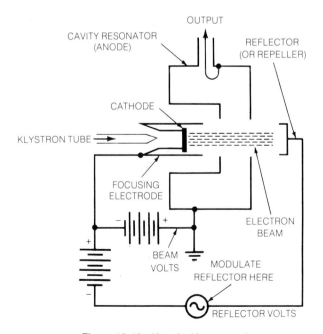

Figure 18-18. How the klystron works.

will be sustained. If the time of transit in the anode reflector region is such as to cause the bunches to arrive at a time when they will be accelerated by the gap voltage, then energy is removed from the resonator, and oscillating will tend to stop. To generate oscillations at a given frequency and fixed anode voltage, it is necessary to vary the transit time in the anode reflector space to a suitable value by means of adjusting the reflector voltage: the more negative the reflector voltage, the shorter is the transit time. Though frequency of operation is determined primarily by the resonator dimensions, a small change in frequency may be obtained by adjusting either the reflector or anode voltage. This adjustment is known as **electronic tuning**. A frequency change of several percent is possible in some tubes.

At one time, the klystron was designed for low-power types of radar units. Now, they are used for medium- and high-power units. Today they are used for radar as well as for many commercial purposes. They are used as part of the earth stations that beam television programs and telephone conversations in the 6-GHz frequency range to satellites.

The Magnetron

This tube is used to produce an extremely high frequency that can be built to very high power levels. It is very much in demand for surveillance radar and other military purposes. It can also be used for commercial purposes.

The magnetron tube, whose basic shape is shown in Figure 18-19A, is made in a number of sizes. It is a cylindrical brass block with a large hole drilled down the center and eight smaller holes drilled between the center and outer edges. Slots are used to interconnect the holes. The smaller holes are called cavities. When the magnetron is operating, electrons take a back-and-forth path along the walls of the cavities, as indicated by the arrows in one cavity. Actually a cavity oscillates in a manner similar to a lower-frequency coil-and-capacitor circuit. The walls of the cavity form the inductance, and the capacitance across the cavity opening forms the capacitor.

A cathode is placed down the middle of the magnetron. It has an internal heater wire. A small hook in one of the cavities acts as a pickup loop. This takes the RF energy from the cavity when it is oscillating and feeds it to the transmission lines. The **waveguide** is used in microwaves to handle the movement of the RF energy. A wire is unable to handle the microwaves, since they travel on the outside (skin effect) of any conductor. They are contained inside the waveguides, which are made to a particular size to accommodate the particular frequency being generated.

In some cases, a spark gap is used to generate the pulses needed to get the magnetron to operate or start oscillating. When the tube is pulsed, the cathode is driven negative by 10,000 to 20,000 volts. This makes the plate relatively positive, and the electrons from the hot cathode start moving toward it. However, a strong, external horseshoe-shaped magnet, with its north pole at one end of the cathode and the south pole at the other end, produces an intense magnetic field down the center hole. The electrons are deflected at right angles to the lines of force through which they are passing. This results in an elliptical path for the electrons as they progress toward the anode areas. The positive potential of the anode accelerates the electrons toward it. That is, the electrons pick up energy from the difference of potential. As the electrons move past the slots between the anode areas, they induce voltage between the slot faces. This drives currents into oscillation along the surfaces of the cavity walls. In this way, the energy of the cathode electrons is transferred to the oscillating currents in the cavities. All the cavities are of the same size and oscillate at the same frequency. The magnetrons become hot and have to be cooled. Air-cooled fins are usually attached to the unit, making it appear much larger than its work area really is. Some kilowatt and megawatt units have to be water-cooled.

The Gunn Oscillator

A semiconductor device called the Gunn diode operates at lower power levels than do the klystron and the magnetron, Figure 18-20. The Gunn diode is best for particular devices such as for the hand-held radar guns the police use for traffic control. The output of the diode is about 1 watt maximum.

The **Gunn diode** is not really a diode. It does not have junctions. Instead, it has some of the peculiar characteristics of semiconductors that produce negative-resistance effects. These are called active-area or bulk devices. The microwave- and millimeter-band Gunn diode consists of a thin slice of N-type gallium arsenide between two metal conductors, Figure 18-20. It is assembled in a cylindrical metal and ceramic body. The diode is fitted into a hole inside a cavity and is fed the DC required to make it oscillate. If the voltage across the diode is increased from 300 V per millimeter to 400 volts per millimeter, at some voltage the electrons from the outer, partly filled energy ring of the atoms of the semiconductor crystal jump across the narrow "forbidden" energy gap of gallium arsenide and actually decrease their mobility in the crystal. This produces a negative-resistance effect in the device. A further increase in voltage causes the current to begin to increase in proportion to the applied DC voltage. This produces a result similar to that of the tunnel diode.

The Gunn-diode oscillator can be tuned or the frequency changed by movement of the cavity end.

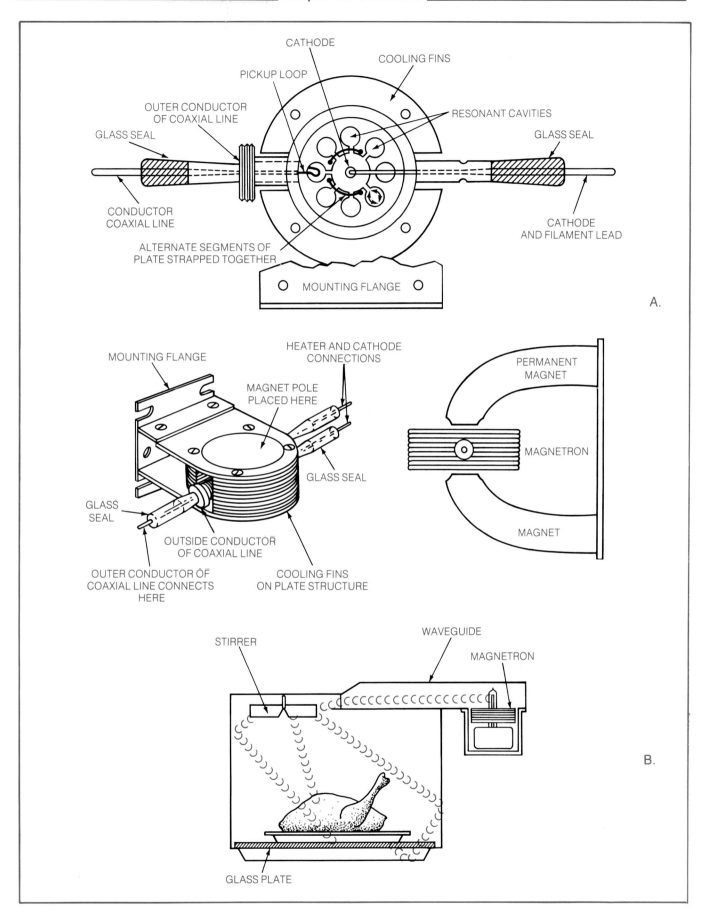

Figure 18-19.
A. Magnetron. **B.** Microwave oven; note location of the magnetron.

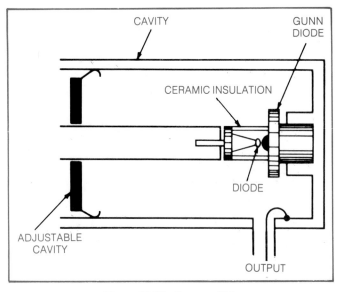

Figure 18-20. Gunn oscillator.

REVIEW QUESTIONS FOR SECTION 18.5

1. How do microwaves get their name?
2. At what frequency does a microwave oven operate?
3. How are microwaves generated?
4. What is a klystron?
5. What is a magnetron?
6. What is velocity modulation?
7. For what is a Gunn oscillator used?

Safety Tips

1. Magnetrons use high voltage to generate microwaves. Keep away from a magnetron when the circuitry is exposed.
2. Microwave ovens should be shielded to prevent radiation. Use a microwave detector to see whether your microwave oven is within set Federal standards.
3. If an oscillator being used requires high voltage, be aware of this and stay clear of areas with exposed connections.
4. Do not replace a fuse without making sure it is of the correct value.

SUMMARY

The oscillator is a type of circuit that produces oscillations of a desired frequency. It can be an audio- or a radio-frequency oscillator. There are a number of types of oscillators, such as the Armstrong, Hartley, Colpitts, Clapp, crystal, voltage-controlled, tuned-base, Wein-bridge and Gunn types. Each has its particular identifiable characteristics and each can be used for specific functions.

An oscillator needs feedback to operate. That means it feeds part of the output back to the input to keep the oscillations going. The proper phase relationship between the signal and feedback must be maintained or the oscillator will stop oscillating.

The Hartley is easily identified by its tapped inductance. The Colpitts is easily identified by its tapped capacitance. The crystal oscillator utilizes a quartz crystal in its circuitry to establish the oscillating frequency.

Op-amps (operational amplifiers) can be utilized in the production of oscillations.

Microwave frequencies are generated by special types of oscillators. The main types are the klystron, the magnetron, and the Gunn oscillator. The klystron can be frequency-modulated and utilized for broad-band FM transmissions. Police radar and microwave ovens, as well as satellite communications systems, utilize microwave frequencies.

USING YOUR KNOWLEDGE

1. Draw an LC tank circuit.
2. Draw the schematic for an Armstrong oscillator.
3. How do the Hartley and Colpitts oscillators differ?
4. What is feedback? Why is it necessary in an oscillator?
5. What is a damped wave? Draw one.
6. Draw a multivibrator circuit.
7. What is a klystron?
8. What three types of devices are used to generate microwaves?
9. Who was responsible for developing the heterodyning effect? Where is it used?
10. How is the Wein bridge different from the Wheatstone bridge?

KEY TERMS

oscillator	magnetron
Armstrong oscillator	Gunn oscillator
Colpitts oscillator	waveguide
crystal oscillator	free-running multivibrator
Hartley oscillator	RC phase-shift oscillator
Clapp oscillator	voltage-controlled oscillator
tuned-base oscillator	op-amp Wein-bridge oscillator
multivibrator	microwaves
tickler coil	X band, Ku band, S band
degenerative feedback	velocity modulation
regenerative feedback	electronic tuning
Wein-bridge oscillator	Gunn diode
klystron	

Chapter 19

AMPLIFIERS

OBJECTIVES

After studying this chapter, you will know

- *how an amplifier works.*
- *the classes of operation of an amplifier stage.*
- *the difference between a JFET and MOSFET transistor.*
- *how a preamplifier is used.*
- *how to draw a schematic of a push–pull amplifier stage.*
- *why an op–amp can be used for so many purposes.*
- *the difference between negative and positive feedback.*
- *the difference between enhanced and depletion modes of operation.*
- *how a PM speaker works.*
- *what a crossover network does.*

INTRODUCTION

It has been said that the amplifier is the workhorse of electronics. Amplifiers are classified into two large groupings: the audio-frequency and the radio-frequency types. The audio amplifier is of most interest to the general public inasmuch as it is used to produce music for entertainment purposes. The radio-frequency amplifier is used in broadcasting equipment and in other communications and data-transmission systems.

19.1 AUDIO AMPLIFIERS

Amplifiers are used to amplify weak signals. However, in the process of increasing the amplitude of a signal, they may also introduce some of their own inherent problems. **Distortion** and **hum** are two of the problems associated with the amplifiers.

Distortion

Distortion is the type of change a signal undergoes from the time it enters the amplifier stage until it comes out. The amount of distortion depends on the linearity or the dynamic characteristics of the semiconductor device used in the amplifier circuit. The amount of distortion depends on the amplitude of the input signal. The input signal may swing into the nonlinear region of the transistor characteristic family of curves and produce severe distortion. Figure 19-1 shows examples of six types of distortion.

Hum

Another type of distortion is **hum**. Hum may be generated by a number of sources. The amplifier power supply is one source. The hum occurs when the power is not properly filtered. The variations in power are then amplified and produce a very objectionable hum level in the amplifier output. Hum may be caused by stray electromagnetic or electrostatic fields. It is usually a low-frequency disturbance.

IM (Intermodulation Distortion)

This is one of the most frequently discussed types of distortion when the purchase of an audio amplifier for home use or for reproduction of recorded music is contemplated. **Intermodulation distortion** is caused by the presence of two or more sine waves beating against one another. A heterodyning effect is produced. This means the new sine wave is the resultant (sum or the difference) of the two sine waves. That means that four instead of two sine waves may now be present. This may cause some very unwanted side effects in the amplifier's output. Intermodulation is found in all amplifiers. However, it is most generally found at the lower end of the frequency range. A distortion analyzer is used to measure the amount of intermodulation of an amplifier.

Classes of Operation

Transistor amplifiers are classified according to their operating point as Class A, Class B, or Class C.

Class A Amplifier

The Class A amplifier conducts all the time. Inasmuch as audio frequencies need both the negative and positive halves of the waveform amplified, this type of amplifier produces a high quality of amplification. A Class A amplifier is shown in Figure 19-2, with the resultant waveform. Efficiency is approximately 15%.

Class B Amplifier

This type of amplifier is used to drive a speaker in an audio amplifier. It conducts only when a positive signal

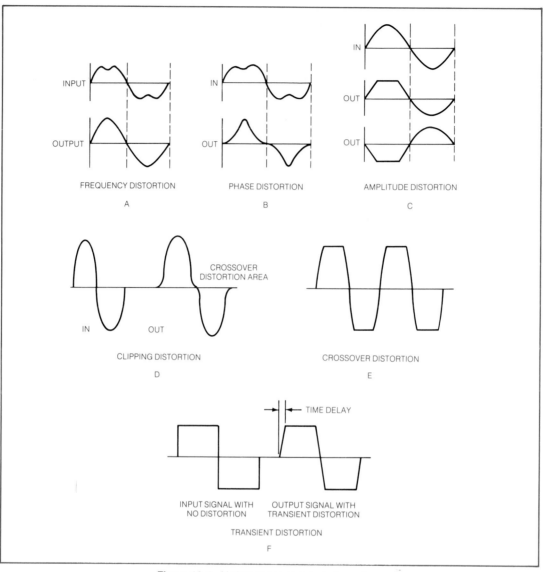

Figure 19-1. Various types of distorted waveforms.

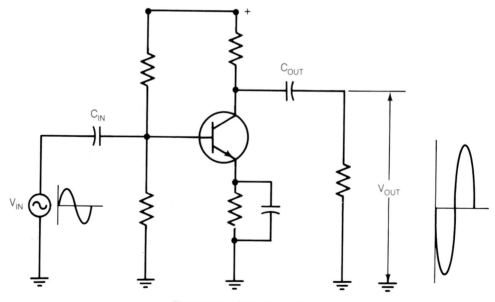

Figure 19-2. Class A amplifier.

is applied. A typical Class B push–pull amplifier is shown in Figure 19-3. This push–pull amplifier arrangement means that one transistor conducts on the positive half of the signal and the other conducts on the negative half. That means the transistor can be driven at a higher power output since it conducts only half of the time. Efficiency is approximately 75%.

Class C Amplifier

This type of amplifier circuit, shown in Figure 19-4, conducts only during a small portion of the input signal. It is used in conjunction with tuned circuits to restore the rest of the signal. The flywheel effect serves to put back the other part of the signal. It is designed so that the bias

operating point is below cutoff. With a sine wave signal applied to the input, the output is less than half of one alternation. The Class C amplifier is one in which the collector signal flows for less than one-half of the input-signal cycle. Today, this type of amplifier is used primarily as a radio-frequency amplifier and for providing energy to oscillators or switching circuits. The operational efficiency of this amplifier is quite high (95%). It consumes energy for only a small portion of the applied sine-wave signal.

Note the capacitor C_N. This is a *neutralizing* capacitor needed in Class C amplifiers to prevent them from becoming oscillators and creating frequencies independent of those being amplified. The capacitor is adjusted to compensate for the internal capacitance of the transistor.

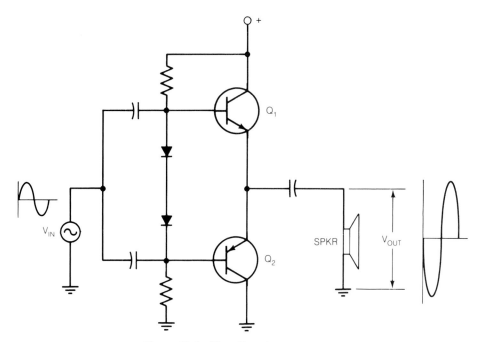

Figure 19-3. Class B push–pull amplifier.

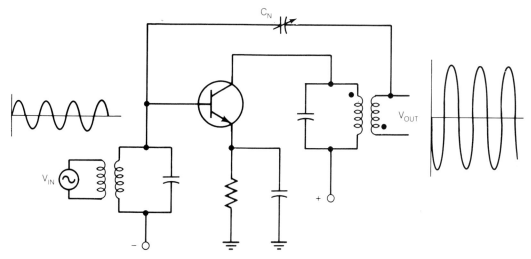

Figure 19-4. Class C amplifier.

Other Classes of Operation

Type AB operation is used in amplifiers adjusted half-way between Class A and Class B operation. The input–output waveforms of Class AB_1 and AB_2 are shown in Figure 19-5. This class of operation is often used in small portable transistorized radios to drive a speaker that is small and not too discerning of the waveform shape.

Class D operation is the use of two amplifiers as switches in push–pull operation. The stage goes between saturation and cutoff. This produces an efficiency of over 100%.

Class E operation uses a high impedance load. The load may be an RF choke. This allows the device to be in saturation for 180° of the input signal and improves the efficiency of this single-stage amplifier.

Class F operation is a single-stage amplifier that acts mostly as a switch. That means it has almost 100% efficiency. The output resembles a square wave. Two sets of tuned circuits are used. One tuned circuit removes the third harmonic and the other passes on the fundamental frequency.

Class S is used in switching regulators.

REVIEW QUESTIONS FOR SECTION 19.1

1. What is distortion?
2. What is hum?
3. Define *intermodulation distortion*.
4. How efficient is a Class A amplifier?
5. How efficient is a Class B amplifier?
6. Where is the bias point in a Class C amplifier?
7. What is neutralization?
8. Where is Class AB used?
9. Where is Class E used?

19.2 TRANSISTOR AMPLIFIERS

There are three types of transistor amplifiers in general use: the common-emitter, the common-base, and the common-collector amplifier. Each type has its own characteristics and advantages in terms of usage in power amplifiers.

Common-Base Amplifier

The connections made to the transistor shown in Figure 19-6 show that it is a common-base-type amplifier stage. The current flow through the NPN transistor is shown by arrows. About 95% of the emitter current reaches the collector. In practical circuits, the flow is 92 to 98%. The use of 95% is a compromise. Note that the remainder (5%) flows through the base.

Common-Emitter Amplifier

The common-emitter amplifier is shown in Figure 19-7. Note the current flow indicated by the arrows. The input signal of the common-emitter amplifier stage is 180° out of phase with the output signal that appears across R_1.

Common-Collector Amplifier

The common-collector is shown in Figure 19-8. This circuit is useful in impedance matching. It does not amplify, but acts as an impedance-matching device. Note that the signal is not reversed but is in the same phase as the input. This can be visualized easily by noting that the output signal is taken from the emitter resistor R_1.

Coupling of Transistor Amplifier Stages

The RC coupling network can be used on almost any size amplifier, from low-level preamps to high-level

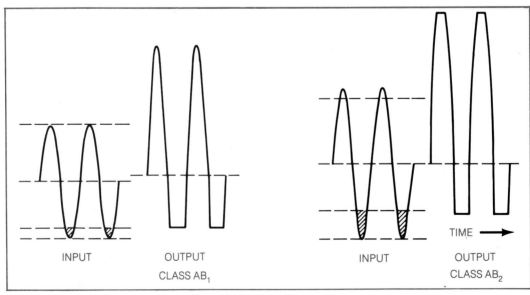

Figure 19-5. Class AB operation.

INPUT   OUTPUT

CLASS AB_1

TIME ⟶

INPUT   OUTPUT

CLASS AB_2

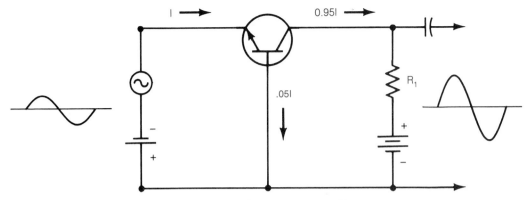

Figure 19-6. Common-base amplifier.

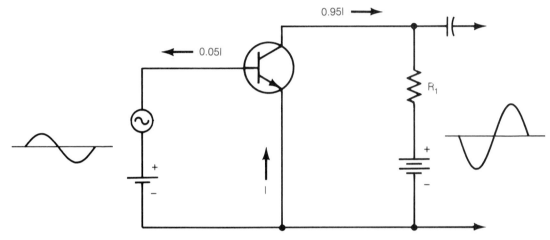

Figure 19-7. Common-emitter amplifier.

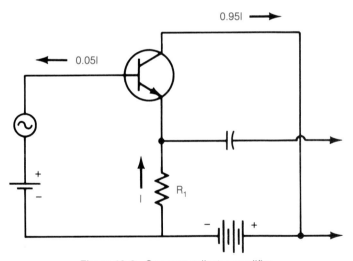

Figure 19-8. Common-collector amplifier.

Transformer coupling is another type of coupling used to connect stages in transistor circuits, Figure 19-10. There are certain advantages to this type of coupling. The very low resistance in the base path aids temperature stabilization of the DC operating point. With a swamping resistor in the emitter lead, the current stability factor is ideal. Because there is no collector load resistor to dissipate power, the power efficiency of the transformer-coupled amplifier approaches the theoretical maximum of 50%. For this reason, the transformer-coupled amplifier is used extensively in portable equipment in which battery power is used.

There are many disadvantages to the use of a transformer as a coupling device. For instance, it is heavy, bulky, and expensive. It also experiences frequency response reduction as compared with the RC coupling. The transformer is rapidly being replaced in audio-amplifier circuits because of expense, frequency-response falloff, and weight and size.

The impedance-coupled amplifier in Figure 19-11 uses L_1 as the load for the input transistor. This inductive load is shunted by the capacitor–resistor combination C_1R_1. The main advantage of this arrangement is that it provides

amplifiers, Figure 19-9. The use of RC coupling in battery-operated equipment is usually limited to low-power operation to limit battery drain. RC coupling is used extensively with junction transistors because of high gain, economy of component parts, and good utilization of board space.

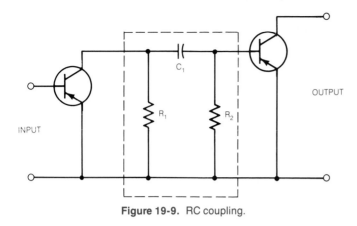

Figure 19-9. RC coupling.

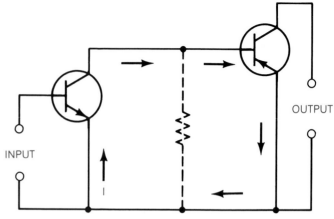

Figure 19-12. Direct coupling.

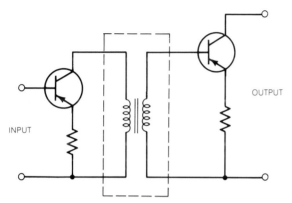

Figure 19-10. Transformer coupling.

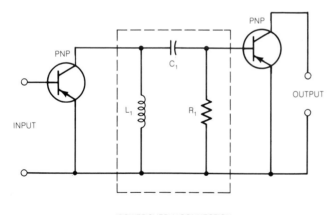

(POWER SUPPLY CONNECTION
NOT SHOWN FOR SIMPLIFICATION PURPOSES)

Figure 19-11. Impedance coupling.

fier is better than that of the transformer-coupled amplifier, but not as good as that of the RC-coupled amplifier.

The direct-coupled amplifier, Figure 19-12, is used to amplify low frequencies of DC signals. Two types of transistors are used in the circuitry. The NPN is connected directly to the PNP-type transistor. Current flow is shown by the arrows. If the collector current of the first stage is larger than the base current of the second stage, a collector resistor (shown in dotted lines) must be used.

Very few parts are used in the directly coupled amplifier, and this makes its production less expensive. There are, of course, limitations on the number of stages that can be directly coupled. Inasmuch as the temperature variations of the bias current in one stage is amplified by all the stages, there is severe temperature instability with this type of coupling.

REVIEW QUESTIONS FOR SECTION 19.2

1. In common-base circuitry, how much of the emitter current reaches the collector?
2. For what is the common-collector circuit used?
3. Define *RC coupling*.
4. What is the main advantage of the impedance-coupled amplifier?
5. Where is the direct-coupled amplifier used?

19.3 INTEGRATED-CIRCUIT AMPLIFIERS

Integrated circuits are also known as **chips** and **ICs**. The chip is a circuit that is integrated, in most instances, with other components to provide a particular arrangement for a specific purpose. An entire amplifier can be made on one chip. A number of transistors and diodes can be arranged, along with the proper capacitance and resistance, to produce a complete circuit without having to

high power efficiency since the DC voltage is not dropped across a load resistor.

Low-frequency response is reduced by the shunt reactance of the inductor. The high-frequency response is reduced by the collector capacitance. Unlike the transformer-coupled amplifier, the impedance-coupled amplifier suffers no loss of high frequencies by leakage reactance. The frequency response of the impedance-coupled ampli-

make the circuit on a board. By chemically etching the silicon, it is possible to place all of the components on a spot no larger than the head of a pin. One type of chip is the op-amp.

Operational Amplifiers (Op-Amp)

One of the most common linear circuits is the operational amplifier. The **op-amp** is an integrated circuit that is classified as a linear or digital amplifier. Several characteristics make the op-amp useful in consumer products. These are its high open-loop gain, high input impedance, low output impedance, and ability to reject unwanted signals.

Differential Amplifiers

The op-amp can be used as a **differential amplifier**, Figure 19-13. This op-amp has inverting and noninverting inputs. Keep in mind that the ideal op-amp has infinite gain, infinite input impedance (open circuit), and zero output impedance. The op-amp unit is basic to such analog circuits as the summing amplifier, integrator, and inverter. Note the symbol for the op-amp with two inputs shown in Figure 19-14. Polarity of the inputs is designated by the use of − and +. A signal applied to the plus input operates in phase and is amplified at the output, whereas that applied to the minus input is amplified but inverted at the output. Although the basic op-amp circuit has very

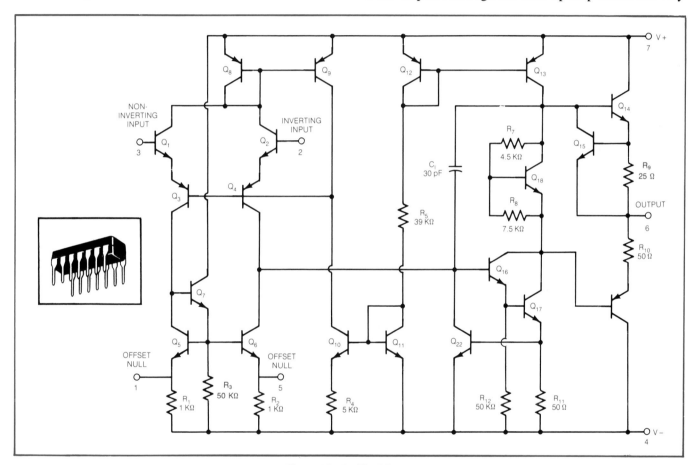

Figure 19-13. The 741 op-amp.

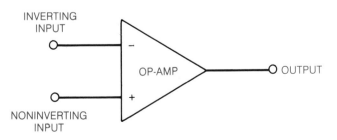

Figure 19-14. Op-amp symbol.

high gain, the useful connection of an op-amp circuit has a gain that is set by an external resistor.

With a few added resistors and capacitors, the op-amp can be used as a final audio-output amplifier in radio receivers. Radios made today utilize an op-amp for the audio-output stage. This amplifier section is capable of delivering, in some cases, 5 watts to the speakers. There must be heat sinks on the IC to prevent overheating, Figure 19-15.

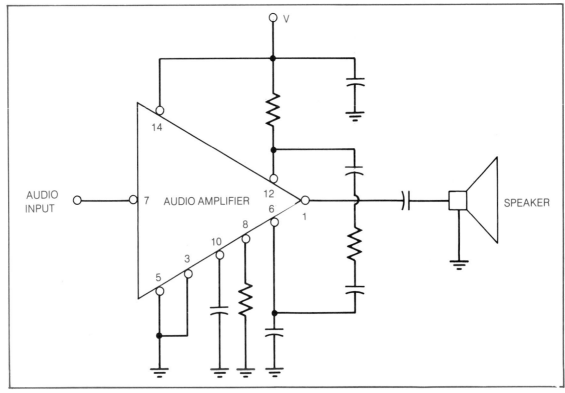

Figure 19-15. Linear amplifier as a final audio amp.

Different Forms of Op-Amps

Analog computers use a basic form of the op-amp circuit. It is an inverting constant-gain multiplier, shown in Figure 19-16A. The op-amp is also capable of operating as a noninverting constant gain amplifier, Figure 19-16C. It is also capable of being used as a unity follower that has a gain of 1, or unity. This means it undergoes no polarity reversal and the output is of the same polarity and magnitude as the input. The circuit acts the same as the emitter-follower circuit in transistors except that it has the advantage of the gain being set much closer to unity, Figure 19-16B. This means the circuit can be used for impedance matching. In an integrator circuit, the summing unit alone allows addition or subtraction operation. An integrator circuit is needed in solving differential equations, Figure 19-16C. The constant-gain multiplier can be converted to an integrator circuit by use of a capacitor as a feedback element, rather than by use of a resistor, Figure 19-17.

19.4 OTHER TYPES OF AMPLIFIERS

Amplifiers are used in many electronic devices. They may be utilized to amplify audio frequencies or radio frequencies. Each type of amplifier has its own characteristics and each is made to operate within its physical and electrical limitations in order to produce the quality of amplification and signal strength needed for a particular application. Radio-frequency stages are usually coupled in a manner somewhat different from that of the audio amplifer.

Tuned-Circuit Coupling

Radio-frequency stages are sometimes coupled by tuned stages. The IF amplifier in receivers use this type of circuit, Figure 19-18.

The induced voltage in the secondary circuit is considered to be in series, since it is generated by the reaction of the coil to the magnetic field. The secondary circuit is not tuned to resonance. The current that flows is in phase with the induced voltage. The voltage and current induced is 180° out of phase with those in the primary. Transformer coupling has some advantages. The resonant conditions of the tank circuit result in a gain in signal voltage that is very selective. This type of coupling is frequently used for intermediate-frequency (IF) amplifiers in receivers and in output stages of trans-

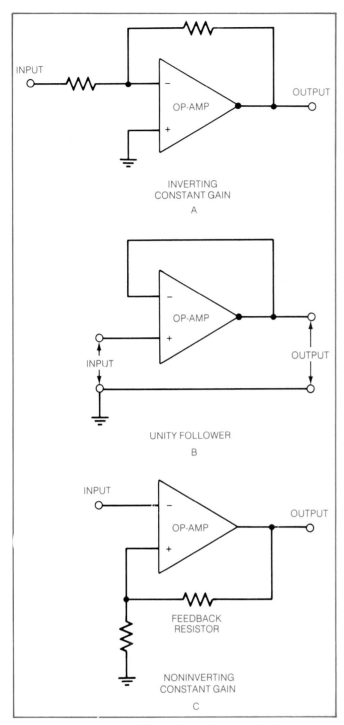

Figure 19-16. Different forms of op-amp circuitry.

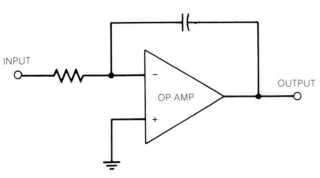

Figure 19-17. Op-amp used as an integrator.

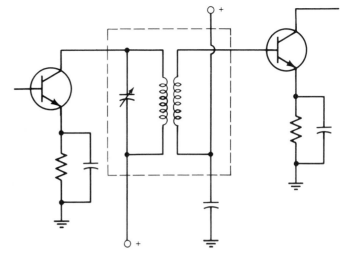

Figure 19-18. Transformer coupling.

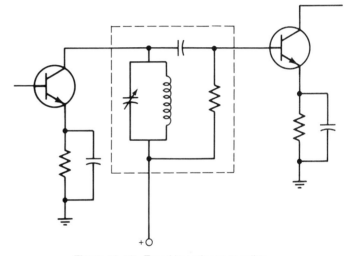

Figure 19-19. Tuned-impedance coupling.

mitters. Figure 19-19 shows a single-tuned stage with RC coupling to reduce the number of tunable capacitors and inductors. This is known as impedance coupling since it uses a resistor, capacitor, and inductor.

Voltage Amplifiers

RF amplifiers are primarily users of tuned circuits. This means that either one or two tuned circuits are used in each of the RF stages of a radio receiver. In a transistor circuit, such as that shown in Figure 19-20, you can see the tuned circuits in both the input and output of the stage. This type of circuit must develop maximum signal at the tuned frequency and must also match the input and output impedances. Note how the tuned circuits are tapped for connection to the base and collector. These are not always center taps, but are placed where they will cause an impedance match for the particular transistor being used.

Power Amplifiers

Power amplifiers are usually needed in transmitters when a weak signal must be boosted to sufficient strength

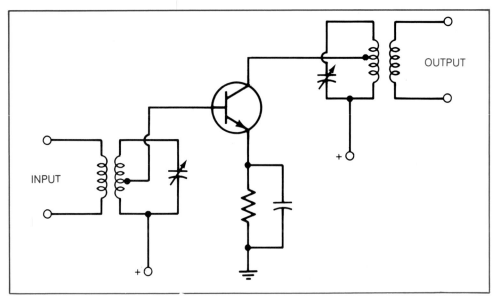

Figure 19-20. Tuned circuits.

to drive an antenna and be sent over many miles. That calls for classes of operation that can handle the large amounts of power. This is where Class C and Class D amplifiers are utilized to their best advantage. Inasmuch as they do not operate at all times during the input signal, they have the ability to handle more power for a short time. Components can also be obtained to operate under these conditions at a lower cost than if full-time operation were required.

Power amplifiers usually require larger input signals. Inasmuch as they usually operate as Class B or Class C, current flows for only part of the input signal. The output is sinusoidal, however, since the tank circuit has a coil whose magnetic field will collapse after the first portion of the signal flows through it. The magnetic-field collapse puts back the energy in almost the same amount required to make the field. Therefore, the output is of the same shape as the input signal.

Class D amplifiers are used with pulse-width modulation (PWM), Figure 19-21. The efficiency of this type of amplifier is as high as 95%. When used as an RF amplifier, it is usually operated as Class C. This means that for most of the input cycle only a small amount of cutoff current flows in the collector circuit. In Figure 19-22, the circuit shown is a grounded-emitter type. It conducts only during the positive peaks of the input signal, and then only in bursts. The bursts of input signal are converted in the tank circuit to sinusoidal output by the nature of the tank. Note that the transistor is biased in reverse of what would normally be the case. This is done to achieve a bias point considerably beyond cutoff and to make the device operate as a Class C amplifier. The emitter is biased positive with respect to the base.

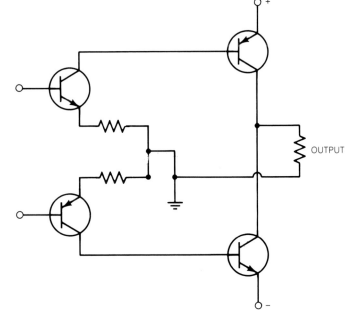

Figure 19-21. Class D amplifier.

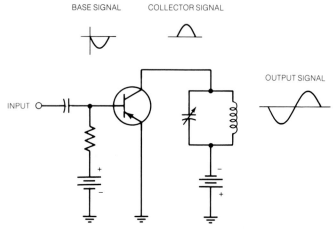

BASE SIGNAL   COLLECTOR SIGNAL

OUTPUT SIGNAL

INPUT

Figure 19-22. Reverse biased base-emitter circuit.

19.5 FEEDBACK IN AMPLIFIERS

There are two types of **feedback** used in amplifiers: regenerative and degenerative, which are sometimes called positive and negative feedback. In some instances, it is desirable to feed back part of the output of an amplifier to the input. This fed-back signal can take two forms— either negative or positive. *Negative and positive* refer to the phase of the signal in reference to the input signal. If the feedback *aids* the input signal or adds to it then it is called *positive* feedback. If the fed-back signal is out of phase with the input signal, it causes a decrease in the amplitude of the input signal and is called degenerative or negative feedback.

Figure 19-23 shows how the regenerative and degenerative signals are affected. To produce the regenerative feedback, the original feedback signals must be in phase with each other. Adding these two waveforms produces the regenerative signal. It is larger in amplitude than the original signal. The original and feedback signals are in opposite phase in the degenerative-feedback configuration. The waveform is smaller in amplitude than the original signal.

Positive feedback produces an increase in the amplifier-stage output since it added to the input signal. This, however, does have some negative aspects. The greater amplification increases the amount of distortion and noise in the amplifier. In some instances, the amount of regeneration is so great it produces sustained oscillations. In negative feedback, the voltage gain of an amplifier is decreased because the effective input voltage is decreased. Practical applications of this type of feedback make use of this characteristic to reduce the effects of distortion. Degenerative feedback also improves the frequency response and stability of amplifiers.

Neutral feedback (or neutralization) is obtained by use of a capacitor to feed back part of the output signal. This capacitance across the transistor has a tendency to cancel out the effect of the capacitance between the elements of the device. Without neutralization, the amplifier becomes an oscillator, since it can obtain a path of feedback through the transistor or amplifier device itself.

19.6 FET TRANSISTORS

The field-effect transistor (FET) has high input impedance when compared with the low input impedance of the bipolar transistor. The FET operates on low DC supply voltages. This type of transistor is used in many consumer devices because it is lightweight, rugged, and very small.

The FET has three terminals labeled **gate**, **drain**, and **source** (as opposed to the *base*, *collector*, and *emitter* of regular transistors). The drain (D) is also called the anode, and the source (S) is also called the cathode. The gate (G) is the current-controlling connection, Figure 19-24.

The FET can be further classified as the junction FET (**JFET**) and the metal-oxide semiconductor (**MOS**). In some cases, the **MOSFET** is also referred to as the insulated gate FET, or IGFET.

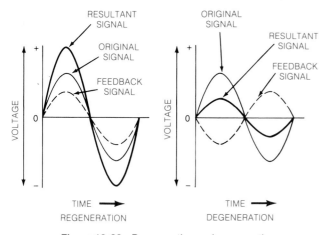

Figure 19-23. Degeneration and regeneration.

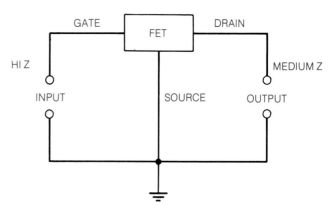

Figure 19-24. FET circuit input and output connections.

The junction FET is classified as to the type of material used to make its channel. The channel can be of either P-type or N-type material. However, the MOSFET is classified according to its *mode* of operation—that is, its **depletion mode** or **enhancement mode**. Figure 19-25 shows how the depletion-mode and the enhancement-mode MOSFET are biased.

The MOSFET is the most commonly used type of FET. It was originally designed to replace the vacuum-tube-type circuit. The MOSFET has the characteristics of the tube in regard to input and output impedances. The MOSFET also has a drain, source, and gate, and operates in much the same way as the JFET. The drain is connected to the positive-voltage power source and the source is connected to the negative-voltage source. The substrate that makes up the transistor is connected to the source voltage supply. The device is biased to set up an electron flow between the source and the drain. Current flows through the narrow channel created by the substrate, as shown in Figure 19-26.

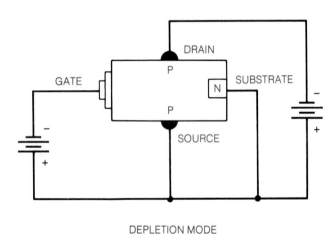

DEPLETION MODE

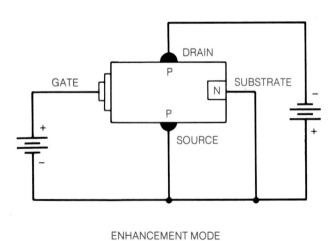

ENHANCEMENT MODE

Figure 19-25. MOSFET depletion and enhancement modes.

Enhancement Mode

Examine Figure 19-25 and note that the gate of the MOSFET is insulated from the channel. That means that a negative or positive voltage can be applied to the gate. A positive voltage is applied in the enhancement mode of operation, and a positive voltage is applied to the gate. In the depletion mode, the gate, insulator, and channel act as a capacitor. Note, however, that in this case the gate has developed a positive charge. This means that the channel has a negative charge. Negative charges that develop in the channel are current carriers in the N-material. They improve the conditions, so more electrons reach the drain. That means the current increases and the current flow in the channel is *enhanced*. As the gate voltage becomes more positive, it increases the current flow through the drain.

Depletion Mode

Keep in mind how a capacitor functions when examining the FET mode of operation, Figure 19-25. The gate and N-type material are similar to the two plates of a capacitor. The metal oxide is the dielectric material between the plates. When a negative voltage is applied to the gate, a negative charge is developed on the gate. Note that this assembly acts as a capacitor, and that the other plate develops a positive charge. That means the positive plate creates a *depletion* area. That in turn restricts current through the narrow channel. The more negative the gate voltage, the wider is the depletion region. When enough negative voltage is applied to the gate, current between the source and the drain can be cut off. The N-channel is depleted of electrons by this action of the positive charge. This is the type of operation that gives the MOSFET its depletion mode of operation.

Some FET Circuits

Transistor circuits are classified as **common-base**, **common-emitter** and **common-collector** types. FET circuits are classified as **common-source**, **common-gate**, and **common-drain** types, Figure 19-27, page 308.

Common-source is the most common type. It has high input impedance, a medium-to-high output impedance and a voltage gain greater than one.

The *common-gate* type has medium input impedance. It also has a high output impedance. It can operate at high frequencies. This type of amplifier offers a low gain to the signal. It does not need neutralizing.

The *common-drain* type is also called a source follower. The name comes from the vacuum-tube-type circuit that was a cathode follower and was used for impedance-

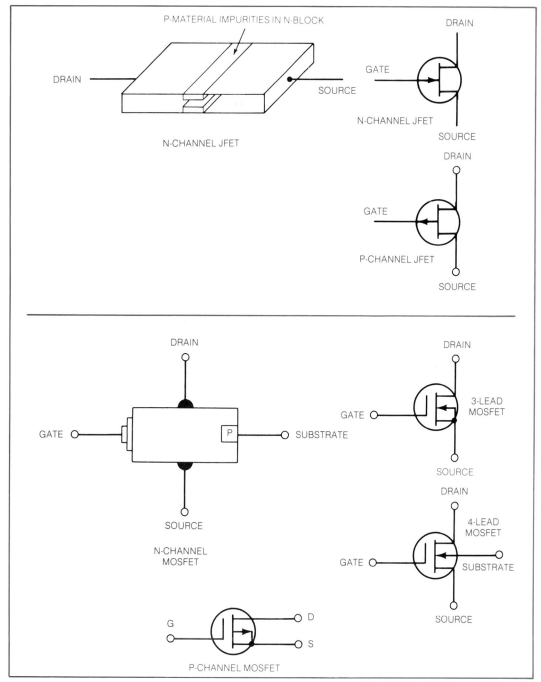

Figure 19-26. N-channel JFET and N-channel MOSFET.

matching purposes with no amplification. Input impedance of this type of circuit is high—much higher than for the common source. The output impedance is very low. There is no signal shift between the input and output. Voltage gain for this type of amplifier is less than one, and it too is used for impedance-matching purposes.

1. What does *FET* stand for?
2. Where is the FET transistor used?
3. What does *MOS* mean?
4. What does *JFET* mean?

5. What does *MOSFET* mean?
6. What is the enhancement mode?
7. What is the depletion mode?
8. List three classifications of FET circuits.

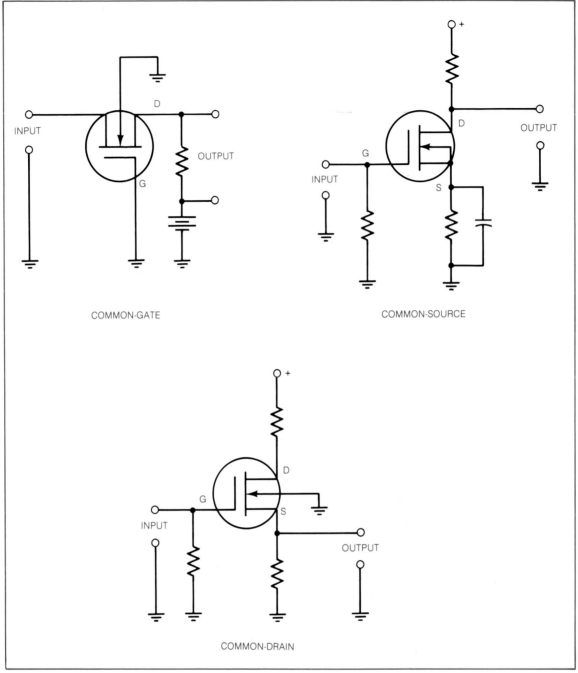

Figure 19-27. Common-gate, common-source, and common-drain configurations.

19.7 PREAMPLIFIERS

Turntable signals, tape-head signals, and microphones need to be amplified *before* they can be heard or used. Because this amplification is required before the signals are applied to an amplifier stage, the signal-boosting stages are referred to as *pre*amplifiers. The preamplifier amplifies weak signals so they are strong enough to drive a regular-power-amplifier stage, or does so before such signals can be used to drive a speaker, headsets, or some other device.

For a good example of the preamplifier, examine Figure 19-28. The schematic shows Q_1 and Q_2 as the two transistors used to amplify the input from the tape head, microphone, or auxiliary input. Q_3 is the phase-splitter or driver stage, since it drives the transformer that causes part of the input signal to Q_4 to be different from that presented to the base of Q_5. The first two transistors are needed to amplify the signal sufficiently to cause Q_3 to operate properly. The push–pull operation of Q_4 and Q_5 are sufficient to drive the speaker connected to the transformer T_2. This complete circuit is typical of a portable

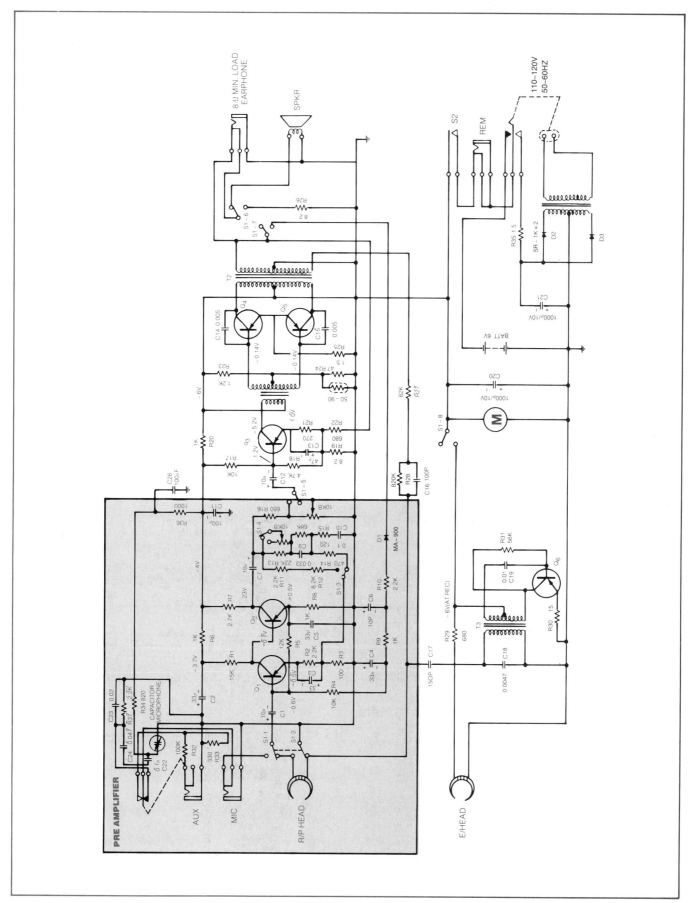

Figure 19-28. Portable-tape-recorder schematic.

tape recorder. Note how the switches S1-1 and S1-2, are attached to the same shaft and are moved at the same time. As shown, they are in the tape position. When they are up, they put the microphone jack into the circuit. When no outside microphone is attached, the capacitor microphone is in the circuit.

This schematic shows a *single* preamplifier with one tape head, one microphone jack, and one auxiliary jack. A stereo recorder has twice the equipment this one does. It requires two tape heads, two microphone jacks, and two auxiliary jacks for other inputs, such as turntables, and a duplication of the five transistors and their associated parts. There will, of course, be the need for two speakers to produce the separate signals to give the stereo effect.

Also keep in mind that one chip can serve as the preamplifier and the power amplifier, and that, in low-power applications, they are all located in one integrated circuit.

___ REVIEW QUESTIONS FOR SECTION 19.7 ___

1. What is the purpose of a preamplifier?
2. Where are preamplifiers used?

19.8 POWER AMPLIFIERS

Power amplifiers are available in many variations for various functions. They may be used to drive speakers, relays, and other equipment. This discussion is limited to the amplifiers used for stereo. At least two amplifiers are needed for stereo, and they have to be separate units. Two or more sources are needed to produce the stereo effect. Some stereo systems use up to seven or eight amplifiers, with a corresponding number of signals and speakers.

Today, integrated circuits (ICs or chips) are used to make the power amplifiers more compact and less power-hungry. They are able to produce large amounts of output power with a higher efficiency than ever before. However, some circuits still call for heat dissipation at such a level that heat sinks have to be added to remove the heat before it damages the IC or individual transistors, Figure 19-29.

Push-Pull Power Amplifier and IC Amplifier

Figure 19-30 shows two types of **push-pull amplifiers**, one with transformer and the other without. In push-pull operation, the input transformer T_1 receives the sinusoidal voltage from a low-level source. That means the signals applied to the two transistors are 180° out of phase. The output transformer T_2 delivers to the load a current that is proportional to the difference of the two collector

currents. Any even-harmonic distortions tend to cancel, and the only distortion is that due to odd harmonics. Also, the performance of the transformer T_2 is improved because the DC components of I_{C1} and I_{C2} are canceled. Magnetic-core saturation and the accompanying nonlinearity are avoided. The push-pull circuit is particularly useful for Class B operation. Since the heat losses in a transistor cannot exceed the rated heat dissipation, two transistors in Class B can supply nearly six times the power output of one similar transistor in Class A. The expensive and heavy transformer can be eliminated by use of a PNP and an NPN transistor in a complementary circuit, such as that shown in Figure 19-30B. In *IC*'s of this type, the resistors R_1 and R_2 are replaced by diodes whose forward voltages *track* the base-emitter voltages of the transistors, Figure 19-3, page 297. This means the high efficiency of Class B can be achieved at low cost.

An operational amplifier IC that has 20 transistors and 10 resistors all in one package is shown in Figure 19-13, page 301.

___ REVIEW QUESTIONS FOR SECTION 19.8 ___

1. What is the smallest number of amplifiers and speakers needed for the stereo effect?
2. Describe what *push-pull* means.

19.9 SPEAKERS

Speakers are transducers that are employed to change electrical energy to sound energy so that it can be heard by the human ear. Many improvements have been made

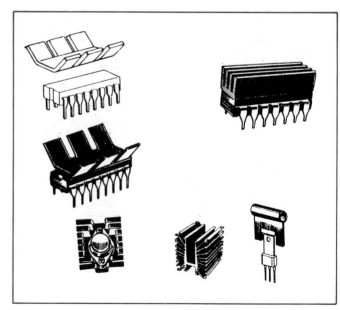

Figure 19-29. Heat sinks for ICs and transistors.

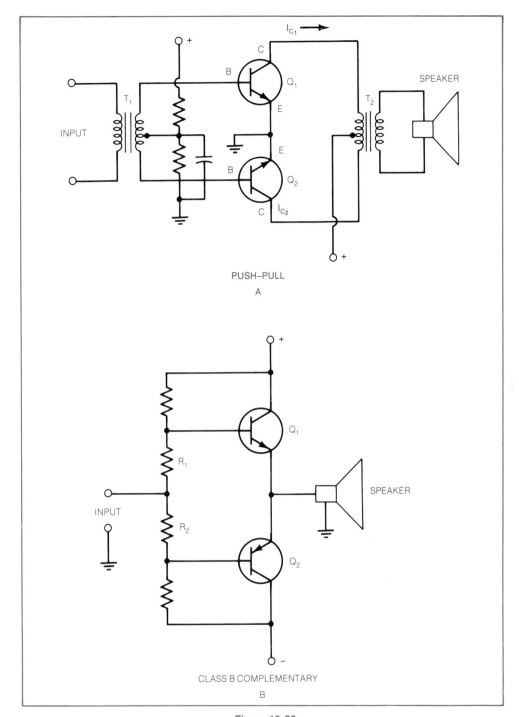

Figure 19-30.
A. Push–pull amplifier. **B.** Class B complementary amplifier.

on, the basic permanent-magnet (PM) speaker in recent years, with some attention being given to various other methods of reproducing sound. The permanent-magnet speaker consists of a large permanent magnet that sets up a north-and-south polarity. The voice coil has audio-frequency energy passing through its coils of wire. This varying energy produces a varying magnetic field in step with the frequency variations. That means that the voice coil is attracted or repelled by the instantaneous polarity produced by the energy passing through it. Inasmuch as the voice coil is attached to the speaker cone, it causes the speaker cone to vibrate or be moved in and out according to the attractions and repulsions. The moving cone com-presses the surrounding air according to its movements.

This *compression and rarefaction* of the air by the cone produces variations that can be heard by the human ear as it reacts to the variations in air pressure.

Note the *spider*, Figure 19-31. It keeps the voice coil centered around the permanent magnet, but allows back-and-forth motion of the voice coil and cone. It also returns the coil to its original resting position when no current is flowing through the coil.

Enclosures

Speaker enclosures come in many sizes and shapes. There is the flat baffle, the **infinite baffle**, and the ducted-port or bass reflex. Others are constantly being developed. Visit a local stereo store to keep up with the latest. Magazine articles are released periodically when new advances or new sounds are possible with speaker enclosures. It is the enclosure that makes the speaker behave the way you want it to. If you want to accentuate the highs or lows or the in-between frequencies for a particular application, you must use the proper enclosure and **crossover network** to make sure the right range of frequencies reaches the right speaker in the enclosure.

Crossover Networks

Crossover networks are just what their name implies. The *crossover* refers to speakers in an enclosure. If there are 4-inch, 8-inch, and 12-inch speakers in an enclosure, each speaker should be fed the frequencies it is best at reproducing. The 12-inch one can handle the low frequencies (of about 30 to 300 Hz). The 8-inch one can better handle the mid-range of frequencies (300 to 3000 Hz). The 4-inch one is better equipped physically to handle

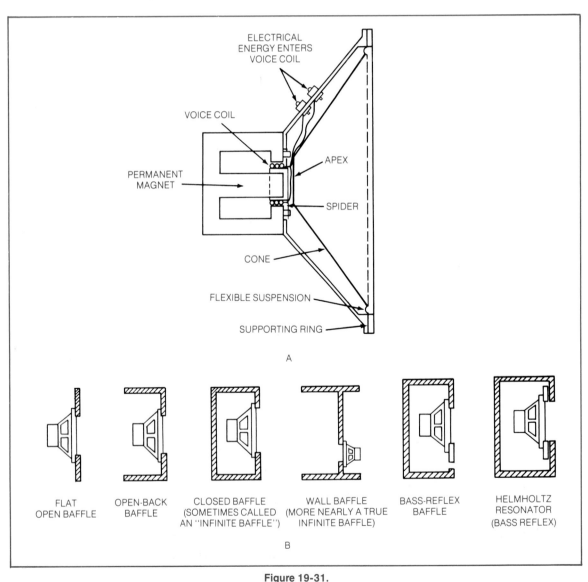

Figure 19-31.
A. Cutaway of PM speaker. **B.** Speaker enclosures.

those frequencies above 3000 Hz. The physical nature of the speakers is such that many of these frequencies will naturally be reproduced by each. However, for higher efficiency of each, there is a possibility of making sure each receives its best range of frequencies. That is where the crossover network is utilized, Figure 19-32.

There are a number of ways to hook up a stereo amplifier to its outputs and inputs. Each should come with its own labeled instructions.

REVIEW QUESTIONS FOR SECTION 19.9

1. What is the spider in a speaker?
2. For what is the speaker enclosure used?
3. What is a crossover network?
4. Where is a crossover network used?

Safety Tips

1. Keep the volume down on amplifiers. The output may cause damage to the ears if it is not properly controlled.
2. Speaker enclosures can be large and bulky. Be sure to lift using the knees instead of the back muscles.
3. Power amplifiers used for public-address systems have a higher voltage output than do other self-contained units. Do not connect remote speakers unless the amplifier is turned off.
4. Do not operate speakers under water unless the proper type is obtained and the wiring installation meets with the *National Electrical Code®* specifications.

SUMMARY

Amplifiers are used to boost or amplify weak signals. Distortion is one of the results of amplification. Distortion is the type of change a signal undergoes from the time it enters the amplifier state until it comes out. Hum is another type of distortion. It may be generated by a number of sources. Intermodulation distortion is produced by a heterodyning effect. This means the new sine wave is the resultant of the two sine waves present or is the difference of the two sine waves. It is found at the low end of the frequency range.

When transistor amplifiers are classified according to their operating point, they are referred to as Class A,

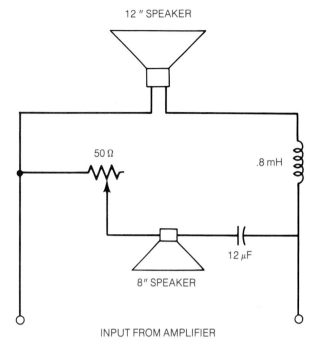

Figure 19-32. Crossover network.

Class B, or Class C. Other types of operation include AB and D, as well as E, F, and S.

There are three types of transistor amplifiers: common-emitter, common-base and common-collector. Each type has particular characteristics of operation due to the arrangement of its elements in a circuit.

Integrated-circuit amplifiers are also known as ICs and chips. An entire amplifier can be made on one chip. Many transistors, diodes, resistors, and capacitors can be placed on a single chip.

Operational amplifiers (op-amps) are among the most common linear circuits. They are packaged in dual-inline packages. The op-amp can be used as a differential amplifier for inverting or noninverting purposes. There are different forms of op-amps.

Tuned-circuit coupling can be used in radio-frequency amplifier stages to make sure the signal is properly coupled from one stage of amplification to another. Transformer coupling, impedance coupling, and RC coupling may also be used to couple the output of one stage to the input of another.

Power amplifiers are usually needed in transmitters where a weak signal is boosted to sufficient strength to drive an antenna and be sent over many miles. Class C and Class D operation is usually needed for power amplifiers. Power amplifiers usually require larger input signals than do voltage amplifiers. Class D amplifiers are used with pulse-width modulation (PWM).

There are two types of feedback in amplifiers: negative and positive.

FET transistors (field-effect transistors) have high input impedance compared with the low input impedance of the bipolar transistor. The terminals are labeled *drain, source,* and *gate.* The FET can be classified as *JFET,* or *junction FET,* or *MOSFET,* which means *metal-oxide semiconductor FET.*

FETs may operate in the enhanced mode or the depletion mode. Each has its own characteristics. FET circuits are classified as common-drain, common-source, and common-gate.

Preamplifiers are used to boost signals so they can drive a power amplifier. Power amplifiers utilize the push–pull configuration or a transistorized version of the arrangement.

Speakers are used to reproduce signals that amplifiers have boosted to the level at which human ears can hear them. Crossover networks are utilized to make sure the right frequencies reach the correct speaker in an enclosure.

_____ USING YOUR KNOWLEDGE _____

1. Draw a distorted waveform.
2. Draw a common-collector transistor amplifier circuit.
3. Draw a common-emitter transistor amplifier circuit.
4. Identify the leads from a FET transistor.
5. Draw a schematic for a push–pull amplifier circuit.

_____ KEY TERMS _____

amplifier	MOSFET
distortion	op-amp
hum	differential amplifier
intermodulation distortion	tuned circuit
chip	voltage amplifier
integrated circuit (IC)	power amplifier
common-base	feedback
common-emitter	enhancement mode
common-collector	depletion mode
common-drain	push–pull amplifier
common-source	crossover network
common-gate	speaker enclosure
JFET	infinite baffle

Chapter 20

ELECTRONIC COMMUNICATION SYSTEMS

After studying this chapter, you will know

- *how an AM receiver works.*
- *how an FM receiver works.*
- *the difference between AM and FM.*
- *how heterodyning operates.*
- *how to draw a block diagram of AM, FM, and TV receivers.*
- *how to draw the signals utilized in PWM, FSK, and PPM.*
- *how satellite TV works.*
- *the value of antennas in both receivers and transmitters.*
- *how telephone, telegraph, and facsimile systems are used to communicate.*
- *how computers are connected to phone lines.*
- *where amateur radio fits into the communications system.*
- *how light waves are used to communicate by means of fiber optics.*

INTRODUCTION

Electronic communication systems are constantly being upgraded, changed, and improved. The rapid development of electronics is well documented and evident in the new devices on the market each year. The advent of the transistor and the integrated circuit caused a great leap forward, bringing the world closer together. Whatever happens anywhere in the world can also become public knowledge within minutes of its occurrence by means of satellite communications coupled with telephone, radio, and television broadcasting.

20.1 RADIO FREQUENCY

Frequencies that can produce sound waves that can be heard by humans are considered to be audio frequencies (AF). These usually range from 15 Hz to 20 kHz. Frequencies that can be fed to antennas and will radiate electromagnetic and electrostatic waves are called radio frequencies (RF).

Radio frequencies (RF) are typically upward of 16,000 Hz. RF can be used to carry other frequencies from one location to another. Information of all types can thus be transmitted on the carrier waves. Television utilizes radio frequencies to send its pictures from the transmitter to the receiver in the home or office. Radio waves are used to send any number of messages and data from point to point. Inasmuch as radio frequencies are so important in electronic communications systems, we will take a closer look at this phenomenon in this chapter.

RF Spectrum

In 1857, Heinrich Rudolf Hertz (for whom the frequency unit was named) was born in Hamburg, Germany. He began an engineering career, but was fascinated by the appeal of pure science and went on to become a physicist. He was a professor of physics at Karlsruhe Polytechnic between 1885 and 1889 and experimented with electromagnetic wave theory there. Hertz was able to show that light was an electromagnetic wave. Such waves travel at a speed of 186,000 miles per second. (The metric equivalent is 300,000,000 meters per second.) Figure 20-1 shows the electromagnetic-frequency spectrum.

The spectrum is divided into slow oscillations (ranging from no frequency to 20 Hz), audio frequencies (from 20 Hz to 20,000 Hz) and radio frequencies (from 20 kHz to 1 terahertz).

Giga = 1,000 Mega	1×10^3 — Kilo
Tera = 1,000 Giga	1×10^6 — Mega
	1×10^9 — Giga
	1×10^{12} — Tera

Bands of radio frequencies are further divided into the following:

very low frequency, VLF	3 kHz–30 kHz
low frequency, LF	30 kHz–300 kHz
medium frequency, MF	300 kHz–3 MHz
high frequency, HF	3 MHz–30 MHz
very high frequency, VHF	30 MHz–300 MHz
ultrahigh frequency, UHF	300 MHz–3 GHz
superhigh frequency, SHF	3 GHz–30 GHz
extremely high frequency, EHF	30 GHz–300 G MHz

Amplitude modulation (AM) radio broadcasts on 535 kHz to 1605 kHz, whereas **frequency modulation** (FM) uses 88 MHz to 108 MHz. Commercial television uses

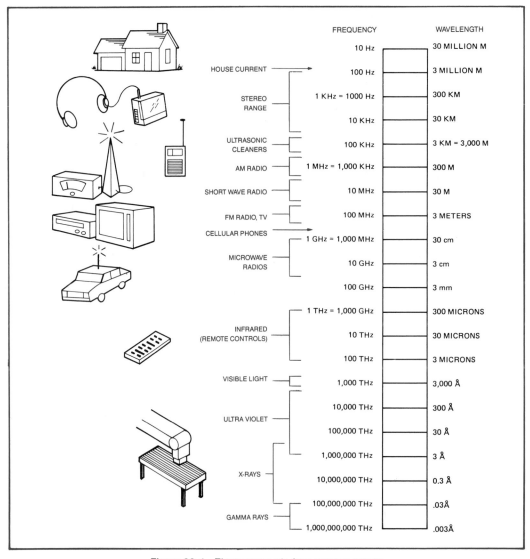

FREQUENCY	WAVELENGTH
10 Hz	30 MILLION M
100 Hz	3 MILLION M
1 KHz = 1000 Hz	300 KM
10 KHz	30 KM
100 KHz	3 KM = 3,000 M
1 MHz = 1,000 KHz	300 M
10 MHz	30 M
100 MHz	3 METERS
1 GHz = 1,000 MHz	30 cm
10 GHz	3 cm
100 GHz	3 mm
1 THz = 1,000 GHz	300 MICRONS
10 THz	30 MICRONS
100 THz	3 MICRONS
1,000 THz	3,000 Å
10,000 THz	300 Å
100,000 THz	30 Å
1,000,000 THz	3 Å
10,000,000 THz	0.3 Å
100,000,000 THz	.03Å
1,000,000,000 THz	.003Å

Device labels (left to right on the spectrum):
HOUSE CURRENT
STEREO RANGE
ULTRASONIC CLEANERS
AM RADIO
SHORT WAVE RADIO
FM RADIO, TV
CELLULAR PHONES
MICROWAVE RADIOS
INFRARED (REMOTE CONTROLS)
VISIBLE LIGHT
ULTRA VIOLET
X-RAYS
GAMMA RAYS

Figure 20-1. Electromagnetic-frequency spectrum.

a VHF and a UHF band. VHF television is allotted 54 MHz to 216 MHz, with some exceptions. UHF has been assigned 470 MHz to 890 MHz.

The microwave range at 2450 MHz is also used by microwave ovens. Police radar uses 10.525 GHz and 24.15 GHz.

REVIEW QUESTIONS FOR SECTION 20.1

1. What does *RF* stand for?
2. What is another name for radio waves?
3. What is a carrier wave?
4. Who was Heinrich Hertz, and what contribution to science did he make?
5. What are the frequencies in the SHF range?
6. On what frequency does a microwave oven operate?

20.2 TRANSMITTERS

Transmitters are units used to convert messages in code, computer language, voice, or music into electrical impulses for transmission either on closed lines or through space from a radiating antenna.

Various types of transmitters are utilized for specialized purposes. There are amplitude-modulated and frequency-modulated transmitters. Each has its own characteristics and performance abilities. There are also radar transmitters that operate in the microwave range. Of primary interest here are the AM and FM transmitters used for commercial broadcasting.

AM Transmitters

The purpose of a transmitter is to carry the audio code from one location to another without the use of wires. In most instances, it is used to broadcast information that can, in turn, be picked up by a receiver and decoded or demodulated to obtain the intelligence thus transmitted. See Figure 20-2 for a simple transmitter block diagram.

Amplitude of the radio frequency wave may be modulated by means of a signal of constant frequency that is varied to send coded messages. This results in what is called **modulated continuous wave** (MCW). In this case, a tone is used to do the modulating.

The amplitude-modulated transmitter may also have voice or music impressed on the RF carrier wave. The received radiations are demodulated and can then be heard directly without the need of an operator to decode the message. See Figure 20-3 for a block diagram of a transmitter with waveforms found in each stage.

Oscillator

Every transmitter needs an oscillator to establish its operating frequency. The frequency of a transmitter can be stabilized by the use of a crystal oscillator. In CB (citizens' band) transmitters, the phase-locked loop is used to generate a number of frequencies for various channels, using only one crystal to establish the basic frequency.

The oscillator is assigned the task of establishing and maintaining a frequency to be used as a carrier. This frequency is then doubled, tripled, or quadrupled in the case of FM, and then sent on to an intermediate power-amplifier IPA stage that boosts its strength. Once it is amplified in the IPA, it is sent to the power amplifier, where it is amplified again to reach the power required. Note in Figure 20-3 that the audio amplifier is amplifying the audio signal. Once the signal from the microphone, tape head, or phono pickup is amplified, it is fed into a **modulator**, where it is boosted again to sufficient level to modulate the RF carrier in the power amplifier (PA).

Buffer

A buffer is needed to prevent variations in load at the antenna from affecting the oscillator frequency. This purpose is served by the IPA, which is located between the PA and oscillator. This particular stage may also serve as a doubler or tripler to make the frequency of the oscillator meet the requirements of the transmitter. As you can see, the buffer can serve many purposes.

Power Amplifier

Take a closer look at the location of the power amplifier in the transmitter block diagram, Figure 20-4. Note that in this diagram the IPA mentioned before is now called a driver and it is used as a doubler. This simply

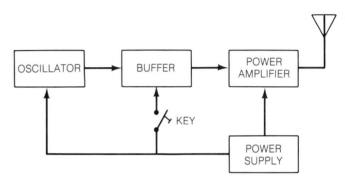

Figure 20-2. Master oscillator/power-amplifier transmitter with keying in the buffer stage.

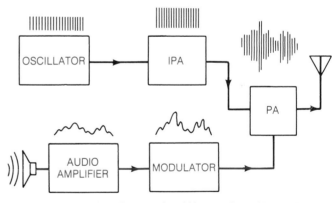

Figure 20-3. Block diagram of an AM transmitter with waveforms.

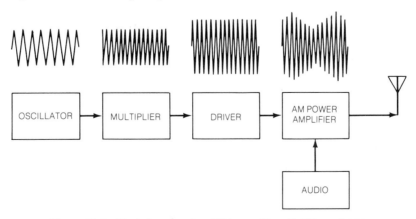

Figure 20-4. Block diagram of an AM transmitter with RF waveforms.

means it is a stage used to drive the power amplifier, which, since it is a high-power amplifier, needs a strong signal in order to operate. The purpose of the power amplifier stage is to increase the signal strength sufficiently to cause the signal to travel to its intended destination. The PA stage is connected to the antenna directly or through some type of coupling.

Sidebands

Sidebands are produced when a modulating frequency modulates a carrier frequency. The modulation process produces four different sine waves:

the original *modulating* frequency
the original *carrier* frequency
the *sum* of the two frequencies
the *difference* between the two frequencies

When the modulating frequency is a much lower frequency than the carrier, then the low-frequency modulating signal is never transmitted and as a result only three of the above are received by the receiver.

Figure 20-5 shows how the lower and upper sidebands are placed in reference to the carrier frequency. This can be seen if a spectrum analyzer is used to check the transmitted AM wave. The sideband that is at a higher frequency than the carrier is called the upper sideband and the sideband that is at a frequency lower than the carrier is called the lower sideband.

Some transmitters utilize only one sideband. They can use either the lower or the upper sideband. Specially designed receivers are required to receive this single-sideband transmission. This is one way two signals can be transmitted on the same frequency without interference. Most commercial broadcasting is done on double sidebands.

FM Transmitters

Direct and indirect means are utilized for producing frequency modulation. Both involve changing either the frequency or phase of an oscillator in accordance with some modulating signal. In the direct method, the modulating signal is injected into a modulator whose output varies the frequency of the oscillator in accordance with the original modulating signal. In the indirect method, the modulating signal is passed through a correction network to a phase modulator. The correction network changes the phase of the modulation in such a manner that, when the output of a crystal oscillator is passed through the modulator, the oscillations are frequency-modulated in accordance with the modulating signal. See Figure 20-6 for a block diagram of the FM transmitter.

Frequency Multipliers

In FM transmitters, frequency multiplication of the FM signal performs two functions. It increases the frequency of the signal to the value desired for transmission, in this way acting the same as a frequency multiplier in an AM transmitter. It also increases the effective frequency deviation of the FM signal.

Frequency Deviation

The amplitude of the modulating wave causes the change in frequency of the FM carrier. The frequency of

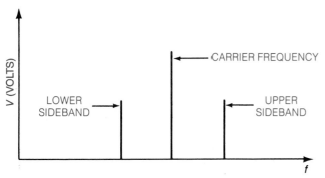

Figure 20-5. Upper and lower sidebands for an AM signal.

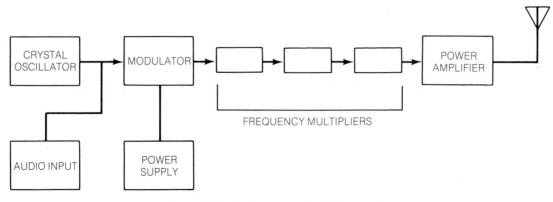

Figure 20-6. Block diagram of an FM transmitter.

the modulating signal affects how rapidly the FM wave changes its frequency. For example, if there is an audio frequency of 200 Hz, the FM wave will deviate from its resting frequency 200 times each second. Just how far the FM wave deviates is determined by the amplitude of the audio signal. How often the FM signal deviates from its resting frequency is determined by the frequency of the audio signal, Figure 20-7.

Modulation Index

The FM carrier is influenced by both the amplitude and the frequency of the modulating wave. This is measured according to what is called the **modulation index**. The modulation index is found by dividing the deviation of the FM wave, in hertz, by the frequency of the modulating signal, in hertz.

Power Amplifier

The requirements of the FM power amplifier are somewhat different from those of the AM transmitter. Since the FM power amplifier has no connection with the modulating process, the only losses that are involved are those inherent in the transistor and circuit when amplifying an unmodulated carrier.

FM Transmitter Uses

FM transmitters are found in many applications other than commercial broadcasting. They broadcast on 88 to 108 MHz for the commercial band. This provides static-free reception of music and voice communication or entertainment. The FM transmitter is also used on airplanes for ground-to-air, air-to-ground, and air-to-air communications. The U.S. Army uses FM transmitters and receivers to set up temporary communications links between the battlefield and headquarters. Any number of uses for FM have been found and more are being utilized all the time.

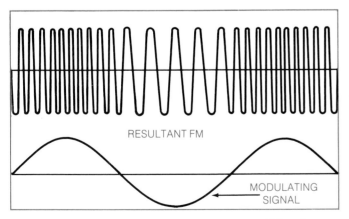

RESULTANT FM

MODULATING SIGNAL

Figure 20-7. Using an audio signal to modulate an FM carrier.

Television uses FM for transmitting the sound portion of each channel.

REVIEW QUESTIONS FOR SECTION 20.2

1. What do transmitters do?
2. What is the difference between AM and FM transmitters?
3. Why is a crystal oscillator used?
4. What is a modulator?
5. What does a buffer do?
6. What does a PA do?
7. What are sidebands?
8. What are the two ways frequency modulation is produced?
9. What is the purpose of a frequency multiplier?
10. What is meant by the term *modulation index*?
11. How does the PA in an FM transmitter differ from that in an AM transmitter?
12. What are some uses, other than broadcasting, for FM transmitters?

20.3 RECEIVERS

In its early days, radio was called *wireless telegraphy.* This is because the only signals carried were dot-and-dash codes.

The first major development toward wireless was the theory of J. Clark Maxwell. In 1865, Maxwell discovered that waves of energy could travel through air.

In 1888, Heinrich Hertz showed that wireless signals were related to electric circuits.

The first useful wireless came out of the laboratory of Guglielmo Marconi. This occurred in 1901. Marconi was able to demonstrate transmissions of code messages over distances of more than 2000 miles [3218.6 km]. Wireless code communication was completed between England and Newfoundland, across the Atlantic Ocean.

The first voice message was transmitted by radio in 1915. This traveled from Arlington, Virginia, to Hawaii.

Early radio receivers were crystal sets. They used large crystals of lead sulfide (galena) as rectifying elements. A "cat whisker" (small piece of wire) had to be adjusted carefully. This made contact with a sensitive part of the crystal. These early devices proved the practicality of radio as a means of communication.

How the Radio Signal Is Received

The radio receiver must be tuned to the same frequency as the transmitter. This is done through use of resonant

Activity: RADIO RECEIVER

OBJECTIVES

1. *To study the process of radio detection.*
2. *To study the simple crystal-diode receiver.*
3. *To identify the transistor amplifier.*

EQUIPMENT NEEDED

1 Headphone, 2000 Ω
1 Tuning coil, RF, ferrite core
1 Capacitor, variable, 0–365 pF
1 Diode, 1N34 or equivalent
1 Capacitor, 100 pF
3 Resistors, 100 K, 10%, 1/2-watt
1 Capacitor, 0.01 μF any working voltage
2 Resistors, 10 K, 10%, 1/2-watt

1 Capacitor, 10 μF, 10 working voltage DC
1 Potentiometer, 1 Meg, audio taper
1 Transformer, 2000 to 8 ohms
1 Speaker, 3", 8 ohms, 100 mW
2 Transistors, 2N1048 or equivalent
1 Power supply, 6 volts DC
10 Wires for connections
1 Long wire for antenna

PROCEDURE

1. Connect the circuit shown in Figure 1. This is the simple crystal-diode detector. It will give good reception for local stations with the aid of a headphone.
2. Plug in the headphone. Carefully tune across the broadcast band (535–1605 kHz). Do this by turning the knob on the tuning capacitor. How many stations can you hear?
3. Disconnect the circuit you are using and connect the circuit shown in Figure 2. This is the same diode detector used in Step 1, with a two-stage transistor amplifier used to amplify the detected signal. If you are in an area close to a radio station, connect 5 to 10 feet of wire to the short antenna terminal (see Figure 2). If you are some distance from a radio station, connect the long antenna to the terminal marked L in the schematic. How many stations do you hear now?
4. Disconnect the transformer from the speaker and connect the transformer to the headphone. Again, tune across the broadcast band. Are the stations louder than they were in Step 2? How many stations can you receive?
5. Disconnect all circuits that you have been using and return all equipment and materials to their proper places.

SUMMARY

1. What is the purpose of the diode detector?
2. What is the purpose of the transistor amplifier?

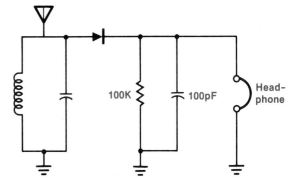

100K   100pF   Head-phone

Figure 2

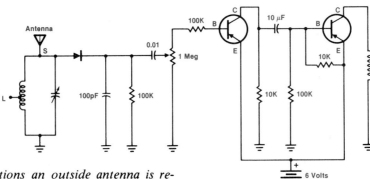

NOTE: *In some locations an outside antenna is required for the circuit to function.*

Figure 1

circuits. You learned about resonant circuits in Chapter 10. The receiver detects radio waves at the frequency for which it is set. Then it converts those waves back into sound.

The patterns of transmission between transmitter and receiver are shown in Figure 20-8. Signals follow one of two paths from transmitter to receiver. *Ground waves* move directly from the transmitter to the receiver. *Sky*

waves rise to the ionosphere and are bounced back to earth. The ionosphere, in effect, is a barrier, or boundary. It marks the end of the earth's atmosphere and the beginning of what we call outer space. Waves of radio frequency reflect from the ionosphere.

Figure 20-9 shows a schematic of a radio receiver. The radio wave is picked up by the antenna, which is sensitiv to radio signals in the air. The antenna picks up signal

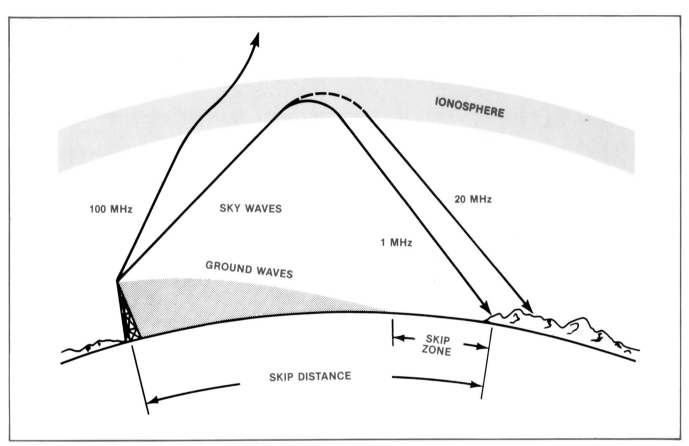

Figure 20-8. Transmission patterns of radio-frequency waves.

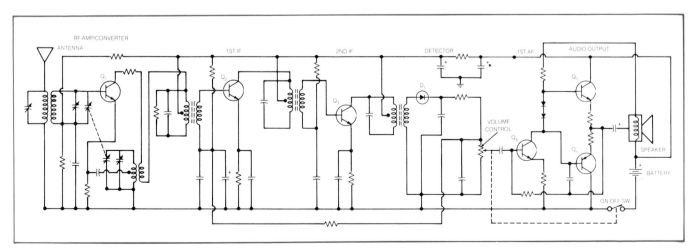

Figure 20-9. Schematic diagram of radio receiver.

at all frequencies. They pass through the primary of the antenna coil to ground. There is a secondary antenna coil that is a tuned circuit. This circuit presents a high impedance to current at the frequency to which it is tuned. The tuned frequency only is amplified through the radio receiver.

The signal is carried to a radio-frequency (RF) _amplifier._ Here the signal is increased. Then the signal moves through the IFs and on to the detector. The detector converts the signal from RF to an audio frequency. The signal is then processed through an audio-frequency amplifier. The audio-frequency amplifier increases the sound level. In effect, the signal is made stronger than when it came from the detector. Finally, the audio signal goes to the speaker, where it is changed to sound.

AM Radio on a Chip

Virtually an entire AM radio can now be obtained on one integrated-circuit chip. This is done with one 14-pin DIP, as shown in Figure 20-10. This particular chip does not contain the audio output stages. That means another chip of whichever audio power output is desired can be connected to cause this to operate as a radio. Note that the antenna, oscillator coil, and IF cans are mounted externally. In some more-advanced chips, the IF cans are replaced by ceramic filters. The dotted lines with numbers

in circles make up the chip, with its pins identified for external-connection purposes. This chip operates well with 12 volts DC, making it ideal for use in automobile radios.

Frequency-Modulation Receivers (FM)

The FM receiver is more complex than is the AM receiver. Methods employed to produce FM and especially stereo are somewhat more complicated than those involved with the AM receiver. Some interesting devices involving integrated circuits are used to decode the stereo signal and to produce an output sufficient to drive a couple of speakers. Improved receivers are being released for public use every day. Newer sets are more selective, more sensitive, and less noisy.

The FM band (88–108 MHz) is used to transmit music and programming for home and commercial use, in what is called wideband transmission. Narrowband FM uses only voice transmission and is used for many communications channels. FM for the commercial broadcast band covers the 88-to-108-MHz band with 200-kHz channel sidebands. Television audio signals use 50-kHz channel sidebands at 54 to 88 MHz, 174 to 216 MHz, and 470 to 890 MHz. The narrowband amateur radio channels are at 29.6 MHz, 52 to 53 MHz, 146 to 147.5 MHz, 440 to 450 MHz, and in excess of 890 MHz for experimental purposes.

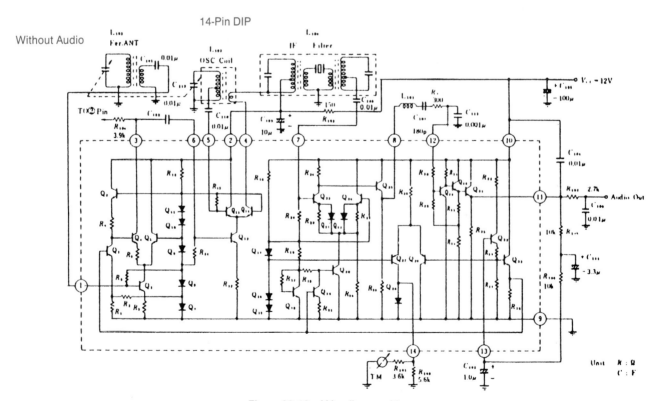

Figure 20-10. AM radio on a chip.

Frequencies below 30 MHz are not used for FM. The earth's ionosphere introduces phase distortion for FM signals at frequencies below 30 MHz. Line-of-sight transmission is used for signals above 30 MHz because they are not reflected and can pass through the ionosphere. The earth's curvature limits the communications range of FM to about 80 miles maximum. This is especially true of the narrowband public-service channels that operate on 108 to 175 MHz, between channels 6 and 7 on TV. The narrowband is also assigned to frequencies in excess of 890 MHz. Output power from a television station's FM transmitter is about 50 kW, whereas the amateurs use some walkie-talkies with only 130 mW of power.

FM Stereo

Most of today's FM receivers are made for stereo. For the receiver to receive and decode this stereo signal, several stages have to be added to the regular monaural FM receiver.

The Federal Communications Commission authorized FM stereo in 1961. This made it possible for home receivers to obtain complete information from records and tapes that were already available. Stereo uses two separate signals to produce a spatial dimension to music or speech. This called for addition of another channel to the FM single-channel transmissions. Stereo high fidelity requires two channels for the 30-Hz-to-15-kHz signals to modulate the carrier frequency in such a way that the receiver can separate them and reproduce the outputs in left and right speakers, Figure 20-11.

More efficient use of the 200-kHz bandwidth was the answer to the stereo problem. This was done through the process of **multiplexing**. Multiplexing is the simultaneous transmission of two different signals on one carrier. It is also possible to broadcast more than two signals under the right conditions. The FCC approved a compatible system for stereo broadcasts. This means that the stereo and the FM monaural signals can be received on receivers that

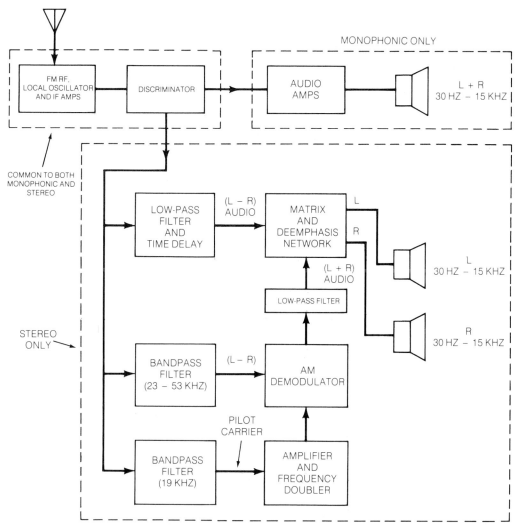

Figure 20-11. Comparison of stages needed for mono and stereo FM.

normally pick up the monaural. Alternatively, the stereo receiver can receive the monaural signal and reproduce it properly.

The stereo signal is a composite. The sum of the left and right (L + R) signals extends from 30 Hz to 15 kHz, Figure 20-12. The left-minus-right (L − R) channel extends from 23 to 53 kHz. Note the placement of the 19-kHz pilot subcarrier on the line. This 19-kHz pilot subcarrier is included in the composite stereo signal that modulates the transmitter. That means two different signals are multiplexed together by placement in two different frequency ranges.

Stereo Demodulation

The FM receiver and the FM stereo receiver are the same up to the discriminator stage. At this point, the stereo signal must be detected and processed properly to add the missing channel.

FM on a Chip

The entire FM receiver can be placed on one or two chips. This is done by using a 14-pin DIP, Figure 20-13. A preamplifier for the audio is also furnished on the same chip. This feeds directly to the amplifier of choice to drive a speaker or speakers. By using the chips, it is possible to make car radios very inexpensively. (That does not necessarily mean they are inexpensive when bought as part of the car, however.)

___ **REVIEW QUESTIONS FOR SECTION 20.3** ___

1. What was an early name for radio?
2. What are ground waves?
3. What does a detector do?
4. How many chips does it take to make a complete AM radio today?
5. What frequencies does the FM band cover?
6. Why is FM considered "line-of-sight" communications?
7. When was stereo FM first authorized in the U.S.?
8. What is the difference between FM and FM stereo?
9. How is stereo demodulated, or taken from the transmitted signal?
10. How many transistors can you identify on the FM IF, the Detector, and the AF preamplifier chips?
11. How many diodes can you identify on the FM stereo demodulator chip?

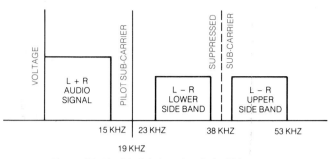

Figure 20-12. Modulating signals for FM stereo.

FM Stereo Demodulator

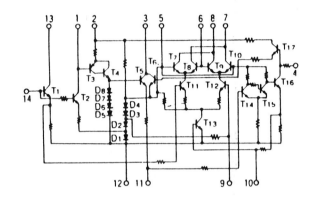

FM IF Amplifier/Detector/AF Preamplifier

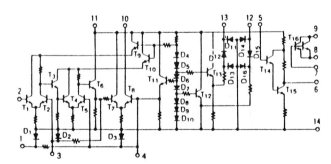

Figure 20-13. FM stereo on a chip.

20.4 TELEVISION

Television is an advanced form of radio transmission. As you know, images as well as sounds are transmitted. The TV camera converts images into electronic signals that are transmitted along with sounds.

How a TV Camera Works

Figure 20-14 diagrams a television camera. At the front of the camera is a lens system. The lens focuses the image of the subject being photographed. TV cameras use the same kinds of lenses as do motion-picture cameras. However, the image is not focused on film. Instead, the

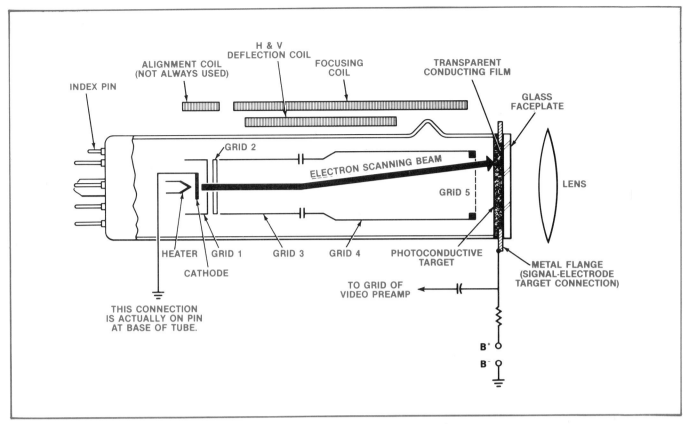

Figure 20-14. Diagram of vidicon camera tube.

TV camera has the face of an electronic tube behind the lens. This is known as a **vidicon** tube.

The camera senses and transmits the image from the face of the vidicon tube. The face of the tube is scanned by an electron beam. The rate of scan is 30 times per second. In one-thirtieth of a second, the TV set creates a full picture, or *frame*. This has 525 scan lines. Each frame has two *fields*, even and odd. Each of these fields contains 262.5 scan lines. The scanning electron beam senses light and dark areas of the picture. A series of grids within the tube acts as a color controller. The signal from the vidicon tube is then fed into a video preamplifier.

Liquid-Crystal-Display (LCD) Television Pictures

The liquid-crystal display has been used for a number of years for such things as calculator displays and signs. However, it is still in the development stages for use in television pictures. There are a number of problems to be solved before it can be made to function as a color picture using the same circuitry and information available in the present television system.

One design being developed in London at the Imperial College of Science and Technology does show promise. It makes use of a flat panel with glass optical waveguide strips in vertical columns covered by a glass panel with rows of nearly invisible electrodes, Figure 20-15.

The LC material is sandwiched between the strips and glass sheet. Three colored-light-emitting diodes (LEDs) at the bottom edge of each strip are fired in a horizontal sequence by a red, green, and blue video signal. Simultaneously, a shift register circuit applies AC pulses to the horizontal electrodes one row at a time. The electric field at each electrode-strip crossover realigns the LCs, allowing light to escape. LED selection here creates pure red, green, and blue picture elements, plus white (three LEDs fired at once).

Small-screen TV receivers utilizing the LCD screen are now commercially available, Figure 20-16. There is an advantage to this type of screen. The size of the receiver is greatly reduced, inasmuch as it does not need the depth that the picture tube requires. The future of this type of receiver depends upon its ability to be mounted flat on a wall with very thin cables supplying the power and signals needed to operate the display.

How a TV Transmitter Works

The picture signal from the camera is only part of the TV signal. Audio, or voice, signals are also picked up and

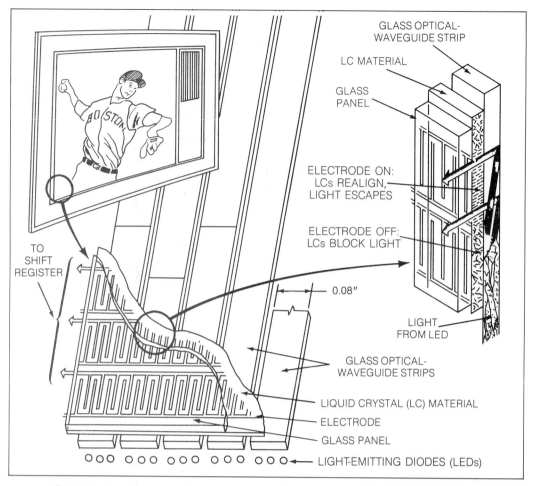

Figure 20-15. Diagram of a liquid crystal display. (Popular Science and Eugene Thompson)

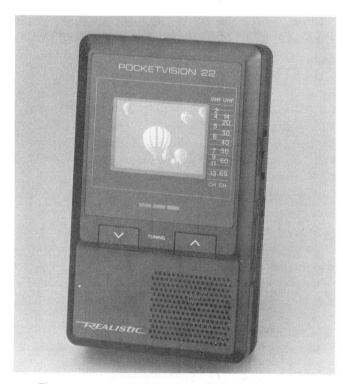

Figure 20-16. Two-inch LCD TV receiver. (Tandy Corporation)

recorded or transmitted. Most programming materials broadcast by TV stations are prerecorded. Recording is done on tape equipment. However, some programs, particularly news broadcasts, are transmitted live. A live broadcast is one that is transmitted as the action takes place.

Usually, two or more cameras are used to cover action for a broadcast. The cameras are connected to a control console. The director at the console decides which picture and sound content are broadcast. The console then creates a broadcast signal that is sent through the air by a transmitter.

A TV signal is composed of three elements: picture, sound, and synchronization signals. The synchronization signals are also called timing signals. They coordinate the picture scanning and sound patterns from the camera or tape. The timing signals ensure that complete pictures and usable sound are transmitted.

Figure 20-17 shows the pattern of TV-signal transmission. Notice the difference between these patterns and those for radio. TV signals move in straight lines. This is called line-of-sight transmission. To transmit TV signals over long distances, special techniques are needed. One method is to send the signal on communication cables. Another is to relay the signal from one tower to another. This requires that a series of towers be built across the countryside. Both techniques are used.

How a TV Receiver Works

Figure 20-18A is a block diagram of a TV receiver. The signal is received on the antenna. From there it moves to the tuner section. The tuner selects the signal to pick the desired channel. The tuned signal is converted to intermediate frequency (IF).

From the tuner, the signal moves to an IF amplifier. The IF amplifier increases the signal to a usable level. Next, the signal goes to a video detector. This separates the three elements of the video signal. The video signal is transmitted to the video amplifier. The audio signal goes to the audio amplifier. And the synchronization signals move to create the raster (scan pattern).

The audio amplifier is connected directly to a speaker. The video amplifier is connected to the picture tube. The

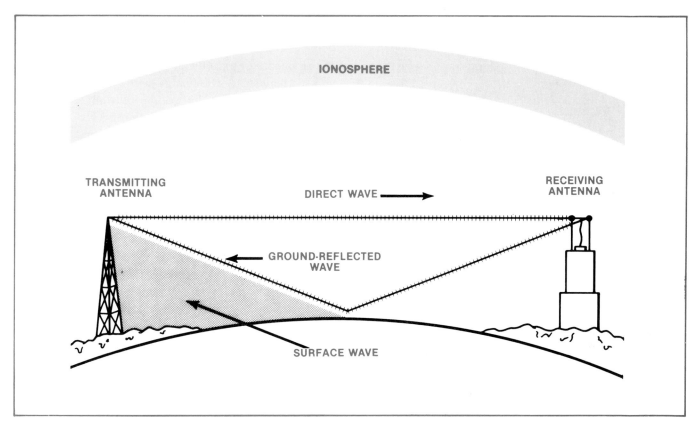

Figure 20-17. Transmission pattern for TV and FM waves.

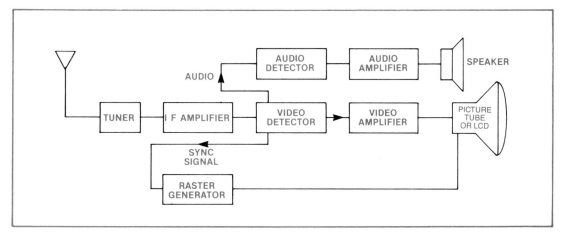

Figure 20-18A. Block diagram of black-and-white-TV receiver.

Figure 20-18B. Color-TV receiver with push-button tuning. (Tandy Corporation)

raster generator controls the scanning patterns of the electron beam that produces the picture. The **raster** is the scan pattern. The raster generator reacts to the synchronization signals from the camera.

Figure 20-18B is a photograph of a TV receiver with push-button tuning.

REVIEW QUESTIONS FOR SECTION 20.4

1. How many lines make up a TV picture?
2. How many complete pictures do you see per second on television?
3. What purpose does a synchronizing signal serve?
4. What does *LCD* stand for?
5. What is replacing the vidicon in television cameras?
6. What is the advantage of using LCD picture screens instead of picture tubes?
7. What is line-of-sight transmission?
8. What is a raster?
9. How does scanning produce a picture?

20.5 OTHER TYPES OF COMMUNICATIONS AND DETECTION CIRCUITRY

As mentioned previously in this chapter, other methods are used to detect FM signals. The quadrature-detection method is one of these.

Quadrature Detection

Quadrature detection involves amplification and phase shifting of the input IF signal. That means it is more easily accomplished by means of a chip with those capabilities, Figure 20-19. The chips shown are used in most television sets and FM receivers made today.

A few external components are required to obtain quadrature detection. Figure 20-20 shows how the IC is used to produce the method of detection. The IF signal is amplified and limited and then applied to the IC at pin 4. The IF output is amplified and then leaves the chip at pin 10. This signal is also coupled internally to the phase comparator. The capacitor attached to pin 9 serves to couple some of the IF signal to the external quadrature circuit, enclosed in the dotted lines. The center frequency for the quadrature circuit is 10.7 MHz or the IF frequency. The quadrature signal is developed so it will lead the IF signal by 90°. The leading quadrature signal is sinusoidal in shape. It is applied to pin 12 of the chip. Pin 12 is the other input to the phase comparator. The IF signal at this point is varying above and below 10.7 MHz in step with the modulation of the transmitter signal. Upward deviation of the IF frequency makes the quadrature circuit become capacitive, and the quadrature signal then lags its former phase angle. Downward deviation of the IF frequency makes the quadrature circuit become inductive, and the quadrature signal leads its former phase angle. The configuration of the phase comparator is such that it compares the phase angle of the quadrature signal with that of the IF signals, develops an output voltage that varies inversely with the phase angle, and outputs it at pin 1.

Frequency-Shift Keying, or FSK

Frequency modulation, or FM, can provide a better method of broadcasting coded information. FM has a number of advantages, one of which is the elimination of noise that can cause misinterpretation of messages. **Frequency-shift keying** (FSK) is a form of frequency modulation. The modulating wave shifts the output between two predetermined frequencies. These are usually referred to as the *mark and space frequencies*. This system uses a carrier frequency that is located halfway between the mark and space frequencies and that is modulated by a rectangular wave.

Figure 20-21 shows how this type of FSK keying causes a shifting in the output signal. This type of modulation allows for a number of narrowband FM channels to be placed in a small portion of the spectrum. The wideband FSK uses a 10-to-20-kHz channel, whereas the narrowband utilizes a channel that is less than 10 kHz.

FM & TV SOUND IF AMP/DETECTOR/LIMITER
FOR PORTABLE & VEHICULAR EQUIPMENT

FM & TV SOUND IF DETECTOR & ELECTRONIC
ATTENUATOR

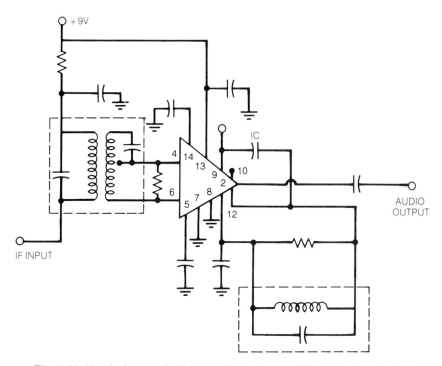

AUDIO OUTPUT — 1
DETECTOR INPUT REFERENCE — 2
N.C. — 3
I.F.-INPUT — 4
DECOUPLING — 5
I.F.-INPUT REFERENCE — 6
GND. — 7

14 — DE-EMPHASIS
13 — +V_{CC}
12 — DETECTOR INPUT (QUADRATURE NETWORK)
11 — TEST POINT
10 — AMPLIFIER HIGH OUTPUT
9 — AMPLIFIER LOW OUTPUT
8 — N.C.

IF-INPUT — 1
IF-INPUT — 2
GND. — 3
GND. — 4
+V_{CC} — 5
VOLUME CONTROL — 6
BYPASS — 7

14 — AF-INPUT
13 — TONE CONTROL CIRCUIT
12 — AF-OUTPUT
11 — N.C.
10 — QUADRATURE NETWORK
9 — QUADRATURE NETWORK
8 — AF-BUFFER OUTPUT

A

B

Figure 20-19. Quadrature detection by a 14-pin chip.

Figure 20-20. Hookup needed for operation of a 14-pin DIP for quadrature detection.

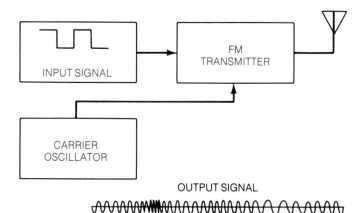

Figure 20-21. FSK, or frequency-shift keying, modulates an FM transmitter.

Telephone lines can transmit information, using the FSK system. This is one of the better ways to transmit digital information or data over long-distance lines.

Pulse-Width Modulation (PWM)

Pulse-width modulation is more like a pulse time modulation; the time varies instead of the amplitude. It can also be referred to as PLM, or pulse-length modulation. A phase-locked loop can be used in the generation of both PWM and PPM. The phase-locked loop (565 IC) can be used. By varying the VCO, it is possible to cause the phase detector to increase or decrease its output. The output then is a phase-shifted signal that can be applied to a carrier and transmitted. The advent of ICs and the phase-locked loop have made it possible to obtain phase shifting easily, Figure 20-22.

Pulse-Position Modulation (PPM)

Pulse-position modulation is generated from PWM. PPM is usually used since it has better noise characteristics. It is possible to generate PPM by inverting PWM, Figure 20-23. Once inverted, the signal is differentiated and produces PPM with a very short pulse. This pulse can be used to modulate an AM carrier. Its improved quality at reception is due to the pulse's very short duration, and the information content is not contained in either the pulse amplitude or width.

Time-division multiplexing or PCM transmission for the telephone has proven its ability to cram more messages into short-haul cable than frequency-division multiplex analog transmission can. PCM is the most noise-resistant transmission system available. Much work is being done with the transmission of data. Other methods of transmission may be developed as newer ICs with the necessary characteristics to handle the requirements of various situations are made available.

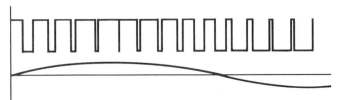

Figure 20-22. PWM, or pulse-width modulation.

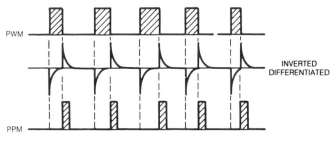

Figure 20-23. PPM signal made by inverting and then differentiating the PWM signal.

REVIEW QUESTIONS FOR SECTION 20.5

1. Where is quadrature detection used?
2. Why are chips required for quadrature detection?
3. What is FSK?
4. What is the advantage to the use of FSK?
5. What is PWM? Where is it used?
6. What is PPM? How is it utilized?
7. Where is PCM transmission most often utilized?
8. What is the most noise-resistant system available?

20.6 ANTENNAS

AM Receiver Antennas

A transmitted wave passes a receiving antenna and induces a current in the antenna as a result of electromagnetic induction. This weak induced current varies with the frequency of the wave. The current is then amplified and processed to obtain the intelligence it contains.

Requirements for receiving antennas are much less demanding than those for transmitters. If the different antennas are used for transmitter and receiver, then simple receiver antennas may consist of a piece or coil of wire or some more sophisticated designs for special purposes. Most portable receivers used commercially and in homes are made with a ferrite core. The core is wound with a few turns of very fine wire, Figure 20-24.

FM Receiver Antennas

The antenna is usually a piece of wire in areas in which signals are very strong. A folded dipole is commonly used when any type of signal strength is presented at the front end of the receiver. The front end is made up of the first two stages. The front end usually includes the RF amplifier and mixer. A good antenna improves the signal-to-noise ratio and results in the receiver's being able to pick up weak signals while still permitting a stereo effect in the output of the receiver. Figure 20-25 shows a folded dipole. More-sophisticated antennas are available for better reception. In practice, the folded dipole may be very simple, but it can be improved and become directional if a reflector and director are used. Figure 20-26 shows a typical single-bay yagi antenna with five elements that can be used for TV or FM reception.

TV Antennas

The folded-dipole antenna is usually recommended for outside TV reception. It has 300 ohms impedance, which makes it a wideband receptor, wide enough to handle the 2-through-13-channel frequency allocation. The dipole is

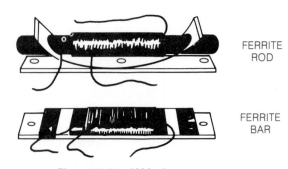

FERRITE ROD

FERRITE BAR

Figure 20-24. AM ferrite-core antennas.

about one-half wavelength. The folded-dipole is directional, and the major lobes are at right angles to the elements. Dipoles may be stacked for higher gain in areas of weak signals. Outside antennas usually have directors in front of the dipole and reflectors in back of it. The directors make the antenna a little more directional, and the reflectors reject the unwanted waves from the back of the antenna. The directors are shorter than the dipole or driven element, and the reflectors are longer in physical size than the driven element. Figure 20-26 shows the five-element antenna that can also be used for directional reception of FM signals. Figure 20-27 shows the loop or

bow-tie UHF antenna, which can also be used for better reception of weak signals.

REVIEW QUESTIONS FOR SECTION 20.6

1. Where are antennas used?
2. What type of antennas use ferrite cores?
3. Where are folded-dipole antennas used?
4. Sketch a single-bay yagi antenna with five elements.
5. Why can a TV antenna be used for FM reception?
6. What is the impedance of a folded-dipole antenna?

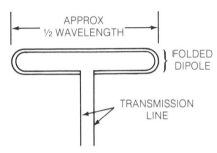

Figure 20-25. Folded-dipole antenna.

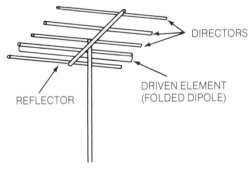

Figure 20-26. Yagi five-element TV or FM antenna.

20.7 THE TELEGRAPH

Samuel F. B. Morse invented the first practical electrical communication system, the **telegraph**, in 1832. However, it was 1835 before he was able to demonstrate his invention. Morse, an American artist, sent his first telegram in 1844. This traveled 40 miles, from Washington, D.C., to Baltimore, Maryland. The telegraph spread rapidly throughout the United States. By 1851, there were 50 separate telegraph companies.

How the Telegraph Works

The principle behind the Morse system had existed for many years. The idea is to transmit signals based on the presence or absence of electric current. The circuit is closed or opened through use of a telegraph key, Figure 20-28. In effect, the telegraph key is a push-button switch. When pressed down, the key completes the circuit. When the operator releases the key, the spring forces ⁻ upward. This breaks the circuit.

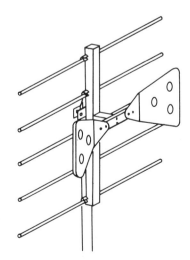

Figure 20-27. Bow-tie UHF antenna with a reflector.

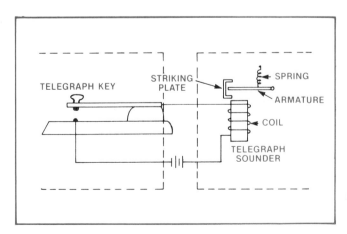

Figure 20-28. Schematic of Morse telegraph set.

The telegraph key is connected to a device known as a sounder. The sounder contains an electromagnet. When the circuit is closed, the coil of the magnet is energized. At this time, an armature is brought down to hit a striking plate. This creates a loud click that is really the result of an electrical impulse. When the coil is de-energized, a spring lifts the armature. The armature then returns to its original position.

The clicks created in the sounder are transmitted over wires. They are then reproduced on a telegraph instrument at a distant point. Patterns for making and breaking the electrical current represent letters and numbers. Messages are built from these letters and numbers.

Telegraph Codes

Messages are composed through control of the patterns of clicks (electrical pulses). This means that the clicks must have the same meaning to the sender and the receiver. Morse developed the **Morse code**, which is based on clicks called dots and dashes. A dot is a short signal. It is created by depressing the telegraph key, then lifting the key quickly. A dash is three times as long as a dot.

Morse's original code was modified for efficiency as telegraphy spread. Before the end of the nineteenth century, there was an **international code**. This is now known and respected all around the world. To illustrate, the letters *S O S* represent an urgent message all over the world. This is an abbreviation for the words "Save our souls." In international telegraphy, the letters are coded dot-dot-dot, dash-dash-dash, dot-dot-dot. This message has priority over all airwaves and telegraph lines. A SOS message mobilizes rescue efforts anywhere in the world.

The signals that make up the Morse and international codes are shown in Figure 20-29. The international telegraphic code is still in use. All ham radio operators must know and be able to use this code. Telegraphy is still taught to communications technicians in the military services. All shipboard radio operators must also know this code.

Computer Monitors

A computer **monitor** resembles the TV picture screen, Figure 20-30. The TV screen can be used to display computer information. However, it was not designed to produce the detail that the computer monitor can.

The monitor is slightly different, in that it does not have a tuner section. It displays the numerals and letters used in everyday "typewriter" communications. It also is required to reproduce graphics with a great deal of resolution, which is not possible in a TV picture. By taking out the radio-frequency section that deals with the recep-

tion and detection of picture information as well as audio, it is possible to concentrate on reproducing and displaying the digital information presented to it. Circuits capable of handling the digital information are included in the monitor. The monitor also has an audio section capable of alerting the operator under certain conditions. This audio capability usually consists of audio oscillator circuits and an audio amplifier to drive a small speaker.

High-Speed Printers

New printing devices have also been introduced into communication systems. Today, communication printers

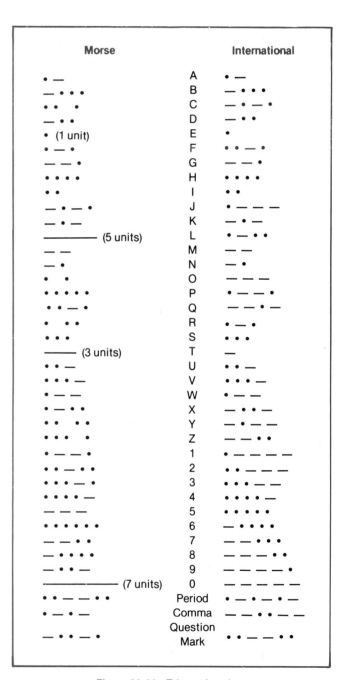

Figure 20-29. Telegraph codes.

regularly operate at speeds of 660 to 1250 words per minute. Figure 20-31 shows a computer-connected printer used in a modern communication system.

Facsimile

Another method for sending graphic information over information lines is *facsimile*. A facsimile is a reproduction of an original document. Graphic information is anything written, printed, or drawn on paper.

A facsimile (FAX) machine reads, or scans, documents photoelectrically. A light beam scans an original document. The light values it senses are converted into electrical signals. The signals are then transmitted to a receiver. The receiver reproduces the document.

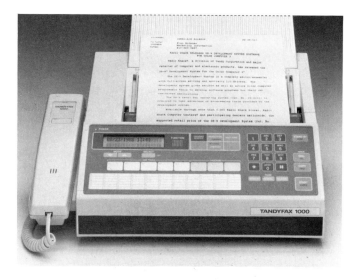

Figure 20-32. Facsimile machine. (Tandy Corporation)

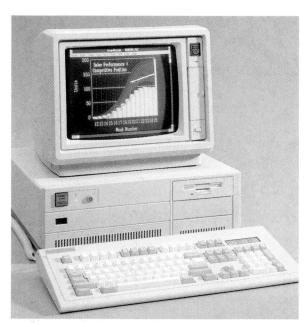

Figure 20-30. The Tandy 4000 microcomputer can handle 32 bits at a time. This high speed allows the user to produce exceptional graphics. (Radio Shack, a division of Tandy Corporation)

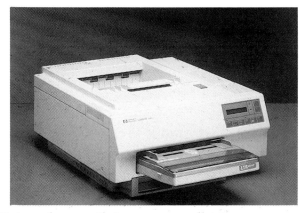

Figure 20-31. This small desktop laser printer produces very-high-quality printed copies at a high speed. (Hewlett-Packard Company)

Machines of this type are in use in many offices. Modern units send signals over telephone lines, Figure 20-32.

Police departments use the same type of machine. The facsimile units can send and receive such documents as criminal records. Suppose a suspect is arrested in one city. The police suspect he is wanted in another city. Fingerprints and photographs can be transmitted in minutes. These are used to establish identification.

Banks also use facsimile machines. They transmit account information and signature records. These also establish identification at points where the information is needed.

REVIEW QUESTIONS FOR SECTION 20.7

1. Who made the first practical telegraph?
2. Explain how the telegraph works.
3. What is the Morse code for the letter *X* and what is the International code for that letter?
4. What does *SOS* stand for?
5. How fast can communications printers operate?
6. How are TV monitors different from TV screens?
7. What are facsimiles? Where are they used?
8. How does a FAX operate?

20.8 THE TELEPHONE

Even as the use of the telegraph was spreading, scientists were busy devising the next communications breakthrough. Throughout the world, the scientists were striving for a system that would carry the human voice electrically. In effect, there was a race to invent the telephone.

Alexander Graham Bell was the first to succeed. He built his first telephone instrument in 1875. By today's standards, it was crude, Figure 20-33. This instrument was not a telephone as we now know it. It could transmit the basic sounds of voices, but quality was a problem. The spoken word could not be understood at the receiving end. However, this instrument did establish a technology. The principle was sound enough so that Bell was awarded a patent. Then he continued to work on his invention. He was able to transmit understandable speech at a public showing on May 10, 1876.

Today, millions of individual telephones are linked into local exchanges. The exchanges are connected in a network, Figure 20-34. The network forms what is probably the most automated, highest-quality-service system in the world. Semiconductor devices do the switching and have completely replaced the relay in the exchanges. That means as soon as you dial the last number of a long-distance call, the phone on the other end will ring. The Touch-Tone® system produces tones that are quickly identified by filters and sent on their way to the proper location within the exchange. Dial phones are a little slower. The pulses have to be converted to tones before they can be utilized to operate the switching gear in the automated exchanges. The push-button phone is shown in Figure 20-35.

Cellular Phones

Newer means of telephone communication rely on **cellular phones** and radio transmission in addition to wires. Such phones operate from cars or any other site, Figure 20-36.

Modems

Telephone lines are the least expensive way to transmit digital data over long distances. One of the most popular schemes to transmit a series of pulses over telephone lines is FSK, which was discussed earlier. Inasmuch as the phone lines were designed for the transmission of voice, they are not necessarily ideal for the transmission of data.

Figure 20-34. This digital central-office switch in Dallas, Texas, enables telephone companies to offer new services to customers. (S.W. Bell Corporation)

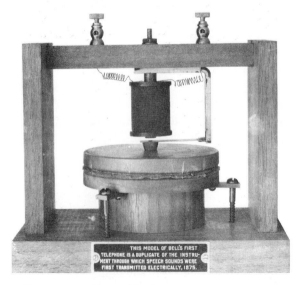

Figure 20-33. Model of Bell's first telephone. (A. T. & T.)

Figure 20-35. Touch-Tone® telephone. (A. T. & T.)

Figure 20-36. Cellular phone. (Tandy Corporation)

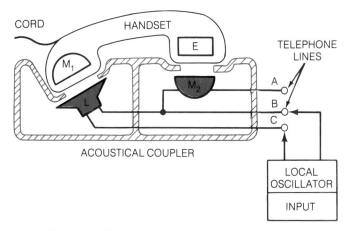

Figure 20-37. Acoustical coupler or early-type modem.

90 volts at 20 Hz. The 90 volts AC can ruin most electronic circuitry. Most telephone lines have 48 volts DC across them and the electronics in the modem must be protected from these excessive voltages.

REVIEW QUESTIONS FOR SECTION 20.8

1. When did Alexander Graham Bell invent the first telephone?
2. How does the Touch-Tone® phone operate switching gear?
3. Why are dial phones slower than Touch-Tone® phones?
4. How do cellular phones work?
5. What is a modem?
6. Why can't a computer be connected directly to telephone lines?

FSK encoding can be used to transmit digital data over phone lines. Low- and medium-speed transmission are possible using an acoustical coupler, Figure 20-37. This is a simple way of connecting the telephone to the computer. Tones are fed into the circuit and then modulated in pulses as the data are generated. The rate at which this is done is limited to the method being used to encode the tones used as carriers.

A couple of integrated circuits are used to code and decode the transmissions. A **modem** is a system that modulates and demodulates signals so that they can be transmitted over various communications systems. Directly wiring into the phone line makes what is called the baud rate somewhat faster, Figure 20-37. *Baud* is the time it takes for the shortest signaling element to operate. Audio tones from 300 to 3000 Hz are fed to the L (loudspeaker) in Figure 20-37, and the telephone earphone signals are picked up by the coupler microphone, M_2. That means the digital data are coded onto the 300-Hz tone and sent along the two wires of the telephone system to a receiver at the other end that will decode or separate the tone from the digital data.

Direct coupling to the lines requires protection of the electronic circuitry from the ringing voltage of about

20.9 CITIZENS' BAND RADIO

A number of radio channels have been set aside for private use by average citizens. These are called **citizens' band**, or *CB*, channels. Many CB radios are mobile (movable) units. They are installed in cars and trucks. There are also miniature CB units that can be carried in a person's hand. These are sometimes called "walkie-talkie" radios.

Some CB sets, called base units, are also located in homes or offices. These units are often used for communication with mobile operators. Part of the use of CB sets is for exchanging information. Drivers talk to each other on CB channels about road conditions. They call each other for help in emergencies. They also hold friendly conversations. Having people to talk to helps keep drivers alert and awake. So, CB makes some contribution to highway safety.

Figure 20-38. CB transceiver, or transmitter–receiver.

There are four classes of CB frequencies. Class-A transmitters produce signals between 460 and 470 MHz. This band can have input power to the antenna of 60 W. Class-B operators can use the 460-to-470-MHz band. But power is limited to 5 W. Class C is limited to 72 to 76 MHz. It is reserved for use by hobbyists for radio control of model airplanes and boats. Other Class-C band frequencies are 27.23 MHz to 29.96 MHz. These are used for remote control of other devices. Class-D frequencies range from 26.96 MHz to 27.93 MHz, with maximum outputs limited to 4 W.

To operate on the lower frequencies of Class-C bands, persons must be 12 or older. For the other bands, users must be 18 or older.

The great majority of CB use is on Class-D frequencies. These frequencies have been divided into 40 channels. Of these, one (Channel 9) is set aside specifically for emergencies. Channel 9 operates at 27.065 MHz.

CB equipment is illustrated in Figure 20-38. Because of the limit of 4 W on power, CB units are usable only for short distances. Persons who want to communicate over longer distances by radio need different equipment. They must also qualify for special licenses.

REVIEW QUESTIONS FOR SECTION 20.9

1. What is the purpose behind allocation of CB radio channels?
2. How many CB channels are on Class-D frequencies?

20.10 AMATEUR RADIO

Worldwide communication through code and voice is possible over amateur radio equipment. Persons who use this equipment are generally called *ham operators* or *hams*. Specific frequencies have been set aside for their use.

To qualify to use these airwaves, individuals must pass tests and receive licenses. In the United States, licensing is handled by the Federal Communications Commission (FCC). All ham operators must qualify in both code and voice transmission. Hams may operate fixed stations, usually in their homes. They may also have mobile units in cars or trucks. There are six classes of ham licenses. These are based on qualifications and skills of the operators.

The lowest, beginning level is the *novice* class. A novice must be able to send and receive five words per minute in code. There is also a test about basic radio theory and operation. Other classes are *technician*, *conditional*, *general*, *advanced*, and *amateur extra*.

An extra-class operator must be able to send and receive in code at 20 words per minute. Special channels are set aside for extra-class operators. These are 3.5 to 3.525 MHz. This is part of the 80-meter band, which includes 3500 to 4000 kHz. Other popular bands include the 40-meter band, which covers 7000 to 7300 kHz. The 20-meter band covers 14,000 to 14,350 kHz. Higher bands are also available. The 2-meter band covers 144 to 148 MHz.

REVIEW QUESTIONS FOR SECTION 20.10

1. What are the requirements for a novice ham license?
2. What code speed is required for an amateur extra-class ham license?

20.11 SATELLITE COMMUNICATIONS

As demand for communication channels has increased, shortages have developed. Demand is expected to exceed supply for cable-connected and radio channels. One method for expanding the capacity of people to communicate is the **satellite**.

A communication satellite is equipped with electronic equipment. This equipment is designed to receive and relay high-frequency radio signals. The satellite is placed in what is called a *stationary orbit*. This means that it travels in space at a speed equal to the rotation of the earth. Thus, the satellite is always at a specific position above the earth. Figure 20-39 shows how satellites can beam signals down to large areas or to smaller targeted areas. Figure 20-40 is a photograph of a satellite dish and receiver for a television system.

Ground stations transmit signals to the communication satellites, Figure 20-41. The relayed signals are received at ground stations with special antenna units. Traffic transmitted can include the full range of electronic communication. Examples include television programming, military and government messages, ship navigation, and ordinary telephone calls.

Most satellite-relayed messages travel in part through telephone networks. The ground lines carry the signals to the transmitters. Then other ground lines move the messages from the receiving stations to their destinations. Some large organizations have their own sending and

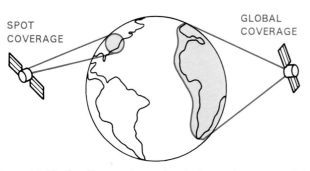

Figure 20-39. Satellites can beam signals down to large areas of the earth (global coverage) or to smaller, targeted areas (spot beams). (Reprinted from *Communication Technology* by Barden and Hacker, © 1990, Delmar Publishers Inc.)

Figure 20-41. Satellite communication ground station. (A. T. & T.)

Figure 20-40. Satellite dish and receiver. (Radio Shack, a division of Tandy Corporation)

receiving stations. These relay messages from one location to another entirely via satellite. Systems are also being introduced that transmit directly to homes from satellites. Satellite communication is expanding rapidly. This expansion is expected to continue for many years.

20.12 CABLE TELEVISION

A cable-television system is usually referred to as a *CATV*, which stands for *community antenna television*. Cable started as a means of bringing television to those located in valleys and remote areas where commercial or public television signals were not available or were of such low quality as to be useless.

The main objective of a cable television system is to deliver high-grade TV signals to a subscriber's home. Note the word *subscriber*, for it designates that the person pays for the privilege of watching a number of channels other than just those available from stations located nearby. The cable TV system is a system that is well established. It is possible to shop and watch movies

without leaving home. Special-interest programming is available inasmuch as many channels are possible.

Figure 20-42 shows a complete television cable system. It also includes the cable link. The cable system deals with the radio-frequency signal from the TV station. The *head end* is where the television channel is received and then processed for distribution through the cable network. The heart of the cable TV system is the coaxial cable transmission line. The other components of the system are used to get signals into or out of the cable or to overcome some of the basic limitations of the cable itself.

Some signal losses occur in the cable system. Amplifiers have to be used to amplify the signal so that every receiver or subscriber receives a good-quality picture. Three categories of amplifiers are used in cable-TV systems: (1) the *trunk*, or mainline, amplifiers, spaced along the main trunk lines of the system to compensate for the cable attenuation; (2) the *bridger* amplifiers, located either in the same case with the trunk amplifiers or in a separate place, and used to take off the trunk line for distribution in feeder cables to the subscriber's neighborhood; and (3) the *line extenders*, are used along the feeder system where required. Some amplifiers can be used for any of the three purposes.

Some CATV systems provide more than 24 channels. Cable operators also produce their own programs for local interests. Satellite-TV links are another innovation that is

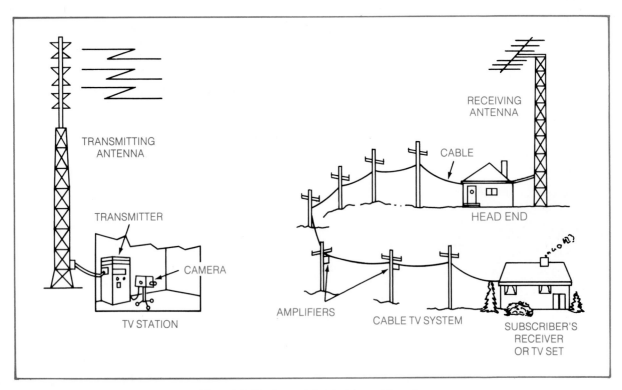

Figure 20-42. A complete television cable system.

expected to develop rapidly, and have, in some locations, replaced cable TV.

20.13 FIBER-OPTIC COMMUNICATIONS

Another approach to expanding communication channels is through use of light. The principle used is similar to that in radio transmission. The major difference is that the frequencies used are in the visible (light) ranges. Two methods have been used to send communication signals via light.

One is the *laser* beam. A laser is a powerful beam of light transmitted through air or special tubes. A laser uses "coherent" light. A laser beam has light waves that are extremely close together in their wavelengths. Laser beams can be used to encode and transmit communication signals.

The other light-wave communication carrier is **fiber optics**. This uses a strand of glass about the same thickness as human hair. Fiber optics is illustrated in Figure 20-43. Transmission of information is at high frequencies. Because of this, fiber optics can handle more messages than a large electrical cable can. A communication message conductor using fiber optics is illustrated in Figure 20-44.

The capacity of fiber optics is greater than that of anything used in the past. For example, one of the first uses of this technique, in Chicago, involved use of hair-thin fibers. Each fiber-optic strand had a capacity to carry the equivalent of 576 telephone calls. These carriers are supplementing the use of communication cables. They are handling telephone calls and other commercial traffic.

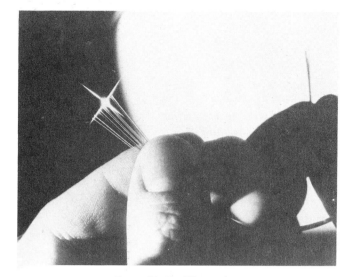

Figure 20-43. Fiber optics.

Figure 20-44. Fiber-optics conductor. (A. T. & T.)

SUMMARY

Radio frequencies cover a range above 20 kHz. The RF spectrum was studied in 1857 by Hertz, who extended the findings of Maxwell. There are bands of frequencies classified as VLF, LF, MF, HF, VHF, UHF, SHF, and EHF, each occupying a range of frequencies.

AM transmitters utilize amplitude modulation to send music and voice transmissions from one point to another. The transmitter has an oscillator, usually a buffer, and an intermediate power amplifier as well as a power amplifier to drive the antenna. There are audio amplifiers to boost the signal to sufficient levels to modulate the RF carrier in the output stage of the transmitter.

Sidebands are produced when a modulating frequency modulates a carrier frequency. There is an upper and a lower sideband. In some instances, single-sideband transmissions are utilized so that two channels can be used on one frequency.

FM transmitters use frequency modulation to transmit music, voice, or data. This type of transmitter utilizes stages such as frequency multipliers and involves modulation index and frequency deviation. FM broadcasting uses 88 to 108 MHz or line-of-sight transmission.

Radio receivers date to 1901, when Guglielmo Marconi transmitted a signal without wires across the Atlantic Ocean from England to Newfoundland. Voice transmissions were accomplished in 1915 when a signal was sent from Arlington, Virginia, to Hawaii.

Receivers take the transmitted signal, amplify it, detect or separate the audio from the radio-frequency carrier and then amplify the audio until it is sufficient to drive a speaker. The AM broadcast band covers 535 kHz to 1605 kHz.

The stages in an AM receiver are the RF amplifier, mixer, oscillator, IF amps, detector, audio amps, speaker, and power supply. Both AM and FM can now be purchased on a chip or integrated circuit. All the essential parts of the receivers are brought down to the size of a chip.

Television cameras convert the scene being photographed into electrical signals that can be impressed on an RF carrier wave and broadcast to receivers in a wide area. Liquid-crystal displays have now been utilized to replace television picture tubes. So far, LCDs have been limited to very small displays. They do show promise for flat TV pictures in the near future.

Quadrature detection is utilized in both FM and TV receivers. This uses an IC or chip to accomplish the job of separating the video or audio information from the carrier wave.

Data can be transmitted on telephone lines or sent through the air on FM carriers. FSK, PPM, and PWM are utilized as methods of accomplishing data transmissions without errors or noise problems.

Receivers and transmitters need antennas. AM receivers have the simplest types of antennas. They range from a piece of wire to many turns of small-diameter wire on a ferrite core. FM can also use a receiving antenna made of a short piece of wire. However, for better reception, both FM and TV need antennas that are cut to fit the frequencies being received. TV and FM can both use the folded-dipole antenna while the bow-tie type is used for reception on the UHF band.

The telegraph dates back to 1832 and Samuel F. B. Morse's experiments in sending messages along a wire for long distances. Improvements have been made, and today we send data rather routinely over long distances by way of wire and FM transmissions, as well as utilizing the satellites parked in orbit to relay messages.

The computer monitor has a slightly different picture tube than does the television set. It does not need a tuner to separate out various channels. The resolution is better on made-for-computer monitors than on TV sets adapted to produce the numbers and letters so often associated with computer readouts.

The facsimile has taken on new uses recently and has begun to replace overnight-letter-delivery services. The facsimile can transmit a whole letter or picture by use of telephone wires.

Alexander Graham Bell made his first telephone instrument in 1875. The telephone has moved a long way since the first crude instruments, and today 742 parts are used to make a telephone.

Telephone lines are the least expensive way to transmit digital data over long distances. One of the popular schemes to transmit a series of pulses over telephone lines is FSK, or frequency-shift keying. A couple of integrated circuits are used to code and decode transmissions. A modem is a system that modulates and demodulates signals so they can be transmitted over various communications systems.

Citizens' band radio has lost some of its mass appeal but is still utilized in many ways by persons who travel the interstate highways. It is an inexpensive way of communicating with others who have the same type of transceiver. CB operates on 40 assigned channels produced by a crystal and a phase-locked loop IC.

Amateur radio has certain bands assigned for use by licensed operators who develop these bands as a hobby.

Satellite communications is expanding constantly as the demand for more channels becomes obvious. It is possible to connect a radio network by way of satellite and not use phone lines to each location. Each radio station or television station has its own receiving dish to pick up the satellite feed.

The laser beam is being channeled through fiber optics to produce clear and noise-free telephone communications with less expense than by wire.

Cable television is usually referred to as CATV, or community antenna television. One location may generate its own programs or pick up stations not easily obtained by homeowners, and provide the service to everyone who pays a monthly fee.

USING YOUR KNOWLEDGE

1. Send a message in International Morse Code. See whether you can tap it out and have a partner understand your message.

2. Place two radios close together. Try to tune both to the same frequency. Then slightly move the dial on one of them. Do you hear the beating of the oscillator signals? What produces this squealing sound? What frequency is it?

3. Try de-tuning an FM receiver (one with a sliding dial) so that on the high end you can pick up a local television channel's audio. If you have a sliding-dial TV receiver you can get an FM station between channels 6 and 7.

4. Look through a Radio Shack catalog and try to find the number needed to order a chip with an AM radio on it.

5. Tune in an amateur-radio receiver until you can hear the tone transmission of messages using one of the pulse-modulation systems.

6. Connect your personal computer to a color or black-and-white TV to display its output. Note the difference in quality in the display on the monitor and that on the TV screen.

7. What channel does the State Police monitor on CB for emergencies? Look for the signs on the interstate highways to see which is monitored in the particular locales you are in at the time.

8. Why won't a TV set receive local stations on their proper assigned location on the dial when the set is connected to a cable system?

KEY TERMS

radio frequencies (RF)
modulated continuous wave
modulator
modulation index
multiplexing
quadrature detection
frequency-shift keying
pulse-width modulation
pulse-position modulation
telegraph
Morse code
international code

facsimile
transmitters
amplitude modulation
frequency modulation
vidicon
raster
satellite
citizens' band
modem
monitor
cellular phone
fiber optics

Chapter 21

DIGITAL CIRCUITS

OBJECTIVES

After studying this chapter, you will know

- *how logic functions such as AND, OR, selection, and switching circuits are performed in electromechanical circuits.*
- *how variable resistors can be used to control current in both AC and DC circuits.*
- *how variacs and saturable reactors are used to control current.*
- *the principles behind digital computers.*
- *how to use truth tables to determine functions and values of digital computer circuits.*
- *how AND, OR, NOT, NAND, and NOR circuits function.*
- *the principles and methods of binary coding.*
- *the basic differences between microprocessors and microcomputers.*
- *the purpose of and the operation of the central processing unit (CPU).*
- *the difference between bits and bytes.*
- *how the CPU works.*
- *how various types of memory are used for storage and retrieval.*
- *the purpose of software.*
- *the purpose of and the operation of the joystick and mouse.*
- *the difference between a mainframe and PC.*

INTRODUCTION

Circuits hold the key to the operation of calculators and computers. These are known as **digital circuits**. The term comes from the fact that the circuits are used to calculate. That is, the circuits process digits, or numerical information. Digital circuits also make simple decisions based on the signals they process. For example, this chapter helps you learn about circuits with names like NOT and OR.

In this chapter, you learn some practical applications of relays and switches. You learn how to make a switching tree. Other content deals with controlling current with diodes and using transistors for switching functions.

Relays and switches are easily understood. Contacts open and close. The process is visible. With transistors and diodes, there are no visible actions to watch. So, circuit analysis can be more difficult.

You already understand basic switching circuits. This is your starting point for learning about digital circuits. Digital circuits switch current. The major difference lies in the type of switch used: In computers, switches are electronic rather than mechanical. Keep this relationship in mind as you build your digital knowledge.

21.1 ELECTROMECHANICAL CIRCUITS

Relays and switches are classified according to use. A single switch controlling a light bulb is called a **direct-switch circuit**, Figure 21-1. This diagram shows the elements of a complete circuit.

AND Circuit

In some situations, it is necessary to place several switches in series, Figure 21-2. If this is done, the result is a **series switching circuit**. This is also known as an **AND circuit**. In Figure 21-2, both switches—*A and B*—must be closed to light the lamp. The light is not energized if either switch is open. A single condition must be met: Both switches must be closed.

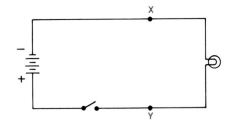

Figure 21-1. Diagram of direct-switch circuit showing a single switch controlling the complete circuit.

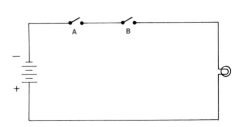

Figure 21-2. Diagram of circuit with switches in series controlling lamp.

OR Circuit

Another situation in which the closing of two switches lights a lamp is shown in Figure 21-3. In this circuit, the light is energized if either switch *A* OR switch *B* is closed. The lamp will also glow if both *A and B* are closed at the same time. There are two conditions under which the lamp will glow. So, this is known as an **OR circuit**.

Combined AND/OR Circuit

It is possible to combine the AND circuit and the OR circuit. Such a circuit is shown in Figure 21-4. To energize the lamp, switch *C* must be closed *and* either *A or B* must also be closed. Thus, *A* and *B* form an OR circuit. Combining these switches with *C* produces an AND circuit. For the light to glow, *AC* or *BC* must be closed. If only *A* and *B* are closed, the lamp is off.

These are the basic direct control circuits for switching action. Many switches can be connected to form AND circuits. The same is true for OR circuits. Many combinations involving both types of circuits are also possible.

Selection Circuits

Circuits can be selected one at a time. Figure 21-5 demonstrates how this is done. A stepping switch is used. In this circuit, any one of seven points may be selected. Selection is done by causing the switch to step around to a particular contact. This is the prime function of a stepping switch.

Another type of selection circuit is built around the use of type *C* contacts. This is illustrated in Figure 21-6. In this circuit, outputs 1, 2, 3, or 4 may be selected by operating switches *A* and *B*. Output 1 is connected with both switch *A* and switch *B* in their upper positions. Note that *B* is a double-pole, double-throw (2C) switch. If *B* is in the upper position and *A* is in the lower position, current flows to output 3.

This circuit is usable for the selection of several outputs. But it still uses a minimum number of switching contacts. In fact, it is possible to build this circuit out to provide many more outputs. This is done in Figure 21-7.

The circuit in Figure 21-7 has four switches. *A* is single-pole, double-throw. *B* is double-pole, double-throw. *C* is four-pole, double-throw. And *D* is eight-pole, double-throw.

These could be relay contact springs just as well as manual switches. If relay springs were used, this circuit would be called a relay selection tree. Selections may be made from 16 separate circuits through use of only four relays. This simplifies circuit selection. Information coded to operate the four relays can result in selection of outputs to as many as 16 separate circuits. Such selection trees were widely used in electromechanical telephone switching centers. They were also popular in early computers.

In computers, the circuit in Figure 21-7 performed an important function. This type of circuit can produce a decimal notation from a binary code. Switches are given values: *A* is 8; *B* is 4; *C* is 2; and *D* is 1.

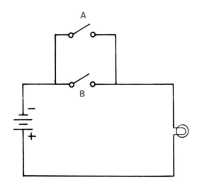

Figure 21-3. Diagram of circuit in which parallel switches control lamp.

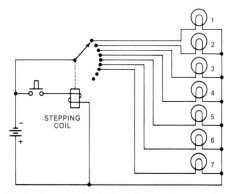

Figure 21-5. Diagram of selection circuit using stepping switch.

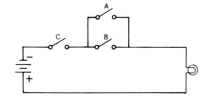

Figure 21-4. Circuit diagram showing combination AND/OR circuit.

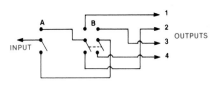

Figure 21-6. Diagram of selection circuit using type C contacts.

Rearrangement of this circuit for computer output is shown in Figure 21-8. Output is delivered to the circuit that represents the sum of the value of the selected relays. Thus, if *A* and *C* are energized and pulled down, the output goes to circuit 10. This is the sum of the values (8 and 2) assigned to the switches. If all four relays were energized and pulled down, output would go to circuit 15. This is a combination of all values — 8, 4, 2, and 1.

Through relay choices, the transfer of current takes place within the selection-tree circuit. This makes it possible to add numbers by using designations assigned by the relays.

Any two-condition device can be used as a binary counter. These devices may be diodes, relays, or transistor switching circuits.

Transfer Circuits

Relays and switches may be arranged for transfer purposes. This application is shown in Figure 21-9.

The illustrated circuit has three relays. This allows a transfer function to take place. Each relay has *1-C* contacts. Assume the contact to be up, as shown. When the relay is energized, the contact drops. When relay *A* is energized, the output transfers to terminal *A'*. If relay *A* is not energized but relay *B* is energized, the current goes to point *B'*. Now suppose relays *B* and *A* are not energized, but *C* is energized. The current goes to *C'*.

If none of the relays is energized, the output goes to point *X*. This is called a transfer function. With this type of circuit, it is possible to transfer current to any of several outputs.

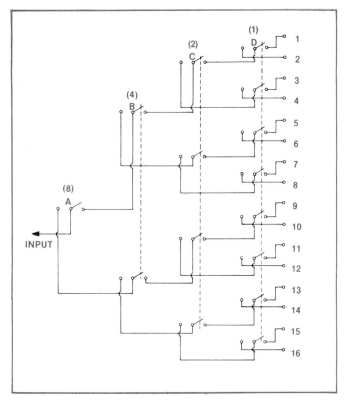

Figure 21-7. Circuit diagram of switch selection tree.

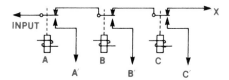

Figure 21-9. Diagram of transfer circuit.

Typical Switching Circuits

A relay can be used to change two devices from a series to a parallel hookup. This is possible with the contacts of a double-pole, double-throw switch or relay.

This type of circuit is shown in Figure 21-10, page 246. If the relay contacts are in the upward position, the two devices would be connected in series. When the relay contacts are in a downward position, the series circuit is broken. With the relays down, the lamps are connected in parallel.

This type of circuit is used in photography. Two flood lamps are arranged so they have less brilliance during focusing. This occurs when they are connected in series. Then, full light is needed to take the picture. Full brilliance is achieved by switching to a parallel circuit.

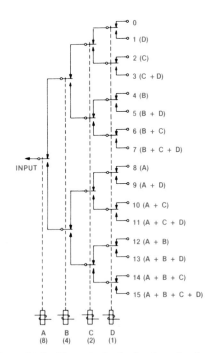

Figure 21-8. Diagram of selection-tree circuit used for computer output.

Activity: AND, OR, AND SELECTION CIRCUITS

OBJECTIVES

1. *To observe the operational characteristics of the AND type switching circuit.*
2. *To observe the operating characteristics of the OR type switching circuit.*
3. *To observe the action of a selection circuit using switches.*

EQUIPMENT NEEDED

1 Power supply, 6 volts DC	4 Lamps, 6 volts with sockets
2 Switches, DPDT, toggle	14 Wires for connections

PROCEDURE

1. Connect the circuit shown in Figure 1. This is the AND circuit, using two switches. Connect the DPDT switch so that only one section is used. Connect to the middle terminal on one side and the top terminal on the other side. Do the same with the other DPDT switch. This way, they can be used as SPST switches as needed in this circuit.

2. Move both switch A and switch B to the down position. *Turn on the power supply.* Adjust the output to 6 volts DC. Does the lamp light? Move the switch A to the up position. Does the lamp light? With switch A in the up position, move the switch B to the up position. Does the lamp light? Use a pencil to trace the circuit on the schematic drawing in Figure 1. Do it with the assumption that both switches are in the up position.

3. *Turn off the power supply.* Disconnect the circuit you have been using. Connect the circuit shown in Figure 2.

4. This is the OR circuit, using two switches. *Turn on the power supply.* Adjust the output to 6 volts DC. With both switches in the down position, does the lamp light? Move switch A to the up position. Does the lamp light? Return switch A to the down position and move switch B to the up position. Does the lamp light? On the schematic, trace the circuit with each switch in either up or down position. Is the circuit complete with either switch in the up position?

5. *Turn off the power supply.* Disconnect the circuit you have been using. Connect the circuit shown in Figure 3. This is the two-station control circuit.

6. *Turn on the power supply.* Adjust the output to 6 volts DC. With both switches in the down position, does the lamp light? Move switch A to the up position. Does the lamp light? With switch A in the up position, move switch C to the up position. Does the lamp light? On the schematic, trace with your finger or a pencil the path for current flow to the bulb and back to the power supply. Why must they both be in the same position for the lamp to light?

7. *Turn off the power supply.* Disconnect the circuit you have been using. Connect the circuit shown in Figure 4. This is the switch selection circuit.

8. *Turn on the power supply.* Adjust the output to 6 volts DC. With switch B up and switch C up, which lamp lights? With switch B up and switch C down, which lamp lights? With switch B down and switch C up, which lamp lights?

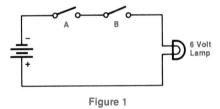

Figure 1

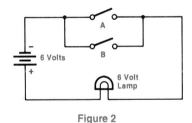

Figure 2

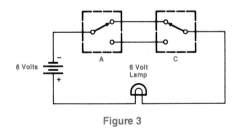

Figure 3

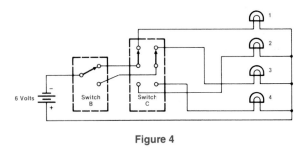

Figure 4

9. Disregard all lamps but #1. What must the positions of the two switches be for this lamp to light? Are the two circuits an AND circuit or an OR circuit?

10. *Turn off the power supply.* Disconnect all circuits you have been using and return all equipment and materials to their proper storage places.

SUMMARY

1. What is the difference in the action of the AND circuit and the OR circuit?
2. Describe a practical use of the two-station control circuit. Discuss its use in the home.
3. In what way does a selection circuit differ from the AND circuit?

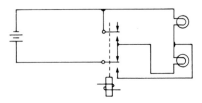

Figure 21-10. Diagram of typical switching circuit.

REVIEW QUESTIONS FOR SECTION 21.1

1. What is a digital circuit and what is its purpose?
2. What is an AND circuit and how does it operate?
3. What is an OR circuit and how does it operate?
4. What is an AND/OR combination circuit and how does it operate?
5. What are selection circuits and how do they operate?
6. What are transfer circuits and how do they operate?
7. What are switching circuits and how do they operate?

21.2 RESISTIVE-REACTIVE CONTROL OF ELECTRIC CURRENT

A resistor within a circuit limits the flow of current. This can be a very valuable function. Using resistors to limit current works well in either AC or DC circuits.

A variable resistor can be used to control the amount of current that passes through the load. A circuit demonstrating this capability is shown in Figure 21-11. Resistance in this circuit is adjustable. Maximum resistance occurs when the arm of the resistor is moved to the far right. At this point, very little current passes through

the circuit. As the arm is moved left, current flow increases. With the arm at the far left, current flow is at its maximum. In power circuits, such resistances are called rheostats.

Within circuits such as those used in electronics, different designs are used. The variable-resistance element may be connected as shown in Figure 21-12. This circuit can be called a voltage divider. Current to the lamp is still limited. However, the entire current passes through the resistance. Source voltage divides across the entire resistance. Moving the arm up or down takes off more or less voltage. If the arm is at the top of the resistance, the lamp receives full voltage. With the arm at the bottom of the resistance element, the reverse is true. The lamp receives no voltage. Therefore, it does not glow.

Current can also be controlled with electronic devices. This is done with a diode—a silicon controlled rectifier, or SCR. SCRs are used as light dimmers and motor speed controls. SCRs are solid-state electronic devices. They are discussed in Chapters 16 and 17.

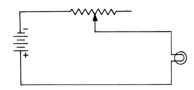

Figure 21-11. Diagram of circuit using variable resistance in series with lamp.

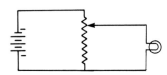

Figure 21-12. Diagram of voltage-divider circuit.

346

Activity: *SWITCH TRANSFER CIRCUITS*

OBJECTIVES

1. *To observe the action of a transfer circuit using switches.*
2. *To observe the action of a transfer circuit using relays.*

EQUIPMENT NEEDED

1 Relay, 6 volts, 150 ohms coil resistance, DPDT contacts, 2 sets
1 Push button, normally open
4 Lamps, 6 volts with sockets

3 Switches, DPDT, toggle
1 Power supply, 6 volts DC
10 Wires for connections

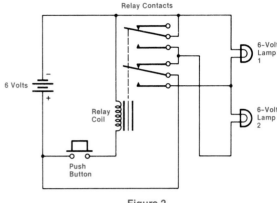

Figure 2

PROCEDURE

1. Connect the circuit shown in Figure 1.
2. With switch B in the down position, *turn on the power supply* and adjust its output to 6 volts DC. Are the lamps dim or bright? Use a pencil to trace the circuit in the schematic through the switch in its down position. Are the lamps connected in series or in parallel? How would this account for their level of illumination?

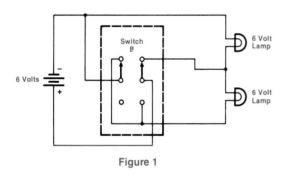

Figure 1

3. Move switch B to the up position. Are the lamps dim or bright? Use a pencil to trace the circuit in the schematic. If the switch is up, are the lamps connected in series or in parallel? How would this account for their level of brightness?
4. *Turn off the power supply.* Disconnect the circuit you have been using. Connect the circuit shown in Figure 2.
5. *Turn on the power supply.* Adjust to 6 volts DC. This is the same circuit you had in Step 1, except that the

relay contacts have been substituted for the switch. Are the lamps glowing dimly or brightly? Would this indicate that the lamps are connected in series or parallel?

6. Press the push-button switch. Does the relay operate? While the push-button switch is held depressed, are the lamps dim or bright? With the relay energized, are the lamps connected in series or parallel?
7. *Turn off the power supply.* Disconnect the circuit that you have been using and connect the circuit shown in Figure 3.

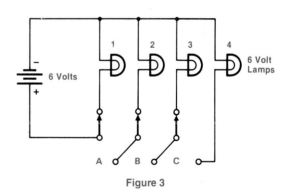

Figure 3

8. Place switches A, B, and C in the down position. *Turn on the power supply,* with its output adjusted to 6 volts DC. Which lamp lights?
9. Move switch A to the up position. Which lamp lights? Move either switch B or C to the up position. With switch A in the up position, do either of the other two switches have any control over the circuit?

Use a pencil, or your finger, and trace the circuit on the schematic. Why does only switch A control the circuit when it is in the up position?

10. Return switch A to the down position. Move switch B to the up position. Which lamp lights?
11. Return switch B to the down position. Move switch C to the up position. Which lamp lights? In this circuit, which switch has priority over the others?
12. *Turn off the power supply.* Disconnect all circuits you

have been working with and return all equipment and materials to their proper storage places.

SUMMARY

1. What is a transfer circuit? How does it operate?
2. Explain how the series-to-parallel switching circuit operates.
3. Where could this type of series-to-parallel switching be used in a practical everyday application?

REVIEW QUESTIONS FOR SECTION 21.2

1. What is a variable resistor?
2. What is a rheostat?
3. Describe the operation of a voltage-divider circuit.
4. How does the use of resistance affect the load in a simple circuit?

21.3 AC CURRENT CONTROL

AC current flow may be controlled by an inductor. Figure 21-13 shows a circuit with a variable inductor connected in series with a lamp. Varying the inductive reactance in this circuit controls the amount of current to the lamp.

Inductors are made variable by fitting them with removable cores. Inserting or removing the core from the coil controls the current.

Another current-control device is the **variac**. This is a tapped auto transformer, Figure 21-14. This type of

variable transformer is used in laboratories. Its value lies in the ability to deliver precise AC voltage. The tapped transformer works on much the same principle as the voltage divider. It is applicable only for AC circuits.

A third device for AC control is the **saturable reactor**, Figure 21-15. The saturable reactor is built on an E-I-type transformer core. The center leg has a control winding. The two outer legs have AC-controlled coils. Current passes through these AC coils, which are connected in series. If there is no current in the control winding, current flow is limited. Inductive reactance is the limiting factor.

Now, suppose a small amount of current is passed through the control winding. The DC saturates the magnetic path. The effect is the same as though the cores were removed from the two coils. Inductive reactance is greatly reduced. So, more alternating current flows to the lamp. The amount of current passing through the DC coil determines the degree of magnetic saturation. This leads to the name *saturable reactor.*

Sometimes, the saturable reactor is called a *magnetic amplifier.* This name comes from the nature of the device: A small amount of DC applied to a magnet controls a large amount of AC. This is possible because a magnetic flux pattern of DC is set up easily. Because DC is used, the pattern is not subject to the constant change of AC devices. AC magnetic flux fields repeatedly build up, then collapse, as current changes directions. Therefore, when a DC flux is established, it maintains control over AC fields.

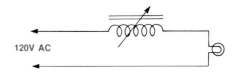

Figure 21-13. Diagram of circuit in which variable inductance determines circuit current.

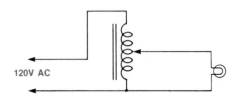

Figure 21-14. Diagram of circuit in which variac controls lamp voltage.

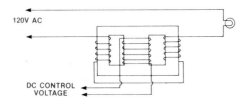

Figure 21-15. Diagram of saturable reactor.

21.4 THE GATE CIRCUIT

Electricity can also be controlled by determining the direction in which it flows. A device used to control electron flow is the diode.

A diode acts much the same as a one-way gate, Figure 21-16. The diode is connected to an AC source. Despite this, current is allowed to flow in only one direction. In this circuit, current from the AC line flows first through the diode to terminal B. Note that this direction of flow is opposite to the direction of the arrow used in the symbol of the diode. From terminal B, the current continues through the lamp to terminal A. Then it moves out of the circuit to the AC source. The other half of the cycle cannot flow through the diode. This is one way in which the direction of current flow is controlled.

Many circuits depend upon current direction control for their operation. An example is shown in Figure 21-17. In this circuit, box 1 is connected by two wires to box 2. Box 1 has two push buttons. Note that box 2 has two light bulbs. To light these bulbs with two switches, three wires would be needed.

The key to making this circuit work lies in the nature of AC. The current flows first in one direction, then in the opposite direction. In this circuit, current enters at the bottom terminal of the AC line. The current passes through light B. Then it moves through the diode connected to light B. From there, the current moves out through switch B and its diode. The current is carried back to the AC line.

If only switch A is closed, lamp B cannot light. This is because the two diodes are working in opposition. Lamp A lights because the current has a path. The current can flow first through the switch. Then it moves through the diode and down the line. The current can pass through the diode above lamp A. Then it moves through the lamp and back to the line.

To summarize: If switch A is pressed, light A will light. If switch B is pressed, light B will light. If both switches are pressed, both lamps will light. This is easy to do with diodes. Diodes are often used in this type of circuit. They control the direction of current flow. Computers use thousands of such diodes to control the direction of current flow.

21.5 DIGITAL-COMPUTER CIRCUITS

Computers have become vital tools for business, industry, and government. They are also playing increasingly important roles in school and in the home. Many millions of persons require some knowledge of computers to hold their jobs. So, an understanding of how computers function can be valuable knowledge to have. A good place to start is to keep building your knowledge of logic circuits. Computers depend on logic circuits for their basic operations.

Units of Information

Computers perform their functions on a binary basis. Binary values are 0 and 1. Binary functions of computers are based on switch conditions. Switches are either on or off. An *off* condition represents a 0. An *on* condition represents a 1.

Information in a digital computer is represented by strings of digits. These digits assume one of the two values 0 or 1. These units of information are called **bits**. The

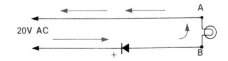

Figure 21-16. Diagram of gate circuit.

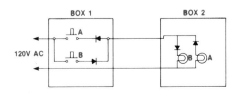

Figure 21-17. Diagram of circuit in which diodes control current.

term *bit* is a contraction for "binary digit." Bits are used internally in the computer to represent both numerical and nonnumerical information. Bits can be converted to magnetic form and stored on memory devices, tape, or discs. In some memory systems, small bubbles are used to store bits of information. Magnetic tapes or discs can be used to put in pulses needed to cause the circuit to operate.

To increase efficiency when dealing with information, a fixed number of bits are grouped together and referenced as a discrete unit. These units are used to encode and format the information that can be processed by the computer. Units of 8 bits are common and are called a **byte**. Bytes are used to encode the basic symbolic characters that provide the input–output (I/O) information such as the alphabet, decimal digits, punctuation marks, and special characters. The byte size is not universally adopted on all machines but is perhaps the most popular basic unit of information.

The larger bit groups are organized into units called **words**. These units are used to encode the basic instruction repertoire of the machine and format the numerical data. The common word sizes used on various commercially available computers are 16, 24, 32, 36, 48, and 60 bits.

AND Circuit

Switches, relays, and diodes can be used to create logic circuits. You have already learned about several examples. In addition, the transistor can be used as a rapid, reliable switch. This chapter shows transistors used in some circuits. You learned about transistors and how they work in Chapter 16.

Figure 21-18 shows a basic AND circuit. The lamp is the output. Switches *A and B* must both be closed for the lamp to glow.

Figure 21-19 is a basic AND circuit with a transistor. When switches *A and B* are both closed, the transistor is energized. Conduction through the transistor causes current through *R*. This is in the same relative location as the lamp in Figure 21-18. The flow of current through *R* produces a voltage drop. This voltage can then be passed along to the next processing stage in a computer

system. In Figure 21-19, current can also be fed to another circuit, *C*. That circuit can also perform a function needed by a computer.

Figures 21-18 and 21-19 both show AND circuits. The symbol for an AND circuit is shown in Figure 21-20.

Truth Table

A truth table shows how the inputs and outputs of a circuit are obtained. For instance, in Figure 21-18, if *A* and *B* are off, this means that *C* must be off. Since this is an AND circuit, both *A and B* must be closed for *C* to have an output.

Look at the truth table in Figure 21-21. The first line shows that $A = 0$, $B = 0$, and $C = 0$. On the second line, values indicate that $A = 0$, $B = 1$, and $C = 0$. Again, the lamp will not light. Remember that this is an AND circuit. So, both switches must be closed. Now look at the third line on the truth table. Both *A* and *B* have values of 1. Therefore, $C = 1$. Both switches are closed. Thus, current is delivered to the lamp. The fourth condition shows $A = 1$, $B = 0$, so $C = 0$.

Transistors and diodes are extremely small. So, it is possible to have these switching functions carried on in small spaces. Complex problems can be solved by having a circuit either conduct or not conduct at the proper time.

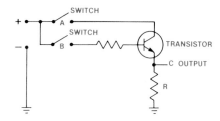

Figure 21-19. Diagram of basic AND circuit using a transistor.

Figure 21-20. Symbol for AND circuit.

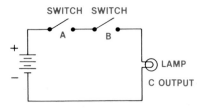

Figure 21-18. Diagram of basic AND circuit with lamp as output.

TRUTH TABLE		
INPUTS		OUTPUT
A	B	C
0	0	0
0	1	0
1	1	1
1	0	0

Figure 21-21. Truth table for AND circuit in Figure 21-19.

OR Circuit

Another computer building block mentioned earlier is the OR circuit. Figure 21-22 shows an OR circuit with two switches in parallel and a lamp. If either *A or B* is closed, the lamp will glow. This feature identifies an OR circuit.

Figure 21-23 is an OR circuit with a transistor. Closing either *A or B* will complete the transistor's base bias circuit. With either switch closed, the transistor receives and can conduct current. The transistor used in OR circuits is a sensitive control over the current in the load resistor *R*. With the base open, there will be little if any current through the transistor. When the base circuit is made positive, current (at a maximum or saturation level) will flow. The current flows through the transistor and the load resistor is connected in series.

The symbol for an OR circuit is shown in Figure 21-24. Notice that it is shaped a little differently from the AND symbol.

A truth table for the OR circuits in Figures 21-22 and 21-23 is shown in Figure 21-25. If both switches, *A* and *B*, are open and have values of 0, then the output value, *C* is also 0. If *A* = 0 and *B* = 1, then *C* = 1. If *A* = 1 and *B* = 1, then *C* = 1 as well. If *A* = 1 and *B* = 0, *C* = 1. Note that there are three conditions in this table in which *C* = 1. That is because it is an OR circuit. It can operate with either *A or B* closed. Remember, *1* means *on* or *closed*. *Zero* means *open* or *off*.

NOT Circuit

Another building block vital to the operation of computers is the **NOT circuit**. Figure 21-26 illustrates a basic NOT circuit. *A* is the control switch. *B* is the output, a lamp. There are two power sources. One source controls the relay. The other controls the lamp by furnishing power when the relay contact is closed.

The relay contacts are important. Notice that the relay contacts are closed when the switch is open. This means that *A* = 0 and *B* = 1. When switch *A* is closed, the relay energizes. This causes the relay contacts to open. Opening the contacts causes the lamp to go out. This is a

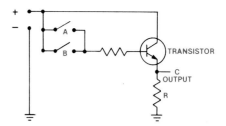

Figure 21-23. Diagram of basic OR circuit using a transistor.

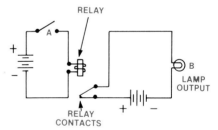

Figure 21-24. Symbol for OR circuit.

TRUTH TABLE		
INPUTS		OUTPUT
A	B	C
0	0	0
0	1	1
1	1	1
1	0	1

Figure 21-25. Truth table for OR circuit in Figure 21-23.

Figure 21-26. Diagram of basic NOT circuit with lamp as output.

condition represented by 0 in the logic circuits. That means this is an *inverting* circuit. The closing of switch *A* causes lamp *B* to go out. This is the opposite of what usually happens in a circuit. Therefore, it is called an inverting function. This circuit is also referred to as an *inverter*.

Now look at the circuit in Figure 21-27. This is a NOT circuit with a transistor. Note that there are two power supplies. When switch *A* is open, no current flows in the circuit. When switch *A* is closed, current flows through the transistor to output *B*. From *B*, the output moves

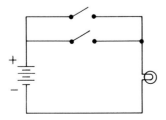

Figure 21-22. Diagram of basic OR circuit with lamp as output.

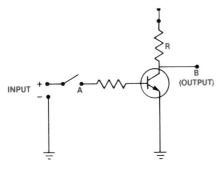

Figure 21-27. Diagram of basic NOT circuit using a transistor.

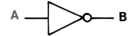

Figure 21-28. Symbol for NOT circuit.

TRUTH TABLE	
INPUT	OUTPUT
A	B
0	1
1	0

Figure 21-29. Truth table for NOT circuit in Figure 21-27.

across a resistor, *R*. In this type of circuit, the batteries have to be connected with proper polarity. This connection of polarity determines whether or not the transistor conducts. Figure 21-28 shows the logic symbol for a NOT (inverter) circuit.

Figure 21-29 is a truth table. This shows only two conditions for this type of circuit. When $A = 0$, $B = 1$. That is, when *A* is off, *B* is on. The other condition is that when $A = 1$, $B = 0$. This means that when *A* is on, *B* is off. There are only two conditions because the circuit has only one switch. The switch can be either on or off.

NAND Circuits

A **NAND circuit** is an AND circuit with a negative structure. In effect, a NAND circuit does the same thing within a computer system as an AND circuit. The only difference is in output. Outputs for a NAND circuit are exactly opposite from those of an AND circuit.

These relationships are shown in Figure 21-30. Included are the symbols for both the AND and NAND circuits. It also presents truth tables for both types of circuits. Notice that the only difference in symbols lies in the circle added to the output side. This identifies a negative circuit structure.

In designing computer circuits, NAND structures are used because they are easier and less costly to make. Since NAND circuits can substitute easily for AND circuits, they are used in the majority of cases.

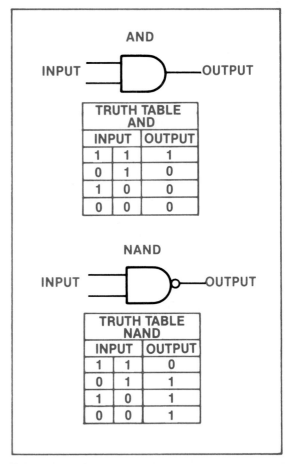

Figure 21-30. Symbols and truth tables for AND and NAND.

NOR Circuits

A **NOR circuit** is a negative OR circuit, Figure 21-31 on page 355. This illustration shows that OR and NOR symbols are similar. The only difference is a negative circle on the output. This indicates a negative circuit structure. Truth tables are also opposite for OR and NOR circuits.

The reasons for using NOR circuits are the same as those for using NAND circuits. That is, negative circuits are easier and less costly to manufacture. So, where an OR function is needed, most computer manufacturers use negative circuits. Both OR and NOR circuits are available from semiconductor manufacturers. Using negative circuits saves money without sacrificing performance.

In summary, five circuits have been reviewed in this

Activity: *THE NAND GATE*

_____ OBJECTIVES _____

1. *To understand the operation of the NAND gate.*
2. *To prove the NAND gate truth table.*
3. *To observe the conversion of the NAND gate to an AND gate.*

EQUIPMENT NEEDED

1 Circuit strip
  (See Figure 2)
1 7400 integrated circuit
  (Quad 2-input NAND gate)
1 7404 integrated circuit
  (Hex inverter)

3 LED indicator lamps
3 270 ohm resistors
1 Power supply, 5 volts DC
12 Wires for connections

PROCEDURE

1. Figure 1 shows the internal connections of the 7400 Quad 2-input NAND gate integrated circuit used in this experiment. It should be mounted on the circuit strip in the approximate position shown in Figure 2. The circuit strip will need to have the LED driver and LED indicators connected, as shown in Figure 2.
2. Three-inch lengths of wire should be connected to the connector strips, as indicated for inputs A and B. Plug the ends of A and B into the ground connector strip and energize the circuit strip. *Turn on the power supply* and adjust to 5 volts DC. Which of the LED's lights?

3. With A and B both connected to ground, they in effect are zeros (0) and would be indicated in the truth table as a zero. When they are connected to the positive connector strip, they have a high voltage input and would be indicated as a 1 in the truth table. Remove the input A from the ground connector strip and plug it into the positive connector strip. What happens to the LED indicator A?

4. You are now ready to develop the truth table of the NAND gate by connecting the inputs of the gate to the different input voltage levels. For each of the four input conditions listed in the truth table, Figure 3, indicate the condition of the output as shown by the LED indicator. Note that when the C LED is lit, it indicates a high output (1).

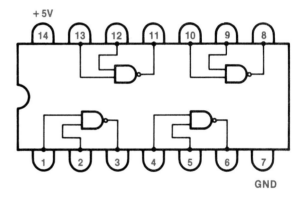

Figure 1

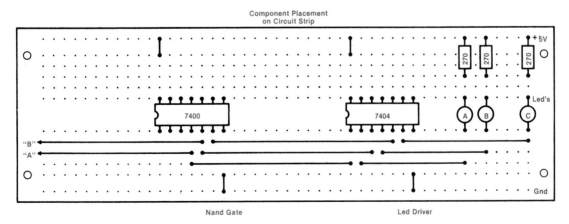

Figure 2

NAND Gate Truth Table

Input A	Input B	Output
0	0	
1	0	
0	1	
1	1	

Figure 3

5. Compare the truth table that you have just developed for the 7400 with the truth table of the NAND gate in your textbook. Are they the same?

6. The NAND gate (7400) is a versatile device and, with its four separate gates, can be connected to produce all of the basic gates used in digital electronics. In the next part of the experiment, you will convert the NAND gate that you have been using into an AND gate, by simply inverting the output of the NAND gate. The diagram for this circuit is shown in Figure 4. Study the diagram and change your circuit board connections to make this circuit.

7. Energize the circuit strip. Fill in the truth table in Figure 5. Use the same procedure you used before.

8. Compare the truth table of the AND gate that you have just completed with the truth table of the NAND

gate. You are interested in the comparison of the output tables. Is the output of the AND gate inverted, as compared with the output of the NAND gate? What part of the circuit in Figure 4 caused this change in the output?

9. There is an alternative method of making an inverter from a two-input NAND gate, rather than connecting the inputs together as shown in Figure 4. Study the truth table of the NAND gate and determine the connections for the input of the NAND gate that causes it to behave as an inverter. What change in the connections should be made?

10. Change the connections on your circuit strip and prove that you have indeed constructed an inverter using a method other than that shown in Figure 4.

11. *Turn off the power supply.* Remove the connections and integrated circuits from the circuit strip. Return all equipment and materials to their proper storage places.

SUMMARY

1. Describe the action of the NAND gate, as shown in the truth table that resulted from your experiment.

2. Explain how the addition of an inverter permits the construction of an AND gate from NAND gates.

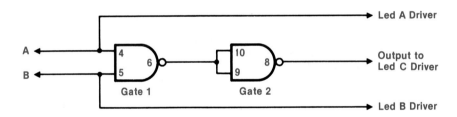

Figure 4

AND Gate Truth Table

Input A	Input B	Output
0	0	
1	0	
0	1	
1	1	

Figure 5

section—AND, OR, NOT, NAND, and NOR. These represent the basic building blocks for computers and calculators. Various data can be stored on tapes, discs, or cards to cause pulses to operate the transistors. In all cases, the information is placed into the circuit as either a 1 or a 0. This binary coding makes it easy to direct a machine to make decisions. A computer is preprogrammed to react to commands from binary signals. The grouping of the circuits can cause a computer actually to make selections and comparisons.

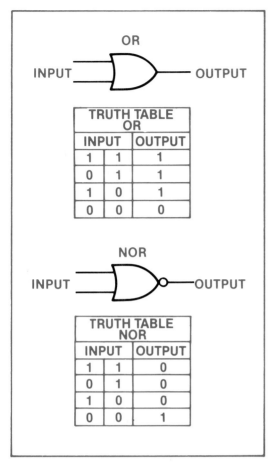

Figure 21-31. Symbols and truth tables for OR and NOR.

The OR and NOR symbols shown with their truth tables:

TRUTH TABLE OR

INPUT		OUTPUT
1	1	1
0	1	1
1	0	1
0	0	0

TRUTH TABLE NOR

INPUT		OUTPUT
1	1	0
0	1	0
1	0	0
0	0	1

Normally, we work with a base-10, or digital, numbering system. It takes 10 pennies to make a dime. It takes 10 dimes to make a dollar. The numbering system uses 10 digits—0 through 9.

The binary system uses only two digits. In both the base-10 and the binary system, the position of the number is important. Consider the numbers 10 and 01. Both use a 0 and a 1. The positions of the numbers determine value.

The scientific notation, or *powers of 10*, are based on the positions of numbers. These offer a good example of the importance of the positions of numbers.

$$10^0 = 1 \quad 10^1 = 10 \quad 10^2 = 100 \quad 10^3 = 1,000$$

Now take the numeral *2*. Raise it to various powers. This demonstrates the relationship between binary and decimal numbers.

$$2^0 = 1 \quad 2^1 = 2 \quad 2^2 = 4 \quad 2^3 = 8$$

There are some rules you should remember about this method of numbering. One rule is that any number raised to 0 power is 1. Also, any number raised to the first power is that number. Thus, $2^0 = 1$, $3^0 = 1$, $10^0 = 1$; and $2^1 = 2$, $3^1 = 3$, $10^1 = 10$.

Through use of this principle, the binary numbering system can represent any quantity, Figure 21-32 on page 356. Note how the relationships between decimal and binary values follow the same principle.

To convert binary notations to decimal values, just add the powers. For example, look at the decimal value 56. This is represented by *1* entries under the binary columns for 32, 16, and 8. Just add the values: 32 + 16 + 8 = 56.

One of the advantages of binary coding is its simplicity. There would have to be 10 circuits to represent the decimal numbering system. In the binary system, only four circuits are needed to represent values of 0 through 9. With the binary system, computers can be smaller and less expensive than if the decimal system were used.

REVIEW QUESTIONS FOR SECTION 21.5

1. Define the terms *bit* and *byte*.
2. What is a truth table?
3. Describe the function of an AND circuit.
4. Describe the function of an OR circuit.
5. Describe the function of a NOT circuit.
6. Describe the function of a NAND circuit.
7. Describe the function of a NOR circuit.
8. In what form does a digital computer process data?

21.6 BINARY CODE

Uses for **binary coding** in computers have a number of names. The term *binary* is applied to a *bistable condition*. This simply means that computer circuits have two conditions. The bistable condition may be called *on* or *off*. It may also be described as either *conducting* or *nonconducting*. Still another set of descriptions is *energized* or *de-energized*. These terms are used to describe functions within a binary numbering or processing system.

REVIEW QUESTIONS FOR SECTION 21.6

1. What is binary arithmetic?
2. Why does a digital computer use binary coding?
3. How are binary notations converted to decimal values?

21.7 COMPUTERS

Central Processing Unit

A computer system consists of various components or devices needed to perform computer functions. **A central**

Decimal Numbers	Decimal Value of the Binary Digit					
	2^5 (32)	2^4 (16)	2^3 (8)	2^2 (4)	2^1 (2)	2^0 (1)
0	0	0	0	0	0	0
1	0	0	0	0	0	1
2	0	0	0	0	1	0
3	0	0	0	0	1	1
4	0	0	0	1	0	0
5	0	0	0	1	0	1
6	0	0	0	1	1	0
7	0	0	0	1	1	1
8	0	0	1	0	0	0
9	0	0	1	0	0	1
10	0	0	1	0	1	0
20	0	1	0	1	0	0
30	0	1	1	1	1	0
40	1	0	1	0	0	0
41	1	0	1	0	0	1
42	1	0	1	0	1	0
50	1	1	0	0	1	0
56	1	1	1	0	0	0

Figure 21-32. Conversion table of decimal and binary values.

processing unit (CPU) controls the computer system. This system needs some kind of central control. A CPU can be built on a single chip. This chip is called a **microprocessor**. A **microcomputer** is built around a single microprocessor, Figure 21-33.

The control unit (CU) directs the computer so it will be able to do the work expected of it. It directs and coordinates data moving throughout the CPU. Programming determines what is to be done with the input data. Programs may be BASIC, Pascal, Fortran, or any of a number of others.

The size of the CPU varies from small computers to very large computers that serve as a computing facility for many users. Memory attached to the system creates the ability for the computer to extend its utility. The speed of the memory and CPU subsystems depends on the speed of the electronics (chips) used and on the number of activities that can occur at the same time. It usually suffices to compare a family of machines—that is, a set of machines that have the same internal organization and instruction sets but that vary in speed—on the basis of their memory-cycle time. The parameter is the time required to read or write a unit of information between the memory and the CPU.

Memory

The ability to store and recall data is a very important function of a computer. Individual elements or cells in a memory unit have the ability to store a single bit of binary data. Total capacity of a memory unit depends on the number of cells in the structure. In most instances, data are placed into memory in binary words.

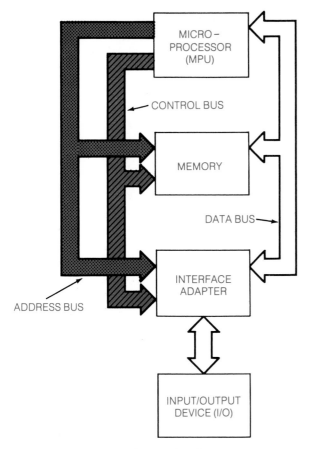

Figure 21-33. Block diagram of a microcomputer system.

Two terms are associated with memory: *writing* and *reading*. Writing is the placing of binary data in the memory for storage. Reading is the sensing of the stored information already in the memory. Figure 21-34 illustrates this **read/write** concept.

In most computers, both reading and writing capabilities are needed. This type of read/write memory is called **RAM**, or **random-access memory**. Another capability is that of the **ROM**, or **read-only memory**. Firmware instructions are placed in the ROM. ROM are usually found in the form of chips. There are two classifications of ROM memory: **volatile memory** and **nonvolatile memory**. Nonvolatile memory stores data without the need of electrical power, whereas volatile memory uses electrical energy to store the data. Disc memories store data on discs that are magnetized. Magnetic tape can also be used to store data. Each type of storage has its advantages. Various types of equipment are used to store and retrieve the information from memory.

The ROM is an IC chip that has memory built into it. Its memory is of the nonvolatile type. The information is "burned in" and cannot be changed. This is the type of information that consists of math tables and **firmware**, or

instructional data built into a system. It is placed in the chip by the manufacturer.

A **programmable read-only memory (PROM)** chip is programmed by the user. Each time something is stored in its memory, it is there to stay. The emitter resistor in the chip is burned out by the storage of data in memory. That means the information is permanently there once the chip has been programmed.

An **EPROM** is an *erasable programmable read-only memory* chip. EPROMs can take the form of MOS transistors with isolated gates. A charge deposited on the gate is retained and interpreted as a 1 bit; a no charge as a 0. It has been estimated that the charge will drop by less than 30% over a 100-year period. This is certainly a nonvolatile form of storage. To erase the charges, one must expose the memory cells to ultraviolet light for 30 minutes to an hour, the light discharging them by means of the photoelectric effect. This creates a blank that can then be reprogrammed. Other types of read-only memory include EAROMs, which are *electrically alterable* ROMs that can be programmed and erased electrically.

These memories are located on a microprocessor chip. The chip arrangement of memory provides a much more rapid access time than does an off-board memory. Such a memory is often termed a **register**, and microprocessors are divided into two broad categories depending on their reliance on such registers. Register-oriented microprocessors (such as the 8080 and the Z80) have a larger number of internal registers that can be manipulated by the programmer. They tend to require more complex timing and control. Memory-oriented microprocessors (such as the 6800 and 6500) have few internal registers but have instruction sets that allow easy manipulation of memory content.

Software

A set of operating instructions is needed to make a computer operate in a logical fashion to perform the tasks assigned. Whereas firmware is the instructional data that are built into the system and that tell the computer how to operate in general, **software** is the instructional data utilized to tell the computer what to do during a specific operation. This software can be altered to conform to a wide range of different procedures. Programs are designed to make the computer conform to this range of operations.

The more real-world statements used in the programming, the more memory is required. The firmware of a high-level language computer requires little knowledge of the computer on the part of the operator, but more memory space must be used to translate the commands into the computer functions that involve binary coding.

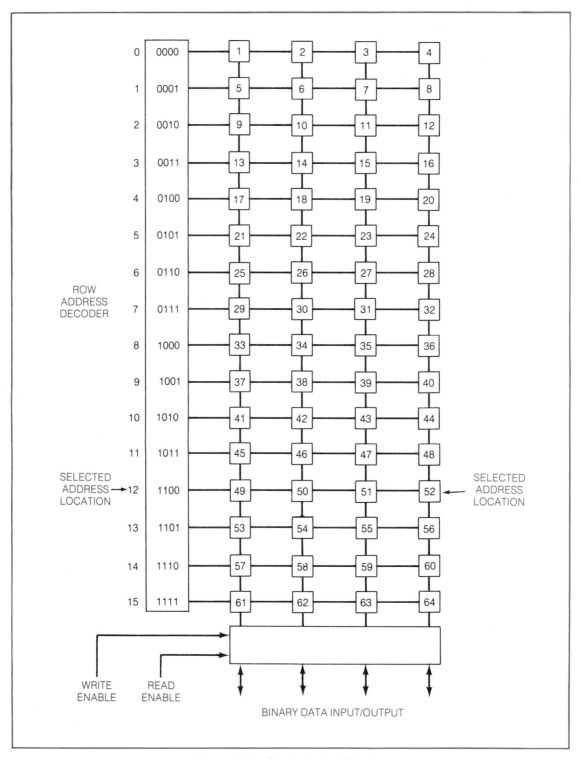

Figure 21-34. Read/write simplification.

Computer Operation

The **language** is the set of symbols and the rules that are used to compose statements acceptable to the computer. Computer languages are more precise and have a simpler structure than do natural spoken languages. A clear separation is made between what is written to com-

pose an acceptable statement (syntax) and the meaning of the statement (semantics). For a program to be executed on the computer, it must ultimately be composed of a sequence of basic instructions to the computer, Figure 21-35. The statements written by the user are put onto some input medium such as a keyboard device and entered

AMERICAN STANDARD CODE FOR INFORMATION INTERCHANGE (ASCII)								
	000	001	010	011	100	101	110	111
0000	NULL	DC_0	b	0	@	P		
0001	SOM	DC_1	!	1	A	Q		
0010	EOA	DC_2	"	2	B	R		
0011	EOM	DC_3	#	3	C	S		
0100	EOT	DC_4	$	4	D	T		
0101	WRU	ERR	%	5	E	U		
0110	RU	SYNC	&	6	F	V		
0111	BELL	LEM	,	7	G	W		
1000	FE_0	S_0	(	8	H	X		
1001	HT SK	S_1	)	9	I	Y		
1010	LF	S_2	*	:	J	Z		
1011	V_{tab}	S_3	+	;	K	[		
1100	FF	S_4	,	<	L	\		ACK
1101	CR	S_5	–	=	M	]		②
1110	SO	S_6	★	>	N	↑		ESC
1111	SI	S_7	/	?	O	←		DEL

Definitions of control abbreviations:

ACK	Acknowledge	ESC	Escape	SO	Shift out
BELL	Audible signal	FE	Format effector	S_0-S_7	Separator (space)
CR	Carriage return	FF	Form feed	SOM	Start of message
DC_0-DC_4	Device control	HT	Horizontal tabulation	V_{tab}	Vertical tabulation
DEL	Delete idle	LEM	Logical end of media	WRU	"Who are you?"
EOA	End of address	LF	Line feed	②	Unassigned control
EOM	End of message	RU	"Are you...?"	SYNC	Synchronous idle
EOT	End of transmission	SK	Skip		
ERR	Error	SI	Shift in		

Example of code format:

B_7 B_1

1000100 is the code for D

three-bit group   four-bit group

Figure 21-35. The ASCII code.

into the machine. The translator, or compiler, is a computer program that accepts these statements as input and translates (or compiles) them into a sequence of basic computer instructions, which form a version of the program that the computer can handle.

Much has been written on computer operation, languages, chips, and the wonders of the computer in general. Every month more information is available as new chips are developed and faster times are achieved that allow some computers to do more work in a shorter time.

Computer applications are everywhere—for example, in automobiles to control combustion and emissions and in robots that make the automobiles. Mainframe computers, microcomputers (PCs), and microprocessors all make production easier and can create a higher standard of living for everyone.

Microcomputers

The microcomputer integrates a number of machines for the control or production of complete jobs. For exam-

Figure 21-36. This student is studying at home using a personal computer and a modem. (Courtesy Racal-Milgo)

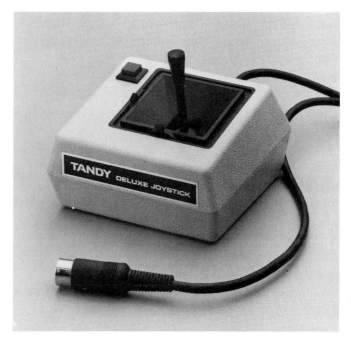

Figure 21-37. Joystick. (Tandy Corporation)

ple, microcomputers are often used to control automatic electronic assembly equipment. In offices, microcomputers are used to process paperwork. Microcomputers, also called PCs or personal computers, are used in modern word-processing systems, Figure 21-36. These systems generate letter and other business documents.

The microcomputer is one that has a CPU, or central processing unit, on one chip or IC. This type of computer is portable—it can be moved around with little effort, usually by one person. Data analysis is but one of its possibilities. Word processing, graphics creation, and game play make use of this versatile piece of electronic equipment.

The Joystick. The **joystick**, Figure 21-37, is a device that can be used with the microcomputer to perform certain operations. In many computer games, the joystick is used to move characters around on the monitor. In robotics, the joystick is very useful. If a joystick is incorporated during programming, an interpolation of the arm axes can be performed, thus causing the robot to move in a straighter path between points. The joystick can also help reduce robot programming time.

The Mouse. The **mouse**, Figure 21-38, is another device that improves the quality of work the computer can do. It is a device that digitizes information and feeds it into the computer for processing.

*C*omputer-*a*ided *d*esign (CAD) utilizes the mouse with good results. It feeds digitized information into a computer being used to make the original drawings for a part. It is then possible to study the part from many design perspectives. The object is often transformed into a solid model by the computer from these two-dimensional drawings. The solid model can then be studied in minute detail for stress and various pressures that might occur, and then be put through a variety of simulated tests. These tests allow the design to be changed until the best design is developed. Once the best design for all the projected factors is available, the product can be manufactured. The significant features of this computer-aided-design process are its speed and the interactive capabilities it gives the design engineers. It allows many potential designs to be considered and evaluated before any are built.

Output Devices

In order to allow use of the processed binary data, a method must be available to make such data usable. An output unit converts the binary data into some type of usable energy or form. Output devices include printers or plotters that make hard copy, Figure 21-39, cathode-ray

Figure 21-38. Mouse for use with computer. (Tandy Corporation)

Figure 21-39. A plotter is one type of output device. (Courtesy Cal Comp Plotter Products Group)

tubes that display information or data, and digital-to-analog converters.

Output data come from the MPU (microprocessor unit) over the data bus. The interface adaptor accepts these data and changes them to an appropriate output signal for a particular type of device. A specific output device is selected by the MPU. This is done through the input/output address bus. During operation, the input/output devices are alternately switched into operation by the system.

A complete knowledge of how the computer operates is not necessary to operate a microcomputer (PC). They have been designed so that a limited amount of instruction is needed to get results. Software packages are designed to make it easier for everyone to operate a computer.

Mainframe Computers

Mainframe computers are designed for use by large commercial and industrial firms. They can have many output terminals and printers. Many people may access them at the same time. Permanent storage of most of the data is usually effected through use of magnetic tape.

In some cases, a large computer may be located thousands of miles away from an operator. Telephone lines are usually then utilized to access the computer. In some instances, however, the computer is accessed by way of satellite communications. Some large banks use a microwave dish to beam a signal to an orbiting satellite. Such

information can also be provided by the swipe of a credit card's magnetic tape across an input device. The information on the card account number is sent to a central processing center and checked with information stored there on tape, and then the permission or rejection is sent back to the store clerk by way of satellite communications.

Future of Computers

There is almost no limit to what can be done with computer systems. The popularity of the inexpensive home computer is such that millions of people now have access to them and use them from time to time. In the future, such computers will become less expensive, faster, and more widely available, as well as easier to use.

Many other improvements in the field will also be made. Chips with megabyte storage capabilities are already in development. The silicon in them may be replaced by other materials, such as gallium arsenide. Laser printers have already improved the quality of computer output, and will continue to do so. Flat screens for all monitors (instead of just for lap-tops) will probably become standard. Hand-held computers will probably replace the less versatile calculators.

In short, one need only think about what has happened to hand-held calculators in the last 25 years in order to project what can be done with computers, which can intrinsically do much more. The future of computers is limited only by human imagination.

REVIEW QUESTIONS FOR SECTION 21.7

1. What is a microcomputer?
2. Describe the purpose and main functions of a microprocessor.
3. Describe some uses for microcomputers.
4. What is a CPU?
5. What are the 2 types of ROM memory?
6. What is meant by the term *burnt in*?
7. What is an EPROM?
8. What are registers?
9. What purpose does computer software serve?
10. What is meant by the *language* of the computer?
11. What is the difference between a mainframe computer and a PC?
12. What does CAD stand for?
13. How does the future look for computers?

SUMMARY

Relays and switches are easily understood. They involve contacts that close and open. The process is visible. Processes involving transistors and diodes involve no visible actions. So, circuit analysis can be more difficult.

Relays and switches are classified according to use. A single switch controlling a light bulb is called a direct switch circuit.

An AND circuit is a circuit with two switches in series. That means both switches A and B must be closed for the circuit to be complete. An OR circuit is a circuit with two switches in parallel. That means both switches do not have to be closed for the circuit to be complete; it can be complete when either switch is closed.

Selection circuits can be used when a stepping switch is utilized. Any one of several points may be selected. Selection is carried out by causing the switch to step around to a particular contact. Relays and switches can be arranged for transfer purposes.

A resistor within a circuit limits the flow of current. This can be a very valuable function. Using resistors to limit current works well with either AC or DC circuits. A variable resistor can be used to control the amount of current that passes through the load.

Current can also be controlled with electronic devices. This is done with a diode or silicon controlled rectifier. SCRs are solid-state electronic devices.

AC current can be controlled by the use of an inductor. A variable inductor can be used to vary the light output of a circuit containing a lamp. The variac is a tapped auto transformer. A saturable reactor is also a control device. The saturable reactor is built on an E-I type transformer core. The center leg has a control winding. The two outer legs have AC-controlled coils. Current passes through the AC coils, which are connected in series. If there is no current in the control winding, current flow is limited. Inductive reactance is the limiting factor. That means a small amount of DC can control the higher-voltage AC. The saturable reactor is sometimes referred to as a magnetic amplifier.

A diode acts much the same as a one-way gate. It allows current to flow in only one direction. Computers use thousands of diodes to control the direction of current flow.

Computers perform their functions on a binary basis. Binary values are 0 and 1. Binary functions of computers are based on switch conditions—either *on* (1) or *off* (0). A truth table shows how the inputs and outputs of a circuit are obtained. A computer or calculator operates on pulses of electrical energy. These pulses are called bits. Units of 8 bits are common and are called bytes. The pulses can be converted to magnetic form and stored on memory devices, tape, or discs. The OR, NOT, NAND, and NOR circuits are used in the computer. The binary system uses only two digits. In both the base-10 and the binary system, the position of the number is important.

Microprocessors are individual printed-circuit chips that perform specific processing functions. Each processor has a specialized job to do. Microcomputers are generally systems built around microprocessors and other printed circuits. The microcomputer integrates a number of machines for the control or production of complete jobs.

The bit is used by a computer to represent both numerical and nonnumerical information. Units of 8 bits are common and are called a byte.

A computer system consists of a CPU, memory, programming devices and output devices. Chips are used for memories in some cases. They may be PROM, EPROM, ROM or other types of memories. EPROM's are eraseable by using light, magnetic fields or by using electrical means. The 8080 or Z80 chips are memories with registers that can be manipulated by a controller.

Software is a set of operating instructions for a computer. Programs are designed to make the computer conform to a variety of operations. Computer language is made up of rules that are acceptable to the computer. They are simpler than our spoken languages.

The microcomputer integrates a number of machines for the control or production of complete jobs. Microcomputers are also called PC's or personal computers. They may also be controlled by a joystick or mouse. The mouse is used in Computer Aided Design (CAD).

Mainframe computers are designed for use by large

commercial and/or industrial firms. They can be operated from a number of remote locations all at once. The future of the computer looks bright. It is limited only by human imagination.

USING YOUR KNOWLEDGE

1. Draw a schematic for a variac.
2. Draw the symbol for an AND circuit.
3. Draw the symbol for an OR circuit.
4. Draw the symbol for a NOT circuit.
5. Draw the symbol for a NAND circuit.
6. Draw the symbol for a NOR circuit.
7. Write out the truth table for the number 32.
8. Express the following number in terms of powers of 10: 10,251.
9. Sketch a block diagram for a microcomputer system.

KEY TERMS

digital circuits
series switching circuit
direct-switch circuit
AND circuit
OR circuit
variac
saturable reactor
NOT circuit
NAND circuit
NOR circuit
binary coding
central processing unit (CPU)
microprocessor
microcomputer
RAM (random-access memory)
ROM (read-only memory)
register
bits
bytes
words
joystick
mouse
read/write
volatile memory
nonvolatile memory
PROM (programmable read-only memory)
EPROM (erasable programmable read-only memory)
software
firmware
language

Chapter 22

ROBOTICS

OBJECTIVES

After studying this chapter, you will know

- *how a robot works.*
- *how each part of a robot functions.*
- *the limitations and advantages of robots.*
- *the meaning of "degree of freedom."*
- *the robot motion capabilities.*
- *the uses of robots in CIM, CAD-CAM, and JIT systems.*

INTRODUCTION

What is a robot? That is the question that has to be answered before any discussion can be properly directed toward coverage of the topic in general.

There are a number of definitions attributed to the word *robot*. A **robot**, for our purposes, is a programmable machine. It has a manipulator that is designed to move material, parts, tools, or specialized devices through a variety of programmed motions.

A robot can be classified as a system that simulates human activities from computer instructions. Figure 22-1 shows such a robot system. The computer is the secret of the system since it is, in effect, the system's brain. The computer is thus an integral part of any robot and must be taken into consideration whenever the robot is studied as a device, a system, or a means of eliminating human effort.

Some nonintelligent robots do not use electronics for brains. Many of the pick-and-place robots are cam controlled. They simply pick up their load and place it elsewhere. Loading and unloading tasks usually are performed by this type of robot, Figure 22-2.

22.1 COMPUTER PROGRAMS

Special computer programs designed for specific jobs are used to control robots. Industrial robots, for instance, are designed to carry out a particular operation. This one operation may be done over and over again, but the robot, unlike humans, does not become fatigued or bored. A program is written to take into consideration the exact tasks to be performed. In some instances, it may take years to analyze the moves needed to perform a particular job. This information then has to be fed into a computer in terms that it understands, and the computer then sends signals to the robot so it will perform exactly as desired.

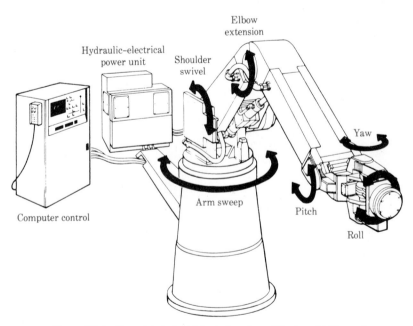

Figure 22-1. Complete industrial robot system. (Cincinnati Milacron)

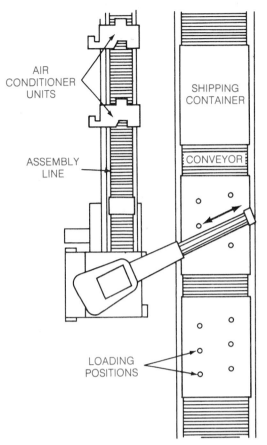

Figure 22-2. Robot used to pick up finished air-conditioning units and pack them in shipping containers.

Languages

Robots use a number of computer **languages**. These languages are designed for specific operations. Some of the languages used with robots are: AL (Stanford's Artificial Intelligence Lab language), VAL, AML (developed by IBM), Pascal, and ADA.

Applications for Industrial Robots

Examples of the use of industrial robots include the following applications:

- Machine loading and unloading: placing parts where they are needed for machining or shipping
- Materials handling: packing parts or moving pallets
- Fabrication: making investment castings, carrying out grinding operations, deburring, doing water-jet cutting, wire-harness manufacturing, applying glues, sealers, putty, and caulks, and for drilling, fettling, and routing
- Spray painting: painting cars, furniture, and other objects (see Figure 22-3)
- Welding: welding cars, furniture, and steel structures
- Assembly: assembling electronic devices, automobiles, and small appliances
- Inspection and testing: conducting quality-control inspection for surface and interior defects, using vision sensors and feelers

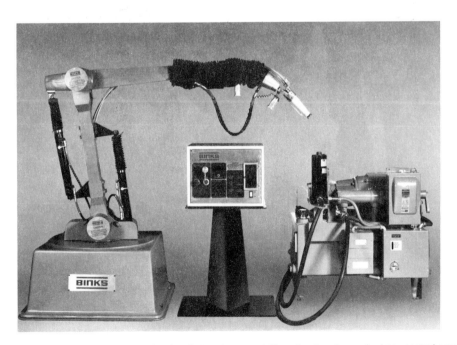

Figure 22-3. Spray-painting robot: a basic robot system consisting of a six-axis manipulator, control console, and hydraulic power supply (Binks Manufacturing Company)

22.2 INDUSTRIAL ROBOTS

Industrial robots have arms with grippers attached, Figure 22-4. The grippers are fingerlike and can grip or pick up various objects. They are used to pick up and place. They pick up an object and place it elsewhere or move materials from one place to another. These robots can be programmed and computerized. A "teach box" is used to program the microprocessor used as the computer brain. Sensory robots, welding robots, and assembly robots usually have a self-contained microcomputer or minicomputer.

There are, of course, other types of robots for different jobs. There are laboratory robots, explorer robots, hobbyist robots, classroom robots, and entertainment robots, Figure 22-5A–D.

The Manipulator

There are about 250 manufacturers of robots in the United States, Europe, and Japan, making it very diffi-

Figure 22-4. Industrial robot used to pick up and place.

cult to identify all the parts used in the available robots. However, there are some common components that may be examined for a better perspective on how a robot works.

The **manipulator** is one of the three basic parts of the robot. The other two are the controller and the power source. In order for the robot to do work, each of these three components must be operational.

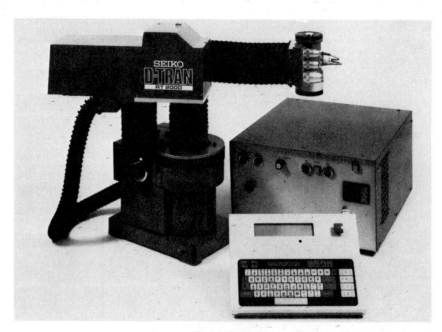

Figure 22-5.
A. Laboratory robot used to handle dangerous materials.
(Seiko Instruments USA, Inc.)

The manipulator is classified by certain arm movements. Four coordinate systems are used to describe the arm movements: polar coordinates, cylindrical coordinates, Cartesian coordinates, and articulate (joint-arm, spherical) coordinates.

B. Explorer robot. (National Mine Service Company)

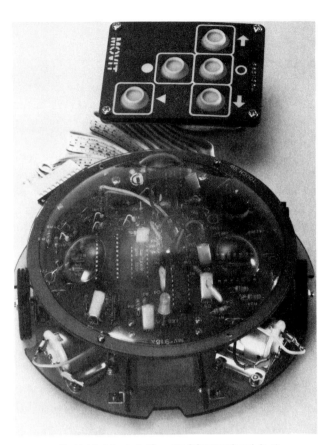

C. Hobbyist robot. (Greymark International, Inc.)

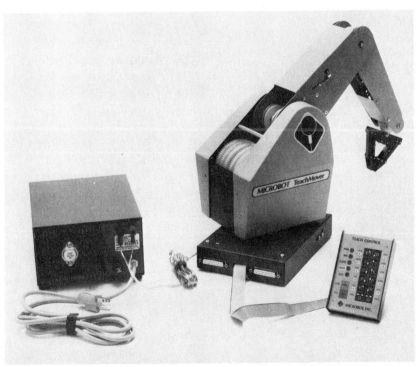

D. Classroom robot for teaching basic functions. (Microbot®)

Base

The base of the robot is its anchor point, Figure 22-6. The base may be rigid. It is usually designed as a supporting unit for all the component parts of the robot. The base does not have to be stationary, since it may become part of the operational requirements of the robot. It may be capable of any combination of motions, including rotation, extension, twisting, and linear motion. Most robots have the base anchored to the floor, but if floor space is limited they may be anchored to the ceiling or suspended from support systems overhead. A track or conveyor system may be used to move the robot along as needed.

Arm

Some type of arm is found on most industrial robots. It may be jointed and resemble a human arm or it may be a slide-in/slide-out type used to grasp something and bring it closer to the robot. A jointed arm consists of a base rotation axis, a shoulder rotation axis, and an elbow rotation axis. This type of arm provides the largest working envelope per area of floor space of any design thus far. If this is a six-axes type of arm, some rather sophisticated computer control is required. Most arms have some type of joint. From one to six jointed arms may be attached to a single base for special jobs. The expense of controlling this complex type of movement is rather high.

Wrist

Figure 22-7 shows how the wrist is attached to the jointed arm, together with the three axes of rotation. The wrist is similar to a human wrist and can be designed with a wide range of motion, including extension, rotation, and twisting. This aids the robot in reaching places that are hard for the human arm to reach. It comes in handy especially when spray painting the interior of an automobile on the assembly line. It is also helpful in welding inside a pipe. This type of flexibility can improve the manufactured products we now enjoy and add others we were unable to fabricate earlier.

Grippers

The **grippers** are at the end of the wrist. They are used to hold whatever the robot is to manipulate, Figure 22-8. Pick-and-place robots have grippers to move objects from one place to another.

Some robots have end-of-arm tooling instead of grippers. In such cases, the robot is used primarily for one type of operation, such as spray painting or welding. If a tool is attached, it is unnecessary to have a gripper on the end of the arm. A pneumatic impact wrench can be fitted at the end of the arm just as easily as the grippers can.

Grippers are made in a variety of sizes and shapes, Figure 22-9. They are designed for special applications by the manufacturers of robotic equipment. The simplest type of gripper is a motion-producing device that is joined by two fingers. The fingers open and close to grasp an object. In most instances, only the finger mounts are purchased, and the designing and building of the fingers are carried out in the local machine shop to fit the job being done. Variations in design account for much of the design time spent on robots.

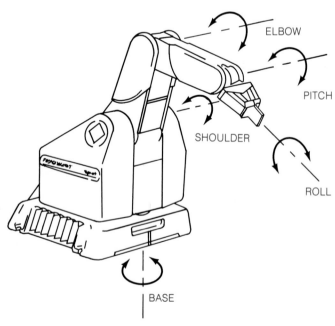

Figure 22-6. Operating envelope for Alpha II. (Microbot, Inc.)

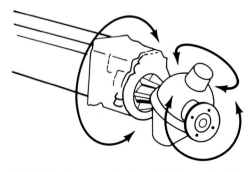

Figure 22-7. The three axes for the wrist of a manipulator arm. (Cincinnati Milacron)

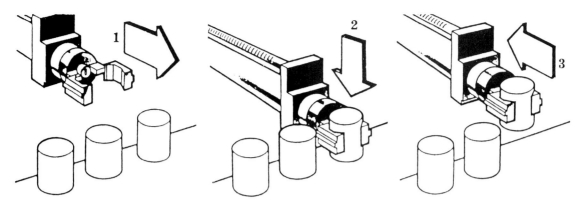

Figure 22-8. Robot arm (manipulator) with grippers for reaching out and picking up an object on a line and bringing it back to place it elsewhere. (Cincinnati Milacron)

Figure 22-9. Different types of grippers. (Cincinnati Milacron)

Work Envelope

The **work envelope** is also referred to as a sphere of influence or work area (the latter since it is not always spherical). This is the space a robot occupies when it swings around and up and down to do the work for which it is designed, Figure 22-10 on page 270.

Articulation

Three articulations are needed for a robot to move its arm, wrist, and hand to any place within its work envelope. The articulations are: (1) extension and retraction of the arm, which can be the simple motion of the arm out and in; (2) swinging or rotation of the arm, which is motion of the arm left and right; (3) elevation of the arm, which is lifting or depression (lowering) of the arm.

Wrist Motions

The three types of motion for a robot's wrist are: (1) bending or rolling forward and backward; (2) yawing (spinning) from right to left or left to right; and (3) swiveling, which is rolling down to the right or left, Figure 22-11.

Degrees of Freedom

The robot has six degrees of freedom if it can move its wrist three ways and its arm three ways, Figure 22-1, page 364. This is very limited when you compare it with human shoulder, arm, and wrist motions. Humans have forty-two degrees of freedom. As you can see, the robot arm requires technical improvement if it is to be as versatile as the human arm. There is some question as to

369

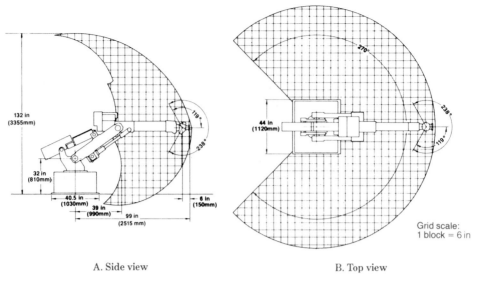

A. Side view   B. Top view

Figure 22-10. Tear-shaped work envelope.

Figure 22-11. Wrist action known as yaw. (Cincinnati Milacron)

whether the human arm and wrist should serve as the model in this respect. Such emulation would call for some rather involved engineering. At the present time, it is at least possible for the robot hand to do basically what the human hand can do: grip, push, pull, grasp, and release.

Robot Motion Capabilities

Robots have four basic motion capabilities: (1) linear motion, (2) extension motion, (3) rotating motion, and (4) twisting motion. These four motion capabilities are referred to collectively in the LERT classification system. (L stands for linear motion; E stands for extensional motion; R stands for rotational motion, and T stands for twisting motion.) The superscript used with the classification system indicates the number of times the robot is capable of a particular motion.

Keep in mind that most robots are mounted to the floor on a base. However, it is possible for them to be mounted to the ceiling or onto a mobile platform. Each axis is listed in order as it is mounted to the first component or base. For instance, L^3 indicates there are three linear

motions. R^2L^3 indicates two rotational motions and three linear motions.

Moving the Manipulator

In order for the manipulator to do work, it must have some power source. Industrial robots work on 220-volt and 440-volt systems available within the plant or factory. Entertainment and smaller robots may work with batteries that need recharging occasionally. Robots use electricity to give them their basic power. However, hydraulic, pneumatic, and purely mechanical methods can also be used to produce motion and do work.

REVIEW QUESTIONS FOR SECTION 22.2

1. What are grippers?
2. What is the name of the anchor point for a robot?
3. Where are grippers located on a robot?
4. What is the simplest type of gripper?
5. What does *work envelope* mean?
6. What does *degrees of freedom* mean?
7. What does *LERT* stand for?
8. What voltage does a robot use in industry?

22.3 THE FUTURE OF ROBOTS

The **computer-integrated manufacturing (CIM)** factory of the future will look quite different from the factory of today. It will be based upon the integration of the traditional or process-based technology of today, with the emerging software- or systems-based technology of tomorrow, Figure 22-12.

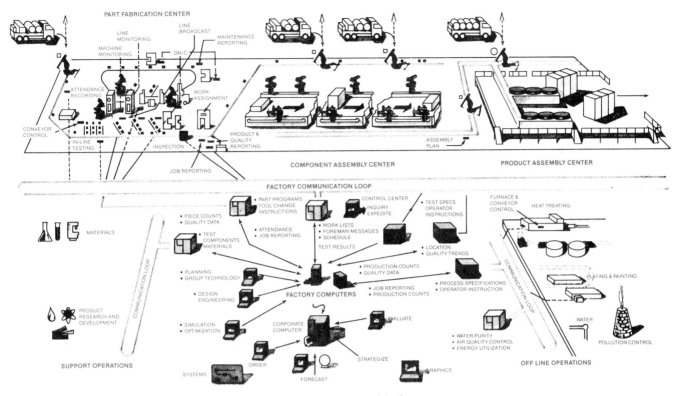

Figure 22-12. Factory of the future

The factory of the future will utilize robots and programmable controllers to make sure all machines are doing what they are supposed to be doing at the right time. Timing is very important in making any product on the assembly line. Robots have been used for a few years now in the manufacture of automobiles. They weld (Figure 22-13) and paint. They are also used to insert seats and other routine jobs of which humans tire easily.

In order to coordinate the manufacture of a large product, such as an automobile, there must be checkpoints and inspections with feedback information. The CIM system is ideal for this type of operation, Figure 22-14. Assembly is done in work cells where parts related by their design and their manufacturing characteristics are grouped together for the most efficient operations.

The total process of manufacture can be controlled from a central location. In a properly designed CIM system, there are work cells integrated with other shop-floor and factory-management computer systems. There is a central database management system that utilizes the information fed to it by other systems. All these sub-systems communicate with one another and the central computer system. From one location, management has complete control over the manufacturing process.

Robots can be used in work stations for the handling of materials. They can pick and place parts that humans quickly tire of dealing with. Less floor space is required

Figure 22-13. Robots are utilized in production of automobiles in up-to-date factories. (Ford Motor Company)

for robots, and pallets are emptied and gravity-feed parts delivery systems are utilized to the maximum. Delivery of parts in as efficient a manner as possible, with the smallest space occupied by the process, is the goal of computer-integrated manufacturing.

A **CAD-CAM** (computer-aided design and computer-aided manufacturing) system is needed for integrated design and manufacturing. The system is integrated so that

Figure 22-14. The entire manufacturing system is in the control of workers who look at the computer monitors. (International Business Machines)

Figure 22-15. Computers control the operation of machines in a computer-assisted manufacturing (CAM) line. (Cincinnati Milacron)

the design of the parts allows for efficient manufacture. The process of manufacture is also flexible enough to accommodate a variety of parts, Figure 22-15.

In **JIT** (just-in-time) manufacturing, parts are ordered to arrive just in time for assembly. This way there is no need for costly inventories or expensive warehousing of parts. JIT is possible only by the use of the computer and other electronic communications systems. The system also relies on trains being on time and trucks delivering needed parts at just the right time so the assembly line is not halted. Materials procurement and handling are the largest cost in manufacturing. JIT manufacturing is used to lower the cost of manufacturing any item. The automobile industry utilizes it to its fullest degree.

REVIEW QUESTIONS FOR SECTION 22.3

1. What does *CIM* stand for?
2. List at least three objectives of CIM.
3. What is a CAD-CAM system?
4. What does *JIT* stand for?

22.4 HEAVY-DUTY INDUSTRIAL ROBOTS

The GCA/DK B3200 robot is designed for heavy-duty industrial tasks, such as material handling, loading and unloading pallets, and long-reach spot welding.

The robot has an end-of-arm capacity of 220 pounds. It has a reach of more than nine feet, and a work envelope of about 650 cubic feet, Figure 22-16. All-electric DC servo-drives are on all axes, and a unique, automatic counterbalancing system is located on the arm articulation axes. This assures rapid, smooth, and responsive manipulation of maximum payloads at maximum speeds everywhere in the work envelope.

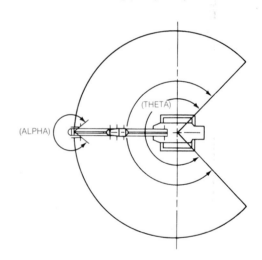

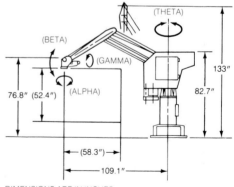

DIMENSIONS ARE IN INCHES

Figure 22-16. Work envelope for the GCA/DKB Robot. (GCA Corporation/Industrial Systems Group)

Pneumatic counterbalancing facilitates the optional lead-through teaching and reduces power consumption to less than 10 percent of that required by some robots with similar payload capacities.

Up to six degrees of freedom are available with the B3200 robot. Four degrees are standard: horizontal travel (*x*-axis), vertical travel (*z*-axis), horizontal sweep (*theta*-axis), and wrist yaw (*alpha*-axis). Wrist pitch (*beta*-axis) and wrist roll (*gamma*-axis) are optional. See Figure 22-17 for specifications.

This B3200 robot uses a dual-coordinate system (cylindrical and Cartesian) to maximize dexterity. The robot is offered with the standard robot controller for point-to-point applications, or with the new CIMROC-2 robot controller where either six-axis, simultaneous continuous-path, or similar integrations are required.

An externally mounted, *theta*-axis motor makes maintenance easy. This motor can be changed without suspending the robot's body by a crane or other lifting device.

The B3200 robot can be furnished with a variety of optional accessories, including: mechanical gripper hands; vacuum-cup lifters for flat material such as glass, non-ferrous metal sheets, or plastic items; magnetic lifters for iron, steel, and ferrous alloy materials; forklifts for pallets; lifters for cartons and boxes; and complete systems for welding.

REVIEW QUESTIONS FOR SECTION 22.4

1. What is the reach of the GCA/DK B3200 robot?
2. What is this robot's end-of-arm capacity?
3. How many degrees of freedom does this robot have?
4. What is another name for the wrist pitch?
5. What is another name for the wrist roll?
6. Why is it easy to do maintenance on the B3200 robot?

SUMMARY

A robot is a multifunctional manipulator designed to move material, parts, tools, or specialized devices through variable programmed motions for the performance of a variety of tasks. A robot can be classified as a system that simulates human activities from computer instructions.

MODEL		DKB3200 ROBOT
Degrees of Freedom		6 (Beta and Gamma optional)
Payload, max including end effector weight		220 lb
Repeatability		±0.04 in.
Range (speed)	**Axes**	
	X (Horizontal, in and out)	58.3 in. (35.4 ips)
	Z (Vertical travel)	52.4 in. (35.4 ips)
	Theta₁ (Horizontal swing)	270° (75°/sec)
	Theta₂ (Main arm vertical rotation)	—
	Theta₃ (Forearm vertical rotation)	—
	Alpha (Wrist yaw)	300° (75°/sec)
	Beta (Wrist pitch)	120° (75°/sec)
	Gamma (Wrist rotation)	180° (75°/sec)
	Composite axes speed	—
Robot weight		3300 lb
Electric power requirement		AC 200/220v, 50/60 Hz, 2.5 KVA
Pneumatic counterbalancing (where applicable)		Pneumatic counterbalancing on X and Z axes

— All **GCA**/DK pedestal robots have a DC servomotor drive system
— Pneumatic power requirement (where applicable) is 85 psi.

Figure 22-17. Spec sheet for GCA.

Some nonintelligent robots do not use electronics for brains. Many pick-and-place robots are cam controlled. Special computer programs are designed for specific jobs. Robots use a number of computer languages. These are designed for specific operations. They include AL, VAL, AML, Pascal, and ADA.

Robots can be used for loading and unloading, materials handling, fabrication, welding, spray painting, assembly, and inspection.

Industrial robots have arms with grippers attached. The manipulator is one of three basic parts of a robot. The other two are the controller and the power source. In order for the robot to do work each of these three components must be operational. The manipulator is classified according to certain arm movements or coordinates.

The base of the robot is the anchor point. It is usually rigid. The arm is found on most industrial robots. It may be jointed and resemble a human arm or it may be a slide-in/slide-out type used to grasp something and bring it closer to the robot. The wrist is attached to the jointed arm and is similar to a human wrist, with a wide range of motion.

The grippers are at the end of the wrist and are used to hold whatever the robot is to manipulate. They are made in a variety of shapes and sizes.

The work envelope is also referred to as a sphere of influence or work area. This is the space a robot occupies when it swings around and up and down to do the work for which it is designed.

Three articulations are needed for a robot to move its wrist, arm, and hand to do work in a work envelope: (1) extension and retraction of the arm; (2) swinging or rotation of the arm; and (3) elevation of the arm.

There are three types of wrist motions: bending or rolling forward or backward; yawing from right to left or left to right, and swiveling.

The robot has six degrees of freedom if it can move the wrist three ways and the arm three ways. Humans have forty-two degrees of freedom. Robots have four basic motion capabilities.

Robots use pneumatics, hydraulics, and electricity for power sources.

Computer-integrated manufacturing (CIM) utilizes robots and programmable controllers.

_____ USING YOUR KNOWLEDGE _____

1. Draw a robot. Label the parts of the robot and show its six degrees of freedom.
2. What is the work envelope of a robot?
3. Sketch a factory of the future, using blocks as units for manufacturing processes.

_____ KEY TERMS _____

robot
languages
manipulator
grippers
work envelope

computer-integrated
    manufacturing (CIM)
CAD-CAM
JIT

Chapter 23

CAREERS

OBJECTIVES

After studying this chapter, you will know

■ *the nature of the electrical engineer's work.*
■ *what an entrepreneur does for a living.*
■ *the employment outlook for the jobs that interest you.*
■ *how to locate information on positions for which you may eventually apply.*
■ *all aspects of at least one career that appeals to you.*

INTRODUCTION

In completing your study up to this point, you have achieved a basic level of knowledge about electricity and electronics. This chapter, which deals with career opportunities, helps you to look ahead. The information in it will help you decide whether you want to go further in your study of electricity and electronics. To help you decide, the chapter previews a number of job opportunities. These are opportunities open to people who have a background in electricity and electronics.

One or more of the positions and careers described in this chapter may interest you. If so, the opportunities you select should guide you in necessary further study of electricity and electronics.

The content that follows reviews a number of opportunity areas. In each area, there is a review of the requirements and the nature of the work. In addition, there are names and addresses of sources for further information.

You should note that the purpose of this chapter is different from that of the chapters you have completed so far. The content in it is for your information and guidance only. Do not try to memorize the content. Do use this information to help you think about career decisions and future studies. In other words, you alone are in control of how you use the information covered. Because of this emphasis, there are no review questions between major topics in this chapter.

23.1 THE ENTREPRENEUR

Some people choose not to work for someone else. They are people who want to work for themselves and want to be their own boss. A person who starts, organizes, and manages a business is usually called an entrepreneur. This person also assumes the risks of the business and is responsible for paying the business expenses. All responsibility for the success of the business rests with the entrepreneur.

A person who is self-employed, or in business for himself or herself, is an entrepreneur. All entrepreneurs face four similar problems. These problems are

• identifying a need,
• finding a product to satisfy the need,
• financing the business, and
• selling the product, Figure 23-1.

In every town, there are people who fit the definition of *entrepreneur*. Some businesses are small and some are large. Some of the small businesses may be owned and managed by one person. Others may be medium-sized, employing several people. There may also be large businesses, such as major manufacturing plants or, the headquarters of large companies. Such businesses could employ hundreds—or even thousands—of people.

Most large businesses started small. Some of the largest corporations started as one-person operations. The building of a business requires skill and hard work. Anyone starting a small business faces a variety of tasks. Money is the first concern. This money may have to be borrowed. Then the person must deal in a product or service that can be sold at a profit. Next, there must be a way of distributing the product or service to the public. Small businesses have only the owner–operator to do these jobs. The larger a business is, the more departmentalized it is. One department may be in charge of making the product. Another may be in charge of advertising it. A third may be in charge of delivering it. Usually, the larger the business is, the greater is the number of people needed for each of these duties.

Keep in mind that all successful businesses have one thing in common. They were started because someone observed that people needed or wanted an item or service. Once an entrepreneur has noticed a need, a search for a product to fill the need is the next step. There are many

Figure 23-1. All entrepreneurs must know how to sell their products.
(Radio Shack, a Division of Tandy Corporation)

electronic devices that are yet to be designed and that are waiting for the right entrepreneur to develop them to the sales level. Some entrepreneurs sell a service, such as the servicing of electrical and electronic products.

Getting started in business is not easy. But, succeeding in business is even harder. In other words, anyone who starts a business is an entrepreneur, but not all entrepreneurs are successful.

Take a look at some of the jobs listed in the next few pages and see whether you will fit into one of them or whether you have an idea for a different one. You may need more training or education to achieve your goal. Now is the time to start.

23.2 ELECTRICAL ENGINEERING

The generation, transmission, and use of electricity are the concerns of the electrical engineer. Electrical engineers design, develop, and supervise the manufacture of electrical and electronic equipment. Equipment items include electric motors, generators, communications equipment, electronic devices, and computers and other systems.

Nature of Work

The major areas of work in this branch of engineering include electronics and electrical-equipment manufacturing. Other opportunities lie in communications, power, illumination, and transportation. Electrical engineers usually specialize in subdivisions of one of these broad areas. For example, many electrical engineers specialize in computer systems. Others deal only with special aspects of communications.

Electrical engineers have many types of responsibilities, including research, design, and development. Other opportunities lie in manufacturing and administration. In addition, many engineers are engaged in selling jobs.

Opportunities are open in many types of organizations. Electrical and electronics-equipment manufacturers employ many engineers. So do the electric light and power, aircraft and missile, telephone, and broadcasting industries. Many electrical engineers work for government agencies, colleges, universities, or consulting firms. Some are self-employed as consulting engineers.

Most jobs are located in industrial centers. These areas tend to be heavily populated. However, there are opportunities in rural areas with some employers, including electric light and power companies, telephone companies, and broadcast stations, Figure 23-2.

Employment Outlook

Rapid growth can be expected in this profession. Increased demands for qualified engineers can be expected in the computer and fiber-optics fields. Also growing is the area of satellite communications. Newer types of power generation are expected to produce many more opportunities in the solar area, through use of semiconductor materials.

Finding More Information

For additional information about careers in the electrical engineering field, contact: Institute of Electrical and Electronics Engineers (IEEE), 1111 Nineteenth Street NW, Washington, DC 20036.

23.3 MANUFACTURING OCCUPATIONS

Persons with backgrounds in electricity can find many opportunities in companies that manufacture electrical and electronic equipment. Jobs exist in such areas as assembly, equipment operation, inspection, and testing.

Nature of Work

Electrical and electronics companies offer opportunities that are somewhat different. In the electrical field, emphasis is on production-type equipment. Typical specialties include motors, generators, appliances, machine tools, and automotive electrical equipment.

In the electronics field, emphasis is on the formation and testing of circuits. Completed circuits are then assembled into electronic devices. These include consumer products such as radios, TV sets, burglar alarms, calculators, and electronic games. Also included are products used by businesses. These include electron tubes, semiconductors, computers, communication equipment, and defense systems. Figure 23-3 shows automatic laser equipment used to produce integrated circuits.

Actual job assignments include assembly work, testing, and inspection. Testing and inspection begin when raw materials or components are received. Inspections continue through all stages of manufacturing.

About three of every five workers in electrical and electronics manufacturing work in plants. These jobs are in production, maintenance, transportation, and service. The rest are administrative and management positions. These include engineering, scientific, technical, administrative, clerical, and sales positions. Approximately one-half of administrative and management workers are in research and development.

Figure 23-2. On-the-job training is offered by many companies.
(Niagara-Mohawk Power Corporation)

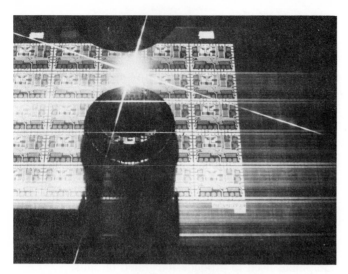

Figure 23-3. Laser equipment used in manufacturing of integrated circuits. (Western Electric)

Employment Outlook

Opportunities exist in several areas of electrical and electronics manufacturing. Some of these are summarized below.

White-Collar Positions. White-collar workers are those in professional, technical, sales, office, and managerial positions.

Managerial workers in this field include engineers, sales executives, and self-employed business operators.

Among technical workers are engineering aides, electronic technicians, drafters, and laboratory technicians.

These technical positions assist the managerial staff. Many electronics technicians help engineers design and build experimental models. Electronics technicians are often responsible for setting up and supervising assembly jobs, Figure 23-4. They may also handle some of the more complex assembly work themselves. Technicians may also assemble into complex systems subassemblies prepared by other workers.

People who want to seek advanced or specialized training can move ahead without restriction. Many women have succeeded in engineering, scientific, and technical positions.

Plant Occupations. Some 60 percent of the employees in electrical and electronics manufacturing are plant workers. Their jobs include assembly, machining, fabricating, processing, inspecting and testing, and maintenance.

Assembly Occupations. Assemblers make up the largest group of plant workers. Both electronics end-product and component manufacturing firms employ assemblers. Most assemblers are classed as semiskilled.

Machining Occupations. Many of the plant jobs in electrical and electronic companies require metal-machining skills. These employees operate power-driven machine tools. They produce metal parts for their companies' products. Toolmakers construct and repair jigs and fixtures used in making and assembling parts. Diemakers specialize in making metal forms (dies) used in power presses to shape metal parts.

Fabricating Occupations. These workers handle sheet metal. They also build frames, chassis, and cabinets.

Figure 23-4. Manufacturing engineers review photographic masters for integrated circuits. (Veeco Instruments)

Others work with glass and ceramics. These tend to be specialties. The number of persons in each occupation is not large.

Processing Occupations. Processes, in general, are continuous, relatively automatic functions. They often involve use of chemistry or chemical reactions. Examples include the plating, tin-coating, or anodizing of parts to prevent corrosion. Also included in this field are the persons who work on printed circuits. The processes involved are photographic. The etching of boards is also a process.

Testing and Inspection. Testing and inspection jobs are designed to check quality and assure that standards are met. These jobs begin with the receipt of raw materials at a plant. Responsibilities extend to all manufacturing or assembly work done within a plant. Many electronic parts and assemblies are now tested on computerized equipment. Inspectors usually specialize. For example, there may be separate people checking incoming materials, plating, machined parts, or electronic assemblies.

Finding More Information

For more information, contact the Electronic Industries Association (EIA), 2001 Eye Street NW, Washington, DC 20006. Also contact the International Brotherhood of Electrical Workers, 1125 Fifteenth Street NW, Washington, DC 20005.

23.4 ELECTRIC UTILITIES

There are approximately 3700 electric utilities in the United States. Some of these are privately owned. Others may be owned by federal, state, or local governments. Each utility has the ability to generate, transmit, and deliver power to users. In addition, there are power-distribution networks that make it possible for utilities to buy power from each other.

Employment Outlook

There were 582,000 people working for utilities in 1980. Many thousands of new employees enter this field every year. Some of these jobs open because of people leaving the industry. Other jobs open as a result of continuing growth. Because of changes taking place, excellent job potential should continue.

Many older generating plants are converting to new power sources. Most of these have been fired by oil or natural gas. Some will shift to coal as their energy source. Many nuclear plants are expected to open in years to come. In addition, experimentation continues for the use of new energy sources. In the future, electricity may come

from solar power. Other potential sources of energy include wind, geothermal wells, and the tides.

About 10 percent of this industry's jobs are in generating plants. About 40 percent are related to the transmission and distribution of power, Figure 23-5. About 20 percent of the workers are involved in repair and other maintenance jobs. Some 15 percent are in customer service, Figure 23-6. An estimated 9 percent of the employees are in engineering and technical positions.

Power-Plant Positions. These employees are responsible for operating the plants that generate power. Responsibilities include operating control panels, checking equipment, and keeping records. Specific job titles include *boiler operator, turbine operator, auxiliary-equipment operator, control-room operator,* and *watch engineer.*

Figure 23-5. Maintenance worker at power substation. (Niagara-Mohawk)

Figure 23-6. Utility service personnel are responsible for maintaining lines, sometimes under difficult conditions. (Niagara-Mohawk)

Figure 23-7. Installing power lines. (Niagara-Mohawk)

In this field, promotions usually result from experience. Beginning employees start with cleanup and other simple jobs. They advance as they become familiar with the equipment and with responsibilities of other jobs.

It takes from one to three years to become a fully qualified auxiliary-equipment operator. From four to eight years of experience are required to operate boilers, turbines, or control consoles.

Transmission and Distribution. Positions in this area include those of load dispatchers, substation operators, line installers, line maintenance crews, trouble shooters, and cable splicers, Figures 23-7 and 23-8.

Some people in these jobs begin their careers in power plants. For example, console operators at power plants may become load dispatchers. They may also move into positions as substation operators. Typically, it takes seven to 10 years of console operation experience to qualify for one of these positions. Even then, substation personnel usually begin as assistant or junior operators. Assignment as a substation operator may take three to seven years of additional experience. Substation operators often begin at small facilities. Assignment to larger substations comes with proven performance.

Line-installation and maintenance employees usually build their skills through on-the-job training. Many utilities have formal apprenticeship programs to cover this training.

Figure 23-8. Power-line cable splicers. (Niagara-Mohawk)

The same is true for most cable splicers. They learn on the job. It takes about four years to become fully qualified. New employees usually start as helpers and are promoted as they gain experience.

Consumer-Service Occupations. Persons who work in these fields perform a number of technical, managerial, and office jobs. Included are technicians who install, test, and repair meters. In some areas, utility employees also repair electrical appliances and machinery. Administrative and office jobs include the preparation of work orders. These work orders can involve installation, maintenance, or termination of service. Other office workers are involved in billing, collection, and record keeping.

Technicians who install and repair meters usually learn on the job. About four years of experience are usually required to qualify as a meter repairer.

Meter readers also learn on the job. The necessary skills for these positions are acquired quickly.

Finding More Information

For additional information about job opportunities in the electric-utility field, contact the International Brotherhood of Electrical Workers, 1125 Fifteenth Street NW, Washington, DC 20005. Also contact the Utility Workers Union of America, 815 Sixteenth Street NW, Washington, DC 20006.

23.5 TELEPHONE INDUSTRY

Telephone communications is a giant industry. There are a number of companies providing telephone service: more than a thousand independent telephone companies exist today. Deregulation of the telephone industry has led to the manufacture of much more telephone equipment. Extensive networks of fiber-optic cables are being installed and the quality of service is being improved constantly.

About one-fourth of the workers in telephone companies are involved in installation and service positions. They install, repair, and maintain telephones. They also set up wires and cables, switching equipment, and message accounting systems. Positions in this area include those of central office installers, line construction crews, and installation/repair specialists, see Figure 23-9. These occupations are described further in the discussions below.

Employment in telephone companies and related industries has grown rapidly in the past and is expected to continue a substantial growth rate in the future. The number of telephones in use expands by about 5 to 10 percent per year.

Telephone Operators. Operators work with the latest switching equipment through use of computer terminals. Information can be pulled up quickly and efficiently. There are long-distance operators, information operators, service assistants, and several categories of supervisors. With increasing automation and continuing expansion of direct-dialing systems, the demand for operators is likely to decrease.

Central-Office Installers. Persons in this category are employed by the companies that make telephone switching equipment. Inasmuch as most of the equipment is now made in factories and needs little work at the site of

Figure 23-9. Telephone installation and maintenance specialist.

installation, this area does not present much of a picture of growth in the near future. Some need will be experienced with the conversion to satellite and fiber optics as means of establishing private communications networks.

Line Construction and Maintenance. People in this category set up and maintain the lines that carry communications between users and central offices. Responsibilities include the setting up of cables on poles or in underground conduit. This work is done by line installers, Figure 23-10.

Another specialty is cable splicing. Splicers are responsible for all connections within the system of telephone lines. Included are the connections between cables and the service lines to individual homes or offices.

Most employees are trained on the job. Minimum requirements for employment usually include a high-school education. Job candidates should also have some background in electricity or electronics. A physical examination is required. Physical requirements include good eyesight and hearing. Applicants must also be skilled at working with their hands.

Opportunities should continue to grow as the communications field expands. The work may become increasingly specialized as satellites and light-wave carriers come into greater use.

Figure 23-10. Telephone line installer.

Installation and Repair Specialists. These employees install and maintain equipment for telephone users. Most telephone equipment is owned by the utility providing service. Users lease their switchboards and telephones. These leases include maintenance and repair services. Even if companies or individuals buy their own telephone equipment, a service contract is usually included.

Personnel in this field have "outside" jobs. That is, they work outside of the telephone company offices. They visit homes, offices, or factories where telephone equipment is used. Some of the work involves installing lines to which telephone equipment is connected. Responsibilities also include the connection, installation, testing, and maintenance of equipment.

Many thousands of people are employed in telephone installation and repair. Most individuals are hired as beginners and trained on the job. High-school graduation is usually a requirement. A physical examination is necessary. In addition, these employees must have a neat appearance. They must be skilled at meeting and dealing with people.

As with other communications areas, opportunities should continue to expand in this field.

Central-Office Operations. Remember that central-office equipment is usually installed by employees of the manufacturers. Therefore, separate personnel are needed to operate and maintain these installations. Responsibilities include monitoring of equipment to detect problems. Routine maintenance may be required on some systems. In addition, repair work is necessary when equipment malfunctions, Figure 23-11.

Some persons in these jobs are former equipment installers. Others are hired for these positions and trained

Figure 23-11. This technician is using data communication test equipment to remotely repair telephone circuits. (Fiber Optic Services)

on the job. Individuals require a high-school education and some background in electricity and electronics. Opportunities should continue to expand as telephone utilities grow. In addition, there are many positions open with independent or private communications companies.

Finding More Information

One place to go if you want more information is to your local telephone-company office. Every telephone business office has a personnel department. People there will be glad to give you information on job openings and career opportunities.

23.6 CONSTRUCTION ELECTRICIANS

Wherever construction takes place, you will find electricians on the job. Types of construction include housing, offices, commercial facilities, factories, airports, oil wells, and refineries. Anyplace people will live or work will almost certainly require electricity. Construction electricians will therefore always be on the job.

These specialists lay out, assemble, install, and test electrical fixtures, equipment, and wiring. The systems they install provide power, heat, light, air conditioning, and refrigeration. Electrical construction crews also install the motors for manufacturing equipment or conveyors. In tall buildings, they install the electricity and equipment that operates elevators.

The majority of workers in this field are members of the International Brotherhood of Electrical Workers. Most serve in four-year apprenticeship programs. High-school graduation and excellent physical health are usually required for entry. These programs require on-the-job experience as well as course work at night. Successful apprentices become journeymen. Continuing experience can lead to positions as foreman or superintendent.

Workers in this field are expected to own and maintain their own hand tools. Power tools, meters, and instruments are furnished by employers.

Employment opportunties vary in geographic areas. The best way to tell whether there are opportunities in your area is just to look around. If there is a great deal of building and development, construction electricians are needed. There are usually waiting lists for apprenticeship programs.

Finding More Information

For further information, contact the International Brotherhood of Electrical Workers, 1125 Fifteenth Street NW, Washington, DC 20005.

23.7 MAINTENANCE ELECTRICIANS

Maintenance of electrical equipment requires growing numbers of skilled people. All mechanical and electrical equipment requires continuing maintenance. In effect, persons in this field maintain the equipment installed by construction electricians. Sometimes construction electricians also do maintenance work.

In addition to maintenance, this work may also involve modifying or modernizing existing equipment. Assignments can include work on motors, transformers, generators, circuit breakers, controls, and lighting equipment. These installations may be in industrial, commercial, public, or residential facilities. Much of this work is done on a contract basis. That is, companies contract to inspect and maintain electrical equipment for regular, fixed fees.

When equipment breaks down, the effects can be costly. A tall building might be without an elevator. A production line might have lost its conveyor. Or an important piece of equipment may be out of service. The electrician must find the trouble and repair the equipment quickly. If the electrical problem presents any dangers, management must be advised quickly.

A maintenance electrician may travel regularly and work in many locations. Assignments may involve any of a full range of electrical skills. Individual jobs may call for wiring, splicing, parts replacement, or even installation of new wiring. A maintenance electrician must be able to use virtually all test equipment. It may also be necessary to read and work from construction drawings.

Maintenance electricians usually learn on the job. They generally go through the same type of apprenticeship program as described above for construction electricians. They usually own their own hand tools. Individual employees may be subject to emergency calls when problems develop.

Employment opportunities in this field should increase in the future. This is hard work that requires skilled people. The need for maintenance never ends.

Finding More Information

For further information, contact The National Joint Apprenticeship and Training Committee for the Electrical Industry, 9700 East George Palmer Highway, Lanham, MD 20802.

23.8 TV, RADIO, AND APPLIANCE REPAIR

Many thousands of persons are employed at shops that service TV sets, radios, stereo equipment, and a wide range of home appliances.

Almost every assignment in this field begins with a need to find out why a piece of equipment does not operate. Once the problem is diagnosed, repair may involve simple replacement of an electronic part. Rewiring or other repairs may also be necessary.

Many people in this field are self-taught. After basic education in electricity and electronics, they have worked through manuals on their own. Many have learned by buying kits and assembling their own radios, TV sets, stereo equipment, or even small computers. It is customary for all persons to own their own hand tools. Employers usually furnish needed instruments and meters. However, many technicians own these tools as well.

Formal training for TV and other electrical repair work is readily available. There are many public and private vocational and trade schools. Many correspondence schools also offer training courses. Many young persons also learn basic electronic repair skills in military schools. These skills can be valuable for individuals returning to civilian life.

In general, it takes two to three years of on-the-job or vocational-school training to qualify as a TV repair technician. In some states or cities, written tests are required. In some areas, individuals or businesses must be licensed to offer these services.

The outlook is for continued growth in this field. TV sets and appliances are becoming increasingly complex. As they do, service requirements grow.

Finding More Information

Any of the dozens or scores of vocational schools that operate in any medium-sized or large city can provide some career information. So, the Yellow Pages of your telephone directory is one source of more information.

23.9 BROADCASTING

Broadcasters, basically, are in the business of providing entertainment and information to listeners. Many technical skills are necessary to deliver the broadcasters' program materials. Included are sound engineers, lighting specialists, and camera operators. In addition, technicians must operate the control-room consoles for radio and TV production studios.

Broadcasters hire many technicians to install, operate, and maintain broadcast transmitters. At least one person on duty at every station must have a license. To qualify for the license, the person must take a test given by the Federal Communications Commission (FCC). Thus, some special education is usually required to qualify as a broadcast technician. This usually involves a high-school

education that is followed by work in a vocational or technical school.

Employment Outlook

At present, there are 550 UHF and 550 VHF commercial television stations in the United States. And, there are 226 UHF and 124 VHF noncommercial television stations in this country. Radio stations are broken down into AM and FM stations. There are 4980 AM radio stations and 4292 FM stations on the air presently. These broadcast facilities were all licensed to transmit signals through the airwaves. In addition, the cable TV industry is still growing rapidly. Thousands of cable TV systems are serving almost every community in the country. Satellite communications have opened new areas to electronics technicians, Figure 23-12. Fiber optics have a future in the cable and telephone service. People at home, with the addition of a fiber-optics cable hookup, have unlimited communications, entertainment, banking, and commercial services.

Thousands of people are employed in broadcasting at the various levels. Radio and television stations, both commercial and noncommercial, often affiliate with networks to provide them with all or some of their programming. These networks are no longer limited to ABC, NBC, and CBS. Networks such as the Associated Press News, Business Radio, and Fox also serve television with special programs.

Although thousands of workers are employed in these networks, both in the entertainment segment and at the

Figure 23-12. Satellite communications have opened up more jobs in the broadcast industry for electronics technicians.
(Photo by Al Marlin/WRGB-TV)

technical end, opportunities tend all the same to be somewhat limited in broadcasting. This is considered a "glamor" industry. So, there are usually many more applicants than there are jobs available. In addition, employment in large stations usually requires membership in at least one union. To "break into" broadcasting, many individuals start at small stations in remote areas. These positions tend to be low paying. As experience is gained, qualified people often try to move to larger stations or networks.

Finding More Information

The easiest way to learn more about broadcasting is to visit a radio or TV station in your area. You will usually find someone willing to discuss your interests and potential opportunities.

Figure 23-13. Microelectronics assembly. (PCB Piezotronics, Inc.)

23.10 NEW ELECTRONICS OCCUPATIONS

In the past ten years, the electronics industry has grown very rapidly. Many of the jobs now in demand did not exist just a few years ago. Most of the titles are self-explanatory. However, more-detailed information is available in the *Dictionary of Occupational Titles*, printed by the U.S. Government Printing Office in Washington, DC.

Note how the rapid growth and development of computers and digital electronics have generated many of these job titles.

Jobs Directly Related to the Production of Semiconductors

Assemblers. There are a number of job titles with which the name *assembler* appear. The microelectronics processor has a particularly distinct type of assembly duty. Semiconductor devices, components, and subassemblies are assembled according to drawings and specifications, using a microscope, bonding machines, and handtools, Figure 23-13. This person may have to perform many duties, such as reading work orders and assembly drawings to determine the operations to be performed. Semiconductor wafers under the scribing-machine microscope have to be aligned and scribed, or marked.

As the demand for more digital equipment and electronic devices using semiconductor chips increases, the demand for assemblers who can do this type of work is expected to increase also. Foreign competition may moderate the demand at present, but new markets will accelerate the demand toward the end of the decade. Many of these jobs are located in California, with some in Pennsylvania and Texas.

Inspectors. The production of semiconductor wafers for use as chips demands many types of inspections all along the line. Duties of an inspector include inspecting, measuring, and testing semiconductor wafers for conformance to specification. Defects, such as scratches, chips, stains, burns, or haze, must be sought out and eliminated in the quest for a perfect product. The diameter and flatness of the wafers must also be checked. Electronic gauges are utilized in checking for wafer bow and flatness.

Crystal evaluators or inspectors are also utilized in the production of semiconductor ingots before they are cut into wafers. Such workers must be able to use the microscope to detect any variations in specifications.

Testers. There are a number of job titles related to testing semiconductor wafers. One person may test the electrical characteristics of the circuits on the wafer, another may check the substrate with a probe tester, spectophotometer, and/or curve tracer. These testers try to determine the electrical characteristics of a particular chip or wafer.

The job of packaging tester is one that calls for tending the automatic equipment that tests the functions of the semiconductor packages. This person reads production documents to determine test specifications and inserts a specified test program (magnetic card) into test equipment to program it.

The leak tester tests the voltage of dies that contain individual devices, such as diodes and transistors, using a computerized testing system to ensure that the dies meet company and manufacturer specifications. Other duties may include reading production sheets and computer code books to determine test instructions and codes, and keying instructions and codes into the computer, using the computer terminal.

As the demand for more electronics equipment increases, the demand for persons to test wafers, finished chips, and transistors will also increase.

Jobs Directly Related to the Making of Printed-Circuit Boards

Printed-circuit boards (PCBs) are a necessary part of any electronics product today. The PC board is used to mount chips, transistors, diodes, and various other electronic parts. It has the ability to hold any number of parts or chips, according to the design of the circuitry or package.

Assemblers. A printed-circuit-board assembler places components by hand, where needed, on a board, Figure 23-14. The assembler performs a combination of tasks, including using handtools, reading worksheets and wiring diagrams, taking verbal instructions, inserting color-coded wires into designated holes, and clinching wire ends. The job also calls for the recognition of various electronic components and the ability to place them in their proper locations. However, much of the component insertion is now done by machines. The assembler becomes more of a machine attendant or operator in larger firms.

A test-fixture assembler (electronics) assembles fixtures used to test electrical circuitry of printed-circuit boards according to diagram instructions. This person is responsible for positioning test-fixture boards on a work table and inserting copper sleeves into designated holes of the boards.

Many assembler jobs are being eliminated, with the advent of the component insertion machine. However, many of the smaller tasks still demand personal attention to the printed-circuit boards before the boards are utilized in a finished product.

Operators. One of the important jobs in the making of printed-circuit boards for electronics equipment is that of the wave-soldering-machine operator. This person controls and monitors operation of the wave soldering machine, a system used to solder electronics components onto PCBs. The ability to read production schedules and operations manuals and to follow verbal instructions regarding sequential start-up and operation of the machines or systems is important. The operator controls the machine and activates the flux, preheater, wax stabilizer, chiller, lead-wire cutter, wave soldering unit, and conveyors, as well as PCB washing and drying machines.

Figure 23-14. Electronics assembly. (PCB Piezotronics, Inc.)

As long as there is a demand for PCB equipment, there will be a need for wave-soldering-machine operators. This position in the production of completed boards will always be needed, since it is vital that all components be properly mounted and soldered before they become part of any piece of electronics equipment.

Testers. The electronic-circuit tester for printed-circuit boards uses computerized test equipment. This person reviews test specifications to determine the program disks and test fixtures required to test the boards.

Testing of printed-circuit boards will always be necessary. This job title has many specific descriptions, according to the exact nature of the work performed. For instance, a PCB component tester (chemical) will test electronic components for compliance to company standards. Chemicals are placed on surfaces in petri dishes to prepare for lead testing. Testing wire leads in acid and observing the reactions is part of this particular job description. Freon is used to remove traces of the acid and then the component is tested to check for corrosion. Many tests of this type are performed to assure quality in components. Small electronics manufacturing firms are located all over the country. Check the local newspapers for ads concerning this type of operation and their localized need for personnel.

Reworkers. It is nearly impossible to turn out PCBs all day long without producing some defective ones. These defects have to be checked for and repaired. The reworker repairs defective surfaces and circuitry on the PCB, using power and handtools. This person utilizes knowledge of electronic repair techniques. The work calls for reading to determine the number of PCBs to be repaired, type of repairs required, and the method and tools to be used in reworking. The utilization of a microscope to examine circuitry may become necessary. The use of knowledge acquired through years of experience, from high school through vocational trade school, comes in handy in this type of repair. A variety of tests and repairs must be handled by this person. That means some sort of background must be acquired before the job is attempted. However, some employers have enough demand for a specific type of repair that a person can be trained in a very short time to do a specific repair function. As the demand for more electronics equipment increases, the need for this job increases.

Inspectors. Inspectors are needed everywhere along the line during production of printed-circuit boards. There are job titles such as Printed-Circuit Boards (electronics) Board Inspector; Circuit-Board Inspector; and Touch-up Inspector. These positions may involve inspecting PCBs for conformance to specifications and touching up defects. The conformance tests may require insertion of plug gauges into holes to verify dimensions and measurement of plated areas of the board.

Making the printed-circuit board itself requires inspection to make sure the board meets specifications. *Inspector of circuitry, negative film* is also a job title. This is a job that requires inspection of the negatives or artwork for conformance to specifications and the touch-up of defects on the negative. The job requires inspection under a magnifying glass for holes, breaks, and bridges (connections) in the photoresist circuit image. This person may also be required to use a knife to remove excess photoresist. Other tasks include touching up holes and breaks in the photoresist and using photo-resist ink pens.

A mechanical inspector must be able to operate coordinate-measuring equipment to measure dimensions of printed-circuit boards. This person reads blueprints to determine dimensions and tolerance specifications for the boards and records the information on inspection sheets.

Supervisory Positions

Group Leader. The function of direct supervision falls under the job-title classification of *group leader*. This person assists the supervisor in coordinating activities of workers engaged in the assembly of printed-circuit boards, applying knowledge of assembly techniques, specifications, and production scheduling. In conferences with the supervisor, this person reviews production schedules and specs and priorities. The job also includes planning departmental work assignments and usually involves responsibility for requisitions and distribution of supplies and materials. The group leader also assigns duties to assembly workers and oversees departmental activities.

The position of group leader is usually one that a person earns after years of experience and on-the-job training. Requirements for this position vary with the organization of a company. They may include a two-year college or specialty-school degree.

Assembly Supervisor. This supervisory position usually requires a higher level of responsibility than does that of group leader, in that supervisory personnel report directly to the assembly supervisor. The person in this position usually supervises and coordinates activities of the workers engaged in assembling, testing, and inspecting, Figure 23-15. The training of workers in interpretation of shop specifications, component recognition, equipment operation, and performance of job duties ensures that assembly standards are met. This person is also responsible for planning and coordinating assembly assignments to ensure that production goals are met.

Figure 23-15. Assembly supervisors will oversee the activities of those workers involved in the assembly of printed-circuit boards. (Pitney Bowes, Inc.)

This is sometimes considered a middle-management position and as such requires a four-year college degree with a major in industrial engineering or industrial technology. The future looks bright for this particular specialty, inasmuch as the need for electronics equipment will steadily increase over the next decade.

Testing Supervisor. This supervisor oversees and coordinates the activities of workers engaged in repairing and testing the performance of electronic equipment during and after assembly, applying a knowledge of electronic theory and circuitry, test procedures, and documentation. The position includes direction of training of workers in areas such as test procedures and equipment operation, troubleshooting and repairing, and interpretation of specifications and documentation. The job calls for coordinating, testing, and repair to ensure completion of work assignments.

Job opportunities in this field appear to be increasing with the advent of production increases and the possibility of new markets for electronics products opening worldwide.

Training Needed

In most instances, a high-school education is sufficient for the above jobs. Highly developed manual dexterity is demanded for the assembler jobs. Vocational-school training in reading blueprints or working drawings is helpful. Often, training is done on the job, through work with experienced personnel.

For supervisory positions, the level at which a person is expected to perform will determine the educational requirements. Larger operations will require the services of someone with a four-year college degree with some experience in electronics repair and assembly. Engineering-technology and industrial-technology programs are usually slanted toward providing the middle-management type of supervisory individual.

Finding More Information

Information on these job titles may be found by consulting the *Dictionary of Occupational Titles* located in your local library or school guidance office. Some companies operate their own training program. Checking the ads in your local newspapers will assist you in locating a company that has a program and is in need of applicants. Two-year and four-year technology programs are operated by many colleges throughout the country. It is possible to obtain more information by writing to the technology department of your state college or university.

23.11 OTHER ELECTRONICS AND ELECTRONICS-RELATED JOBS

There are other jobs in electronics not directly related to the production of PCBs and the components mounted on them. There are also thousands of jobs that are related to electronics or that depend on electronics for their equipment or tools. Some of them have developed recently, due to the demands of an advancing technology.

The future for those types of jobs is dependent on the demand for new electronic equipment. With the worldwide market improving, it appears there will be a need for additional personnel in the electronics or electronics-related fields.

Employment Outlook

Assemblers. Electronics assembly requires skilled technicians to produce a wide variety of signal-conditioning electronics for use in transducer systems, including voltage, charge, and computer-controlled electronics, Figure 23-16.

Operators. Many production machine shops have precision lathes that are computer controlled and completely automatic in operation. They utilize the skills of an operator familiar with the demands of close tolerances and the precision requirements of the electronics field, Figure 23-17.

Welders. Laser and electronic-beam welding have not been available for long, but both can do a precision job. Figure 23-18 shows high-technology laser welding that complements electron-beam welding, assuring reliable, hermetically sealed sensors for critical applications.

CAD/CAM. A CAD (computer aided drafting or design) person is expected to be able to draw complicated electronics as well as produce mechanical drawings to specifications, using the computer to increase efficiency and cut the time involved. CAD and CAM (computer-aided manufacturing) are linked together to provide better-quality products at a much faster pace than previously possible.

CAD/CAM, for instance, aids in the design and production of quartz sensors and electronic circuits, providing a higher-level technology product, and facilitates modifications to suit the needs of the user, Figure 23-19.

CNC Operator. Computer numerical control has increased the efficiency of many high-technology factories. The spindle lathe shown in Figure 23-20 is automatic, thanks to electronics. The machinist sets up the completely

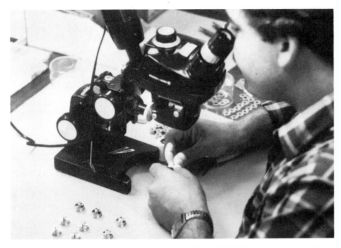

Figure 23-16. Electronics assembly. (PCB Piezotronics, Inc.)

Figure 23-17. Production machine shop. (PCB Piezotronics, Inc.)

Figure 23-18. Laser and electron-beam welding.
(PCB Piezotronics, Inc.)

Figure 23-19. Computer-aided design and manufacturing
(PCB Piezotronics, Inc.)

Figure 23-20. CNC lathes. (PCB Piezotronics, Inc.)

automatic lathe to perform both primary and secondary operations. The operator is also expected to be able to program the machine to meet the specifications for a completed product.

Factory Applications Engineer. Technical consultation and answers to technical questions are usually provided by the factory applications engineer. The job is just as the title specifies, requiring a person who knows how to apply the products made at a specific plant or facility. This person is also needed to expedite orders from customers.

Quality-Control Technician. The quality-control technician performs a variety of electronic, mechanical, and electromechanical tests on electronic systems, subassemblies, and parts to make sure the unit functions according to specifications. In the case of a known failure problem, the technician uses a full range of electronic test instruments to check the component. The range of skills must be wide since the person is expected, in some cases, not only to locate a defect, but also to remedy it, Figure 23-21.

Test Engineers. Test engineers use the latest computer technology to perform final tests and National Bureau of Standards traceable calibration in compliance with military standards, Figure 23-22.

Training Needed

These jobs require various degrees of skill and knowledge. For some jobs, a high-school diploma and some experience with electronics or a vocational course are all that is required. Many companies provide on-the-job training. Other jobs may require a two-year or four-year degree program.

Most of these positions require an individual who can be easily trained in new developments inasmuch as the technology is constantly being improved.

Finding More Information

Only a few of the possible jobs in electronics and electronics-related fields have been mentioned here. For further information, consult the *Dictionary of Occupational Titles*.

Information about two-year technical schools and four-year technology programs may be obtained from your guidance counselor or the local state university or two-year college near you.

23.12 LEADERSHIP

Our society has always prized resourcefulness and initiative. These qualities were vital to the westward expansion of the American frontier. They were also important in the early shaping of the infant Republic. Those early days required strong decision-making skills and firm business commitment.

Historically, we have developed a strong propensity toward leading the world in many aspects, and this shows in the standard of living we now enjoy. It comes as no surprise that hard work and great effort can be rewarded in a number of ways. Everyone is urged to develop his or her leadership talents in this society.

Keep in mind that good leaders are characterized by their ability to communicate effectively. They have a firm will and a singleness of purpose, as well as moral integrity. While it seems these qualities are more readily apparent in some, all of us can develop, in some degree, the skills needed for effective leadership.

Figure 23-21. Quality control personnel check a product for quality. (PCB Piezotronics, Inc.)

Figure 23-22. Computerized calibration. (PCB Piezotronics, Inc.)

Membership in student clubs such as VICA or the Technology Student Association can help develop leadership ability. Although a club, in bringing together as it does individuals with a single common interest, is perhaps less diverse than society generally, it can still offer you an opportunity to practice effective communication and learn the skills needed to work within a group, such as a committee.

The most valuable qualities of leadership are often exercised in the most quiet situations. Each of you will be given opportunities to exert the force of leadership to bring about a decision, resolve a crisis, or prompt an action. Leadership is essential to active participation in a democratic society.

SUMMARY

There are many jobs created directly and indirectly by electronics. Thousands of people are employed in this and closely associated industries. The field is constantly changing, and the demands on workers in the field are changing as well. It is necessary to keep updated if you want to progress. Self-education and training, as well as a formal education, are usually required to earn and keep a job in this field.

GLOSSARY

Absolute units These are units that have no prefixes. A good example is volts, amperes, ohms, and watts.

Accuracy Indicates how close to the true or actual value the meter reading actually is.

Acid-core solder The kind of solder NOT TO BE USED in electronics work. It has an acid flux core. Usually 50% tin and 50% lead.

Active component A device whose characteristics can change during normal circuit operation.

Air-core choke Coil of wire with no core. Used in radio-frequency circuits.

Alkaline cell Manganese-alkaline primary cells have a manganese-dioxide positive electrode, a zinc negative electrode, and a potassium hydroxide electrolyte. Nominal output is 1.5 volts. Secondary cells are hermetically sealed units with manganese dioxide cathodes, zinc anodes, and potassium hydroxide as the electrolyte. They also put out 1.5 volts unloaded and may be overcharged without damage, but should not be discharged below 0.9 volt.

Allen wrench Usually a piece of hexagonally shaped rod with one end bent 90.° Used to adjust set screws that have a recessed socket shaped head.

Alternating current Current that periodically changes magnitude and direction; abbreviated AC.

Alternation One-half of a cycle.

Alternator An alternating current generator.

Ammeter Device used to measure current in a circuit. Both AC and DC ammeters are available.

Ampere Basic unit of electric current. One ampere is 6.25×10^{18} electrons flowing past a given point in one second.

Amplification A process used to increase the voltage, current or power of an electrical signal.

Amplification factor A ratio that exists between the base current and the collector current in a transistor.

Amplifier Arrangement of electronic parts used to increase the voltage or power output of a circuit. Amplifiers may be so structured as to operate at different frequencies.

Amplitude modulation Method of modulation used in radio broadcasting; abbreviated as AM. The amplitude of the carrier wave is increased or decreased in step with the audio frequency being impressed on it.

Analog meter Meter whose pointer deflection is proportional to the current flow. The scale is placed on a clocklike dial and the pointer position indicates the value being read.

Analog multimeter Multimeter that displays its measured values by the displacement of a pointer over a calibrated scale. See *Multimeter*.

AND circuit Digital circuits are made up of *gates*. The AND gate is an electronic circuit that will function in some desired manner when one or more high- or low-voltage signal(s) are fed to its inputs. The AND gate requires both inputs to be high in order to conduct.

Anode Positive section of a vacuum tube diode.

Antenna System of electrical conductors used to receive and/or transmit electromagnetic waves.

Armature Rotating part of an electric motor or generator. May also be the moving part of a relay, buzzer or speaker.

Armored cable Usually refers to residential electrical cable that has a sheath of metal wound around it for protection from physical damage.

Armstrong oscillator An older type oscillator that utilizes a tickler coil for feedback.

Atom Smallest part of a chemical element that still retains all the properties of the element.

Attenuation Reduction of circuit voltage by some proportion.

Attenuator Device used to reduce voltage proportionally.

Audio choke Coil or inductor designed for use with audio frequencies.

Audio frequency Range of frequencies that can be heard by the human ear.

Audio-frequency oscillator Circuit or device used to produce sound in the audio range.

Autoranging Term used to describe a digital meter function when it automatically selects the range it needs to measure voltage, current or resistance.

Autotransformer Transformer that utilizes only one coil that is tapped to provide the correct voltage.

Avalanche breakdown Reverse voltage breakdown effect characteristic of zener diodes.

Average value Refers to the average of the instantaneous values of an AC voltage or current waveform. Usually 63.6% of maximum or peak value.

Back emf (counter emf) Force produced in an inductor that is opposed to that which produced it.

Ballast Choke or inductor used in a gaseous discharge (including fluorescent) lamp circuit.

Bandwidth Number that describes the frequency range between the upper- and lower-half power points of a resonant circuit.

Base One of the three terminals on a transistor.

Battery Electrochemical device consisting of two or more cells that serves as a DC power source.

Beam-power tube Vacuum tube that has beam forming plates to increase its power amplification capabilities.

Beta Amplification factor in transistors.

Bias Fixed DC voltage applied to a circuit before a signal is injected. It establishes the operating characteristics of a particular transistor or vacuum tube.

Binary coding BCD is a term often used in digital circuitry. It means *binary coded decimal*. It is the process whereby numbers and letters are coded into a series of 1's and 0's to indicate *on* (1) and *off* (0) conditions for a digital circuit to operate.

Bipolar-junction transistor Transistor whose current between two terminals is controlled by the current through a third terminal; abbreviated BJT.

Bits Unit of digital information. Binary values are usually in groups of four. Each pulse is a *bit* of information. The group of bits is a *word*. The words can have 4, 8, 16, or 32 bits. An 8-bit word is called a *byte*.

Bleeder resistor Resistor connected across the output of a power supply filter circuit for the purpose of discharging the capacitors when the power is turned off. It also aids in the voltage regulation of the power supply. Also used in some instances as a voltage divider.

Branch circuit Refers to a residential electrical circuit that handles a certain number of outlets or lights.

Bridge rectifier Rectifier circuit usually made up of four diodes arranged in a diamond configuration.

Brushes Devices that make contact with the rotating connections to the armature of a motor or generator. Usually made of carbon or metal.

Bytes See Bits. A byte is an 8-bit word.

CAD-CAM Computer aided design and computer augmented manufacturing.

Candlepower Unit of measurement of light.

Capacitance Property of a capacitor that allows it to store an electrical charge; measured in farads (F).

Capacitive reactance Opposition that a capacitor presents to an alternating current; symbol is X_c. Measured in ohms.

Capacitor Device used to store an electrical charge. Consists of two plates and a dielectric. Opposes any change in circuit voltage.

Capacitor-start motor Electric motor that utilizes a capacitor in its start circuit to enable it to start under load.

Cathode Part of a vacuum tube that emits electrons. May also be one terminal of a semiconductor. The cathode of a semiconductor diode is marked with a band around the cathode end.

Cell Device that produces an emf by chemical action. Usually made from two electrodes and an electrolyte.

Cellular phone Mobile telephone system using cells or strategically placed receivers and transmitters connected with wired communications systems to allow persons in moving cars, trains or planes to talk with one another.

Cemf Counter electromotive force. Generator effect in a motor when it is running. It opposes the incoming current that produced it.

Central processing unit (CPU) The central processing unit in a computer system that directs signals to the proper processing circuit.

Ceramic-bonded To be permanently attached to a ceramic substrate.

Chassis punch Device used to cut or punch holes in the chassis of a piece of electronics equipment.

Chip Another name for an integrated circuit or IC.

Choke Another name for a coil or inductor.

Circuit Pathway of conductors and devices designed for current flow. A complete circuit is a path for electrons to flow from the power source to the consuming device and back to the power source. An open circuit is not a complete path for electron flow to and from the power supply. A short circuit has much less resistance than it was originally intended to have.

Circuit breaker Device designed to open a circuit when too much current flows.

Circular mil Measurement unit for cross-sectional area of a round wire; equal to one-thousandth (0.001) of an inch in diameter.

Citizens' band Those frequencies on which anyone can broadcast once they are registered with the FCC. No test required.

Clapp oscillator An improved Colpitts oscillator. It uses an additional capacitor to improve its stability.

Coil Device made by turns of insulated wire wound around a core. A coil sometimes has a hollow center portion. Also called an inductor and/or a choke.

Color code System used to indicate resistance and capacitance. Rings or dots of color mark electronic devices. Values assigned to colors indicate the amount of resistance or capacitance.

Colpitts oscillator Easily identified by its tapped capacitance in its tank circuit.

Combination pliers Pliers that can be used for gripping and for wire cutting.

Common base Type of transistor circuit where the base is common reference point.

Common collector Type of transistor circuit where the collector is the reference point. No amplification, with this type of circuit. Unity power factor.

Common drain A type of circuit configuration on a special type of transistor that has a drain lead.

Common emitter Type of transistor circuit configuration where the emitter is the reference point. The most commonly used type of circuitry for transistors.

Common gate A type of circuit configuration on a special type of transistor that has a gate lead.

Common source A type of circuit configuration on a special type of transistor that has a source lead.

Commutator Device made of segments of copper insulated by mica or some other material. Used to reverse direction of current flow from a generator or to a motor. Brushes are usually placed to make contact with surface of commutator. Segments of commutator are connected to ends of armature coils in a motor or generator.

Composition Act of putting together two or more parts.

Compound motor Motor that uses field coils connected in series parallel combinations.

Computer-integrated-manufacturing (CIM) Controlling manufacturing processes using a computer based instructions unit.

Conductor Solid, liquid, or gas that permits the flow of electrons.

Conduit Usually a piece of pipe — metallic or plastic — used to house electrical wires or cables or both.

Conduit bender Device used to bend conduit to specified contours.

Control grid An element in a vacuum tube used to place the signal into the tube.

Convenience outlet The wall outlet in a residential wiring system. The outlet which allows 120 volts to be connected to an appliance or lamp.

Coulomb Unit of measure for electrons at rest on a surface area. One coulomb includes 6.25×10^{18} electrons.

Crossover network Circuitry that determines the point at which a given frequency will be sent to another speaker or output device.

Crystal amplifier Original name for a transistor.

Crystal oscillator Oscillator that uses a crystal to generate its basic frequency.

Current Movement of electrons along a conductor, usually defined as being from a negative to a positive terminal.

Current lag Current can be said to lag behind voltage in an inductive circuit.

Damped oscillation Oscillation that has a tendency to lose some of its amplitude as each oscillation occurs until it dies out.

D'Arsonval meter movement Portion of a meter consisting of a fixed permanent magnet and a moving electromagnetic coil.

Decibel Unit of measurement for sound energy; abbreviated as dB. One tenth of a bel.

Deflection Usually refers to a meter pointer being deflected.

Degenerative feedback A fedback signal that causes the circuit to break into oscillations.

Delta Greek letter in the shape of a triangle. A method of connecting 3-phase current transformers and motors windings.

Delta to wye. Another method used to connect 3 phase windings in a transformer or to an electrical motor.

Depletion mode Refers to FET transistor operation.

Depolarizer Chemical used to cause a battery to last longer by keeping the carbon rod from becoming covered with impurities.

Deposited carbon resistor A method used to make resistors by depositing vaporized carbon on a form.

Desoldering iron Device used to unsolder component parts.

Detector (modulator) Electronic circuit or stage that changes the form of a radio frequency current. Used in radio and television to allow reception of audio frequencies and picture information from transmitted signals. Usually employs a diode.

Diagonal-cutting pliers Small pliers with ability to cut small diameter wires — usually those encountered in electronics circuits.

Dielectric Insulating material used in a capacitor or other electrical device.

Differential Resultant difference between two voltages or currents.

Differential amplifier A linear amplifier and is abbreviated as diff amp. It does not include capacitors in its composition. Two inputs make it capable of distinguishing between two sources. A difference output is generated from the two outputs.

Diffusion To spread out. Diffused light reduces glare.

Digital circuit Circuit designed to handle digitized electrical pulses.

Digital multimeter A meter that produces a readout in numbers rather than indicated with a moving needle against a scale.

Diode Device used to change AC to DC. Can be either vacuum tube or semiconductor.

Direct control Where a device such as a switch has control over the operation of another device or circuit.

Direct current Abbreviated DC. Current that flows in one direction only.

Discharge When a capacitor looses its charge.

Distortion When the output of a waveform is not the same as it was when it was inputted it is said to be distorted.

Distribution panel Usually refers to a panel whereby the electrical power originates for branch circuits within a house or building.

Domain theory Refers to one theory which states that some magnetic materials have a glass-like, or crystalline, structure. These crystals can be lined up so that their individual magnetic properties add to or build upon, each other. This causes the entire piece of material to become a magnet.

DPDT Double pole-double throw switch.

DPST Double pole-single throw switch.

Dry cell Device that produces electricity through chemical action. Usually made of zinc and carbon with sal ammoniac as an electrolyte.

Dual inline package Abbreviated DIP. One packaging arrangement for integrated circuits, ICs, or chips.

Edison effect Noted by Thomas Edison when experimenting to produce an electric light bulb. A positive plate attracts electrons (as we know now — he didn't know about electrons) when so charged, but will not allow current flow through a vacuum when negatively charged.

Electrical potential Usually refers to voltage or electrical pressure or potential difference between two poles or terminals.

Electric discharge light source Light produced when two carbon arcs sustain an electric arc between them.

Electricity Form of energy present when electrons move through a complete path.

Electrolyte Solution capable of conducting electric current; the liquid part of a battery.

Electrolytic Capacitor that has parts separated by an electrolyte. Thin film formed on one plate provides the dielectric.

Electromagnet Magnet produced by current flow through a coil of wire. Core is usually used to concentrate the magnetic lines of force.

Electromagnetic force Force that causes electrons to move through a conductor; abbreviated emf. Also called voltage.

Electron Negatively charged atomic particle that normally revolves around the nucleus of an atom.

Electron tube Device consisting of two or more electrodes or elements enclosed in a vacuum tube. Contained within a shell of either glass or metal.

Electronics Reference to devices or systems with circuits utilizing the flow of electrons through semiconductors or vacuum tubes.

Electronic tuning The ability to tune a circuit using a voltage difference applied to a diode type material.

Electrostatic The force of attraction or repulsion between two stationary charged bodies.

Energy Ability to do work.

Enhancement mode In a MOSFET a positive or negative voltage may be applied to the gate. In the enhancement mode of operation a positive voltage is applied to the gate.

EPROM Erasable programmable read-only memory chip.

Equivalent resistance Term used when solving parallel resistor circuits.

Extension cord An electrical cable with a plug and jack on the opposite ends; used to extend the source of electrical power.

Farad Unit of measurement of a capacitor; abbreviated F. Represents the amount of capacitance needed to produce 1 ampere of current flow when a circuit has a change of 1 volt per second.

Feedback Amplifiers use part of their output and feed it back to the input to improve the quality of the output.

FET Field effect transistor.

Fiber optics Plastic fibers used to direct the movement of laser light sources from one place to another. Utilized in modern telephone circuitry.

Fidelity Term used to describe the quality of sound reproduction.

Field coil Refers to the coils of a DC motor or generator.

Filaments Small coils of resistance wire in light bulbs and vacuum tubes that heat up to glow either red or white hot. The filament in a vacuum tube boils off electrons from the cathode. The filament in a light bulb glows to incandescence to produce light.

Filter A frequency selective circuit that passes circuit frequencies and rejects all others.

Firmware Permanently installed instructions in computer hardware that control operation.

Floating ground A separate grounding wire or system that is not tied to the existing Edison System utilized in house wiring or building power distribution. Especially helpful in placing digital equipment at ground potential different from that of its electrical distribution system.

Fluorescence To fluoresce or glow like in a fluorescent lamp or tube.

Fluorescent lamp Lamp that produces light through action of ultraviolet rays striking a fluorescent material that is coated on the inside of a glass tube and causes the material to fluoresce or glow.

Flux Refers to the magnetic force field of a magnet or electromagnet.

Flywheel effect Used to describe the charge-discharge of a capacitor through a coil or inductor in a tank circuit arrangement to produce an oscillation equal to the ratio of capacitance to inductance.

Force field A magnetic flux field.

Forward bias Used to describe the forward or conducting condition of a transistor or semiconductor diode.

Four-way switch A switch used in residential wire to allow control of a circuit load from three or more locations.

Free-running multivibrator An astable multivibrator. Astable means it is not stable in terms of the *on* and *off* states for either stage. It does not need an input signal.

Frequency Term describing the number of times an alternating current (AC) completes its cycle in one second.

Frequency-shift keying One way to transmit digital information or data over long distance telephone lines. Abbreviated FSK.

Frequency modulation Abbreviated FM. A method used to cause audio frequencies to be impressed on radio frequencies for broadcast purposes. The rf carrier varies its frequency instead of its amplitude.

Full-wave rectifier Rectifier that uses two diodes to convert AC to DC. Both halves of the sine wave are utilized.

Fuse Safety device designed to open if an excess current flows through a circuit.

Gain Proportional increase in voltage or current produced by an electronic circuit.

Galvanometer Meter used to indicate amount and direction of current flow in circuits.

Gate Terminal of silicon controlled rectifier (SCR).

Gauge (or gage) number Size of a piece of wire in its cross-sectional area. A gage is also used to measure the diameter or number of the wire.

Generator Electromechanical device used to produce electrical power.

Germanium Chemical element used in the manufacture of semiconductor materials such as transistors and diodes.

Grippers Device or tool needed for gripping or holding.

Ground Actual connection to Earth, a common connection point in an electronic device.

Grounding Establishment of a common connection to Earth or ground.

Gunn diode Semiconductor device used to generate radar frequencies.

Half-wave rectification Process that converts only half of the AC voltage cycle to pulsating DC voltage.

Harmonic Even-number multiple of a given frequency.

Hartley oscillator A circuit that utilizes a tapped inductance in its tank circuit to produce oscillations.

Headset, headphones, earphones Alternative terms for device that produces sound in audio-frequency range. Units fit either over or into the ears.

Heat sink Clamp or some other device used to dissipate undesired heat from the leads of heat-sensitive components during soldering. Can also refer to large fins placed on transistors, diodes, and chips to dissipate heat generated during operation.

Helix Used in microwave applications. A coil of wire utilized in directing microwaves through a traveling-wave tube (TWT).

Henry Unit of measurement for inductance; abbreviated H. An inductance of 1 henry is present in a coil when the current through it, changing at the rate of 1 ampere per second, produces 1 volt across the coil terminals.

Hertz Unit of measurement of frequency; abbreviated Hz.

Heterodyning Process of beating two frequencies together to produce the sum, the difference, and the original two frequencies.

Hole A type of free charge that is present only in P-type semiconductors.

Hum When the filters in a power supply do not have enough capacitance, the result is too much of the AC ripple component in the DC output voltage. The effect is excessive hum in the sound output from an audio system.

Hybrid circuit Circuit made with a combination of hard wiring and printed circuits or with integrated circuits.

Hysteresis Magnetic loss in a transformer or inductor caused by residual magnetism of the core.

IC Integrated circuit.

Illumination Light.

Image orthicon Type of television camera tube.

Impedance Total opposition to alternating current; symbol is Z. Measured in ohms.

Impedance matching Use of transformers to match one amplifier stage to another or one device, such as a microphone, to another to allow for the maximum transfer of energy.

Impedance ratio Relationship between the impedances of the primary and secondary of a transformer.

Incandescent lamp Lamp or bulb that uses a filament glowing white hot or to incandescence to produce light.

Induced current Current caused by magnetic lines cutting a conductor.

Induced voltage Voltage caused by magnetic lines cutting a conductor.

Inductance Ability of a coil, choke, or inductor to oppose any change in circuit current. It is represented by the symbol L and is measured in Henrys.

Induction motor A form of AC motor most commonly used to power appliances and heavy equipment. The rotating current in the stator windings induce a magnetic field in the rotor and consequently a rotating motion.

Inductive filter Filter circuit that uses an inductor connected in series with the power-supply load.

Inductive reactance Opposition to current flow in an alternating-current circuit caused by an inductor. It is measured in ohms and abbreviated as X_L.

Inductor Device used to oppose any change in circuit current. Also referred to as a choke or coil.

Infinite baffle Unlimited baffle such as used in loudspeaker enclosures.

Infrared oven Device used to heat articles to drive off excess moisture and speed drying or cooking.

Input Term describing the energy applied to a circuit, device, or system.

Instantaneous voltage Term used to indicate the voltage at a specified point in the cycle.

Insulator Any material that inhibits or slows the flow of electrons.

Integrated circuit (IC) Device made up of a collection of electronic components and their connections. They are manufactured by the doping of a slice of silicon.

Intermodulation distortion Another result of amplitude distortion where the harmonics introduced in the amplifier can combine with each other or with the original frequencies to produce new frequencies that are not harmonics of the fundamental. This effect is referred to as intermodulation distortion.

International code A series of universally accepted electrically produced dots and dashes assigned to each letter of the alphabet, numbers, and punctuation marks.

Inversely Usually used when referring to two quantities that vary in such a way as to be inversely proportional; when one increases, the other decreases.

Inverter Device used to change DC to AC; can make a positive voltage negative.

Jack Device used to make connections or to disconnect points in electrical circuits.

JFET Junction field effect transistor.

JIT *Just in time* manufacturing means no inventory to control — parts arrive just in time to keep the assembly line going.

Joule Metric unit of electrical work done by 1 coulomb flowing with a potential difference of 1 volt.

Joystick Device used to manipulate images on a computer screen.

Junction Point within a circuit where wires meet; also the area where pieces of semiconductor material are joined within a diode or transistor.

Junction diode Semiconductor device made from P and N type materials joined at a junction point. Designed to handle larger currents than the signal diode.

Key Hand-operated device (switch) used to send telegraphic messages.

Kilo Prefix meaning one thousand. (1000). Abbreviated k.

Kilohertz Term used for 1000 hertz. Abbreviated kHz.

Kilohm Term used for 1000 ohms.

Kilowatthour Term used for 1000 watts consumed in one hour. Abbreviated kWh.

Kirchhoff's current law Describes the relationships that determine current at a junction in the analysis of complex circuits. It states that the algebraic sum of currents at a junction equals zero. In other words, the quantity of current flowing to the junction is equal to that flowing away from the junction.

Kirchhoff's voltage law States that the algebraic sum of voltage drops around a complete loop equals zero, or, the sum of the voltage drops around a complete loop equals the applied voltage.

Klystron Tube used to generate frequencies in the microwave range.

Knife switch An open bladed switch that operates similar to the action of a jack knife.

Ku band Refers to a band of frequencies allocated to be used by radar (24.15 GHz is one of those frequencies).

Lag Delay in current or voltage in a circuit with inductance or capacitance.

Laminated Term describing a solid object constructed of a package of identical sheets of material. The core of a transformer, relay, or coil is usually laminated.

Language Words and/or letter combinations used to tell a robot or computer what to do.

Large-scale integration (LSI) Term used to describe a type of integrated circuit (IC).

Laser *L*ight *A*mplification by *S*timulated *E*mission of *R*adiation.

Law of magnets Law that states: like poles repel and unlike poles attract.

Lead A chemical element; a wire leading from a semiconductor device such as a transistor or diode.

Lead-acid cell A cell made up of lead, lead peroxide and uses sulfuric acid for an electrolyte. Automobile batteries use lead-acid cells.

Left-hand rule States that a current-carrying conductor is grasped with the left hand, with the thumb pointing in the direction of current flow, then the fingers will indicate the direction of the magnetic lines around the conductor.

Light-emitting diode (LED) Semiconductor device that

glows when it is forward-biased. It is used to form the segments of a number in a digital display. It can also be used as a pilot light.

Linear scale Meter scale set up in a line marked in even spaces.

Linear taper Usually refers to a potentiometer that has even amounts of resistance change for equal distances moved by the control knob.

Lissajous figure Oscilloscope waveform that shows the relationship between two frequencies.

Load Anything that may draw current from an electrical power source.

Lock-up control When a relay itself enters the control path.

Lodestone Mineral with magnetic properties; a natural magnet; common name for magnetite.

Long-nose pliers Another name for needle-nosed or pointed pliers used to get into tight spots in an electronics chassis.

Loosely-held Electrons in outer orbits that do not have the same attraction to the nucleus as those closer to the center of the atom.

Loudspeaker Term, now used infrequently, for an instrument that reproduces audio frequencies from electrical impulses. Electrical energy is changed into sound energy.

Luminous Property for emitting or reflecting light.

Magnet Object with polarity (north and south) that attracts certain materials, particularly iron.

Magnet wire Copper wire used to wind coils, solenoids, transformers, and motors; usually coated with varnish or other insulation material.

Magnetic chuck Device used to magnetically hold a piece of metal being machined.

Magnetic field Forces present around the ends of a magnet and around a current-carrying wire or coil.

Magnetic inductance The process of making a material magnetic by allowing a strong magnetic field to come in contact with it.

Magnetic lines Invisible lines that leave the north pole and enter the south pole of a magnet. No longer used to describe a magnetic field.

Magnetism Ability to attract nickel, iron, or cobalt.

Magnetite Another name for lodestone.

Magnetron Device used to generate microwaves for a microwave oven.

Magnetic flux Lines of force used to represent magnetic induction.

Magnetic lines Invisible lines that form loops around a magnet. They leave the north pole and enter the south pole.

Magnetism Ability to attract nickel, iron, or cobalt.

Magnetron Power source for radar frequencies; used in microwave ovens.

Manipulator That part of the robot that does the manipulating — the arm.

Mega Prefix meaning one million (1,000,000); abbreviated M.

Megahertz Term meaning one million (1,000,000) hertz; abbreviated MHz.

Megger Device for measuring extremely high resistance.

Megohm Term used to indicate one million ohms.

Mercury switch A switch that uses the element mercury to complete the path between two electrodes to cause an on-off condition when a toggle is flipped.

Metal film resistor Resistance is obtained by a fine metal film being deposited on a core and then coated to protect it.

Meter Instrument used to measure. Electrical meters include ammeters, voltmeters, ohmmeters, wattmeters, and watthour meters.

Mica Insulating material that can withstand high voltages and elevated temperatures. Used in the manufacture of appliances and capacitors.

Micro Prefix meaning one-millionth (0.000001); symbol is μ.

Microampere Term used for 0.000001 ampere. One millionth of an ampere.

Microcomputer A small computer made for individual use at home, school and in small businesses.

Microfarad One millionth of a farad; 0.000001 F.

Microphone Device used to change sound energy to electrical energy.

Microprocessor May be one chip with the ability to do a specific number of operations.

Microswitch An extremely small switch usually employed in computer circuits.

Microwave Those frequencies that behave like light. Extremely high frequencies and extremely short travel between hertz.

Milli Prefix meaning one-thousandth (0.001); abbreviated m.

Milliampere One thousandth (0.001) of an ampere. Written as mA.

Modem Device used to connect the telephone or other device to a computer.

Modulated continuous wave The process of impressing modulation onto a continuous wave or rf carrier; abbreviated MCW.

Modulation Process whereby an audio frequency is impressed upon a radio-frequency carrier.

Modulation index Term used with the modulation process associated with frequency modulation or FM.

Modulator That stage of the transmitter that impresses the audio onto the rf carrier.

Molecule The smallest unit of a compound made of elements.

Momentary-contact switch A switch that has its contacts closed for a short period or turns off when released.

Monitor To view, listen to, or observe the operation of a circuit or electronics device. Can also be a TV picture tube-like device for displaying images.

Morse code A code devised by Samuel F.B. Morse to send messages in electrical dots and dashes.

MOSFET Metal-oxide semiconductor, field effect transistor.

MOS transistor Metal oxide semiconductor.

Motor Device used to change electrical to mechanical energy.

Mouse Computer input device that digitizes information for use by the computer.

Mu (μ) Greek letter M. Used to represent micro or 1,000,000. This symbol indicates gain of a vacuum tube.

Multimeter A test instrument or meter that measures voltage, current, and resistance.

Multiplexing The process whereby more than one frequency is imposed on a carrier wave. Such as with stereo FM broadcasting.

Multiplier Resistor used in a voltmeter to increase its ability to measure higher voltages. It is placed in series with the meter movement and is used to extend the meter range.

Multirange More than one range.

Multivibrator A stage that is important to digital electronics. There are three types: astable, bistable and monostable. Also called relaxation oscillator.

Mutual inductance Inductance that results when there is an interaction of adjacent inductors. It is measured in henrys.

NAND circuit The opposite of AND circuit. It generates a low when both inputs are high.

National Electrical Code® Set of guidelines covering installation of electric wiring and electrical devices; abbreviated NEC®.

Needle-nose pliers Long, thin-jawed pair of pliers used for holding parts in areas where a wider jaw could not enter.

Negative Electrical property. The point in a circuit with excess electrons, such as the negative terminal of a battery or capacitor; symbol is −.

Negative ion Atom with an excess of electrons.

Neon lamp Device with neon gas encapsulated in a small glass container with two electrodes. Presentation of voltage difference between the electrodes causes neon gas to ionize, conduct current, then glow.

Neutron Neutral particle located in an atom's nucleus.

Nickel-cadmium cell A cell made up of the elements mentioned. It can be recharged.

Noninductive load A load with no inductance in the circuit.

Nonmagnetic Term for materials that are not easily magnetized; nor easily attracted to or repelled by a magnet.

NOR circuit The output of the NOR gate is high when both inputs are low.

North seeking pole This is really the way to define the north pole since it is north seeking. It is in reality a south pole since unlike poles attract.

NOT circuit Gate with one input and one output with the function of inverting the input so that the output will be the opposite.

Nonvolatile memory A computer storage function that retains data without the use of electrical power. Examples are disk and tape.

NPN transistor Transistor made with a combination of n and p materials arranged in an n-p-n sequence. This type uses an n-type material for the emitter, a p-type material for the base, and an n-type material for the collector.

N-type material Piece of silicon or germanium that has been doped with donor atoms and has excess electrons.

Nucleus Central part of an atom, containing neutrons and protons.

Nut drivers Screwdriverlike tools with sockets on the end for turning nuts.

Ohm Unit of measure for electrical resistance; abbreviated with the Greek letter omega (Ω). Circuit has a resistance of 1 ohm when 1 ampere flows through it with a pressure of 1 volt.

Ohmmeter Device used to measure electrical resistance in ohms.

Ohm's law Basic electrical law stating the relationships of voltage, current, and resistance: Current in a circuit is equal to the voltage divided by the resistance ($I = E/R$).

Op-Amp Operation Amplifier abbreviation. Another name for linear ICs or integrated circuits. They have a number of uses.

Op-amp Wein-bridge oscillator Oscillator utilizing the operational amplifier in a bridge arrangement to produce oscillations.

Open circuit Circuit that experiences an interrupted flow of current, caused intentionally with a switch or unintentionally.

Orbit Path traveled by electrons around its atom's nucleus.

OR circuit A gate that has two or more inputs, but only

one output. Equated to two switches in parallel. Either one will turn on the circuit.

Oscillation The output of an oscillator. One oscillation is the same as one complete cycle or hertz.

Oscillator A circuit or device that oscillates or generates a frequency.

Oscilloscope Device used to display and measure voltage and time.

Outlet Socket or receptacle that accepts a plug to make electrical contact, usually to provide power.

Output Term describing power or energy produced by a device or system, such as a transformer or an amplifier.

Output current The output current of a transistor circuit.

Parallax error Error caused in reading analog meters because of the offset in the human eyes.

Parallel circuit Circuit in which resistors or capacitors are connected so as to allow current to divide and flow through each individual device, then to combine and flow back to the source; there are two or more paths for electron flow.

Parallel resonant When the coil and capacitor in parallel reach a point where the inductive reactance equals the capacitive reactance.

Passive component Device whose characteristics remain constant during normal circuit operation or, in other words, do not change as a result of power being applied to the circuit.

Peak-to-peak voltage (p-p) Represents the highest positive and negative levels reached during a complete cycle of alternating current. It is represented as p-p on most meters; rms voltage times 2.828 equals p-p voltage.

Peak-to-valley ratio Ratio of peak voltage (high point) to the valley voltage (low point).

Peak value Maximum point or peak value of a voltage or current. RMS × 1.414 produces the peak value.

Peak voltage Highest positive or negative level reached by a voltage alternation.

Pentode Electron vacuum tube with five elements.

Permanent magnet Magnet that retains its magnetism even after the magnetizing force has been removed.

Permeability Ease with which magnetic lines pass through an object.

Phase Relationship between zero-voltage crossing of two signals; is expressed in degrees.

Phase angle The angular difference between voltage and current in an inductive or capacitive circuit.

Phase shift The amount of change in the lead and lag of a circuit.

Phi (ϕ) Greek letter used to indicate phase.

Phillips screwdriver A type of screwdriver that has its points in the shape of a cross.

Picofarad Unit of measurement of capacitance; symbol is pF. Equal to one-millionth of a millionth (0.000000000001) of a farad.

Piezoelectrical effect Process whereby emf is generated by pressure applied to a crystalline substance, usually Rochelle salts or quartz.

Pi filter Filter circuit that uses an inductor in series with the output of a power supply. There are two capacitors across the line, one before the inductor and one after, forming a *pi* configuration.

Plug Device inserted into a jack to make an electrical connection; also a device for obtaining power from an outlet.

PN junction Place in a semiconductor diode or transistor where the *p*- and *n*-type materials are joined.

PNP transistor Transistor that uses *p* and *n* materials in a *p-n-p* order to form the device. *P*-type material is used for the emitter and collector, the base material being *n*-type.

Polarity Term for opposite positions and properties of electricity, such as negative or positive terminals of a battery or capacitor.

Polarization Process of acquiring polarity; also a process by which the electrode in a dry cell becomes inoperative because of hydrogen bubbles deposited on its surface. Also refers to control of the direction of radiant energy, as in polarized light.

Positive Electrical condition in which there is a deficiency of electrons; symbol is +.

Positive ion Atom with a deficiency of electrons.

Potential coil The voltage coil or coil with many turns of small diameter wire in a kilowatt-hour meter.

Potentiometer Three-terminal variable resistor.

Power Equal to voltage times current. Symbol is P; measured in watts (W).

Power amplifier A power amplifier stage produces enough power to drive a speaker. A power amplifier may be a complete amplifier unit that is made to drive an array of output devices.

Power factor Ratio of the power actually dissipated or used compared to what appears to be used. It can be expressed in decimal form or in percentage form.

Power rating Rating of how much heat a resistor can dissipate.

Precision resistor Resistor with a tolerance of less than 1%.

Preventive maintenance Scheduled process of inspections and corrections conducted to make sure that electrical or electronic devices operate properly.

Primary Input side of a transformer.

Primary cell Cell that cannot be recharged.

Primary coil Input coil in a transformer.

Printed circuit Phenolic or fiberglass board with copper lines etched on it. The copper is bonded to the board material and then etched away to allow for the desired connections of component parts.

PROM A chip used for its programmable read-only memory.

Proton Positively charged particle in an atom's nucleus.

P-type material Semiconductor material that has a deficiency of electrons. It has been doped with acceptor atoms that produce holes that accept electrons easily.

Pulsating direct current Current produced when AC is changed to DC and left unfiltered; abbreviated PDC.

Pulse-position modulation PPM is another form of pulse time modulation. The position of the pulse relative to a reference is changed in accordance with the amplitude of the modulating signal.

Pulse-width modulation PWM is where the width of the pulse is determined by a comparator.

Push-pull amplifier Circuit used to obtain maximum output from two tubes or transistors. They are turned on one then the other. Half of each hertz or cycle is amplified by the two devices and their associated components.

Pythagorean Theorem Square of the hypotenuse is equal to the sum of the squares of the legs in any right triangle.

Q Merit of a coil or circuit; ratio of inductive reactance to reistance, also used to represent the selectivity of a tuned circuit.

Quadrature detection Utilization of a chip demodulate FM. It involves amplification and phase shifting of the input IF signal.

Quartz heater Electrical device that uses quartz as the heating element. Produces infrared rays to warm objects directly in line with its radiation.

Radar Acronym meaning RAdio Detection And Ranging. Electromagnetic waves are sent out to strike an object, which is detected by interpretation of the reflected wave.

Radio Device that receives electromagnetic waves and changes them into sound energy.

Radio frequency (RF) Those frequencies above 16 kHz and in the non-audible range. Electromagnetic radiation. (rf) (RF).

RAM IC chip that is used for random access memory in a computer.

Raster Can be seen as a glowing TV picture tube when the set is without a signal from the transmitter.

RC phase shift-oscillator Oscillator that uses an RC network to shift the phase of a frequency 60 degrees. Three networks produce a shift of 180 degrees. This is combined with the 180 degrees shift produced by the amplifier and

oscillator feedback is produced so it produces a set frequency.

RC time constant Time taken for a charging capacitor to reach 63.2% of its maximum value. $T = RC$. It takes five time constants to reach the practical maximum value of voltage in an RC circuit (99.3%).

Read/write Another name for the RAM. The memory has the capacity to have bits of data read from it or to have data written into it.

Receptacle Socket or outlet that accepts a plug to make electrical contact.

Rectification Process of converting AC to DC.

Rectifier Device that changes AC to DC; allows current flow in only one direction; another name for a diode.

Rectifier circuit Circuit designed to convert alternating current (AC) to direct current (DC).

Reflection To return light or sound waves from surfaces.

Refraction Deflection from a straight path suffered by a ray of light, heat, sound, or the like in passing obliquely from one medium to another in which its velocity is different, as from air into water or from a denser to a rarer layer of air.

Regenerative feedback In-phase feedback effect capable of producing oscillation is known as regeneration. Out-of-phase feedback will prevent oscillation and is known as degeneration.

Register A series of interconnected flip-flops is generally called a register.

Relay An electrically operated switch.

Reluctance Opposition to the flow of magnetic lines. It compares with resistance in a resistor.

Residual magnetism Magnetism left after a magnetizing force has been removed.

Resin-bonded Using a resin or chemical to bond or hold together.

Resistance Opposition to current flow.

Resistivity Extent to which a material resists electron flow.

Resistor Device designed to offer a specific amount of opposition to current flow.

Resistor network Any arrangement of resistors.

Resonance Condition of a circuit that occurs when the capacitive reactance is equal to the inductive reactance.

Resonant frequency Frequency at which $X_L = X_C$.

Reverse bias Voltage with polarity the opposite needed to cause a transistor or semiconductor device to conduct.

Rheostat Variable resistor, usually having only two points connected into circuit. Used to control voltage by increasing and decreasing resistance.

Right-hand rule Rule used to find the direction of motion in the rotor conductors of a motor.

Ripple Series of pulses that "ride on top of" the direct current when AC is changed to DC.

Ripple voltage Peak-to-peak voltage variation produced by a rectifier circuit.

RMS Abbreviation of Roots-Means-Square.

Robot A machine that will move an end-of-arm tool, sensor, and/or gripper to a preprogrammed location.

ROM (read only memory) A type of memory chip that allows only reading what is stored in it.

Root-mean-square Type of reading obtained by using a standard voltmeter or ammeter; abbreviated RMS.

Rosin-core solder Solder used primarily in electronics work. The rosin core acts as a flux.

Rotary switch A switch that usually has many connections with small fingers making contact to a sliding strip.

Rotor Moving part of a motor or generator.

Satellite Something orbiting the earth, usually with sophisticated electronics equipment aboard.

Saturation State reached by transistor drawing maximum design current.

S-band A band of frequencies in the microwave range. The microwave oven and some police radar operate on 2455 MHz in the S-band.

Schematic Drawing that uses symbols to show how components in an electrical circuit are connected.

Scratch awl A screwdriver like device that has a pointed end similar to an ice pick.

Secondary Output side of a transformer.

Secondary cell Cell that can be recharged.

Secondary coil The output coil of a transformer.

Selectivity Property of a tuned circuit that permits it to pass or select only one frequency at a time.

Self-inductance In a coil an emf is induced in a conductor when the magnetic field changes. This is called a self-induced emf.

Semiconductor Material that is able to conduct electricity under certain conditions and not conduct it under others. Silicon and germanium are two good semiconductor materials used in diodes and transistors.

Series circuit Electrical connection method in which current flows in sequence through all devices in a circuit. The current has only one path.

Series motor A motor, usually DC or universal type, that has the armature and field coil in series.

Series ohmmeter An ohmmeter used to measure high resistances.

Series-parallel circuit Circuit that has at least one resistor in series with at least two in parallel.

Series resonant When inductive reactance equals capacitive reactance in a series LC circuit.

Series switching circuit A circuit that uses a switch in series with various loads to turn the current on and off.

Service drop In residential electrical wiring it is where

that part of the service from the utility meets the in-house wiring.

Service entrance This is where the electrical service enters the house or building.

Shielded cable Cable with one of its conductors wrapped with a clothlike outer covering made of finely woven wire that acts as a shield.

Short circuit Unintended alternative path of relatively low resistance that increases and bypasses the flow of current.

Shunt Parallel. It is often used to describe a resistor placed in parallel with an ammeter meter movement to extend its range.

Shunt control Placing a resistor in parallel with the relay coil to prevent a short when the relay coil is deenergized.

Shunt motor A DC motor that has its field coil in parallel with its armature.

Shunt ohmmeter Device used to measure low resistances.

Side cutting pliers Large handled pliers that also have side cutters capable of cutting larger diameter copper wire. Usually on every electrician's tool belt.

Silicon Chemical element used to make semiconductor devices.

Sine wave The signature of alternating current.

Single-phase motor An electric motor that uses single phase alternating current for proper operation.

Single-phase power Alternating current that has a single sine wave generated sixty times per second in U.S. power plants.

Slide switch Device with an elongated bar that makes contact with terminals when slid forward and disconnects when slid backwards.

Slip ring Ring of copper mounted on the shaft of a motor or generator through which a brush makes permanent (or constant) contact with the end of the rotor winding. Always used in pairs.

Software A set of instructions for computer operation.

Solar cell Cell that turns light energy into electrical energy; usually made of silicon.

Solder Usually a combination of 60% tin and 40% lead for physically securing electrical connections.

Soldering iron Device used for melting solder and making a proper solder connection.

Solder pencil Small soldering iron with a pencil-like tip.

Solder sucker Device designed to suck up molten solder when a rubber bulb is released. Aids in desoldering joints or components.

Solder wick Copper stranded wire that has the ability to take on solder when heated. It will take up the excess solder generated during the desoldering process leaving the joint free of solder.

Solenoid Coil of wire wrapped around a hollow form,

usually with some type of core material sucked into the hollow. Movement of core is used to move a switch or to open valves.

Solid state Another term used to refer to semiconductor devices.

South-seeking pole Really a north pole, but we call it a south pole and drop the seeking part.

Space charge Electrons boiled off a cathode in a vacuum tube have a charge or group of electrons around the cathode until attracted elsewhere by a positive potential.

SPDT Single-pole, double-throw switch.

Speaker Instrument that reproduces audio frequencies from electrical impulses. Electrical energy is changed into sound energy.

Speaker enclosure The box or enclosing fixture for speakers.

Splice Form of electrical connection in which wires are joined directly to each other.

Split phase motor Single-phase motor that uses a start winding and run winding.

SPST Single-pole, single-throw switch configuration.

Stage Complete electronic circuit with all the needed parts connected to produce a given result within a larger system, such as a portion of an amplifier.

Static electricity Form of energy present when there are two charges of opposite polarity in close proximity. Static electricity is generated by friction.

Stator Fixed element or plate in a variable capacitor.

Steady state A condition that does not change with time.

Stepping switch One that moves and in so doing completes the circuit in a number of locations.

Superconductor Conductors that operate at or near absolute zero.

Sweep circuit Circuit used in an oscilloscope or TV picture tube to produce a voltage that varies at a constant rate.

Switch Device designed to quickly connect or disconnect an electrical circuit.

Symptoms Behavioral characteristics of a circuit that indicate a malfunction.

Synchronous motor Motor that has the ability to synchronize with the frequency of its power source.

Tachometer Device for indicating the speed of rotation.

Tank circuit A parallel combination of inductance and capacitance.

Tantalum Material used to make a special type of capacitor.

Telegraph Means of communicating by sending electrical pulses over wires in coded form.

Telephone Instrument for communicating by sending audio signals over wires, microwaves, light-guides, or radio waves.

Television Method used to transmit and/or receive pictures and sound over radio frequencies or communication circuits.

Terminal Connecting point for wires; present on batteries, cells, switches, relays, motors, and electrical panels.

Terminal strip Series of connection points insulated from each other and providing locations for components to be wired or joined.

Tesla (T) Unit of measurement for the density of a magnetic field. Named for Nicholas Tesla.

Tetrode Vacuum tube with four elements.

Thermocouple Device made of two different kinds of metals joined at one end. When junction is heated, an emf is generated across the open ends.

Thermonic emission Boiling off of electrons from a cathode material by using heat.

Thermostat Device that acts as a switch operated when heat causes two metals to expand at different rates.

Thevenin's theorem Provides for the reduction of a complex circuit to simple terms; provides for reduction to an equivalent voltage and an equivalent resistance in a series circuit.

Three-phase motor Electric motor that uses three-phase power to operate.

Three-phase power Similar to single-phase power but has its three sine waves 120 degrees apart.

Three-way switch A switch that is designed to cause a circuit to be turned on and off in two locations if two switches are used. A single-pole, double-throw switch.

Tickler coil Small coils strategically placed to produce a small amount of feedback in a circuit.

Time constant Time constant equals resistance times capacitance or inductance divided by resistance. One time constant equals 63.2% of maximum current.

Toggle switch Switch operated by a pole-like (toggle) handle.

Tolerance The amount of + and − error between the actual and indicated values of a component.

Transformer Device that can induce electrical energy from one coil to the other by using magnetic lines of force, stepping voltage up or down.

Transistor Semiconductor device used to amplify signals or for switching; derived from *trans*fer res*istor*.

Transmission voltage Voltage at which electrical power is transmitted from the generator to the consumer.

Transmitter Electronic device that generates radio frequency waves; also the mouthpiece of a telephone handset.

Trimmer capacitor Small variable capacitor often used across larger capacitors to make fine adjustments in tuned circuits.

Triode Vacuum tube with three elements, usually a cathode, grid, and plate.

Troubleshooting Process of measurement and analysis used to determine a cause of electrical or electronic malfunction.

True power (TP) Product of voltage and current times the power factor in an AC circuit, to produce watts (W).

Tube tester Device for testing electronic vacuum tubes by simulating certain circuit conditions.

Tuned-base oscillator Utilizes a common-emitter configuration and transformer feedback. Uses a transformer as an inductor for the tank circuit and as a source of feedback.

Tuned circuit A circuit that is resonant to some frequency.

Tuner Circuit of a radio or television receiver used to select the desired frequency.

Tunnel diode Device with a negative resistance characteristic. Also called Esaki diodes. Low power device for microwave frequencies.

Turns ratio Relationship between the number of turns in the primary of a transformer and the number of turns in the secondary.

Underwriters' Laboratories, Inc. Testing laboratory for electrical equipment.

Universal motor Motor that operates on both AC and DC.

Variable capacitor Capacitor that can be adjusted to change its capacitance.

Vector Representation that indicates both direction and magnitude; used to graph functions of AC.

Vidicon TV camera tube that operates on the photoconductive principle.

Volatile memory Memory that is lost when the electrical power is removed.

Volt Unit of measurement of electromotive force; abbreviated V.

Voltage Electromotive force (emf) that causes electrons to move along a conductor.

Voltage amplifier Nearly all input amplifier stages are voltage amplifiers. They are designed to increase the voltage level of the input signal.

Voltage-controlled oscillator VCO. A voltage-controlled oscillator provides an output frequency that is a linear function of a controlling voltage. The 566 chip can be utilized for this function.

Voltage divider Series arrangement of resistances connected to a DC or AC supply used to produce propor-

tionally lower amounts of a source voltage by tapping across the various series resistor combinations.

Voltage doubler Circuitry that utilizes two half-wave rectifiers with the outputs summed by using two capacitors.

Voltage drop Voltage change that results from current flow through a resistor.

Voltage gain Ratio of output voltage to input voltage in a transistor circuit.

Voltage regulation Controlling the voltage so that it is not allowed to vary more than predetermined levels.

Voltmeter Instrument used to measure voltage.

Volume control Resistor used to control volume or output of an amplifier stage.

VOM Abbreviation for an analog multimeter that can measure volts, ohms, and milliamperes.

Wafer switch Rotary switch with contacts mounted on wafers.

Watt Unit of electrical power; abbreviated W.

Wattage rating Amount of power the device is capable of handling before being damaged by heat build-up.

Watthour meter Meter that measures electrical power (usually in kilowatthours) consumed in an hour.

Wattmeter Device used for measuring power consumption.

Waveform Pattern of an AC frequency derived by looking at instantaneous voltage values which occur over a period of time.

Waveguide A piece of equipment used to contain microwaves. They are the conductors (replace wires) for microwaves.

Weber (Wb) Unit of measurement for magnetic flux in the SI system. Same as 100,000,000 (1×10^8) lines or maxwells.

Wein-bridge oscillator Uses a resistance-capacitance network for the development of a sinusoidal output.

Welding Joining of two pieces of material, usually by heat or by melting.

Wet cell Cell made with a liquid electrolyte. It is usually referred to as a secondary cell and can be recharged. Wet cells use many combinations of materials for electrodes and electrolytes.

Wicking Process used to remove solder by pressing a braid against the connection with a hot iron. The wick or braid absorbs the solder so it can be removed from the desired spot.

Wire wrap gun Device for wrapping wire around a terminal instead of soldering the joint until it is tightly wrapped.

Wire stripper Device or tool used to remove insulation from a piece of wire.

Wire-wound resistor Fixed-value resistor with turns of wire wrapped around a ceramic core and usually coated with an epoxy or a glaze.

Words In digital parlance, a grouping of pulses that pass through a computer.

Work envelope This is the area in which a robot arm moves and should be avoided by humans when the robot is in operation.

Wye One of the three-phase wiring configurations. It is so named because it resembles the letter Y when the three coils are connected and schematically drawn.

Wye to delta Three-phase connection.

X Symbol used to represent reactance, such as X_C and X_L.

X-band One of the bands of frequencies utilized by radar. Police radar operates on 10,525 MHz in the X-band.

Z Symbol for impedance. Total opposition to flow of AC due to X_L, R, X_C combinations.

Zener Type of semiconductor diode that breaks down intentionally at a predetermined voltage.

Zener effect Produced by a special type of diode. A zener diode is designed to purposely breakdown without being damaged. Breakdown takes place when the reverse bias exceeds the diode's rating. At this breakdown point the inverse current increases rather rapidly. This is referred to as the zener effect.

Zero adjust A variable resistor used to adjust an ohm-meter (compensates for battery deterioration) to zero before it is used to measure resistance.

Zero-adjusted potentiometer Adjustment used in a VOM to adjust for the steady decline of a battery used in its ohmmeter circuit.

INDEX

Numbers followed by an *i* indicate illustrations; numbers followed by a *t* indicate tables.